Connexin Methods and Protocols

METHODS IN MOLECULAR BIOLOGY™

John M. Walker, SERIES EDITOR

176. **Steroid Receptor Methods:** *Protocols and Assays,* edited by *Benjamin A. Lieberman, 2001*

175. **Genomics Protocols,** edited by *Michael P. Starkey and Ramnath Elaswarapu, 2001*

174. **Epstein-Barr Virus Protocols,** edited by *Joanna B. Wilson and Gerhard H. W. May, 2001*

173. **Calcium-Binding Protein Protocols, Volume 2:** *Methods and Techniques,* edited by *Hans J. Vogel, 2001*

172. **Calcium-Binding Protein Protocols, Volume 1:** *Reviews and Case Histories,* edited by *Hans J. Vogel, 2001*

171. **Proteoglycan Protocols,** edited by *Renato V. Iozzo, 2001*

170. **DNA Arrays:** *Methods and Protocols,* edited by *Jang B. Rampal, 2001*

169. **Neurotrophin Protocols,** edited by *Robert A. Rush, 2001*

168. **Protein Structure, Stability, and Folding,** edited by *Kenneth P. Murphy, 2001*

167. **DNA Sequencing Protocols,** *Second Edition,* edited by *Colin A. Graham and Alison J. M. Hill, 2001*

166. **Immunotoxin Methods and Protocols,** edited by *Walter A. Hall, 2001*

165. **SV40 Protocols,** edited by *Leda Raptis, 2001*

164. **Kinesin Protocols,** edited by *Isabelle Vernos, 2001*

163. **Capillary Electrophoresis of Nucleic Acids, Volume 2:** *Practical Applications of Capillary Electrophoresis,* edited by *Keith R. Mitchelson and Jing Cheng, 2001*

162. **Capillary Electrophoresis of Nucleic Acids, Volume 1:** *The Capillary Electrophoresis System as an Analytical Tool,* edited by *Keith R. Mitchelson and Jing Cheng, 2001*

161. **Cytoskeleton Methods and Protocols,** edited by *Ray H. Gavin, 2001*

160. **Nuclease Methods and Protocols,** edited by *Catherine H. Schein, 2001*

159. **Amino Acid Analysis Protocols,** edited by *Catherine Cooper, Nicole Packer, and Keith Williams, 2001*

158. **Gene Knockoout Protocols,** edited by *Martin J. Tymms and Ismail Kola, 2001*

157. **Mycotoxin Protocols,** edited by *Mary W. Trucksess and Albert E. Pohland, 2001*

156. **Antigen Processing and Presentation Protocols,** edited by *Joyce C. Solheim, 2001*

155. **Adipose Tissue Protocols,** edited by *Gérard Ailhuud, 2000*

154. **Connexin Methods and Protocols,** edited by *Roberto Bruzzone and Christian Giaume, 2001*

153. **Neuropeptide Y Protocols,** edited by *Ambikaipakan Balasubramaniam, 2000*

152. **DNA Repair Protocols:** *Prokaryotic Systems,* edited by *Patrick Vaughan, 2000*

151. **Matrix Metalloproteinase Protocols,** edited by *Ian M. Clark, 2001*

150. **Complement Methods and Protocols,** edited by *B. Paul Morgan, 2000*

149. **The ELISA Guidebook,** edited by *John R. Crowther, 2000*

148. **DNA–Protein Interactions:** *Principles and Protocols* (**2nd ed.**), edited by *Tom Moss, 2001*

147. **Affinity Chromatography:** *Methods and Protocols,* edited by *Pascal Bailon, George K. Ehrlich, Wen-Jian Fung, and Wolfgang Berthold, 2000*

146. **Mass Spectrometry of Proteins and Peptides,** edited by *John R. Chapman, 2000*

145. **Bacterial Toxins:** *Methods and Protocols,* edited by *Otto Holst, 2000*

144. **Calpain Methods and Protocols,** edited by *John S. Elce, 2000*

143. **Protein Structure Prediction:** *Methods and Protocols,* edited by *David Webster, 2000*

142. **Transforming Growth Factor-Beta Protocols,** edited by *Philip H. Howe, 2000*

141. **Plant Hormone Protocols,** edited by *Gregory A. Tucker and Jeremy A. Roberts, 2000*

140. **Chaperonin Protocols,** edited by *Christine Schneider, 2000*

139. **Extracellular Matrix Protocols,** edited by *Charles Streuli and Michael Grant, 2000*

138. **Chemokine Protocols,** edited by *Amanda E. I. Proudfoot, Timothy N. C. Wells, and Christine Power, 2000*

137. **Developmental Biology Protocols, Volume III,** edited by *Rocky S. Tuan and Cecilia W. Lo, 2000*

136. **Developmental Biology Protocols, Volume II,** edited by *Rocky S. Tuan and Cecilia W. Lo, 2000*

135. **Developmental Biology Protocols, Volume I,** edited by *Rocky S. Tuan and Cecilia W. Lo, 2000*

134. **T Cell Protocols:** *Development and Activation,* edited by *Kelly P. Kearse, 2000*

133. **Gene Targeting Protocols,** edited by *Eric B. Kmiec, 2000*

132. **Bioinformatics Methods and Protocols,** edited by *Stephen Misener and Stephen A. Krawetz, 2000*

131. **Flavoprotein Protocols,** edited by *S. K. Chapman and G. A. Reid, 1999*

130. **Transcription Factor Protocols,** edited by *Martin J. Tymms, 2000*

129. **Integrin Protocols,** edited by *Anthony Howlett, 1999*

128. **NMDA Protocols,** edited by *Min Li, 1999*

127. **Molecular Methods in Developmental Biology:** Xenopus *and* Zebrafish, edited by *Matthew Guille, 1999*

126. **Adrenergic Receptor Protocols,** edited by *Curtis A. Machida, 2000*

125. **Glycoprotein Methods and Protocols:** *The Mucins,* edited by *Anthony P. Corfield, 2000*

124. **Protein Kinase Protocols,** edited by *Alastair D. Reith, 2001*

123. **In Situ Hybridization Protocols** (**2nd ed.**), edited by *Ian A. Darby, 2000*

122. **Confocal Microscopy Methods and Protocols,** edited by *Stephen W. Paddock, 1999*

121. **Natural Killer Cell Protocols:** *Cellular and Molecular Methods,* edited by *Kerry S. Campbell and Marco Colonna, 2000*

120. **Eicosanoid Protocols,** edited by *Elias A. Lianos, 1999*

119. **Chromatin Protocols,** edited by *Peter B. Becker, 1999*

118. **RNA–Protein Interaction Protocols,** edited by *Susan R. Haynes, 1999*

117. **Electron Microscopy Methods and Protocols,** edited by *M. A. Nasser Hajibagheri, 1999*

116. **Protein Lipidation Protocols,** edited by *Michael H. Gelb, 1999*

115. **Immunocytochemical Methods and Protocols** (**2nd ed.**), edited by *Lorette C. Javois, 1999*

114. **Calcium Signaling Protocols,** edited by *David G. Lambert, 1999*

Connexin Methods and Protocols

Edited by

Roberto Bruzzone

Institut Pasteur,
Paris, France

and

Christian Giaume

Collège de France,
Paris, France

Humana Press ✳ Totowa, New Jersey

This publication is printed on acid-free paper. ∞
ANSI Z39.48-1984 (American Standards Institute) Permanence of Paper for Printed Library Materials.

Cover illustration: From Fig. 3 in Chapter 2 "Sodium Dodecyl Sulfate-Freeze-Fracture Immunolabeling of Gap Junctions," by Irene Dunia, Michel Recouvreur, Pierre Nicolas, Nalin M. Kumar, Hans Bloemendal, and E. Lucio Benedetti.

Cover design by Patricia F. Cleary.

For additional copies, pricing for bulk purchases, and/or information about other Humana titles, contact Humana at the above address or at any of the following numbers: Tel.: 973-256-1699; Fax: 973-256-8341; E-mail: humana@humanapr.com; Website: http://humanapress.com

Printed in the United States of America. 10 9 8 7 6 5 4 3 2 1

Library of Congress Cataloging in Publication Data

Main entry under title: Methods in molecular biology™.

Connexin methods and protocols / edited by Roberto Bruzzone and Christian Giaume.
        p. cm.—(Methods in molecular biology; v.154)
    Includes bibliographical references and index.
    ISBN 0-89603-658-8 (alk. paper)
    1.Connexins—Laboratory manuals. I. Bruzzone, Roberto. II. Giaume, Christian. III. Series.

QP552.C63 C66 2000
571.6'4—dc21   99-087338

Preface

Direct cell–cell communication is a common property of multicellular organisms that is achieved through membrane channels that are organized in gap junctions. The protein subunits of these intercellular channels, the connexins, form a multigene family that has been investigated in great detail in recent years. It has now become clear that, in different tissues, connexins speak several languages that control specific cellular functions. This progress has been made possible by the availability of new molecular tools and the improvement of basic techniques for the study of membrane channels, as well as by the use of genetic approaches to study protein function in vivo. More important, connexins have gained visibility because mutations in some connexin genes have been found to be linked to human genetic disorders.

Connexin Methods and Protocols presents in detail a collection of techniques currently used to study the cellular and molecular biology of connexins and their physiological properties. The field of gap junctions and connexin research has always been characterized by a multidisciplinary approach combining morphology, biochemistry, biophysics, and cellular and molecular biology. This book provides a series of cutting-edge protocols and includes a large spectrum of practical methods that are available to investigate the function of connexin channels.

Connexin Methods and Protocols is divided into three main parts. In the first part, we have included chapters dealing with common laboratory techniques that have been specifically adapted to the study of connexins and gap junction channels. The second part presents a variety of approaches that are more closely related to functional studies of this form of cell–cell communication. Finally, in the third part, we have grouped chapters that discuss the properties of connexins in relation to their functional role. Most chapters are organized in a very schematic fashion with a step-by-step presentation of the technique to facilitate its introduction into the laboratory. Other chapters are more narrative in style, since they discuss specific theoretical aspects of connexin biology and physiology and are, therefore, not amenable to the same format. We hope that *Connexin Methods and Protocols* will set the standard for assays used to investigate connexins and demonstrate their involvement in intercellular communication. Since there is an existing "connexin link" (connexin-connection@listserv.uni-stuttgart.de) that provides a forum to disseminate information among researchers in this field, we hope that this electronic address may be used to post comments and protocol updates.

This book is intended for doctoral students and postdocs who are starting their work in the connexin field and wish to approach a key biological question involving several techniques. It will be equally useful to group leaders whose research activity either brings them to the study of connexins and cell–cell communication or requires the timely addition of more techniques to their laboratory repertoire. For instance, the cloning of the connexin counterparts in invertebrates, the innexin family, together with the association of connexin mutations to human diseases, has now brought developmental and human geneticists into the field of intercellular communication. It is to be hoped that the new breed of connexin researchers may find this book useful at this time.

At the 1987 Gap Junction meeting held in Asilomar (USA) the scientific community agreed on a nomenclature that distinguishes connexins on the basis of species of origin and appends the molecular mass predicted by cloned DNA sequences to the family name connexin (Cx). For example, the 43 kDa protein first identified in myocardial gap junctions is termed Cx43 and connexin homologs from different organisms can be distinguished with a suitable identifying prefix. Despite its limitations, this convention is currently used by the vast majority of researchers and, therefore, has been adopted throughout this book. The connexin family has now grown well beyond the only four members that were known 12 years ago. Thus, at the 1999 meeting on connexins and gap junctions (Gwatt, Switzerland), a committee was formed to address the issue of nomenclature. The committee has now reached a consensus and has recommended that the current nomenclature be retained.

Finally, we wish to thank all the authors for their patience in dealing with our picky reviews of the manuscripts. We hope that, despite the changes and rigidity in style that we have occasionally imposed, they will be as pleased with the outcome as we have been with their excellent contributions: we are already using some of their protocols in our laboratories. We are also grateful to the series editor, John Walker, for his invitation, since our editorial work has considerably improved our practical knowledge of connexins.

Roberto Bruzzone, MD
Christian Giaume, PhD

Contents

Preface v

Contributors xi

PART I: TOOLS TO STUDY CONNEXINS

1 Investigation of Connexin Gene Expression Patterns
   by *In Situ* Hybridization Techniques
   Magali Théveniau-Ruissy, Sébastien Alcoléa,
   Irène Marics, Daniel Gros, Antoon F. M. Moorman,
   and Wouter H. Lamers *1*

2 Sodium Dodecyl Sulfate-Freeze-Fracture Immunolabeling
   of Gap Junctions
   Irene Dunia, Michel Recouvreur, Pierre Nicolas,
   Nalin M. Kumar, Hans Bloemendal,
   and E. Lucio Benedetti 33

3 Purification of Gap Junctions
   Gina E. Sosinsky and Guy A. Perkins 57

4 Culturing of Mammalian Cells Expressing Recombinant
   Connexins and Two-Dimensional Crystallization
   of the Isolated Gap Junctions
   Mark Yeager and Vinzenz M. Unger 77

5 Connexins/Connexons: *Cell-Free Expression*
   Matthias M. Falk 91

6 Biochemical Analysis of Connexon Assembly
   Judy K. VanSlyke and Linda S. Musil 117

7 Expression and Imaging of Connexin-GFP Chimeras
   in Live Mammalian Cells
   Dale W. Laird, Karen Jordan, and Qing Shao 135

8 Analysis of Connexin Expression in Brain Slices
   by Single-Cell Reverse Transcriptase
   Polymerase Chain Reaction
   Laurent Venance 143

9 Use of Retroviruses to Express Connexins
   Jean X. Jiang 159

10 Spatiotemporal Depletion of Connexins Using Antisense
       Oligonucleotides
       **Colin R. Green, Lee yong Law, Jun Sheng Lin,
       and David L. Becker** *175*

11 Transfection and Expression of Exogenous Connexins
       in Mammalian Cells
       Dieter Manthey and Klaus Willecke *187*

PART II: ASSAYS FOR FUNCTION

12 Assaying the Molecular Permeability
       of Connexin Channels
       Paolo Meda *201*

13 Applying the *Xenopus* Oocyte Expression System
       to the Analysis of Gap Junction Proteins
       **I. Martha Skerrett, Mary Merritt, Lan Zhou, Hui Zhu,
       FengLi Cao, Joseph F. Smith, and Bruce J. Nicholson** *225*

14 Mutagenesis to Study Channel Structure
       Gerhard Dahl and Arnold Pfahnl *251*

15 Dual Patch Clamp
       **Harold V.M. Van Rijen, Ronald Wilders, Martin B. Rook,
       and Habo J. Jongsma** *269*

16 Determining Ionic Permeabilities
       of Gap Junction Channels
       Richard D. Veenstra *293*

17 Fluorescence Recovery After Photobleaching
       **Jean Délèze, Bruno Delage, Olfa Hentati-Ksibi,
       Franck Verrecchia, and Jean-Claude Hervé** *313*

18 Capture of Transjunctional Metabolites
       Gary S. Goldberg and Paul D. Lampe *329*

PART III: PHYSIOLOGY AND BIOLOGY OF CONNEXINS

19 The Study of Connexin Hemichannels (Connexons)
       in *Xenopus* Oocytes
       E. Brady Trexler and Vytas K. Verselis *341*

20. Exploring Hemichannel Permeability In Vitro
       Andrew L. Harris and Carville G. Bevans *357*

21 Inducing *De Novo* Formation
       of Gap Juntion Channels
       Feliksas F. Bukauskas *379*

Contents

22 Recording and Analysis of Putative Direct Electrical
    Interactions in the Mammalian Brain
    **Taufik A. Valiante, Jose L. Perez Velazquez,
    and Peter L. Carlen** 395

23 Intercellular Calcium Signaling and Flash Photolysis
    of Caged Compounds: *A Sensitive Method
    to Evaluate Gap Junctional Coupling*
    Luc Leybaert and Michael J. Sanderson 407

24 Biochemical Analysis of Connexin Phosphorylation
    **Bonnie J. Warn-Cramer, Wendy E. Kurata,
    and Alan F. Lau** 431

25 How to Close a Gap Junction Channel:
    Efficacies and Potencies of Uncoupling Agents
    **Renato Rozental, Miduturu Srinivas,
    and David C. Spray** 447

Index 477

In Memoriam

Bernie Gilula (1944–2000)

The field of gap junction research lost one of its leaders and founders with the death yesterday of Bernie Gilula, who succumbed to lymphoma after more than a year of fighting the disease. For those of us who attended the last gap junction meeting in Switzerland a year ago, his Keynote address now stands as his valedictory remarks. The keynote was scheduled well in advance of the diagnosis and was followed by his early departure from the meeting due to the onset of illness. Treatment was aggressive, and we had hoped for its success.

We and many others have been the beneficiaries of Bernie's mentoring and training. He was a scholar of the highest caliber, and one whose sharp intellect, innate curiosity, and engaging enthusiasm were truly inspirational. On a personal note, we often anguished as Bernie positioned himself in the center of controversy—even when the controversy was not really understood as such at the time. Yet our development as individuals and scientists was profoundly influenced by his knowledge, exuberance, and insight. He provoked us to think deeply, and demanded of himself, and those fortunate to have worked with him, the pursuit of scholarship at the highest standards.

His vision was ahead of its time, and encompassed a vital recognition of the inseparable links among physiology, cell biology, structural biology, molecular biology, and biochemistry. That he brought this expansive platform to our field and to the editorship of the *Journal of Cell Biology* ought not surprise any of us. His loss to science is incalculable. His contributions to the gap junction field are a pillar that will surely stand the test of time.

Bernie's life enriched all of us, and we are grateful to have had his stewardship for what seems like a fleeting moment. We deeply mourn his passing, and offer our deepest heartfelt sympathy to Bessie, Johnathan, and Daniel.

September 29, 2000

Cecilia Lo
Biology Department
University of Pennsylvania

Elliot Hertzberg
Department of Neuroscience and Anatomy and Structural Biology
Albert Einstein College of Medicine

Contributors

SÉBASTIEN ALCOLÉA • *Institut de Biologie du Développement de Marseille, Université de la Méditerranée, Marseille, France*

DAVID L. BECKER • *Department of Anatomy and Developmental Biology, University College London, London, England*

E. LUCIO BENEDETTI • *Institut J. Monod, Université Paris VII, Paris, France*

CARVILLE G. BEVANS • *Department of Molecular Biophysics and Biochemistry, Yale University School of Medicine, New Haven, CT*

HANS BLOEMENDAL • *Department of Biochemistry, University of Nijmegen, Nijmegen, The Netherlands*

ROBERTO BRUZZONE • *Institut Pasteur, Paris, France*

FELIKSAS F. BUKAUSKAS • *Department of Neuroscience, Albert Einstein College of Medicine, Bronx, NY*

FENGLI CAO • *Department of Biological Sciences, SUNY at Buffalo, Buffalo, NY*

PETER L. CARLEN • *Playfair Neuroscience Unit, Bloorview Epilepsy Programme, University of Toronto, Toronto, Canada*

GERHARD DAHL • *Department of Physiology and Biophysics, University of Miami School of Medicine, Miami, FL*

BRUNO DELAGE • *Laboratoire de Physiologie Cellulaire, Université de Poitiers, Poitiers, France*

JEAN DÉLÈZE • *Laboratoire de Physiologie Cellulaire, Université de Poitiers, Poitiers, France*

IRENE DUNIA • *Institut J. Monod, Université Paris VII, Paris, France.*

MATTHIAS M. FALK • *Department of Cell Biology, The Scripps Research Institute, La Jolla, CA*

CHRISTIAN GIAUME • *INSERM U114, Collège de France, Paris, France*

GARY S. GOLDBERG • *Department of Biological Sciences, SUNY at Buffalo, Buffalo, NY*

COLIN R. GREEN • *Department of Anatomy with Radiology, University of Auckland, School of Medicine, Auckland, New Zealand*

DANIEL GROS • *Institut de Biologie du Développement de Marseille, Université de la Méditerranée, Marseille, France*

ANDREW L. HARRIS • *Department of Pharmacology and Physiology, New Jersey Medical School, Newark, NJ*

OLFA HENTATI-KSIBI • *Laboratoire de Physiologie Cellulaire, Université de Poitiers, Poitiers, France*

JEAN-CLAUDE HERVÉ • *Laboratoire de Physiologie Cellulaire, Université de Poitiers, Poitiers, France*

JEAN X. JIANG • *Department of Biochemistry, University of Texas Health Science Center, San Antonio, TX*

HABO J. JONGSMA • *Department of Medical Physiology, University Medical Center, Utrecht, The Netherlands*

KAREN JORDAN • *Department of Anatomy and Cell Biology, University of Western Ontario, Ontario, Canada*

NALIN M. KUMAR • *Department of Cell Biology, The Scripps Research Institute, La Jolla, CA*

WENDY E. KURATA • *Cancer Research Center, University of Hawaii at Manoa, Honolulu, Hawaii*

DALE W. LAIRD • *Department of Anatomy and Cell Biology, University of Western Ontario, Ontario, Canada*

WOUTER H. LAMERS • *Department of Anatomy and Embryology, Academic Medical Center, University of Amsterdam, Amsterdam, The Netherlands*

PAUL D. LAMPE • *Fred Hutchinson Cancer Research Center, Seattle, WA*

ALAN F. LAU • *Cancer Research Center, University of Hawaii at Manoa, Honolulu, Hawaii*

LEE YONG LAW • *Department of Anatomy with Radiology, University of Auckland, School of Medicine, Auckland, New Zealand*

LUC LEYBAERT • *Department of Physiology and Pathophysiology, University of Ghent, Ghent, Belgium*

JUN SHENG LIN • *Department of Anatomy with Radiology, University of Auckland, School of Medicine, Auckland, New Zealand*

DIETER MANTHEY • *Institut für Genetik, Universität Bonn, Bonn, Germany*

IRÈNE MARICS • *Institut de Biologie du Développement de Marseille, Université de la Méditerranée, Marseille, France*

PAOLO MEDA • *Département de Morphologie, Centre Médical Universitaire, Genève, Switzerland*

MARY MERRITT • *Department of Biological Sciences, SUNY at Buffalo, Buffalo, NY*

ANTOON F. M. MOORMAN • *Department of Anatomy and Embryology, Academic Medical Center, University of Amsterdam, Amsterdam, The Netherlands*

LINDA S. MUSIL • *Vollum Institute, Oregon Health Sciences University, Portland, OR*

BRUCE J. NICHOLSON • *Department of Biological Sciences, SUNY at Buffalo, Buffalo, NY*

PIERRE NICOLAS • *Institut J. Monod, Université Paris VII, Paris, France*

JOSE L. PEREZ VELAZQUEZ • *Playfair Neuroscience Unit, Bloorview Epilepsy Programme, University of Toronto, Toronto, Canada*

GUY A. PERKINS • *Department of Neurosciences, University of California, San Diego, CA*

ARNOLD PFAHNL • *Department of Physiology and Biophysics, University of Miami School of Medicine, Miami, FL*

MICHEL RECOUVREUR • *Institut J. Monod, Université Paris VII, Paris, France*

MARTIN B. ROOK • *Department of Medical Physiology, University Medical Center, Utrecht, The Netherlands*

RENATO ROZENTAL • *Department of Neuroscience, Albert Einstein College of Medicine, Bronx, NY and the Institute of Tropical Pathology and Public Health, Federal University of Goias, Goiania, Brazil*

MICHAEL J. SANDERSON • *Department of Physiology, University of Massachusetts Medical School, Worcester, MA*

QING SHAO • *Department of Anatomy and Cell Biology, University of Western Ontario, Ontario, Canada*

I. MARTHA SKERRETT • *Department of Biological Sciences, SUNY at Buffalo, Buffalo, NY*

JOSEPH F. SMITH • *Department of Biological Sciences, SUNY at Buffalo, Buffalo, NY*

GINA E. SOSINSKY • *Department of Neurosciences, University of California, San Diego, CA*

DAVID C. SPRAY • *Department of Neuroscience, Albert Einstein College of Medicine, Bronx, NY*

MIDUTURU SRINIVAS • *Department of Neuroscience, Albert Einstein College of Medicine, Bronx, NY*

MAGALI THÉVENIAU-RUISSY • *Institut de Biologie du Développement de Marseille, Université de la Méditerranée, Marseille, France*

E. BRADY TREXLER • *Department of Neuroscience, Albert Einstein College of Medicine, Bronx, NY*

VINZENZ M. UNGER • *Department of Molecular Biophysics and Biochemistry, Yale University, New Haven, CT*

TAUFIK A. VALIANTE • *Playfair Neuroscience Unit, Bloorview Epilepsy Programme, University of Toronto, Toronto, Canada*

HAROLD V. M. VAN RIJEN • *Department of Medical Physiology, University Medical Center, Utrecht, The Netherlands*

JUDY K. VANSLYKE • *Vollum Institute, Oregon Health Sciences University, Portland, OR*

RICHARD D. VEENSTRA • *Department of Pharmacology, SUNY Health Science Center, Syracuse, NY*

LAURENT VENANCE • *INSERM U114, Collège de France, Paris, France*

FRANCK VERRECCHIA • *Laboratoire de Physiologie Cellulaire, Université de Poitiers, Poitiers, France*

VYTAS K. VERSELIS • *Department of Neuroscience, Albert Einstein College of Medicine, Bronx, NY*

BONNIE J. WARN-CRAMER • *Cancer Research Center, University of Hawaii at Manoa, Honolulu, Hawaii*

RONALD WILDERS • *Department of Medical Physiology, University Medical Center, Utrecht, The Netherlands*

KLAUS WILLECKE • *Institut für Genetik, Universität Bonn, Bonn, Germany*

MARK YEAGER • *Division of Cardiovascular Diseases, Scripps Clinic and Research Foundation, La Jolla, CA*

LAN ZHOU • *Department of Biological Sciences, SUNY at Buffalo, Buffalo, NY*

HUI ZHU • *Department of Biological Sciences, SUNY at Buffalo, Buffalo, NY*

I

Tools to Study Connexins

1

Investigation of Connexin Gene Expression Patterns by *In Situ* Hybridization Techniques

Magali Théveniau-Ruissy, Sébastien Alcoléa, Irène Marics, Daniel Gros, Antoon F. M. Moorman, and Wouter H. Lamers

1. Introduction

The principle of *in situ* hybridization is the specific annealing of a labeled DNA or RNA probe to complementary sequences present in tissues, followed by detection of the probe. The complementary sequences can be present in chromosomal DNA, viral DNA or RNA, transcript RNA, etc. A particular *in situ* hybridization technique corresponds to each type of these nucleic acids. All these techniques, including *in situ* hybridization to transcript RNA, have been described in detail in several books or specialized papers that the reader is invited to consult (e.g., *1–6*). This chapter deals only with *in situ* hybridization to transcript RNA, and it is focused on the localization of connexin (*Cx*) gene transcripts in mammalian tissues. In the text below the expression "*in situ* hybridization" is used only in the restrictive sense of "*in situ* hybridization to transcript RNA".

In situ hybridization and immunohistochemistry are complementary techniques that enable spatiotemporal analysis of gene expression in a topographical context. As a first approach to gene expression, the *in situ* hybridization technique provides results more quickly than immunohistochemistry when the required antibodies are not available. Transcript detection by *in situ* hybridization can be carried out either on intact embryos with nonradioactive probes (whole-mount *in situ* hybridization technique) or on tissue sections with radioactive or nonradioactive probes *(7,8)*. The choice between a radioactive and a nonradioactive approach depends on variables such as the sensitivity and the spatial resolution of detection that is required, and whether or not the detection method should be quantitative so as to allow determination of the cellular content

From: *Methods in Molecular Biology, vol. 154: Connexin Methods and Protocols*
Edited by: R. Bruzzone and C. Giaume © Humana Press Inc., Totowa, NJ

of mRNA molecules. The nonradioactive approach, on whole-mount embryos or on sections, is the best technique if the spatial resolution is the determining factor but in spite of considerable improvements *(8,9)*, its sensitivity remains lower than that of the radioactive approach. The latter technique is the method of choice if the transcript to be detected is present only in low abundance in the biological material investigated.

The choice between a whole-mount and a sectioning technique depends on variables such as the size of the tissue being investigated and the three-dimensional complexity of the structures being stained. The whole-mount approach, which has emerged with advances in developmental biology, is more successful at early developmental stages and is faster than an isotopic technique. A complex staining pattern could limit the usefulness of this approach because of the superimposition of stained features, but this may be solved by analyzing serial sections of embryos after the probe immunodetection step. In addition, in most cases, a whole-mount approach has to be accompanied by an analysis of stained embryo sections to identify the cells that express the gene transcript being investigated. As embryos grow, both their size and the extracellular matrix content increase. Both parameters limit the penetration of probes and hence rule out the use of a whole-mount approach. However, even in a sectioning approach, there is a penetration problem: small molecules, such as substrates for enzyme-histochemical reactions, easily penetrate adult and embryonic tissues *(10)* but macromolecules, such as enzymes, antibodies, and polynucleotides *(11)*, hardly penetrate into adult tissues, irrespective of whether the membranous structures are intact (unfixed or formaldehyde-fixed tissue) or destroyed (precipitating fixatives such as alcohols or acetone, or addition of detergents to the probes). Finally, the localization of transcripts in complex organs or tissues by a sectioning approach requires a three-dimensional reconstruction to fully make the most of the results.

The first report dealing with the localization of *Cx* gene transcripts by *in situ* hybridization was published by Micevych and Abelson in 1991 *(12)*. The radioactive *in situ* hybridization technique on sections used by these authors to investigate the distribution of *Cx32* and *Cx43* mRNAs in rat brain was later repeated, and improved, by several authors, on organs other than brain and for the transcripts of various *Cx* genes *(13–17)*. Application of the whole-mount *in situ* hybridization technique for the localization of *Cx* gene transcripts has appeared more recently and has been used on both zebrafish *(18)* and mammalian *(17,19,20)* embryos.

The techniques described below are based on our experiences with, on the one hand, the localization of different *Cx* gene transcripts in the cardiovascular system of mouse embryos with the whole-mount *in situ* hybridization technique *(17,20)* (*see* **Subheading 3.1.**) and, on the other hand, the localization of a variety of mRNAs, including *Cx* mRNAs, in rat, mouse, and human tissues (embryonic, fetal and adult) with the *in situ* hybridization technique on sections with radioactive probes *(15–17,21,22)* (*see* **Subheading 3.2.**).

1.1. Whole-Mount In Situ Hybridization

The whole-mount *in situ* hybridization technique (*see* **Subheading 3.1.**) consists, first, in incubating fixed embryos of different developmental stages with a digoxigenin (DIG)-labeled riboprobe (cRNA) and then, after hybridization, in detecting this probe with anti-DIG–alkaline phosphatase conjugated antibodies which, in turn, are localized with a chromogenic substrate. The chromogenic reaction stains the cells that express the investigated gene transcripts, and thus enables their visualization in a three-dimensional context throughout development. Many embryos can be processed in a single experiment (manually, or better, using an automate) so that statistically significant results are obtained by comparing the embryos to each other. Recently, this method has been expanded to the simultaneous use of two different probes on the same embryos to perform two-color *in situ* hybridizations *(23)*.

1.2. Radioactive In Situ Hybridization on Tissue Sections

The protocol described below (*see* **Subheading 3.2.**) is based on our experience with the detection and localization of various mRNAs in different tissues initially using ^{35}S-labeled cDNA probes *(29)*. This protocol essentially follows that described by Holland *(30)*, but is laced with minor modifications found over the years to be beneficial or more convenient. We have adapted this protocol for the use of cRNA probes *(15–17,21,22)* and assessed the quantitative aspects of the procedure *(31)*. It was applied in particular to rat and mouse embryo sections to identify the *Cx* genes expressed in specific structures such as the trabeculae or the developing compact myocardium of the heart, or parts of the conduction system such as the nodes *(15–17)*. By exchanging the radioactively labeled probe for a nonradioactively labeled probe, as described in **Subheadings 2.1.** and **3.1.**, a nonradioactive *in situ* analysis can be performed on sections *(8)*.

2. Materials
2.1. Whole-Mount In Situ Hybridization

All the solutions have to be made RNase-free to preserve the integrity of both the riboprobe and the embryo transcripts themselves until the first washes of the embryos to eliminate the unbound riboprobe (**Subheading 3.1.4.1.**) (*see* **Note 1**).

2.1.1. Riboprobe Synthesis

1. DEPC-treated water: 0.1% (v/v) diethyl pyrocarbonate (DEPC; Fluka) is added to distilled water, mixed thoroughly, kept at room temperature (RT) under a chemical hood in an open bottle (to allow vapors to escape) for at least 12 h, and autoclaved (*see* **Note 2**).
2. Phenol and chloroform (chloroform–isoamylalcohol 24:1, v/v).
3. 3 *M* Sodium acetate in DEPC-treated water, pH 5.2.
4. 100% and 70% (in DEPC-treated water) ice-cold ethanol.
5. RNA polymerases (T3, T7, or SP6), 10× solution NTP–DIG (nucleotide trisphosphates–digoxigenin) and transcription buffer, all supplied by the manufacturer (Boehringer-Mannheim).
6. RNase inhibitor (RNasin) and DNase I (RNase-free) (Boehringer-Mannheim).
7. Tris-buffered saline solution (TBS): 0.1 *M* Tris-HCl, 10 m*M* EDTA, 1% SDS in bidistilled water, pH 7.5. Sterilize.
8. 6 *M* Ammonium acetate in DEPC-treated water.

2.1.2. Fixation of the Embryos

1. Phosphate-buffered saline (PBS). A 500-mL 10× stock solution is made up of DEPC-treated water (1.5 *M* NaCl, 0.1 *M* Na-phosphate, pH 7.4). This solution is stored at RT. The 1× solution is made fresh as required from the 10× stock solution with DEPC-treated water.
2. PBT buffer: 0.1% Tween-20 (v/v) in 1× PBS.
3. Fixation solution A: 4% Formaldehyde solution in PBT made from paraformaldehyde powder (Sigma). Paraformaldehyde is more readily soluble in hot PBT (60–70°C) and at basic pH (add one or two drops of 1 *N* NaOH to the solution while stirring). Adjust the pH after cooling (*see* **Note 3**).
4. 100% ethanol and 25%, 50% and 75% ethanol solutions in PBT.

2.1.3. Prehybridization and Hybridization Steps

These solutions are prepared from molecular biology grade products.
1. 30% H_2O_2 solution (Sigma).
2. Proteinase K (Sigma): 10 µg/mL in DEPC-treated water.
3. Glycine solution: 2 mg/mL glycine in PBT.
4. Fixation solution B: 0.2% Glutaraldehyde in fixation solution A. The fixation solution B is made fresh as required from both fixation solution A and an 8% glutaraldehyde stock solution. The 8% glutaraldehyde stock solution is made from a 70% glutaraldehyde solution (Sigma) diluted with DEPC-treated water, aliquoted, and stored at –20°C.
5. SSC (20× stock solution): 3 *M* NaCl, 0.3 *M* Na-citrate, pH 7.2 in DEPC-treated water.
6. Formamide (analytical grade).
7. Hybridization buffer: 50% Formamide, 5× saline sodium citrate (SSC), 50 µg/mL of yeast total RNA (Sigma), 1% sodium dodecyl sulfate (SDS), 50 µg/mL of

heparin (Sigma) in DEPC-treated water. Stocks of yeast total RNA (50 mg/mL) and heparin (50 mg/mL) are aliquoted and stored at –20°C.

2.1.4. Immunodetection of the Riboprobe

1. Washing solution 1: 50% formamide (v/v), 5X SSC, 1% SDS in distilled water.
2. Washing solution 2: 0.5 *M* NaCl, 10 m*M* Tris-HCl, pH 7.5, 0.1% Tween-20 in distilled water.
3. Washing solution 3: 50% formamide, 2X SSC, in distilled water.
4. RNase A (Boehringer-Mannheim), 100× stock solution: 10 mg/mL in distilled water. The stock solution is stored at –20°C and working solution (100 µg/mL) is made as required in distilled water.
5. TBST (10× stock solution): 1.4 *M* NaCl, 27 m*M* KCl, 0.25 *M* Tris-HCl, pH 7.5, 1% Tween-20 in distilled water. This stock solution is stored at RT and 1× solutions are made fresh as required.
6. Normal sheep serum (Sigma).
7. Embryo acetone powder *(24)*.
   (a) Homogenize 20 mouse embryos of 13.5 d (E13.5) in PBS (about 20 mL) at 4°C using a Potter-type homogenizer.
   (b) Add four volumes of cold acetone and mix vigorously.
   (c) Keep on ice for 30 min with occasional vigorous mixing.
   (d) Centrifuge for 10 min at 4°C at 10,000g.
   (e) Discard the supernatant. Repeat steps 2–5. Transfer the pellet onto a clean filter paper and let air-dry. Store the powder in an air-tight container (14 mL, disposable round-bottom plasticware from Falcon) at –20°C. This powder is stable for months.
8. Alkaline phosphatase-conjugated anti-DIG antibodies (sheep Fab fragments) (DIG nucleic acid detection kit, Boehringer-Mannheim).
9. Reaction buffer (NTMT): 100 m*M* NaCl; 100 m*M* Tris-HCl, pH 9.5, 50 m*M* MgCl$_2$, 0.1% Tween-20 in distilled water. This solution must be made fresh as required.
10. Staining solution: 10 mL of reaction buffer plus 45 µL of 4-nitroblue tetrazolium chloride (NBT) and 35 µL of 5-bromo-4-chloro-3-indolyl phosphate (BCIP) (DIG nucleic acid detection kit, Boehringer-Mannheim).
11. Glycerol/PBT solutions: 1:1 (v/v) and 4:1 (v/v) solutions supplemented with 0.02% NaN$_3$ (w/v) as preservative.

2.1.5. Embedding and Sectioning of the Stained Embryos

1. 100% Isopropanol and 50%, 75% and 90% isopropanol solutions in PBS.
2. 0.15 *M* NaCl solution in distilled water.
3. Embedding wax: 98% Paraplast (Monojet Scientific) (melting point: 55–57°C; contains dimethyl sulfoxide (DMSO) for rapid tissue infiltration), 2% yellow beewax (Prolabo). Mix both components and melt at 60°C.
4. Two baths (A and B) of xylene in slide containers (*see* **Note 4**).
5. Mounting solution. Mix 6 g of glycerol, 2.4 g of Mowiol (Calbiochem), and 6 mL of distilled water for 2 h at RT, then add 12 mL 0.2 *M* Tris-HCl, pH 8.5, and

incubate under shaking for 30 min at 50°C. Centrifuge for 15 min at 5000*g* to precipitate small aggregates and particles, aliquot the supernatant, and store at –20°C. This mounting solution is stable for months.

2.2. Radioactive In Situ Hybridization on Tissue Sections

2.2.1. Tissue Processing

To prevent RNase contamination, *see* **Note 1**.

1. DEPC/ethanol: 0.1% diethylpyrocarbonate (DEPC) in 96% ethanol. DEPC-containing solutions should be prepared just before use (*see* **Note 2**).
2. AAS: 2% aminoalkylsilane (3-aminopropyltriethoxysilane from Sigma or 3-triethoxysilylpropylamine from Merck) in acetone (dry, 99.5% pure). AAS should be stored at 4°C in a tightly closed vial containing anhydrous calcium sulphate or another drying agent. Let the container warm to RT before opening.
3. PBS (10× stock solution): 1.5 *M* NaCl, 0.1 *M* Na-phosphate, pH 7.4, in autoclaved bidistilled water.
4. 4% Formaldehyde fixative: dissolve 4% paraformaldehyde in sterile PBS at 60–70°C while stirring; add a few drops of 1 N NaGH; dissolution takes about 1.5 h; cool to RT and check the pH before use. *See* **Note 3** for formaldehyde handling.
5. Ethanol solutions *(96%)* in autoclaved bidistilled water.
6. Embedding medium: Paraplast plus (Sherwood Medical, St. Louis, MO, USA).

2.2.2. Probe Labeling

1. 100 m*M* dithiothreitol (DTT) in autoclaved bidistilled water; store in small aliquots at –20°C.
2. RNA Polymerase: T3, T7, and SP6 polymerase (*see* **Note 5**).
3. RNA Polymerase buffer: Supplied with the RNA polymerase.
4. RNasin (Promega): RNase inhibitor.
5. Ribonucleotides (HT BioTechnology, Cambridge, UK): A, C, G-trisphosphate (for single-label transcription assays) and A, G-trisphosphate (for double-label transcription assays) are mixed from commercial solutions with a neutral pH and diluted in bidistilled water.
6. [^{35}S]CTP, [^{35}S]UTP (1000 Ci/mmol) (Amersham).
7. Alkaline hydrolysis mix (2× stock solution): 80 m*M* NaHCO$_3$, 120 m*M* Na$_2$CO$_3$, pH 10.2; in autoclaved bidistilled water. Store in aliquots at –20°C and discard after use.
8. TE (100× stock solution): 1 *M* Tris-HCl, pH 8.0, 0.1 *M* EDTA; in autoclaved bidistilled water.
9. 10 mg/mL tRNA: Dissolve commercially available yeast total RNA at a concentration of 10 mg/mL in TE (1×), 0.1 *M* NaCl. Extract twice with phenol (TE-saturated) and twice with chloroform. Precipitate with 2.5 volumes ethanol at RT. Centrifuge at 5000 *g* for 15 min at 4°C. Redissolve in TE, measure the concentration

spectrophotometrically (an OD at 260 nm of 1 corresponds to 40 µg of RNA) and store at a concentration of 10 mg/mL at –20°C in small aliquots.
10. 3 *M* Sodium acetate, pH 5.2, in autoclaved bidistilled water.

2.2.3. Pretreatment of Paraffin Sections

1. SSC (20× stock solution): 3 *M* NaCl, 0.3 *M* Na-citrate, pH 7.2, in DEPC-treated water.
2. 2 *N* HCl: Add 20 mL of 36% HCl (12 *N*) to 100 mL of bidistilled water.
3. Pepsin (800–2500 U/mg, Sigma): 10% stock solution (w/v) in bidistilled water, predigest for 2 h at 37°C.
4. Pepsin (0.1% working solution in 0.01 *N* HCl): add 0.5 mL of 2 *N* HCl to 100 mL of bidistilled water; warm it to 37°C and add 1 mL of pepsin stock solution approx 15 min before use. The capacity of this solution is approx 40 slides with sections when used within 3 h after preparation.
5. Glycine: 10% (w/v) stock solution in PBS.
6. 10 m*M* DTT in autoclaved bidistilled water. Store at –20°C and keep on ice during use. The solution can be safely frozen and thawed five times.

2.2.4. Hybridization Solutions

1. SSC: 20X stock solution (*see* **Subheading 2.2.3.**).
2. 50% (w/v) Dextran sulfate in 10X SSC: Add approx 3 mL of bidistilled water to 10 mL of 20X SSC and slowly dissolve 10 g of Dextran sulfate by stirring and heating (60°C). Add Dextran in steps and wait until it is dissolved before adding new powder. Adjust final volume to 20 mL and autoclave for only 20 min.
3. Deionized formamide: Add 5 g of Biorad AG 501-X8 mixed-bed resin (20–50 mesh) to 100 mL of formamide; stir for 1 h. Filter through a Whatman no. 1 filter and store in batches at –20°C.
4. 50X Denhardt's solution: 1% (v/v) Ficoll 400, 1% (v/v) polyvinylpyrrolidone, 1% bovine serum albumin (fraction V) in bidistilled water. Sterilize by filtration and store in batches at –20°C.
5. 10% (v/v) Triton X-100: dilute Triton X-100 in sterile bidistilled water and store at 4°C.
6. 1.25× hybridization mixture: 5 mL of deionized formamide; 2 mL of 50% dextran sulfate 10X SSC; 0.4 mL of 50X in Denhardt's solution; 0.1 mL of 10% Triton X-100; 0.5 mL of bidistilled water (sterile); store in batches at 4°C.
7. DTT: make a 1 *M* stock solution in autoclaved bidistilled water and store in small batches at –20°C.
8. Salmon sperm DNA: Make a 10 mg/mL solution in bidistilled water by stirring it overnight at 4°C. Adjust the NaCl concentration to 0.1 *M*. Extract the solution once with phenol and once with phenol–chloroform (1:1, v/v) *(26)*. Recover the aqueous phase and shear the DNA by passing it 12 times rapidly through a 17-gauge hypodermic needle. Precipitate the DNA by adding two volumes of ice-cold ethanol. Centrifuge, dry the pellet, and dissolve the pellet in bidistilled

water, measure the OD_{260}, and adjust the concentration to 10 mg/mL (an OD_{260} of 1 corresponds to 50 µg of DNA). Boil for 10 min and chill on ice. Store in batches at −20°C.

9. 50% formamide 1X SSC. If the formamide used to prepare this solution is not deionized, check the pH, which should be neutral.

10. 5× RNase buffer: 50 mM Tris-HCl, pH 8.0, 25 mM EDTA, 2.5 M NaCl in autoclaved bidistilled water.

11. RNase A solution (10 mg/mL). To make RNase that is free of DNAse, dissolve RNase at a concentration of 10 mg/mL in 10 mM Tris-HCl, pH 7.5, 15 mM NaCl and heat it for 15 min at 100°C. Cool slowly to RT. Dispense into aliquots and store at −20°C.

2.2.5. Probe Detection

1. Photographic emulsion: Ilford Nuclear Research Emulsion G-5 (*see* **Note 6**).

2. Glycerol–water: Add 1.2 mL of glycerol to 59 mL of bidistilled water. Filter through a 0.22 µm filter and store at 4°C.

3. 10% Potassium bromide (KBr) in autoclaved bidistilled water.

4. Amidol-developer: Prepare a fresh solution just before developing. Dissolve 1.13 g of Amidol (4-hydroxy 1,3-phenylene diammonium dichloride) (Fluka) in 200 mL bidistilled water, add 4.5 g of Na_2SO_3 and bring it to a final volume of 250 mL with bidistilled water. Filter the solution through a regular filter paper (e.g., Whatman no. 1) and add 2 mL of 10% KBr.

5. Fixative: 30% (w/v) $Na_2S_2O_3 \cdot 5H_2O$ in bidistilled water.

6. Nuclear-Fast Red (stock solution): 0.1% Nuclear fast-red (Kernechtrot) and 5% aluminum sulfate; dissolve in bidistilled water and filter. Prepare 250-mL bottles and store at 4°C.

7. Nuclear-Fast Red (working solution): Dilute the stock solution 1:5 with bidistilled water. This solution can be used for several weeks.

8. Stainless steel cuvettes ($105 \times 60 \times 40$ mm), lids (7 mm high), and slide holders (12 slides) were made in our workshop (*see* **Fig. 1**).

9. Malinol mounting medium (cat. no. 2C-242) from Chroma Gesellschaft, Germany.

2.3. Apparatus

1. For RNA probe analysis: Minigel electrophoresis apparatus (Mupid-2, Eurogentec SA) or equivalent, spectrophotometer (Perkin Elmer).

2. Dissection material: Sterilized scissors (heavy model, 14 cm long, two sharp blades; Barraquer Wolff, sharp and delicate blades, 5 mm long) and tweezers of different sizes (Moria, 12.5 cm long, serrated jaws; Moria no. 5 straight extra delicate; Dumont-Moria, straight, tips closing flat on 4 mm), sterile Petri-dishes (35×10 mm, Falcon), a binocular (MZ6, Leica-Wild) equipped with optic fibers (Intralux 4000–1, Volpi).

3. Standard benchtop centrifuges and microfuges.

4. Standard heating and shaking water bath incubators.

5. Screw-capped glass containers (2 mL; 12×35.5 mm, Polylabo).

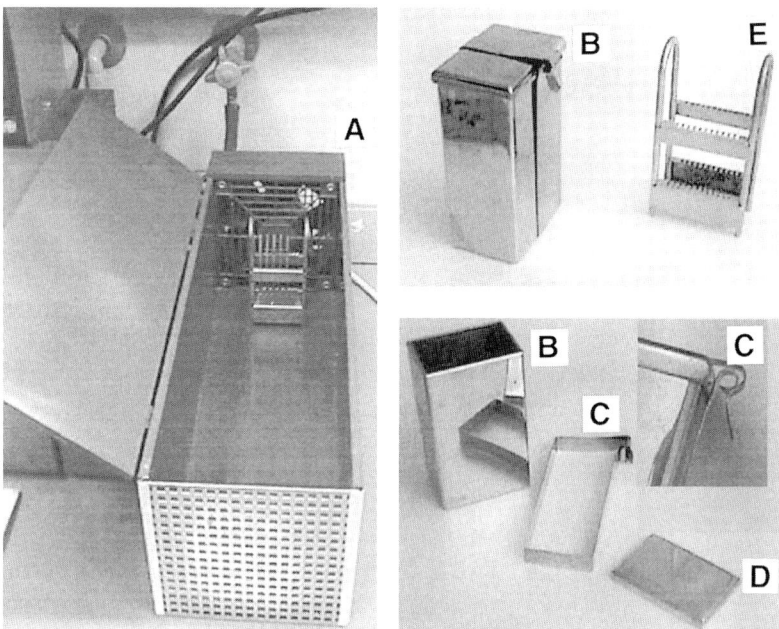

Fig. 1. Homemade equipment used for *in situ* hybridization. **Left panel:** Device to dry AAS-coated slides. Note the fan in rear and the position of the slide holder. **Right upper panel:** Box (*left*) and slide holder (*right*) used for washing. **Right lower panel:** Detail of locking device for the box, which facilitates autoclaving.

6. 14 mL Disposable round-bottom plasticware (Falcon).
7. Multiwell Tissue flat-bottom culture plates (Falcon) (ranging from 6–24 wells and used according to the size of the treated embryos).
8. Precleaned SuperFrost microscope slides.
9. Watch glass (Polylabo) for embedding embryos in wax.
10. Slide containers (Polylabo) for the unwaxing procedure.
11. Observations and photographs of stained embryos: A binocular (M420, Leica-Wild) equipped with optic fibers (Intralux 6000-1, Volpi) and a photoautomate (MPS 48, Leica-Wild).
12. For sectioning of stained embryos and examination of sections: A mechanical microtome, a slide-warming plate, a microscope (Axiophot, Zeiss) equipped with Nomarski optics, and a photoautomate (Zeiss).
13. Examination of sections requires a microscope equipped with dark-field illumination and a conventional or digital camera.
14. A cooled microfuge, equipped with a swingout rotor to centrifuge the labeled probes and prevent spillage of radioactivity.
15. A temperature-controlled stove for the hybridization of the sections. Humidity is maintained by keeping sections in plastic boxes containing wet paper towels.

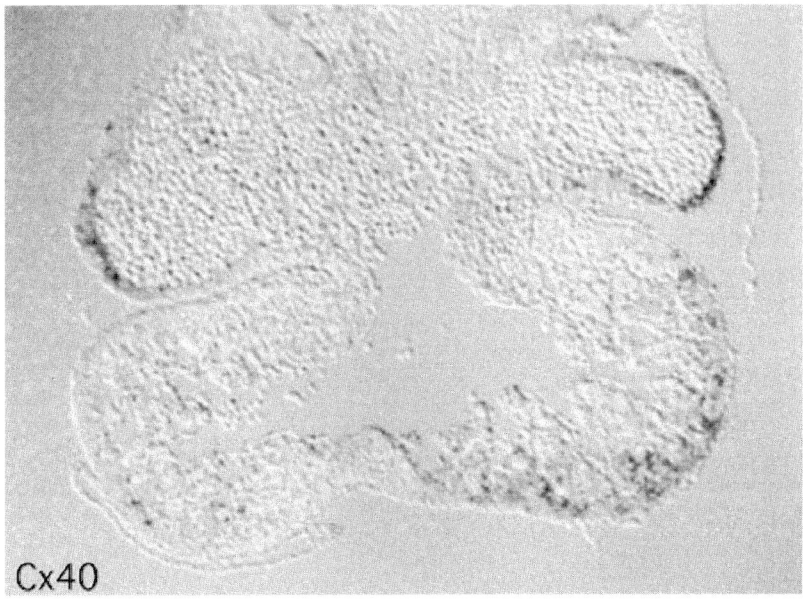

Fig. 2. Section in ED10 mouse embryo hybridized in whole-mount with antisense *Cx40* probe. Note that both left and right atrial walls express *Cx40* transcript; whereas in the ventricles only the left wall shows accumulation of the transcript.

16. Darkroom equipment. Entry to the darkroom is via a light trap to prevent inadvertent access of light. Illumination is provided by a light-proof armature, fitted with an Ilford S 902 filter, and a 15 W lamp. Water baths that are kept at 42°C and at 18°C (cooled), are present to handle the photographic emulsion. Furthermore, a spark-free refrigerator is needed for exposure of sections. The sections themselves are kept in black, hard-plastic slide holders, inside light-tight boxes.

17. A microscope (Axiophot, Zeiss), equipped with a cooled CCD camera (Photometrics series 200 with CAF 1400 chip, Roper Scientific, Bogart, Georgia, USA) (*31*). Image analysis is performed with the NIH Image package (http://rsb.info.hih.gov/nih-image/) or, more recently, the Optimas package (Media Cybernetics, Silver Spring, MD, USA).

3. Methods

3.1. Whole-Mount In Situ Hybridization (see Fig. 2)

To be efficient and reliable, the *in situ* hybridization technique has to comply with certain key requirements, such as: (1) keep the loss of RNA as low as possible; (2) maintain the embryo morphology as close as possible to the native structure; (3) use efficient probes that have high penetration and hybridization capacities, and (4) design specific controls (*see* **Notes 7** and **35**).

3.1.1. Riboprobe Synthesis

We have chosen to work with riboprobes (cRNA) instead of DNA probes, because: (1) they can be synthesized with high efficiency and (2) they form, by hybridization with endogenous transcripts, RNA-RNA hybrids that are more stable than DNA-RNA hybrids *(25)*. The hybridization process can thus be performed at high stringency, and, in addition, the unbound and nonspecifically bound riboprobe can be digested by RNase without affecting the RNA-RNA hybrids. The DNA fragment selected for the synthesis of the riboprobe (*see* **Note 8**) is cloned into a commercial plasmid vector designed for directional probe synthesis. In the vector, the insert of interest is flanked by different RNA polymerase initiation sites (T3, T7, SP6) (*see* **Note 5**).

3.1.1.1. PURIFICATION OF THE LINEARIZED PLASMID DNA

The plasmid containing the DNA insert should be linearized to enable the production of "runoff" transcripts derived from the insert sequence only (*see* **Note 9**).

1. Linearize 10 µg of the plasmid DNA with a restriction nuclease chosen according to the RNA polymerase promoter to be used so that, depending on the orientation of the insert, an antisense or a sense riboprobe will be synthesized.
2. Adjust the reaction volume to 400 µL and purify the linearized plasmid with 1 vol of phenol–chloroform (1:1, v/v) extraction, followed by a second extraction with 1 vol of chloroform *(26)*. At each purification step the mixture is shaken, centrifuged (high speed for 1 min) and the upper aqueous layer is collected into a new RNase-free 1.5-mL Eppendorf tube.
3. Precipitate the plasmid DNA from the aqueous phase by adding 100 µL of a 3 *M* sodium acetate solution plus 1 mL of ice-cold 100% ethanol and incubating that mixture at –70°C for 30 min.
4. Centrifuge for 30 min at 12,000g to pellet the DNA and discard the supernatant.
5. Wash the pellet with 100 µL of ice-cold 70% ethanol in DEPC-treated water, centrifuge for 5 min at high speed, and air-dry.
6. Dissolve the pelletted DNA into 40 µL of DEPC-treated distilled water.

3.1.1.2. SYNTHESIS OF THE NTP-DIG RIBOPROBE

Riboprobes are synthesized as "runoff transcripts" and a non-isotopic hapten, DIG, is incorporated in the probe during its synthesis (*see* **Note 10**).

1. Set up the transcription reaction by adding 1 µg of the linearized plasmid (4 µL of the previous plasmid DNA sample), 2 µL of 10× transcription buffer, 2 µL of 10× solution NTP-DIG, 1 µL (20 U) RNase inhibitor, 2 µL (40 U) RNA polymerase (T3, T7 or SP6) and make up to 20 µL final volume with DEPC-treated water.

2. Vortex-mix and centrifuge.
3. Incubate for 2 h at 37°C.
4. Digest the DNA template with 2 µL (20 U) of DNase I (RNase-free) for 15 min at 37°C. This DNase treatment is optional.
5. Adjust the reaction volume to 50 µL with DEPC-treated water.
6. Add 50 µL TBS, 12 µL of 6 M ammonium acetate solution, 220 µL of ice-cold 100% ethanol and precipitate the riboprobe for 30 min at –70°C.
7. Centrifuge for 30 min at 12,000g and 4°C; wash the pellet with 100 µL of ice-cold 70% ethanol.
8. Carefully remove the supernatant and let the pellet air-dry.
9. Recover the riboprobe in 100 µL of DEPC-treated water. It is stable for several months at –20°C.
10. Estimate the synthesis yield (variable from one template to another and depending on the size of the template) by measuring the optical density (OD_{260}) of the synthesized probe (An OD_{260} of 1 corresponds to 40 µg of single-stranded RNA), then analyze an aliquot by agarose gel electrophoresis. Wash the electrophoresis apparatus and run the riboprobe quickly to avoid RNase digestion. A combination of both procedures enables estimation of the amount of available probe and determination of its size.

3.1.2. Dissection and Fixation of the Embryos

To investigate expression of *Cx* genes in the embryonic cardiovascular system we commonly use embryos ranging from E7 to E12.5. Beyond E12.5, the embryos are relatively large, most of the tissues are compact (especially the outermost layers of the ventricles), and the penetration of the reagents is restricted.

1. Timed pregnant mice are sacrificed, the abdominal cavity is opened, and the uterus is cut with another pair of clean scissors *(27)*. The uterus is dissected under a binocular in a Petri dish filled up with PBS. Remember to keep all dissection material as close as possible to RNase-free conditions.
2. With small scissors remove the muscular wall of the uterus, put embryos in fresh PBS solution, and, if required, dissect and remove the extraembryonic membranes with tweezers. To avoid that dust sticks to embryos or contaminates the samples, each dissection is performed in a different Petri dish with new buffer. Young embryos are transferred from one dish to the other by smooth pipetting with a fire-polished Pasteur pipet with a cut end. When embryos are older than E9 we take care to puncture the heart and brain ventricles so that the riboprobe or the anti-DIG antibodies will not be trapped in the hollow parts of the organs.
3. Dissected embryos are fixed in fixation solution A at 4°C for 45 min (for 8-d embryos) to overnight (for 9.5-d embryos onwards) (*see* **Note 11**). From this point the experiments are carried out either in 2-mL screw-capped containers or in 14-mL disposable plasticware according to the size of the embryos.
4. Eliminate fixation solution A by two quick washes in PBT (3 min each at 4°C).
5. Dehydrate the embryos by passing them sequentially through a graded series of ethanol solutions: 25%, 50%, and 75% ethanol in PBT, 100% ethanol (for 5 min each)

(*see* **Note 12**). Dehydration and rehydration (*see* **Subheading 3.1.3.**) of the embryos should not be skipped because they play a role in the fixation procedure.
6. Embryos can be stored for several months at –20°C in fresh 100% ethanol.

3.1.3. Prehybridization and Hybridization (Day 1)

To obtain reliable results it is recommended to use at least five embryos for each developmental stage investigated and for each probe.

1. Bleach the embryos for 1 h at RT with 100% ethanol–30% H_2O_2 (5:1, v/v). This treatment also inactivates the endogenous phosphatases.
2. Rehydrate through a decreasing series of ethanol solutions in PBT (75%, 50%, 25% for 5 min each), followed by three washes in PBT.
3. Permeabilize with 10 µg/mL of proteinase K (PK) in PBT for 15 min at RT (*see* **Note 13**).
4. Incubate twice 5 min with 0.2% glycine solution (to prevent overdigestion with PK), then twice for 5 min in PBT.
5. Incubate for 20 min at RT with fixation solution B to stabilize the morphology of tissues.
6. Wash twice 5 min at RT with PBT before prehybridization.
7. Prehybridize the embryos with hybridization buffer (in the absence of the probe) for at least 1 h, with gentle shaking in a water bath maintained at 70°C. The prehybridization procedure helps to reduce the background by saturating nonspecific hybridization sites (*see* **Note 14**).
8. Hybridize with fresh hybridization buffer containing a final concentration of labeled probe ranging from 0.5 to 1 µg/mL. The incubation is carried out overnight at 70°C in identical conditions as for prehybridization. Hybridization kinetics is tightly correlated with the time of penetration of the probe. The hybridization time may be modulated according to the length of the probe.
9. Prepare the washing solutions needed for the next day (washing solutions 1 and 3) and prewarm them at 70°C. This will save you a lot of time but wait until the last minute on d 2 before adding RNase to washing solution 2.

3.1.4. Incubation with the Anti-DIG Antibodies (Day 2)

3.1.4.1. Post-Hybridization Washes

Aspirate the hybridization buffer and start the washes.

1. 2×30 min at 70°C in washing solution 1.
2. 1×10 min at 70°C in a mix of washing solution 1/washing solution 2 (1:1, v/v).
3. 3×5 min at RT in washing solution 2.
4. 2×30 min at 37°C in washing solution 2 supplemented with 100 µg/mL of RNase A.
5. 1×5 min at RT in washing solution 2.
6. 1×5 min at RT in washing solution 3.
7. 2×30 min at 70°C in washing solution 3.
8. 3×5 min at RT in TBST.

Washing solutions 1–3 help to reduce the background, not only by washing off the unbound probe but also by destroying the weakly and non-specifically bound probe with the RNase treatment (**step 4**). While the washing procedure is going on, heat inactivate the normal sheep serum and pre-adsorb the anti-DIG antibodies with the embryo powder.

3.1.4.2. HEAT-INACTIVATION OF NORMAL SHEEP SERUM

1. Make 10% normal sheep serum in TBST just before use (*see* **Note 15**).
2. Incubate 30 min at 70°C.
3. Keep at RT until use.

3.1.4.3. PREADSORPTION OF SHEEP ANTI-DIG FABS WITH MOUSE EMBRYO POWDER

This procedure helps to reduce the background due to non-specific recognition of the antibodies.

1. Homogenize 1 mg of embryo powder (*see* **Subheading 2.1.4.**) in TBST ($\cong 1.5$ mL) and incubate for 30 min at 70°C.
2. Centrifuge at high speed for 10 s and remove the supernatant.
3. Add to the pellet: 2.7 mL of TBST, 300 µL of freshly inactivated 10% normal sheep serum (final dilution: 1%), and 3 µL of anti-DIG antibodies (final dilution: 1:1000).
4. Shake for at least 4 h at 4°C, then centrifuge at high speed for 20 s.
5. Collect the supernatant and add 2.7 mL of TBST and 300 µL of inactivated 10% normal sheep serum. The final concentration of anti-DIG antibodies is 1:2000. The total volume of anti-DIG antibodies (diluted to 1:2000) to be prepared depends on the number of embryos treated in one experiment. The volume above can be increased by adjusting the volume used in **step 3**.

3.1.4.4. INCUBATION OF THE EMBRYOS WITH ANTI-DIG ANTIBODIES

1. Pretreat the embryos with inactivated 10% normal sheep serum for at least 60 min at RT with gentle shaking.
2. Incubate the embryos with TBST containing 1:2000 diluted anti-DIG antibodies (*see* **step 5** in **Subheading 3.1.4.3.**), overnight at 4°C, with continuous soft shaking.

3.1.5. Probe Immunodetection (Days 3 and 4)

3.1.5.1. POST-ANTIBODIES WASHES (DAY 3)

The number of washes, the volume of fresh buffer, and the time of incubation increase the sensitivity of the final reaction by decreasing the nonspecific background (*see* **Note 16**).

1. Wash 3×5 min with TBST.
2. Wash 4×1 h with the same buffer.
3. Wash overnight or more with TBST. Two to three days of washing are recom-

mended for embryos older than E9 but the length of this step also depends on the riboprobe used. Check the pH of TBST (7.5) before this step and never wash with reaction buffer, as an alkaline pH will destroy in long washes the activity of the enzyme conjugated to the antibodies.

3.1.5.2. PROBE IMMUNODETECTION (DAY 4)

1. The embryos are transferred to the multiwell plates for easier observation.
2. Preincubate the embryos for 20 min at RT with freshly prepared reaction buffer, pH 9.5, to ensure that ionic concentrations are optimal for the alkaline phosphatase reaction.
3. Replace the reaction buffer with the staining solution. Staining reaction is carried out in the dark in the presence of both substrates, NBT and BCIP, which develop a purple color. Depending on the probe and the abundance of transcript the reaction can last from 30 min to several hours and must be closely monitored to stop the enzymatic reaction before background signals appear (*see* **Note 17**).
4. Stop the reaction with TBST buffer and renew it frequently while keeping the embryos at 4°C.

3.1.5.3. CONSERVATION OF THE EMBRYOS

1. Dehydrate the embryos in a graded series of ethanol solutions in TBST (25%, 50%, and 75%) and in 100% ethanol (5 min for each step at RT). This treatment intensifies the pink-purple reaction product to dark blue.
2. Rehydrate the embryos by passing them through the same solutions inverting the order, from 100% ethanol to TBST (5 min for each step at RT).
3. Clear the embryos in a glycerol-PBT solution 1:1 (v/v) for 1 h at RT.
4. Transfer to fresh glycerol-PBT solution 4:1 (v/v) supplemented with 0.02% NaN$_3$. Embryos can be stored for months at 4°C.
5. Analyze the embryos, including the control embryos, and photograph them (*see* **Note 18**).

3.1.6. Wax-Embedding and Sectioning of the Stained Embryos

Precise identification of labeled tissues and cells requires serial sectioning of the embryos and examination of sections. The detection of a significant signal in sections necessitates that the embryos have developed a strong signal because part of the signal is lost during the unwaxing procedure with xylene. The procedure in the following section describes embedding in embedding wax, but, alternatively, pure paraffin can also be used (*see* **Subheading 3.2.1.2.**).

3.1.6.1. EMBEDDING PROCEDURE

One day before use embedding wax is melted in an incubator at 60°C. Do not overheat.

1. Wash the embryos 3 × 5 min, with PBS at RT.
2. Incubate for 1 h with PBS, then for 1 h with 0.15 M NaCl at RT.

3. Incubate successively in 50% isopropanol in PBS (at least 2 h), 75% isopropanol (at least 3 h), 90% isopropanol (at least 6 h), and twice in 100% isopropanol (at least 6 h × 2).
4. Impregnate the embryos overnight at 60°C with isopropanol/embedding wax 1:1 (v/v), for 6 h with embedding wax, then overnight again with fresh embedding wax.
5. Pour melted embedding wax in molds, then transfer the embryos. Carefully position the embryos using a cut-end Pasteur pipet or tweezers heated in a Bunsen flame. Hot tools are used to keep the wax melted while positioning the embryos under the binocular.
6. Leave to solidify at RT and store until use.
7. Make sections 8–12 μm thick.
8. Lay the sections on a bed of warm water spread on a microscope slide and keep warm on a hot plate until the sections are dried (usually overnight).

3.1.6.2. UNWAXING PROCEDURE

1. Incubate the slides twice in xylene: 10 min in xylene bath A and 10 min in bath B.
2. Rehydrate the slides through a decreasing series of ethanol solutions in distilled water (100%, 90%, 80%, 70%, 40%, 2 min each) followed by two washes in distilled water. Leave to air-dry.
3. Mount the slides with a coverslip by using two to three drops of mounting solution while taking care to avoid air bubbles.
4. Leave the Mounting to air-dry before observation.
5. Observations are made with Nomarski optics.

3.2. Radioactive In Situ Hybridization on Tissue Sections (see Fig. 3)

3.2.1. Tissue Processing (see Note 19)

Embryos are dissected as described in **Subheading 3.1.2.** Small samples (e.g., rat or mouse embryos up to E9) are fixed in 4% formaldehyde for 4 h at 4°C with agitation. Older embryos are fixed overnight at 4°C, but not exceeding 16 h (*see* **Note 20**). From E18, embryos are decapitated and the skin is removed before fixation.

3.2.1.1. PREPARATION OF AS-COATED SLIDES (see Note 21)

1. Put the slides in plastic racks; soak overnight in 1% NaOH (discard after 1 mo); rinse for 15 min in running warm tap water; rinse in demineralized water; soak for 1 h in 2% HCl; rinse for 15 min in running warm tap water; rinse in demineralized water. From here on wear gloves.
2. Place the racks in plastic boxes, filled with DEPC-ethanol for 2 min; repeat this step once. The racks with slides are now RNase free.
3. Put the racks on an RNase-free surface and dry them in a stream of air for approx 1 h. This is generated by a fan, which is placed on one end of a home-made box in which 8 racks are positioned in such a way that slides are parallel to the stream of air (**Fig. 1**). Alternatively, we also use an ordinary fan and

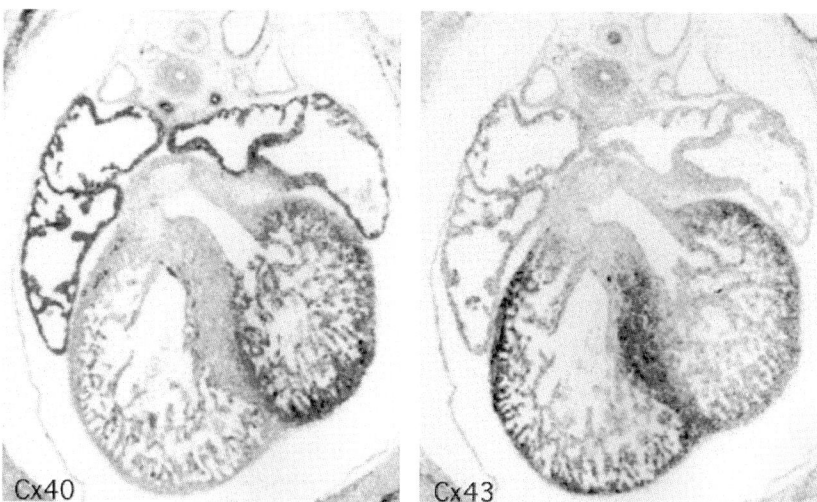

Fig. 3. Adjacent sections in ED16 rat embryo hybridized with radioactive *Cx40* and *Cx43* antisense probes. Note differences in staining intensity between heart compartments.

cutoff chemical containers (e.g., those that contain a kg of NaCl), just to guide the stream of air.

4. Place the racks for 30 s in 2% AAS (250 mL of solution are enough for about 200 slides).
5. Rinse in bidistilled water; move the racks ten times up and down; repeat this step 6×.
6. Place the racks on a RNase-free surface in a stream of air for a few hours or overnight until all slides are dry.
7. Store the slides in an RNase-free box at RT.

3.2.1.2. PARAFFIN EMBEDDING, MOUNTING AND SECTIONING

1. After fixation the tissue is passed through an approx 50-fold volume of a graded series of ethanol solutions (70%, 80%, 90%, 96%, and 100%) with gentle agitation, for about 2 h each, followed by an overnight ethanol exchange in 1-butanol. Use closed containers and handle in fume hood because of explosion hazard.
2. Impregnate the tissue with paraffin at 60°C, 3× for 2 h, the last 2 h under vacuum, if necessary.
3. Pour melted paraffin in a mold and carefully position the tissue.
4. Solidify at RT and store until use. Paraplast-embedded tissue blocks can be stored at RT for several years.
5. Cut 7-μm sections and humidify with bidistilled water if necessary. RNase-free conditions are not necessary during sectioning.
6. Collect the sections on a tray. The sections can be stored at RT for a few weeks before they are mounted onto the slides. Mounting of the sections is done under RNase-free conditions. Wipe the warm plates with an RNase surface decontaminant.

7. Stretch the sections by laying them on sterile bidistilled water on AAS-coated slides placed on a warm plate at 45–50°C; draw off as much of the water as possible by suction, ensuring that no air bubbles are present underneath the tissue and keep the slides at 30°C for a few hours.

8. Dry overnight at 37°C and store the slides.

3.2.2. Probe Labeling

The rationale for working with riboprobes (cRNA), instead of DNA probes, has been discussed in **Subheading 3.1.1.** The protocols for template linearization and purification have been detailed in **Subheading 3.1.1.1.**

3.2.2.1. PREPARATION OF ^{35}S-CALIBRATION SLIDES

1. ^{35}S-Calibration solution: dissolve at 50°C 5% high gel-strength gelatin (Fluka) in 10 mM Tris-HCl, pH 7.8, containing various amounts (87, 375, 750 and 1500 cpm/pg) of ^{35}S-radioactivity. We use [^{35}S]dCTP labeled random-primed cDNA because of its stability.

2. Draw two rows of five circles (7-mm diameter) on the back side of AAS-coated microscope slides.

3. Apply evenly over each circle with a positive displacement pipet, for example. Gilson M25, a 4-μL drop of ^{35}S-calibration solution containing varying amounts of radioactivity (typically 10 spots containing 1000–10,000 cpm) and place the slide immediately on an ice cooled plate for 1 min maximum. Longer cooling results in cracks along the outside of the spots. Repeat the procedure for the preparation of the next calibration slide.

4. Dry the slides in a stream of air for about 30 min and let the spots further fix overnight at 50°C. In theory, one slide/exposure time/development session should suffice, but for superior accuracy two or three calibration slides are recommended.

5. Take 3 × 4 μL-samples for liquid scintillation counting to determine the amount of radioactivity applied on a spot.

6. Fix the drops by immersing the slide in 4% formaldehyde fixative supplemented with 1% glutaraldehyde (4 mL per slide), to prevent swelling of the spots later in the procedure. Glutaraldehyde solution is made from a 25% stock solution (Fluka).

7. Take a 0.5-mL sample of the fixative of each slide for liquid scintillation counting to determine the loss of ^{35}S in the fixation procedure.

8. Dehydrate in 96% ethanol for 10 min (use 4 mL per slide), repeat once, and dry in a stream of air.

9. Process the slides for autoradiography (*see* **Subheading 3.2.5.**).

3.2.2.2. IN VITRO TRANSCRIPTION

The buffer used for in vitro transcription is dependent on the source of RNA polymerase and is provided by the manufacturer of the RNA polymerase. The composition of the final transcription assay is indicated in **Table 1**. Incubation

Table 1
RNA Transcription Assay

Addition	Single label	Double label	Final concentration
Linearized DNA template ~1 μg/mL	0.25 μL	0.25 μL	25 ng/μL
T3/T7 or SP6 buffer (5×)	2 μL	2 μL	
100 mM DTT	1 μL	1 μL	10 mM
RNAsin 40 U/μL	0.25 μL	0.25 μL	1 U/μL
5 mM A, C, G-Trisphosphate	1 μL	—	500 μM
5 mM A, G-trisphosphate	—	1 μL	500 μM
1000 Ci/mmol of [^{35}S]UTP	—	5 μL (freeze-dried)[a]	5 μM
1000 Ci/mmol of [^{35}S]CTP	5 μL	5 μL	5 μM
50 U/μL RNA polymerase	0.5 μL	0.5 μL	2.5U/μL
Total volume	10 μL	10 μL	

[a]The limitations of the final volume may necessitate drying of the label. Label can be dried in a Speedvac within 30 min. The dried label can be stored at –20°C overnight before use. When the probes are used for only one or a limited number of *in situ* hybridization experiments, the assay can be performed with half the volumes.

is for 2 h at 37°C, or for 4 h at 30°C for templates longer than 1500 bp to produce more full length transcripts. Template is removed by the addition of 1 μL of DNase I (RNase-free) at 1 U/μL and incubation for 15 min at 37°C. Ninety microliters of bidistilled water is added and the reaction is put on ice. One microliter is collected to determine by electrophoresis the length of the transcripts (*see* **Note 22**).

3.2.2.3. PROBE HYDROLYSIS

One volume of alkaline hydrolysis mix (2× stock solution) is added to the assay and the RNA is hydrolyzed at 60°C for about 12 min (for a 300 bp probe), 15 min (for a 500 bp probe), 16 min (for a 800 bp probe), 17 min (for a 1500 bp probe), and 18 min (for a 5000 bp probe) (*see* **Note 23**).

3.2.2.4. PROBE PURIFICATION

1. One microliter of tRNA (10 mg/mL) is added to the hydrolyzed RNA, then the solution is vortex-mixed, mixed with 100 μL of phenol (TE saturated), then centrifuged.
2. The (upper) aqueous phase is collected and mixed with 100 μL of chloroform. This solution is centrifuged and the aqueous phase is collected. Two microliters of this phase are counted in a liquid scintillation counter to assess the ^{35}S input value.
3. Add 20 μL of 3 M sodium acetate, pH 5.2, and 440 μL of 100% ethanol (RT) to the aqueous phase, keep overnight at –20°C and then centrifuge for 30 min at 5°C (swinging rotor).

4. Discard the supernatant. Dissolve the pellet in 500 μL of 70% ethanol (RT), keep for 10 min at –20°C, and centrifuge for 10 min. Carefully remove the supernatant, dry the pellet in a Speedvac centrifuge for 10 min, then dissolve in TE (containing 10 mM DTT) to a final concentration of 10^6 cpm/μL. One microliter of this solution is used to estimate the incorporation of radioactivity. Store the probe at –20°C.

3.2.3. Pretreatment of Paraffin Sections

1. Slides are rinsed 3× in xylene for 5 min (see **Note 4**), in xylene–ethanol (1:1, v/v) for 5 min, and twice in 100% ethanol for 3 min and dried for at least 30 min in a stream of air. The pretreatments are carried out in 150 mL stainless steel vessels that can hold one slide tray for 12 slides (**Fig. 1**).
2. Incubate the slides in 2X SSC at 70°C for 10 min and wash with bidistilled water for 5 min. Although the action of this step is unknown, it is essential. After this step the sections are swollen, which may increase the accessibility of the sections for protease and the target RNA for the labeled riboprobe.
3. Pepsin treatment. The time of digestion must be determined empirically for each specific tissue. Each new batch of Pepsin should be tested on a known combination of tissue(s) and probe(s): 0.1% Pepsin dissolved in 0.01 N HCl is used. The incubation temperature is 37°C. When different incubation times are used during one experiment the slides can be left in the bidistilled water wash solution that is used between the SSC step at 70°C and the Pepsin treatment, so that all the slides come together in the glycine stop solution. Typical incubation times are: 20 min for adult and neonatal rat liver and heart; 15 min for E20 rat embryos; 10 min for E16–E18; 5 min for E13–E14; 3 min for E11; and 2 min for E12–E11. The protease reaction is stopped with 0.2% glycine in PBS for 30 s followed by two washes with bidistilled water for 30 s and 5 min, respectively.
4. Incubate in 10 mM DTT for 5 min and dry the slides as quickly as possible to prevent oxidation (see **Note 24**). Hybridize the sections the same day.

3.2.4. Hybridization

3.2.4.1. PREPARATION OF THE HYBRIDIZATION MIX

The composition of 1× hybridization mix (see **Note 25**) is: 50% formamide, 10% Dextran sulfate, 2X SSC, 2X Denhardt's, 0.1% Triton X-100, 50 mM DTT, 200 ng/μL of salmon sperm DNA. Warm the 1.25× hybridization mixture (see **Subsection 2.2.4.**) to RT. For 100 μL of hybridization mix, pipet carefully 80 μL 1.25× hybridization mix (the solution is very viscous). Put the solution on ice. Add 5 μL of 1 M DTT (see **Note 26**). Add 2 μL of denatured salmon sperm DNA (10 mg/mL), preheated for 5 min at 100°C and quenched on ice just before use. Keep the solution (which can be used for 2 h) on ice. Add bidistilled water to the probe to a final volume of 13 μL. Heat the riboprobes for 3 min at 80°C and quench on ice. Add the denatured probe (see **Note 27**) to the hybridization solution and keep on ice until use.

3.2.4.2. HYBRIDIZATION

1. Pipet about 6 μL of hybridization mix onto each section of approx 5 × 5 mm. Larger sections may require more mix. No coverslips are needed (*see* **Note 28**).
2. Put the slides in a tightly closed box containing paper towels soaked in 50% formamide–1X SSC to prevent evaporation of the hybridization mix.
3. Incubate overnight (*see* **Note 29**) at 54°C (*see* **Note 30**).

3.2.4.3. POST-HYBRIDIZATION WASHES

1. Discard paper towels.
2. Wash slides with a few milliliters of 1X SSC, to remove the bulk of the hybridization mix, then by gently shaking in 250 mL (sufficient for 25 slides) of 50% formamide in 1X SSC (2 × 15 min at 54°C), and finally in 250 mL of 1X SSC (1 × 10 min at RT).
3. Repeat wash with 1X SSC, then wash with RNase buffer supplemented with 10 μg of RNase/mL for 30 min at RT.
4. Wash with 1X SSC, and again with 1X SSC for 10 min at RT and then with 0.1X SSC for 10 min at RT.
5. Dehydrate by incubating slides sequentially (3 min each) in 50%, 70%, and 96% ethanol containing 0.3 *M* ammonium acetate.
6. Dry slides in an air stream for at least 1 h and make sure that there are no droplets left on the slides before applying the emulsion.

3.2.5. Probe Detection

The radioactive hybrids are detected by their ability to form a latent image in a photographic emulsion that is applied onto the sections *(32)*. The range of ^{35}S allows the localization of a message at the cellular level within one week and for that very reason is widely used. All work described here is carried out in a darkroom.

3.2.5.1. PREPARATION OF BATCHES OF PHOTOGRAPHIC EMULSION

1. Melt the batch of emulsion at 42°C (it takes at least 20–25 min) (*see* **Note 31**). Carefully mix the emulsion by turning the container upside down several times during melting.
2. Aliquot the batch by transferring 5 mL of the melted emulsion to a 20-mL vial containing 7.5 mL of glycerol–water. Use of an automatic pipet would lead to considerable retention of emulsion in the tip and dark room illumination is insufficient to see the divisions on a syringe. To circumvent this type of problem, a syringe and a block corresponding in size to a 5 mL volume is used. Aspirate the emulsion into the syringe. Put the block between the syringe and the plunger. Push the plunger down. Remove the block and push the plunger completely down, so that 5 mL aliquots are transferred to 20 mL vials.
3. Store batches at 4°C in a light-tight box. Ilford guarantees the emulsion for at least 2 mo without melting. We did not observe elevated background after half a year of storage (*see* **Note 32**).

3.2.5.2. Application of the Photographic Emulsion to the Slides

1. Melt the batch of diluted emulsion for 10 min at 42°C and mix carefully without trapping air bubbles by turning the container upside down.
2. Pour the emulsion into a custom-made 10 mL stainless steel cuvette ($70 \times 30 \times 5$ mm internal diameter) and keep it at 42°C.
3. Slowly lower a slide into the emulsion and slowly move the slide up and down twice. A single slide should be lowered slowly into the cuvette to avoid bubbles and spoiling of emulsion.
4. Hold the slide as horizontally as possible and wipe the back (make sure you have the correct side).
5. Place the slide horizontally at RT for the time needed to dip the following slide.
6. Place the slide on a horizontal ice-cooled glass plate for 10 min.
7. Dry the slides horizontally for 60 min at RT.
8. Put the slides in a horizontal position in a light-tight box at 4°C, containing silica gel as desiccant, and expose them for approx 4–10 d.
9. Immediately after use, pour the remaining emulsion back into a 15 mL glass scintillation vial and store it at 4°C. These vials are also used to store freshly prepared emulsion. The emulsion can be used a second time for normal experiments but if the lowest possible background is required a new batch should be used.
10. Carefully clean the cuvette (any emulsion that remains gets exposed to light and introduces a lot of background in the following experiment).

3.2.5.3. Development of the Autoradiographic Signal

1. Bring all solutions to 18°C.
2. Keep the box with slides for at least 30 min at RT.
3. Develop for 4 min (*see* **Note 33**) in Amidol-developer with agitation.
4. Stop for 1 min in bidistilled water and fix for 10 min in Fixer with agitation.
5. Wash for at least 60 min (the first 30 min in the dark) in running tap water at about 20°C (check temperature frequently).

3.2.5.4. Counterstaining (*see* **Note 34**)

1. Slides are washed for 5 min in bidistilled water and then incubated for 1.5 min in 1:5 Nuclear-Fast Red solution.
2. Rinse in bidistilled water for 5 min, then dehydrate with the following ethanol solutions: 50%, 70%, 96%, 100%, and 100% (5 min each).
3. Incubate 3×5 min in xylene (*see* **Note 4**). The grains will disappear with longer times in xylene.
4. Mount with a cover glass in Malinol, a very old-fashioned, slowly drying mounting medium. Many of the rapidly drying media, such as Entellan (Merck), cause disappearance of the silver grains.
5. Dry for several weeks at 37°C.

3.2.6. Analysis of Sections

Sections are analyzed with bright- or dark-field microscopy. Bright field microscopy enables accurate measurement of signal density; dark-field microscopy is not quantitative and results intrinsically in a loss of resolving power because light is reflected. We recommend, therefore, taking bright field images that can be digitized and converted into "dark field" pictures.

Sections are first assessed qualitatively by the investigator. Pay attention to the histological quality of images, as well as to the specificity and level of staining (hybridization signal). The distribution of the hybridization signal is screened for gradients which are probably not due to regional differences in mRNA concentration. Such gradients usually do not follow anatomical boundaries and often reflect improper fixation. The specificity of the hybridization signal is assessed by including proper controls in the analysis (*see* **Note 35**). Initially, we expose two or three adjacent sections hybridized with the same probes. On the basis of the intensity of the hybridization signal, we decide how much longer adjacent slides, which have been simultaneously hybridized, still have to be exposed to attain a proper signal. It should be kept in mind that if quantification of the signal is required, the optical density of the signals should not exceed 0.8, as there is a linear relationship between OD and cpm per area within the 0–0.9 range. This level of intensity is definitely lower than a visually "nice" signal, which typically exceeds 1.5 OD.

For quantitative analysis (*see* **Note 36**), we use a microscope (Axiophot, Zeiss), equipped with a cooled CCD camera (Photometrics series 200 with CAF 1400 chip, Roper Scientific, Bogart, Georgia, USA) *(31)*. Image analysis is performed either with the NIH Image package (http://rsb.info.hih.gov/nih-image/) or with the Optimas package (Media Cybernetics, Silver Spring, MD, USA).

4. Notes

1. Successful *in situ* RNA detection requires that great care be taken to avoid contamination of specimens, solutions and materials with RNases. Gloves are to be worn during the entire procedure. All materials to be used, glassware, pipet tips, slide trays, and solutions need to be sterilized. "Dry materials" should be heated for 3 h at 140°C and solutions can be autoclaved for 30 min at 120°C, or made with sterile buffer or bidistilled water. The "dry material" that cannot be sterilized by heating and surfaces must be wiped with a fresh solution of DEPC–ethanol or a commercial RNase decontaminant and air-dried. An alternative to the use of glassware is sterile disposable plasticware.
2. DEPC is a strong inhibitor of RNases and a potential carcinogen that has to be used under proper safety conditions. In addition, DEPC must not be used in Tris buffers, as it is known to react with amines and develop toxic fumes.

3. Formaldehyde and glutaraldehyde fumes are toxic. The formaldehyde and glutaral-dehyde solutions should be prepared and handled under a fume hood. Use only freshly prepared fixation solution A. However, the solution may be stored at −20°C and aliquots maintaining its crosslinking properties can be thawed when needed.

4. Xylene has to be handled under a fume hood.

5. SP6 RNA polymerase has a lower enzymatic activity than T3 or T7 RNA poly-merases and its transcription capacity is less efficient. Consequently, the use of T3 and T7 polymerases is preferred to that of SP6.

6. The grade of the emulsion determines its sensitivity. Finer emulsion grades give a better resolution but are less sensitive. Ilford Nuclear Research Emulsion G-5 is generally used for the detection of ^{35}S. The emulsion must be stored at 4°C and protected from radiation.

7. Controls are absolutely necessary, in particular when the riboprobe is used for the first time and its efficiency is unknown. A large panel of controls is avail-able to investigators to check the specificity of either riboprobe or staining, or the proper development of the experiments, etc.

   a. A sense riboprobe is commonly used as negative control. An appropriate sense probe must have the same length as the antisense probe, the same nucleotide composition, and consequently share identical specific activity in terms of background hybridization. However, remember that the absence of signal in a negative control is not necessarily due to the fact that the sense probe did not hybridize. Many other reasons may be responsible for the absence of signal.

   b. Omission of the riboprobe in the hybridization solution or omission of the anti-DIG antibodies in TBST make it possible to check whether or not endog-enous enzymes are responsible for the final signal or contribute to it.

   c. When the antisense probe appears to generate a very low signal that may be confused with the background, one may use Hybridization buffers containing a stable concentration of the labeled probe mixed with increasing concentra-tions of non-labeled probe. The competition of the non-labeled probe with the labeled probe for the hybridization with the transcript should lead to a signifi-cant decrease of the specific signal.

   d. One of the best controls consists in comparing the localization of the *Cx* gene transcript with the localization of the Cx protein, detected by immunohis-tochemistry. However, note that the abundance of Cx protein does not neces-sarily reflect the expression level of the transcript (*see Cx43* gene expression in embryonic rat heart, *[15]*; or *Cx45* gene expression in embryonic mouse heart *[17]*).

8. Because the connexins have emerged as a family of proteins on the basis of their amino acid sequences, thus sharing sequence homologies, care must be taken to raise specific probes which do not crossreact with a nonexpected *Cx* transcript. One must keep in mind that it is not uncommon that organs and cells express multiple *Cx* genes and it is also likely that all *Cx* genes have not been identified yet. Thus, the investigator must choose either to use the whole coding region of the investigated *Cx* or only part of it in order to generate a more specific

riboprobe. Often it is advised to use a probe from the most divergent region of the *Cx* genes, that is, the 3' translated or untranslated regions. Yet by using different *Cx* riboprobes we found that this choice was not the most critical point. We have never seen nonspecific signal using riboprobes complementary with the whole coding sequence, but we have noticed that the optimum probe length is about 1 kb. Probes longer than 1 kb should not be used, because they should be fragmented by a limited alkaline hydrolysis *(28)* (*see* **Subheading 3.2.2.3.**).

9. Before riboprobe synthesis, the plasmid is linearized from the opposite end of the RNA polymerase initiation site to be used to prevent a further transcription of the plasmid sequence. The enzyme used for this linearization must produce a 5' overhang or a blunt end. Use of templates with a 3' overhang can result in erroneous transcripts containing sequences complementary to the expected transcript. The restriction digestion should be complete. Undigested plasmid DNA present in the transcription assay produces unwanted vector sequences.

10. An 11-carbon arm is intercalated between the nucleotide and the hapten for better flexibility and accessibility of the hapten to the antibodies used during the detection procedure. During the synthesis of the probe the concentration of nucleotides is such that one DIG-11UTP molecule is randomly incorporated in the sequence every 20–30 UTP.

11. Fixation is essential to preserve the tissue morphology and to anchor the RNA within the tissue. Over-fixation hardens the tissues making sectioning difficult. Underfixation generally results in weak and/or artificial hybridization. RNA is very sensitive to degradation by RNase and easily washed away, and consequently it is essential to start fixation as soon as possible after dissection of the embryos. Formaldehyde fixation at 4°C prevents artifacts due to too rapid fixation that causes the formation of gradients inside the specimens.

12. In our first experiments we used methanol solutions to dehydrate and rehydrate the fixed embryos as recommended in most of the protocols. Because methanol is toxic we have shifted from methanol to ethanol without any noticeable difference in the results.

13. Proteinase K (PK) treatment helps to reduce the background by releasing RNA from potential surrounding proteins. After this treatment, embryos become very sticky and great care must be taken to minimize deposit of dust on the surface of the embryos. In addition, PK treatment can eventually destroy the smallest embryos. In that case, a shorter incubation time is required. Note also that the activity of PK may vary from one batch to another.

14. According to certain protocols, embryos can be stored before or after heating at 70°C in the hybridization buffer without riboprobe. However, under such conditions, the embryos become very clear and it is very difficult to keep track of the smallest ones. In addition, the embryos, whatever their developmental stages, become very fragile. For these reasons it is much safer to store them in 100% ethanol.

15. Heat-inactivated normal sheep serum in TBST will be used in the next step (**Subheading 3.1.4.3.**), but also in the incubation step of the embryos with anti-DIG antibodies (**Subheading 3.1.4.4.**). Prepare enough serum for both steps.

16. In our first experiments we added 2 mM Levamisole (an inhibitor of tissue alkaline phosphatases) to each solution used after the hybridization step, until incubation with the Reaction buffer. We have detected no noticeable change when omitting it, and consequently, we no longer use it. Nonetheless, before omitting this inhibitor it has to be checked that the targeted tissues do not express endogenous alkaline phosphatases which could interfere with the expected results.

17. Immunostaining is usually performed at RT but when the signal takes a long time to develop, it may be convenient to store the embryos at 4°C for several hours, then to bring them back at RT.

18. The embryos can be photographed just after being rehydrated with TBST but the quality of the staining will be improved (and the photographs also) if the embryos are stored for a few days at 4°C in glycerol-PBT until complete clearing. The embryos are photographed lying in a Petri dish on an agarose bed that can be stained with various kinds of compounds such as bromophenol blue, Coomassie blue, phenol red, etc.

19. Preservation of the cell constituents can be carried out either by direct freezing of the samples in nitrogen-cooled isopentane, or by chemical fixation. We prefer chemical fixation followed by paraffin embedding, as the unfixed frozen tissue is very sensitive to RNase during the process of making cryostat sections. Moreover, the morphology of tissues is better preserved after paraffin embedding of the samples than after freezing.

20. Storage of the fixed tissue for longer periods is possible in 70% ethanol; higher ethanol concentrations make the tissue too hard to cut. The plastic racks/boxes are from Emergo, Landsmeer, The Netherlands. The size of the boxes is $101 \times 88 \times 52$ mm. Each box can contain one rack with 25 slides (75×25 mm).

21. Slide-coating. 3-Aminopropyltriethoxysilane (AAS), also known as 3-triethoxysilylpropylamine (TESPA), reacts with silica glass to produce aminoalkyl groups on the glass surface that may bind either ionically or covalently with aldehyde or ketone groups in the tissue. For both frozen and paraffin sections, AAS-treated slides are far superior than poly-L-lysine-coated slides with respect to their binding characteristics of tissue sections *(33)*.

22. Transcripts should be of high specific activity. Therefore, in the transcription assay [^{35}S]-CTP is not diluted with unlabeled CTP. A specific activity of 1.67×10^9 cpm/µg is obtained with a new batch of [^{35}S]CTP (1000 Ci/mmol). The only way to get higher specific activity of the transcripts is to use more than one labeled nucleotide in the assay. The concentration(s) of the [^{35}S]nucleotide(s) must be at least 5 µM to guarantee full-length transcription. Although the use of two labeled nucleotides results in less full-length transcripts, the signal from hybridizations using double-labeled probes is about twice as strong compared to that obtained from hybridizations using single-labeled probes. We prefer to use [^{35}S]CTP rather than [^{35}S]UTP, as at the optimal nucleotide concentration transcription of poly(U) stretches in the DNA is hampered, thus preventing the synthesis of full length transcripts.

23. Although the optimum size of a probe for *in situ* hybridization is about 50–200 nucleotides it is important to prepare full length transcripts, as the total length of

unique sequences determines the sensitivity of hybridization. Transcripts are subsequently reduced in size by limited hydrolysis.

24. This step, carried out just before hybridization, is extremely important when [^{35}S]labeled probes are used, as it reduces background considerably.

25. The hybridization mixture contains 10% Dextran sulfate, a compound that is strongly hydrated in aqueous solutions. Because of this effect, macromolecules (such as the probe) have no access to the hydrating water, leading to an apparent increase in probe concentration. Furthermore, the mixture contains 50% formamide to lower the melting temperature of the duplex, and Denhardt's, Triton X-100, and competitor DNA to lower background by preventing nonspecific binding of the probe to the tissue or target RNA.

26. To reduce background, it is crucial to hybridize under reducing conditions in the presence of DTT. Use of ^{33}P-labeled probes is worth trying if persistent background problems occur that cannot be overcome by the use of reducing conditions.

27. With a ^{35}S-labeled cRNA probe concentration of 4×10^4 cpm/µL hybridization mix, a good signal is achieved for most probes. For abundant messengers, probes can be used for a period of three months. To get the highest possible signal, especially with double-labeled probes to detect low abundance messengers, the hybridization should be performed as soon as possible after labeling with a new batch of nucleotides. Longer exposure is better than using a higher concentration of the probe. A single-labeled RNA probe made with new radioactivity typically gives a specific activity of 1.7×10^9 cpm/µg. With a concentration of 4×10^4 cpm/µL hybridization mix, the RNA concentration is 24 pg/µL hybridization mix. For double-labeled RNA probes the specific activity is 3.3×10^9 cpm/µg, and at a concentration of 4×10^4 cpm/µL of hybridization mix, the RNA concentration is 12 pg/µL.

28. There is no need to use coverslips to prevent evaporation, provided that a proper moist box is used *(29)*. Drops of hybridization mixture containing the probe can conveniently be applied on the sections and several probes can be easily compared on consecutive sections.

29. Shorter hybridization times (i.e., 4 h), do not improve the signal-to-noise ratio, and lead, in our experience, to a considerably lower signal of hybridization, which is difficult to measure.

30. When a gene family such as that of the connexins is being investigated, it is important to assess the specificity of hybridization. We did compare the use of small, isoform-specific probes with longer probes that include regions of similarity and observed, somewhat unexpectedly, that adjusting the stringency of the hybridization conditions was equally effective as choosing a gene-specific fragment with respect to specificity of hybridization.

31. Ilford emulsions can be handled under safe-light conditions. We use an indirect illumination with a 15-Watt darkroom light fitted with a no. 902 Ilford filter, 1.5 m above the working area. Although this illumination is safe for the duration of the normal procedure, it is advisable to work fast and keep the exposure time of the emulsion to light as short as possible. Switch off the light if you do not need it.

Warming the emulsion will lead to more background grains, as does mechanical stress. So work fast to keep the emulsion's exposure to higher temperatures as short as possible, treat the liquid emulsion gently, and dry the slides slowly. Avoid trapping of air bubbles in the emulsion during melting and dipping, as this will prevent precise allocating of the emulsion and give white spots in the final image.

32. The dilution of the emulsion, and thus the thickness of the emulsion layer, has a considerable effect on the signal and the background. We typically use a dilution of 2.5× that gives a good signal and a good resolution with a low background. Using a less diluted emulsion gives more blackening but less resolution. Undiluted emulsion can give more than 20× the signal of that of a 2.5× diluted emulsion, but the resolution is very poor and, depending on the probe, the background high.

33. The standard developing time is 4 min at 18°C, while agitating to prevent exhaustion of the developer at the emulsion surface. The higher the temperature, the faster the process. To avoid swelling of the emulsion, the temperature is kept at 18°C. The temperature of all solutions should be the same, otherwise swelling or contraction of the emulsion may lead to cracks in or even loss of the emulsion. Longer times of development up to 16 min increase the signal proportional to the time of development, permitting proportionally shorter exposure times *(31)*. It results, however, in lower resolution owing to the development of larger silver grains. Nevertheless it may be extremely convenient in a rapid screen of many different probes. On the other hand, longer exposure times result in sensitive detection with superior resolution. Background signal is acceptable up to about 24 d of exposure.

34. After developing, the sections can be stained to allow localization of the silver grains, indicating hybridization, within the tissue. Great care must be taken with the choice of staining and mounting media, as they can have detrimental effects on the developed silver grains. The image can fade or even completely disappear (negative chemography). Staining with Nuclear-Fast Red gives a reproducible result.

35. Worries about the specificity of the *in situ* hybridization can easily develop into a nightmare for frequent users of the *in situ* hybridization technique who have to accept that absolute specificity does not exist. Several artifacts, possible causes, and controls are listed below. High general background, with grains all over the slide, is most likely due to inaccuracies in one of the steps of the autoradiographic procedures. Check the procedure using blank slides. High tissue background indicates nonspecific binding of the probe. To distinguish whether this is due either to the procedure or to the probe, use a positive control that should be positive in some tissues and negative in others. We have dubbed such a control a "tissue-intrinsic control" *(29)*. Such a control is more informative than negative controls, because it comprises both positive and negative controls in a single section. Moreover, a negative control, such as sense probes, is not a good control, even if it is negative, because many other reasons can be envisioned why a probe is negative. It is the authors' experience that sense probes and probes to low abundance mRNAs tend to display relatively high backgrounds due to binding of the probe

to unrelated sequences. If the tissue-intrinsic control displays uniform background over the tissue, the quality of the probe has to be assessed. Check whether plasmid sequences have been transcribed and whether the probe has the correct length. Check various steps in the procedure such as probe concentration, stringency of hybridization, washing, and RNase treatment. Increase DTT concentration to guarantee that reducing conditions are being used. If the tissue-intrinsic control displays specific hybridization and the probe of interest does not, check the quality of the probe and check whether the sequence contains GC-stretches or T-stretches that may cause artificial hybridization. Isolate mRNA from the tissue of interest and check by RT-PCR whether the probe reacts with a single molecular species. If possible use a probe with more unique sequences and/or double label. If the probe displays a specific pattern of hybridization, this observation can be strengthened by the use of a set of probes that hybridize to different sequences of the mRNA of interest and should yield the same pattern. In all cases the signal has to disappear when the sections are pretreated with RNase.

36. Quantitative aspects. Radioactive *in situ* hybridization is more quantitative than generally thought *(31)*. Thus, relative differences in the expression of mRNA species can be provided and it can be assessed whether changes in the hybridization signal are due either to changes in the cellular levels of a given mRNA species, or to changes in the number of cells that accumulate the same mRNA. Finally, changes in the relative mRNA levels can be related to changes in ribosomal RNA levels, permitting comparison with Northern analyses. Moreover, absolute mRNA levels per cell can be determined, as the quantitative *in situ* hybridization allows assessment of the distribution of the relative mRNA concentrations in a tissue, whereas the quantitative Northern or PCR analyses allow assessment of the mRNA levels. For example, let us assume that we obtain, in embryos, an estimate of 1,000,000 copies of mRNA/mm^3 by Northern blotting or quantitative PCR. Let us further assume that we look at sections (10 μm thick) and find that all the RNA signal comes from the liver and that the liver in a particular section occupies 100 mm^2. The volume of the liver in this section is ($100 \times 0.001 = 1$ mm^3). Therefore there are 1,000,000 mRNA copies in this section or 10,000 per mm^2. Assuming, for the sake of simplicity, that a liver cell in this section is 10,000 μm^3, there are 10 mRNA copies per liver cell. Let us now assume the signal is not homogeneously distributed. In that case we can make a histogram of ODs vs surface area. In such a graph the total area under the curve represents the total amount of OD or the total number of mRNA copies in that section, or 1,000,000. Let's again for simplicity assume that the liver has a low OD for 25% of its area and a high OD (e.g., 3× higher than low) for 75% of its area. The total radioactivity in the low OD area is $0.25 \times 1 = 0.25$ units, whereas that in the high OD area is $0.75 \times 3 = 2.25$ units. Together (2.5 units) this represents 1,000,000 mRNA copies. The low OD area therefore has 100,000 copies, whereas the high OD area has got 900,000 copies. From this, one can again calculate the number of copies per hepatocyte. Reliable OD measurements require homogeneous distribution of the absorbing product in the image. By definition the

autoradiographic image is not homogeneous and OD measurements should, at first sight, necessarily lead to a so-called distribution error. Fortunately, this is not the case. OD values are calculated values based on the transmission observed by a single detector element (in this case a pixel). Non-homogeneous transmissions in that field lead to incorrect calculation of the OD because it is calculated as the negative logarithm of the weighted average of these transmissions, rather than as the sum of the negative logarithms of the individual transmissions, which cannot be measured in the field of a single detector element (pixel). In the case of OD measurements of autoradiographic images, the OD is zero (100% transmission) when no silver grains are present. In that very special situation the OD is proportional to the area covered with silver grains.

Acknowledgments

The investigations dealing with the whole-mount *in situ* hybridization technique were supported by grants from the Association Française contre les Myopathies and the European Economic Community (grant BMH4-CT 96-1427 to D. Gros). In addition, the authors thank Drs. J. M. Ruijter and D. Franco for fruitful discussions about the technique of radioactive *in situ* hybridization on sections.

References

1. Wilkinson, D. G. (1992) *In Situ Hybridization. A Pratical Approach*, IRL, Oxford University Press, Oxford.
2. Wilcox, J. N. (1993) Fundamental principles of *in situ* hybridization. *J. Histochem. Cytochem.* **41,** 1725–1733.
3. Eberwine, J. H., Valentino, K. L., and Barchas, J. D. (1994) *In Situ Hybridization in Neurology,* Oxfor University Press, Oxford.
5. Bagasra, O. and Hansen, J. (1997) *In situ PCR Techniques*, John Wiley & Sons, New York.
4. Jowett, T. (1997) *Tissue In Situ Hybridization*, John Wiley & Sons, New York.
6. Herrington, C. S. and O'Leary, J. J. (1997) *PCR 3: PCR In Situ Hybridization. A Practical Approach*. IRL, Oxford University Press.
7. Emson, P. C. and Gait, M. J. (1992) *In situ* hybridization with biotinylated probes, in *In Situ Hybridization. A Pratical Approach* (Wilkinson, D. G., ed.), IRL, Oxford University Press, Oxford, pp. 45–59.
8. Höltke, H. J., Ankenbauer, W., Mühlegger, K., Rein, R., Sagner, G., Seibl, R., and Walter, T. (1995) The digitonin (DIG) system for non-radioactive labelling and detection of nucleic acids. An overview. *Cell Mol. Biol.* **41,** 883–905.
9. Barth, J. and Ivarie, R. (1994) Polyvinyl alcool enhances detection of low abundance transcripts in early stage quail embryos in a nonradioactive whole mount *in situ* hybridization technique. *BioTechniques* **17,** 324–327.
10. van Noorden, C. J. F. and Frederiks, W. M. (1992) *Enzyme Histochemistry. A Laboratory Manual of Current Methods*. IRL, Oxford University Press.

11. Geerts, W. J. C., Verburg, M., Jonker, A., Das, A. T., Boon, L., Charles, R., Lamers, W. H., and Van Noorden, C. J. F. (1996) Gender-dependent regulation of glutamate dehydrogenase expression in periportal and pericentral zones of rat liver lobules. *J. Histochem. Cytochem.* **44,** 1153–1159.

12. Micevych, P. E. and Abelson, L. (1991) Distribution of mRNAs coding for liver and heart gap junction proteins in the rat central nervous system. *J. Comp. Neurol.* **305,** 98–118.

13. Yancey, B. S., Biswal, S., and Revel, J. P. (1992) Spatial and temporal patterns of distribution of the gap junction protein connexin 43 during mouse gastrulation and organogenesis. *Development* **114,** 203–212.

14. Ruangvoravat, C. P. and Lo, C. W. (1992) Connexin 43 expresion in the mouse embryo. Localization of transcript within developmentally significant domains. *Dev. Dyn.* **194,** 261–281.

15. van Kempen, M. A., Vermeulen, J. L. M., Moorman, A. F. M., Gros, D., Paul, D. L., and Lamers, W. H. (1996) Developmental changes in the connexin40 and connexin43 mRNA-distribution pattern in the rat heart. *Cardiovasc. Res.* **32,** 886–900.

16. Ya, J., Erdtsieck-Emste, E. B. H. W., De Boer, P. A. J., van Kempen, M. J. A., Jongsma, H., Gros, D., Moorman, A. F. M., and Lamers, W. H. (1998) Heart defect in connexin43-deficient mice. *Circ. Res.* **82,** 360–366.

17. Alcoléa, S., Théveniau-Ruissy, M., Jarry-Guichard, T., Marics, I., Tzouanacou, E., Chauvin, J. P., Briand, J. P., Moorman, A. F. M., Lamers, W., and Gros D. (1999) Downregulation of connexin 45 gene products during mouse heart development. *Circ. Res.* **84,** 1365–1379.

18. Essner, J. J., Laing, J. G., Beyer, E. C., Johnson, R. G., and Hackett, P. B. (1996) Expression of zebrafish connexin 43.4 in the notochord and tail bud of wild-type and mutant *no tail* embryo. *Dev. Biol.* **177,** 449–462.

19. Dahl, E., Willecke, K., and Balling, R. (1997) Segment-specific expression of the gap junction gene connexin 31 during hindbrain development. *Dev. Genes Evol.* **207,** 359–361.

20. Delorme, B., Dahl, E., Jarry-Guichard, T., Briand, J-P., Willecke, K., Gros, D., and Théveniau-Ruissy, M. Expression pattern of connexin gene products at the early developmental stages of the mouse cardiovascular system. *Circ. Res.* **81,** 423–437.

21. Moorman, A. F. M., Vermeulen, J. L. M., Koban, M. U., Schwartz, K., Lamers, W. H., and Boheler, K. R. (1995) Patterns of expression of sarcoplasmic reticulum Ca^{2+}ATPase and phospholamban mRNAs during rat heart development. *Circ. Res.* **76,** 616–625.

22. Franco, D., Kelly, R., Moorman, A. F. M., and Buckingham, M. (1997) Regionalised transcriptional domains of myosin light chain 3F transgenes in the embryonic mouse heart: morphogenetic implications. *Dev. Biol.* **188,** 17–33.

23. Jowett, T. (1997) Two-color *in situ* hybridizations, in *Tissue In Situ Hybridization* (Jowett T., ed.), John Wiley & Sons, New York, pp. 39–48.

24. Harlow, E. and Lane, D. (1988) *Antibodies: A Laboratory Manual.* Cold Spring Harbor Laboratory Press, Cold Spring Harbor, NY.

25. Casey, J. and Davidson, N. (1977) Rates of formation and thermal stabilities of RNA:RNA hybrids and DNA:DNA duplexes at high concentrations of formamide. *Nucleic Acids Res.* **4,** 1539–1552.

26. Ausubel, F. M., Brent, R., Kingston, R. E., Moore, D. D., Seidman, J. G., Smith, J. A., and Struhl, K. (1994) *Current Protocols in Molecular Biology,* John Wiley & Sons, New York.

27. Hogan, B., Beddington, R., Constantini, F., and Lacy, E. (1994) *Manipulating the Mouse Embryo. A Laboratory Manual,* 2nd edit. Cold Spring Harbor Laboratory Press, Cold Spring Harbor, NY, pp. 352–368.

28. Angerer, L. M. and Angerer, R. C. (1997) *In situ* hybridization to cellular RNA with radiolabelled RNA probes, in *In Situ Hybridization. A Pratical Approach* (Wilkinson, D. G. , ed.), IRL, Oxford University Press, Oxford, pp. 15–32.

29. Moorman, A. F. M., de Boer, P. A. J., Vermeulen, J. L. M., and Lamers, W. H. (1993) Practical aspects of radio-isotopic *in situ* hybridization on RNA. *Histochem. J.* **25,** 251–260.

30. Holland, P. (1986) Localization of gene transcripts in embryo. Section: *In situ* hybridization with RNA probes, in *Manipulating the Mouse Embryo. A Laboratory Manual.* 1st edit. (Hogan B., Constantini F., and Lacy E., eds.), Cold Spring Harbor Laboratory Press, Cold Spring Harbor, NY, pp. 228–242.

31. Jonker, A., de Boer, P. A. J., van den Hoff, M. J. B., Lamers, W. H., and Moorman, A. F. M. (1997) Towards quantitative *in situ* hybridization. *J. Histochem. Cytochem.* **45,** 413–423.

32. Rogers, A. W. (1979) *Techniques of Autoradiography,* Elsevier, Amsterdam.

33. Henderson, C. (1989) Aminoalkylisane: an inexpensive, simple preparation for slide adhesion. *J. Histotechnol.* **12,** 123–124.

2

Sodium Dodecyl Sulfate-Freeze-Fracture Immunolabeling of Gap Junctions

Irene Dunia, Michel Recouvreur, Pierre Nicolas, Nalin M. Kumar,
Hans Bloemendal, and E. Lucio Benedetti

1. Introduction

By the mid 1960s, pioneering work using high-resolution electron micros-copy, new fixation methods, and negative staining of isolated liver plasma membranes allowed the identification of a geometric subunit pattern likely associated with junctional domains *(1,2)*. Furthermore, the application of tissue impregnation with electron-dense tracers revealed that the minute "gap" (2 nm wide) between the closely adjoining junctional membranes comprised an hex-agonal subunit pattern. This type of membrane–membrane interaction, distinct from tight junctions, adhesion plaques, and desmosomes, was originally called "gap junction" *(3)*.

The connexins are the polypeptides forming the six subunits of a single com-municating oligomer (connexon) *(4–6)*. A large number of connexin isoforms has been sequenced, and several experiments have shown that the gap junction assembly even in one single tissue may display a great variety of expression patterns and that a single communicating channel may be formed by different connexin subunits *(4,5,7)*. Furthermore, the gap junction phenotype is not a static character. During growth, differentiation, and aging the communicating junctional constituents are submitted to many structural and biochemical events such as phosphorylation, proteolysis, assembly, disassembly, and crystalliza-tion *(4)*. Significant advances in gap junction structural biology have been achieved by a new technological endeavor. Among the novel techniques, freeze-fracture and freeze–etching have opened new avenues for the study of the fine structure of gap junctions and their modulation with developmental, functional, and biochemical events (*cf.* **ref. 8**). These techniques produced the

From: *Methods in Molecular Biology, vol. 154: Connexin Methods and Protocols*
Edited by: R. Bruzzone and C. Giaume © Humana Press Inc., Totowa, NJ

most solid and direct evidence of the *en face* view of the subunit organization of the junctional membranes (*see* **Notes 1** and **2**).

In the gap junction a pair of hexameric units, matching one another in perfect register, forms the connecting transmembrane device spanning the lipid bilayers *(4)*. The freeze–fracture simultaneously splits the two opposite junctional membranes in a stepwise fashion. The cleavage passes along one of the bilayers and exposes the inner aspect of the leaflet adjacent to the cytoplasm (protoplasmic fracture face or PF), then the adjacent membrane is cleaved and the exposed fracture face corresponds to the inwardly directed face of the outer membrane leaflet (exoplasmic fracture face or EF). The junctional PF is characterized by a polygonal assembly of 9-nm intramembranous particles (IMPs). The corresponding EF displays a complementary arrangement of pits or depressions *(9,10)* (*see* **Note 2**). However, conventional freeze–fracture and etching do not provide information on the chemical nature of the pleiomorphic features of the gap junctions visualized on the fracture faces. This inherent limitation of the technique prompted several investigators to develop new methods combining freeze–fracture and etching with immunocytochemical labeling. The aim of such technological development was to study in parallel the ultrastructural features and the biochemical nature of the membrane constituents *(11–14)*. Among the different methods of immunolabeling, the sodium dodecyl sulfate-digested freeze–fracture replica labeling (SDS-FL), originally developed by Fujimoto *(15)*, produced consistent information on the organization of the intercellular junctions, particularly during their assembly and functional changes (**Fig. 1**, *see* **Note 3**).

The basic principle of the SDS-FL technique is that the freeze-cleaved membrane halves become physically stabilized by the metal shadowing during replication. These stabilized membrane halves are not dissolved by the detergent, as the apolar inner membrane core positioned against and stabilized by the metal cast became resistant to detergent solubilization. The detergent treatment, however, may unravel and/or expose antigenic polar domains anchored to the stabilized membrane structure and allow immunolabeling thereafter (**Fig. 2**, *see* **Note 4**). A recent modification of SDS-FL, using Lexan stabilized freeze-fracture replicas, has been proposed by Rash and Yasumura *(15a)*.

2. Materials

2.1. Quick Freezing

1. A Dewar flask filled with liquid nitrogen (Balzers, Liechtenstein).
2. A solid metal cylinder comprising a small receptacle at the top (3 cm depth), sitting across the neck of the Dewar flask (Balzers).
3. Propane gas bottle (Balzers) and regulator valve to control the flow of gas coming out (*see* **Note 5**).

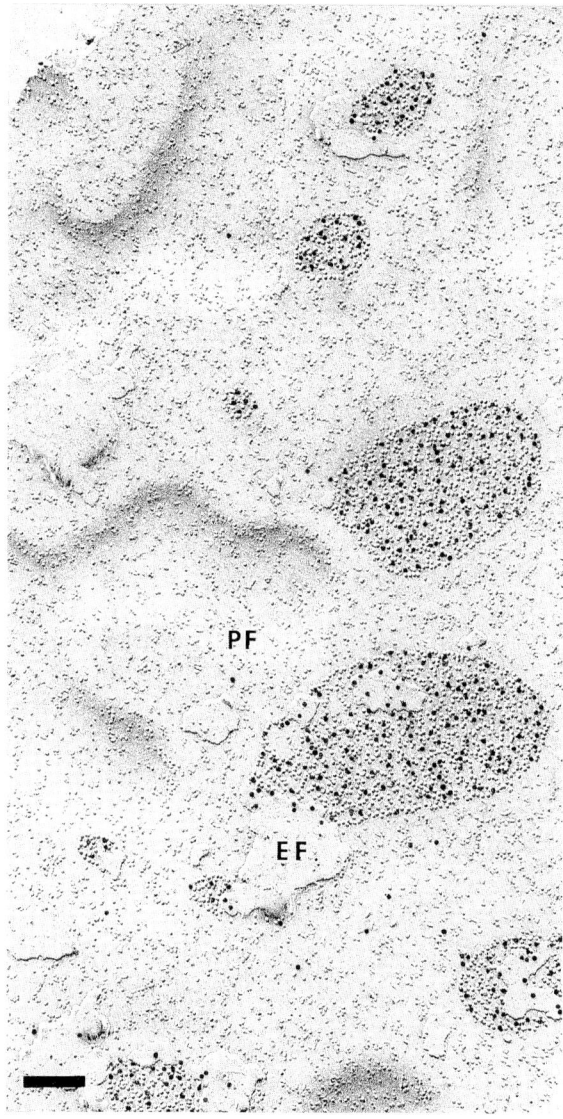

Fig. 1. SDS-FL of the outer lens cortex where the assembly of gap junctions connecting the elongating fibers takes place. Immunogold labeling (10-nm gold particles) using the polyclonal antibody directed against the middle cytoplasmic domain of Cx50, between the second and third transmembrane domains. Freeze–fracture has mainly exposed the membrane leaflet close to the cytoplasm, protoplasmic fracture face (PF). Only small fragments of the exoplasmic fracture face (EF) are exposed. Note the great variety in size of the newly assembled junctional domains, characterized by clusters of identical 9-nm intramembrane particles (IMPs). The immunogold labeling is specifically restricted to the junctional plaques. Bar = 80 nm.

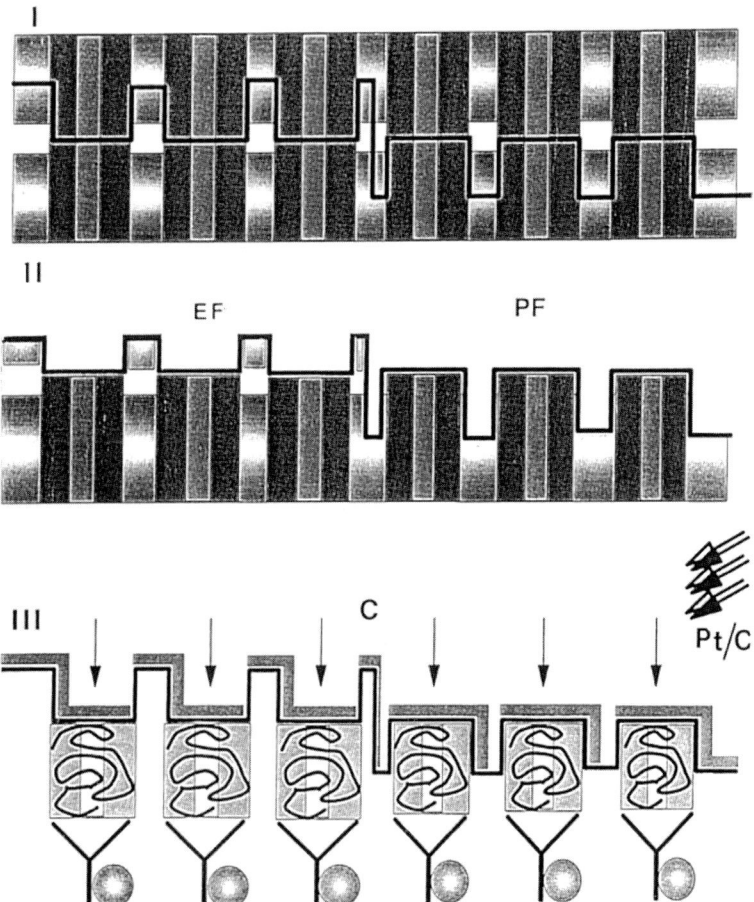

Fig. 2. Diagrammatic representation of the SDS-FL of gap junctions. At low temperature (–150°C) the cleavage plane splits the junctional membranes into two halves, exposing both the pitted exoplasmic fracture face (EF) and the protoplasmic fracture face (PF) where the single connexons are anchored (I and II). SDS solubilization unravels the antigenic sites of connexins exposed at the cytoplasmic surface of the junctional membrane (III). Both fractured and uncleaved junctional bilayers are hold and remain "stabilized " by the platinum/carbon replica (Pt/C and C), allowing the subsequent gold immunolabeling (III).

4. Another Dewar flask filled with liquid nitrogen to store specimens after freezing.
5. Specimen holders: precleaned gold–nickel alloy discs with a central solid flat-topped platform (Balzers), previously scratched lightly in a criss-cross pattern with a scalpel blade, to increase the adhesion of the frozen sample.
6. Fine-tipped forceps.

2.2. Freeze-Fracture and Replication

1. A freeze-fracture apparatus, model 301 or 400 (Balzers), equipped with a turbomolecular pump, liquid nitrogen-cooled trap and electron gun beams ready to evaporate: a platinum and a carbon layer (Pt/C) at an oblique angle of about 45° and a perpendicular layer of carbon respectively.
2. Quartz crystal thin film monitor (Balzers).
3. Appropriate styrofoam box filled with liquid nitrogen to cool the specimen holder injector.
4. Precision tweezers.

2.3. Detergent Treatment of the Replica

1. Phosphate-buffered saline (PBS), pH 7.4.
2. Buffered 2% SDS in 10 mM Tris-HCl and 30 mM sucrose, pH 8.3.
3. Porcelain spotting plaques.

2.4. Immunolabeling

1. PBS, pH 7.4.
2. 1% Bovine serum albumin (BSA) in PBS.
3. 0.2% BSA in PBS.
4. 0.5% Glutaraldehyde in PBS.
5. 50 mM Glycine in PBS.
6. Reagents were obtained from Sigma, unless otherwise indicated.

2.5. Antibodies

1. Affinity-purified rabbit polyclonal antibodies directed against the major intrinsic membrane protein of the lens fibers (MP26) that recognize the native protein *(16)* at a 1:500 dilution.
2. Rabbit polyclonal antibodies directed against MP26 that recognize the native protein and its proteolytic derivatives *(17)* at 1:200.
3. Affinity-purified polyclonal antibodies directed against the Cx46 J peptide *(18)* (*see* **Subheading 3.5.**, **step 3**) at 1:200.
4. Polyclonal antibodies directed against a Cx50 peptide *(19)* at 1:200 dilution.
5. Monoclonal antibody directed against MP70 (Cx50) *(20)* at 1:200 dilution.
6. Protein A conjugated to 15- or 10-nm gold particles (Department of Cell Biology, University of Utrecht, The Netherlands).
7. Mouse-IgG coupled to 15- or 10-nm colloidal gold particles (Amersham).

2.6. Mounting the Replica

1. Distilled water.
2. 300-mesh coated (thin film of Formvar or collodion) grids for electron microscopy.
3. 0.3% Formvar solution (dichloroethane) or 1% collodion (chloroform).
4. Precision tweezers.

3. Methods

3.1. Dissection and Preparation of the Frozen Specimens

1. The Dewar flask filled with liquid nitrogen is placed in the fume cupboard; the solid metal cylinder, comprising a small receptacle at the top, is placed inside. Propane from the bottle is dispensed into the inner receptacle with the regulator valve adjusted to give a slow flow to fill it completely with condensed propane.
2. The material to prepare the samples has to be dissected immediately after the animals are killed. Very small (0.25 mm^3) and thin pieces of the biological material were placed on the specimen holders under a binocular microscope.
3. Once the propane was cooled to $-180°C$ (*see* **Note 5**), each mounted specimen held in a fine-tipped forceps was manually plunged into the liquid propane, at the highest speed possible. The specimens need to be made and frozen one by one (working temperature about 93 K or $-170°C$), to avoid drying and shrinkage of the sample. After standing each sample in the liquid propane for about 15 s to rapidly freeze it, the samples need to be rapidly transferred into the liquid nitrogen contained in the second Dewar flask to store them (*see* **Note 6**). The opening of the propane receptacle must be kept covered with a polystyrene lid between freezing runs.

3.2. Freeze-Fracture and Replication

1. Start the pumping procedure of the freeze–fracture unit once the electron beam guns are ready and the knife (razor blade) of the microtome arm has been changed.
2. Preheat the electron beam guns when vacuum is 10^{-5} Torr. Pt/C gun: 1000 V, 50 mA, preheating for 2 min. C gun: 1500 V, 50 mA, preheating for 5 min.
3. Start the cooling process of the specimen table to $-150°C$ (about 15 min) and the cooling of the microtome arm.
4. Start to cool the injector and the specimen holder plate by immersion in the adapted styrofoam box.
5. Transfer carefully, with a precooled tweezers, one sample (stored in liquid nitrogen) to the precooled specimen holder plate in the liquid nitrogen.
6. When the specimen table is cold ($-150°C$), open the trap of the freeze–fracture unit adapted for introducing the injector holding the specimen.
7. Take out the injector and close the trap as soon as the sample is well fixed in the specimen table.
8. When the microtome arm is cold ($-196°C$) and the vacuum reaches at least 5×10^{-7} Torr (about 20 min after replacement of the sample), adjust at $-140°C$ the temperature of the sample table (requires approx 10 min).
9. Wait till the vacuum is 2×10^{-7} Torr.
10. Fracture the specimen superficially with great care, moving the cold microtome arm and "fracturing" the surface of the specimen with the cold knife. Carefully control these manipulations through the binocular stereo microscope outside the vacuum chamber (*see* **Note 7**).

11. Cover the final fractured surface of the specimen by placing the microtome arm above the sample as soon as the fracture is completed.
12. Start the heating of the electron gun beam for Pt/C evaporation. As soon as the appropriate current values are 1900 V and 90 mA, the evaporation on the specimen can start: uncover the fractured surface, by moving away the microtome arm, and shadowcast the specimen with a 2 nm Pt/C evaporation at an oblique angle of 45°, controlling the thickness of the Pt/C deposit on the quartz crystal monitor.
13. Switch off the Pt/C evaporation and switch on the electron gun beam for carbon at 2300 V and 120 mA. Evaporate a perpendicular 20-nm carbon layer (*see* **Note 8**).

3.3. Thawing and Detergent Treatment

1. Once the replication process has been completed, the trap is vented and the specimen holder is removed with the injector.
2. Manipulate the specimen holder carefully with tweezers. Thawing of the replica should be very delicately achieved to minimize fragmentation. Use a stereomicroscope to control the different manipulations.
3. The replica is detached from the tissue by immersion in PBS, with gentle agitation, using a plastic pipet, to create a movement of the surrounding liquid around the replica. Porcelain spotting plaques are used for all manipulations.
4. As soon as the replicas, complete or in pieces, are detached from the bulk of the biological material, they are transferred to 2% SDS–Tris–sucrose buffer, taking care to immerse the replica completely in the detergent solution. Intermittent agitation, with a plastic pipet, is produced. Change the SDS solution 3–4 times. The total SDS treatment is prolonged for 1–2 h maximum (*see* **Note 4**).
5. Next, wash the replica thoroughly in PBS (6 × 3 min).

3.4. Immunolabeling

3.4.1. Simple Immunolabeling

1. To saturate nonspecific binding sites, incubate in PBS–1% BSA for 10 min.
2. Incubate with a specific antibody at an appropriate dilution in PBS–0.2% BSA for 30 min.
3. Wash 5×, 3 min each, in PBS–0.2% BSA.
4. Incubate with protein A–gold at an appropriate dilution or with the corresponding gold-labeled IgG diluted in PBS–0.2% BSA for 20 min.
5. Wash 5× in PBS, 3 min each.
6. Fix in 0.5% glutaraldehyde in PBS for 5 min.
7. Wash several times in distilled water before mounting the replica on grids previously covered with a thin film of Formvar or collodion.

3.4.2. Double Immunolabeling

1. Incubate in PBS–50 mM glycine (2 × 5 min).
2. Glycine is used to quench free aldehyde groups.

3. For double or triple labeling, repeat all the steps of simple immunolabeling, either once (double labeling) or twice (triple labeling) (*see* **Notes 9–11**).

3.5. Antibodies Raised Against Connexins and MP26

3.5.1. Polyclonal Antibodies

1. Antiserum directed against MIP. Rabbit antiserum purified SDS-denatured calf MIP (MP26) was prepared as described previously (*21*). Lens membranes enriched in junctions were solubilized in SDS, reduced, alkylated, and separated on preparative SDS gels. Gel strips corresponding to proteins with a molecular mass of approx 26 kDa were removed from the preparative gel and crushed. Immunization was done by injection of the crushed gel strips in complete Freund's adjuvant. The resulting antiserum recognized a single component in lens membranes with a molecular mass of 25.5 kDa—MIP—from calf, mouse, rat and chicken by Western blotting and by immunohistochemistry on frozen sections (*16,21*), and did not recognize lower molecular weight, presumably degraded forms, of MIP. The antiserum against MIP was used at 1:500 for immunofluorescence and SDS-FL.
2. Antiserum directed against MP26. This antiserum was produced in rabbits against a chloroform–methanol extract of MP26 (*22*), solubilized from isolated lens fiber plasma membranes, as described (*23*). The resulting antiserum specifically recognized MP26 and its degradation products (MP22, lower molecular weight) by immunoblotting (*17*). The antiserum against MP26 was used at 1:200 dilution for SDS-FL.
3. Antibody directed against a Cx46 (Cxα3) J peptide. Cxα3J peptide corresponds to the intracellular loop (J) based on the putative topology of connexins (*24*) and the amino acid sequence of rat Cx46 (*25*). This peptide was synthesized with a cysteine added to the C-terminus to facilitate attachment to a carrier protein, KLH (keyhole limpet hemocyanin). The sequence of the peptide was as follows: RRDNPQHGRGREPMC, assigned to residue 115–128. The synthetic peptide was coupled to KLH using *m*-maleimidobenzoyl-*N*-hydroxysuccinimide ester (MBS) as previously described (*24,25*). The generation of polyclonal peptide antiserum in rabbits was carried out as previously described (*24,25*). The Cx46 antibodies were affinity purified and characterized by Western blotting and immunofluorescence analysis (*18,26–28*). By these criteria, this antiserum recognizes the full-length form (44.6 kDa) of Cx46, as well as smaller degradation products, but does not cross-react with other connexins. The Cxα3J antiserum was used at a 1:200 dilution for immunohistochemistry and SDS-FL.
4. Antibody directed against a Cx50 peptide. This antiserum was produced in rabbits against a synthetic peptide corresponding to residues 124–136 of Cx50 as described in **ref. *19***. The Cx50 antiserum was characterized by Western blotting and immunofluorescence analysis (*19*). We used a 1:200 dilution of this antiserum for SDS-FL.

3.5.2. Monoclonal Antibodies

1. Antibody to Cx50 (6–4 B2-C6). A mouse monoclonal antibody to Cx50 was generated using urea-extracted membranes isolated from sheep lenses *(20)*. Antibodies from the hybridoma cell line (6–4 B2-C6) were found to be specific for fiber membrane junctional domains as determined by Western blotting and immunohistochemistry *(20,29)*. Hybridoma 6–4 B2-C6 recognizes antigens present in calf, chicken, mouse, rat, and toad *(20)*. Cleavage of Cx50 to a 38-kDa form resulted in a loss of the epitope recognized by this monoclonal antibody *(30)*. Hence, this antibody is likely to recognize a c-terminal domain *(31)*. This monoclonal antibody was used at a 1:200 dilution for immunohistochemistry and SDS-FL.

3.6. Mounting the Replica

This procedure is the final step before the examination of the SDS-FL in the electron microscope. Once the immunostaining is achieved, and after several washes in distilled water where the replicas have been completely immersed, it is necessary to float them at a clean surface of distilled water. Floating replicas may be mounted on electron microscopy grids (Formvar-coated or collodium-coated grids) from above or below, depending on the individual's preference. As cleaning, mounting is always carried out with the aid of a binocular microscope. When mounting from above is done, the grid, held by the tips of fine forceps, is brought face-on into contact with the replica, pushing it down into water, and then, in a single continuous twisting movement, the grid is turned over and out of the water, carrying the replica on top. Any traces of water are carefully removed with pointed strips of filter paper applied to the edge of the grid. Freeze–fracture replicas are chemically inert and reasonably resistant to damage by electron irradiation. If stored in a clean environment replicas will last indefinitely. The resolution of a replica is limited by the dimensions of the metal grains used to generate contrast, being 1.5–2 nm for a Pt/C replica.

4. Notes

1. During freeze-fracture the membrane bilayer is cleaved along its hydrophobic core. Thus the freeze-fracture reveals the internal organization of both the lipid and the protein of the inner core of the membrane bilayer. In addition, freeze-etching exposes the ultrastructural features of the true external and inner cytoplasmic surfaces of the membranes. Therefore, the observation of both fractured and etched exposed faces provide a tridimensional view of the membrane *(32–35; cf.* refs. in *36)*. In a replica of a biological membrane, the inner protoplasmic face (PF) of the leaflet cleaved along its hydrophobic matrix shows a heterogeneous appearance characterized by smooth areas and particulate entities ranging in diameter from 4 to 14 nm (intramembranous particles or [IMPs]) likely comprising the transmembrane proteins. The exoplasmic fracture face (EF) is

characterized by the presence of few IMPs and multiple small pits created by the transmembrane proteins dislocated from the outer half of the bilayer. Thus a distinct asymmetric distribution of IMP is found in freeze-fractured biological membranes. The transmembrane proteins interact more strongly with the protoplasmic half of the bilayer (PF) and/or with extrinsic constituents associated with the inner cytoplasmic surface of the membrane *(37)*. The IMP partition coefficient (K_p) between PF and EF can be calculated as follows: $K_p = CP/CE$, where CP and CE are the concentrations in number of IMPs per unit surface adhering to the PF and to the EF, respectively. During freeze–fracture, integral membrane proteins are asymmetrically partitioned between EF and PF. Therefore, each resulting cleaved membrane half does not contain the entire set of proteins, and probably of lipids, present in the intact membrane *(38)*. So far, it is not easily apparent whether or not connexons, built up by a distinct set of connexins, have a preferential partition coefficient for either PF or EF. The application of SDS-FL using site-directed antibodies will be a useful tool for addressing the preferential partition of connexin isoforms (*cf.* **ref. 24**).

2. We anticipated that in the gap junction the fracture plane extends from one to the other membrane of the same junction. Thus, freeze–fracture of junctional membranes splits the hydrophobic interior of each bilayer, exposing IMPs on the protoplasmic leaflet (JPF) and their complementary pits on the exoplasmic leaflet (JEF). Obviously, junctional IMPs comprise connexins (**Figs. 3** and **4**). It might also be that junctional IMP enclosed not only connexins but also tightly bound lipids and auxiliary proteins *(39)*. These remarks should be taken into consideration when one needs to elucidate the results of the SDS-FL experiments. Another interesting problem concerns the region where the connexon pairs are fractured. Experiments carried out with high-resolution rotary shadowing have provided the evidence that the junctional particles on PF may display different heights *(40)*. This observation suggests that the freeze–fracture of connexon pairs takes place randomly and even along protein domains with prevailing covalent bonding *(37,38)*. We should, however, take into consideration that other data obtained by rapid freezing *(41)*, atomic force microscopy *(42)*, and three-dimensional electron crystallography *(43–45)*, of split or recombinant gap junction membranes demonstrated that the connexon-exposed outer surface is characterized by struc-

---

Fig. 3. SDS-FL of the outer lens cortex using polyclonal antibodies raised against the cytoplasmic exposed Cx50 domain. The freeze-fracture has simultaneously split the opposite bilayers of two adjoining lens fibers. The cleavage opened the inner core of one bilayer and exposed the exoplasmic fracture face (EF). Then, the fracture passed in a step-by-step fashion along the bilayer of the adjoining plasma membrane and discovered the inner aspect of the protoplasmic fracture face (PF). One junctional plaque can be identified on EF by the presence of a round-shaped agglomeration of small depressions (JEF). The gold immunolabeling is positively associated with the pitted plaque. The two junctional plaques exposed on PF, (JPF) are characterized by the assembly of 9-nm intramembrane particles (IMPs). The gold immunolabeling is

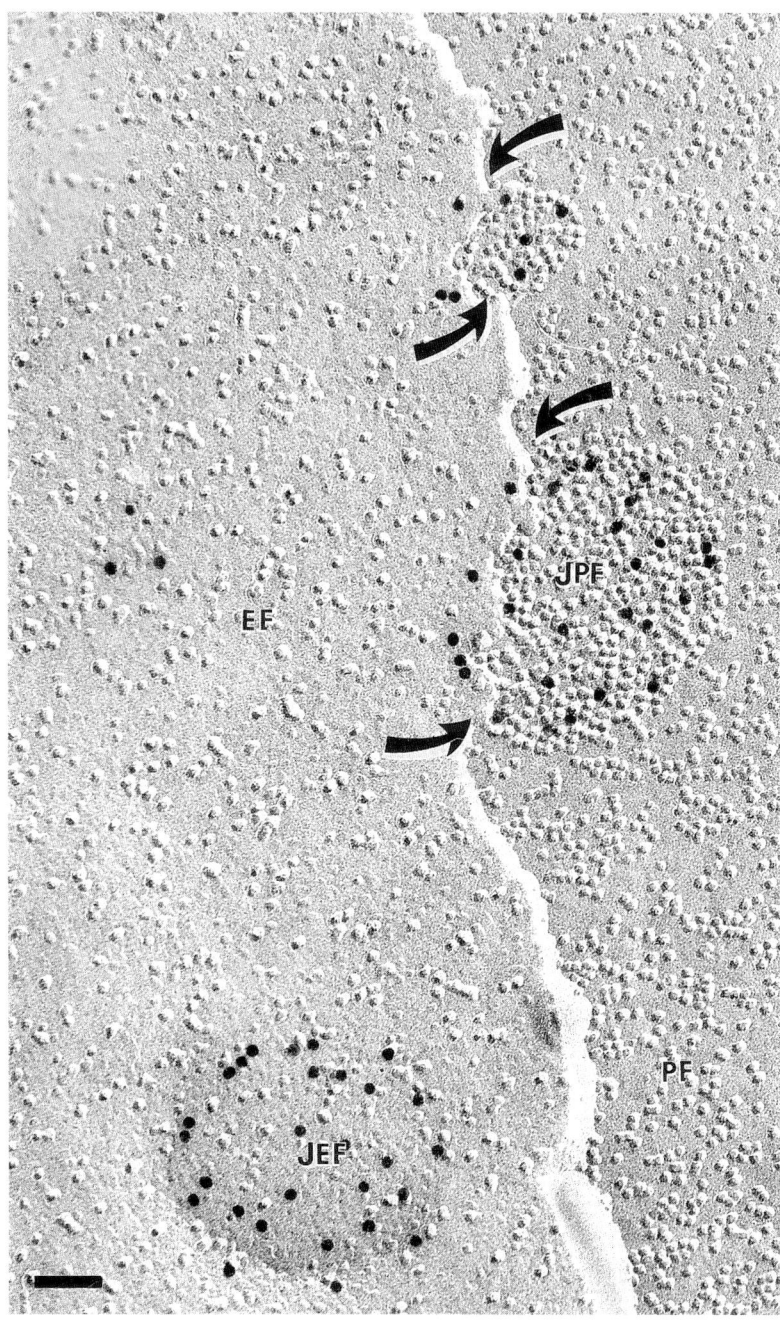

restricted to the junctional domains. The *arrows* point to the sites where the intercellular gap is abruptly reduced. Bar = 50 nm.

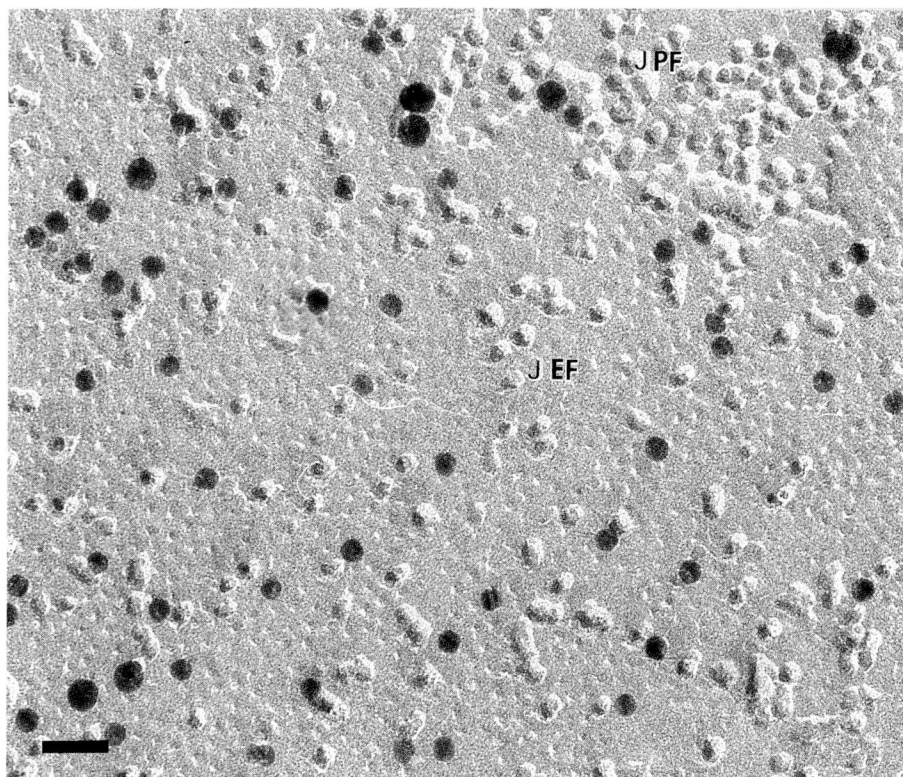

Fig. 4. SDS-FL of cortical lens fiber plasma membranes. Double-immunolabeling
with antibodies directed against Cx46 (15-nm gold particles) and Cx50 (10-nm gold
particles). The double labeling shows that these two lens connexins are codistributed
on the same junctional plaque. JEF, junctional exoplasmic face; JPF, junctional protoplasmic face. Bar = 40 nm. Reproduced from Dunia et al. *(55)*.

tural features comparable to junctional IMPs on PF of conventional replicas of
intact gap junctions. These observations suggest that the IMPs visualized on PF
of gap junctions are primarily single connexons exposed by a freeze–fracture
plane that passes in the middle of the extracellular docking domain of connexon
pairs. The proposed freeze-fracture mechanism of the gap junction and the results
of SDS-FL will be better understood if one takes into consideration the molecular model illustrating the vertical interaction of the connexins forming the transmembrane communicating pair. A model conceives that the two extracellular
loops (E1 and E2) of each connexin connected by disulfide bonds *(46)*, forming a
hemichannel dock with the opposite connexin-exposed segments and inter-
digitates like the two sides of a "zip." The extracellular loop region consists of
stacked intercalated β-sheets resulting in an antiparallel β-barrel motif. The bar-

rel would possess concentric layers, one formed by E1 loops and the other by E2 loops, respectively. The outer surface and the center of the barrel are essentially hydrophilic. Conversely, the interphase between the two concentric layers is expected to be highly apolar, composed mainly of intercalated hydrophobic side-chains of residues originating from two opposed β-sheets (*44–48*). Such a tightly packed hydrophobic wall should confer high stability to the docking of the two connexons forming a pair. It may also explain why conditions used for splitting the gap junctions are usually chemically harsh, involving alkaline pH, chaotropic agents (urea), or a combination of the two treatments (*41,42,48,49*; for additional references *see* Chapter 3 by Sosinsky and Perkins in *this volume*). Furthermore, the molecular features of the docking domain of the connexon pairs will constitute a barrier preventing any leakage from inside the channel to the extracellular space (gap). On the other hand, during freeze-fracture the extracellular domain will be the most fragile site of the connexon pair because at low temperature the hydrophobic bonds are weakened. Therefore, the extracellular domain between two opposed connexons represents the site where the freeze-cleavage can abruptly disjoin the connexon pair (**Fig. 2**).

3. The application of SDS-FL is very useful for the study of the relationship between connexin assembly and cell surface proteins that are implicated in cell–cell recognition and membrane–membrane interactions (*50–53*). SDS-FL will also contribute to the definition of the close lipid environment in the gap junction domains by using specific antibodies against the lipid moiety of the membrane (*54*). In addition, SDS-FL could provide some clues on the potential role of the major intrinsic lens fiber membrane polypeptide (MP26) during gap junction assembly (*55*). This event occurs during the terminal differentiation of epithelial lens cells into elongating fibers (*56*). The results of SDS-FL suggest that MP26 forms transmembrane oligomers randomly distributed in the general plasma membrane that upon freeze-fracture remain anchored to the PF (**Fig. 5**), likely functioning as aquaporin (*57*). On the other hand, MP26 oligomers may also constitute around the gap junctional domain a borderline of transmembrane linked pairs according a model purported for connexon pairs (*see* **Note 2** and **ref. 55**). Hence the application of SDS-FL suggests that a membrane constituent—MP26—may fulfill different functions and is integrated within the lipid bilayer in different conformations (**Figs. 5** and **6**). Furthermore the SDS-FL is a method of general applicability for the identification of spatial distribution of constituents that are involved in the assembly of specialized plasma membrane domains (desmosomes, tight junctions, etc.), and immunochemically characterize the extrinsic membrane proteins that are implicated in cell surface modulation (*58,59*).

4. The assumption of Fujimoto (*15*), addressing the feasibility of SDS-FL, is that the freeze-cleaved membrane halves on the metal shadowing somehow became grasped by the Pt/C mold and physically stabilized. As a consequence the apolar inner core of the cleaved membrane halves are no longer accessible to the SDS binding. Nevertheless, SDS treatment may extract extrinsic membrane constituents and unravel connexon antigenic sites associated with the "stabilized" proto-

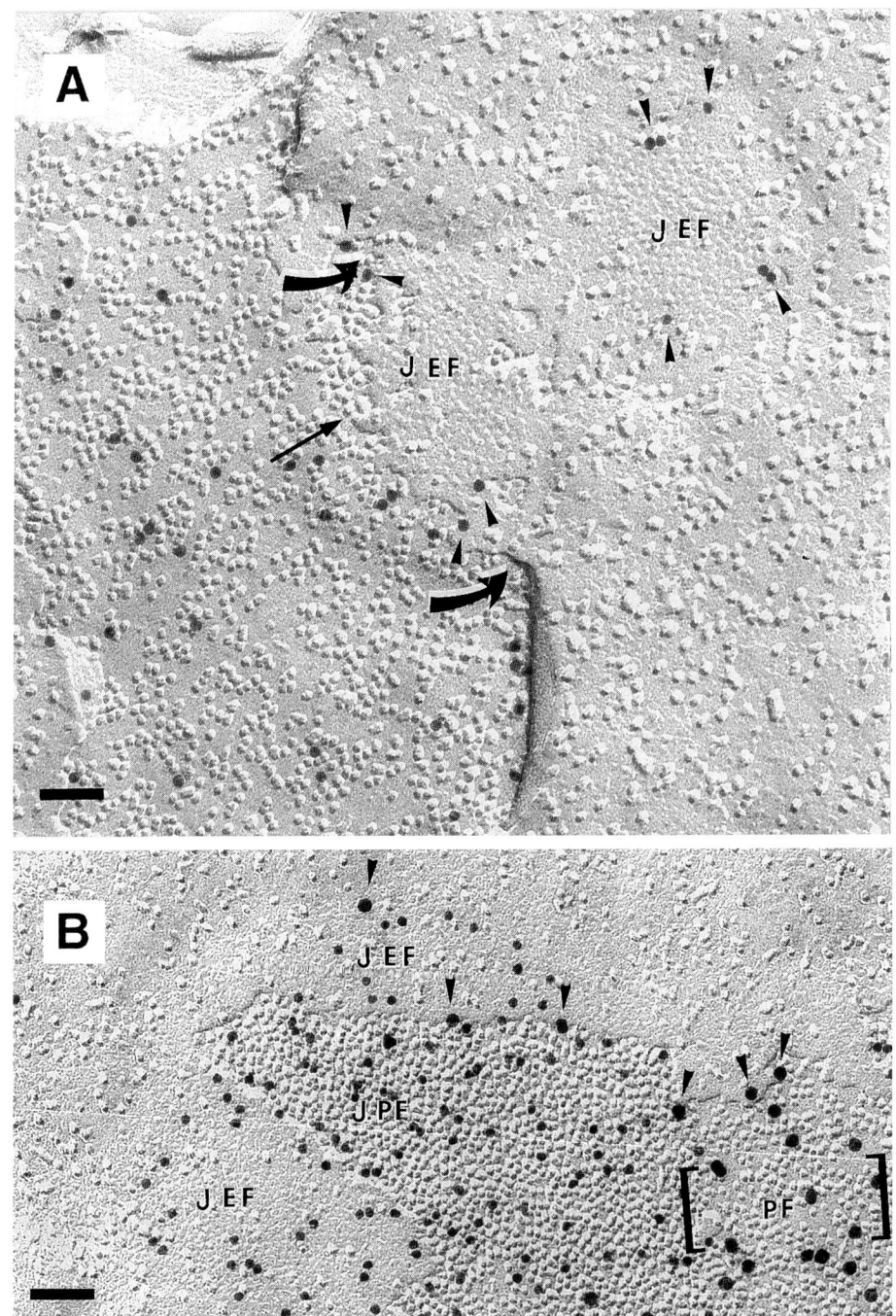

plasmic fracture (PF) halves. The immunolabeling will be further facilitated by washing out the detergent with PBS. An important question concerns the mechanism of gold immunolabeling of the pitted junctional exoplasmic fracture (EF) faces. This question is especially pertinent when the antibody used for immunolabeling recognizes connexin antigenic sites exposed at the cytoplasmic surface of the junctional membrane. One should consider the fact that proteins differ in their intrinsic stability toward SDS solubilization. There exist several examples of protein–lipid membrane domains resistant to SDS denaturation and that may form complexes with the SDS (*60–62*). Therefore one hypothesis could be that the tight protein–protein and protein–lipid interactions between connexons are relatively stable toward the detergent treatment applied for SDS-FL and that SDS instead of denaturate may promote formation of complexes (*63,64*). At this point, it is also noteworthy to recall that gap junctions were originally isolated from the general plasma membrane because this specialized domain is resistant to detergent solubilization (*65,66*). Hence the stability of the gap junction protein–lipid scaffold toward detergent treatment during SDS-FL likely accounts for the persistent association between the connexons within the uncleaved junctional membrane and the freeze-fractured and replicated membrane halves recognized as junctional pitted EF (**Fig. 2**). We assume that the evidence of gold particles apparently labeling the junctional EF in fact correlates with the gold inmunolabeling of the cytoplasmic exposed antigenic sites of connexin within the uncleaved bilayer.

5. Extreme care must always be taken to eliminate any possibility of an explosion hazard when working with liquefied flammable gases in general and particularly with propane. All work must be undertaken within the confines of an extraction fume cupboard suitable for allowing safe escape of flammable vapors and naked flames. Electrical switches that might generate sparks must be excluded from the work area. The liquefied cryogen should be safely discarded after each experi-

---

Fig. 5. (*Opposite page*) (**A**) SDS-FL of cortical lens fibers plasma membranes using polyclonal antibodies raised against MP26 (10-nm gold particles). Both junctional protoplasmic (PF) and exoplasmic faces (EF) are exposed. MP26 immunolabeling remains almost exclusively associated with PF. Conversely, MP26 immunolabeling of EF specifically delimits the borderline between the junctional domain and the general plasma membranes (*arrowheads*). *Curved arrows* point to the site where the intercellular space is reduced. *Black arrows* point to the accumulation of junctional intramembrane particles (IMPs) on PF. (**B**) SDS-fracture double-immunolabeling of cortical fiber plasma membrane with anti-MP26 (15-nm gold particles) and anti-(Cx50 (10-nm gold particles). The immunolabeling of anti-Cx50 (10-nm gold particles) is specifically localized where arrays of 9-nm identical IMPs are closely packed. Conversely, MP26 immunolabeling (15-nm gold particles), is mainly confined to areas where IMPs on PF are randomly distributed (bracket) or along the rim between EF and PF where the intercellular space is reduced (*arrowheads*). Bars = 50 nm.

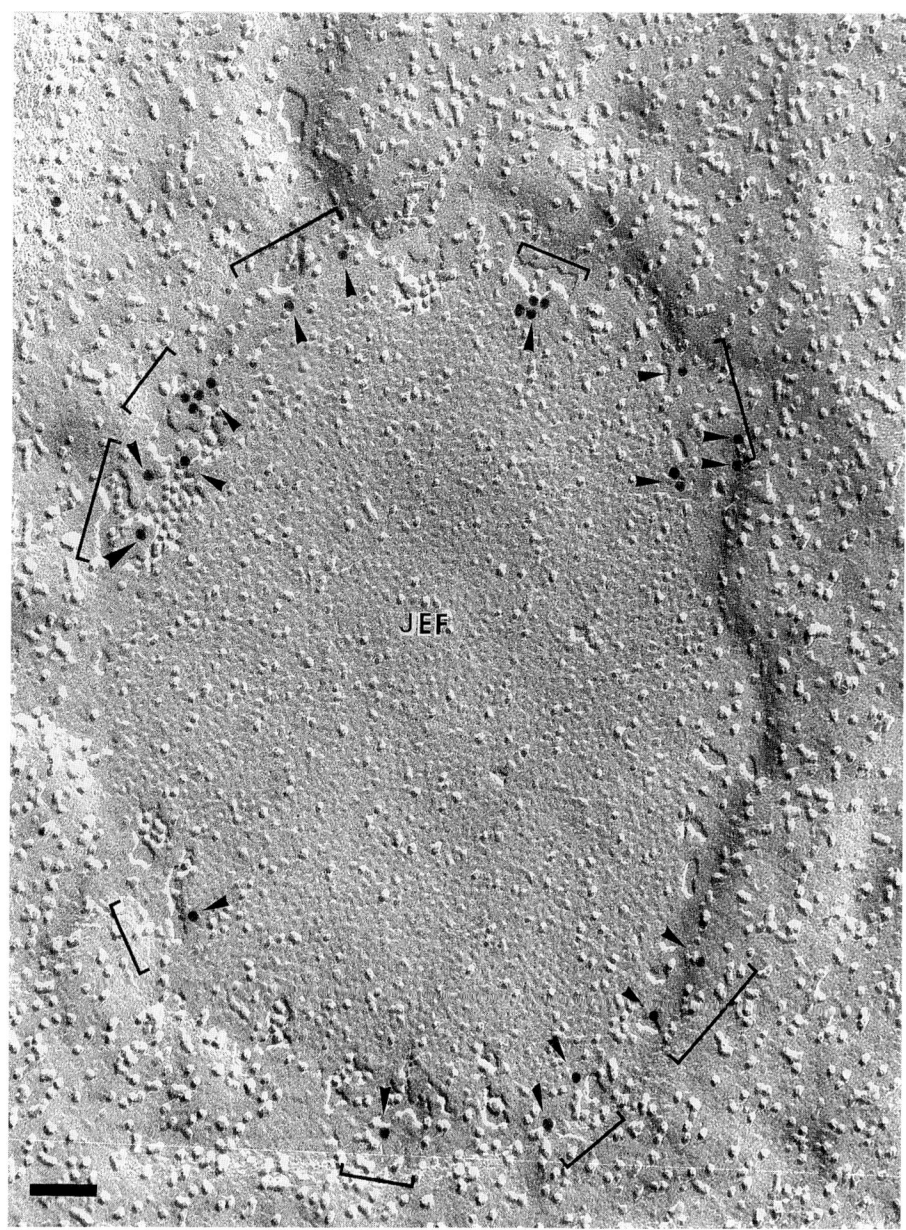

Fig. 6. SDS-FL of cortical lens fibers plasma membrane using polyclonal antibodies raised against MP26. The freeze–fracture exposes a large EF surface. The junctional plaque (JEF) appears as a round-shaped pitted depression. The gold immunolabeling is positively restricted to a belt surrounding the junctional domain (*arrowheads, brackets*). Bar = 50 nm.

ment, either by evaporation within the fume cupboard or by carefully pouring the liquid onto the ground at a distance from people, automobiles, and buildings and other installations.

6. The preparation of the frozen specimens to be fractured and replicated is a crucial step of this technique. The routine method of overcoming the problem of damaging ice crystal formation is the use of a cryoprotectant, with or without previous chemical fixation. For SDS-FL because the material has to be processed unfixed the critical freezing rate of the specimen should be increased to ensure best freezing, to reduce the size of ice crystals compatible with the induction of the so-called vitrification state of the biological specimen. The first sections from the surface of the specimen (measuring 5–20 μm) are properly and quickly frozen. In this "vitrified" state the amount and distribution of the water content of the specimen are preserved. The specimen should be as small as possible and the time interval between sampling and freezing of each sample should be as short as possible to avoid modifications of the structure. Many rapid-freezing methods exist: high-pressure freezing, spray freezing, jet freezing, plunge freezing, and cold-block freezing. The results are satisfactory but they often involve sophisticated and expensive equipment. The entry velocity attainable by manual plunge freezing is not as high or as reproducible as can be achieved using a mechanical device, but the method is simple, has negligible cost, and, with practice, gives consistently good results when used with thin specimens. The detailed methodology, advantages and disadvantages of different techniques are exhaustively described elsewhere (*cf. 67,68*). Once the specimens are frozen they can be used directly for further processing or stored for use at some future time. Frozen specimens must be handled in such a way that temperature increase does not occur. In particular, temperature should not be allowed to rise above about 90 K (–150°C), because some ice crystal growth may occur. Storage in liquid nitrogen is both safest and easiest, and perfectly suitable for specimens to be used for freeze-fracture replication.

7. The fracturing process cleaves the frozen specimen along the plane that offers the least resistance to the applied forces, namely, along the membrane hydrophobic core. Fracturing is done with a microtome knife edge (razor blade). The zone of ideal freezing of the specimen is limited to a thickness between of 5 and 20 μm from the surface depending on the quick-freezing method used (*cf. 67,68*). Thus, the frozen specimen should be fractured superficially.

8. Deposition of the shadowcast (replica) onto the specimen fracture faces must be done in a vacuum coating unit. Excellent vacuum conditions, with a minimum of condensable gases around the specimen during exposure of the fracture planes to the metal deposit, are essential. Several liquid nitrogen or helium-cooled traps have been developed. The aim of these devices is to surround the sample with a cold trap to avoid contamination of the specimen fracture faces particularly with water vapor condensing from either the vacuum unit or parts covered with hoar frost (*cf. 67,68*). The replication process should copy the relief produced by specimen fracturing. This process involves two steps. First, an electron-opaque metal, Pt/C, is evaporated onto the specimen at an oblique angle (approx 45°) to the

average orientation of the fracture planes to provide contrast. Second, the patches of shadowed material are bound together and reinforced by the deposition of a low electron-scattering layer of carbon evaporated perpendicular to the fracture plane. Several parameters increase shadowing resolution (*cf. 67,68*). The user of the freeze–fracture technique is often confronted with the question of whether or not the results of the application of this method are credible. We may assume that when the main physical parameters governing each step of the technique are thoroughly controlled and optimized *(69)*, this method of specimen preparation yields reliable information almost devoid of artifacts. Nevertheless the use of comparative techniques in conjunction with freeze–fracture and etching will provide useful data that will corroborate the interpretation of the replica (*cf. 67,68*). It is noteworthy to recall that the combination of freeze–-fracture and low temperature X-ray diffraction experiments has provided evidence that the crystalline order of complex lipid–water phases remains preserved after freeze–fracture and replication *(70)*.

9. Immunolabeling needs to be performed under extremely clean conditions, with freshly prepared solutions and filtered water. It is important to always keep the replicas wet. Appropriate controls are necessary—for a single immunolabeling, using a comparable nonspecific antibody instead of the specific antibody and testing the protein A–gold conjugate alone without primary antibody. As controls for double or triple labeling, each step needs to be performed separately as a single immunolabeling. For double or triple labeling, one may ask what is the best choice of the size and the order of the gold particles. This should be tested by trial and error. In a multiple labeling procedure the yield of the later gold step is less than in a simple labeling (*cf. 71*). Thus, it is advisable to label the scarcest antigen first, using the largest size of gold particles. The best combination is found by changing the order of the different sizes of gold particles for the same kind of double or triple labeling. Can one gold particle be interpreted as a positive reaction? Any label observed has to be considered specifically if the antibody is characterized by biochemical means as giving a positive reaction. The labeling efficiency of the SDS-FL corresponds to the proportion of an antigen in a replica that is recognized by an antibody, or the number of immunogold particles divided by the known number of antigens that they label. For other methods of immunolabeling, the labeling efficiency is rarely above 10%. Thus, one gold particle can be judged as a positive reaction for an antigen present at very low amounts. Without doubt, this is a handicap for immunocytochemistry at the electron microscopy level, as only antigens present in relatively high amounts can be immunolocalized. When the background is low or nondetectable, more than three gold particles are judged as a positive reaction.

10. Several factors affect the labeling efficiency. First, experimental variables: the quality of the antibody and of the gold and several labeling parameters such as the length of the labeling, dilution of the antibody and gold, temperature, etc. Second, antigen-related variables deal with the different conformations of the antigens to be labeled. Third, the fracture mechanism, detergent treatment, and

particularly the molecular conformation of the constituent (for proteins: number of transmembrane domains, posttranslational modifications, etc.) need to be taken into account. In our experience, prolonged SDS treatment (6–12 h) diminishes the labeling efficiency. Finally, an additional factor to be considered at high antigen concentration is the possibility of steric hindrance, that is, the presence of one bound antibody and /or gold particle may hinder the access of further antibody/gold particles to closely adjacent antigens.

11. The semiquantitative evaluation of immunogold labeling can be applied to SDS-FL, including basic stereology. This methodology has been extensively reviewed (*cf. 72–74*). Quantitative estimation of the immunolabeling of connexins on the gap junction domain, respectively on PF and EF, has been thoroughly illustrated by *(50)* and *(53)*.

Acknowledgments

We gratefully acknowledge the stimulating discussions and support of Pr. N. B. Gilula during our work. We wish to thank Dr. J. G. Dunia for his friendly help and expertise in computer-assisted drawing. The original studies of the authors have been supported by a Collaborative Research grant from NATO International Scientific Exchange Programs on Lens Plasma Membranes and Cataract, by the Alcon Research Institute Awards to E.L. Benedetti and H. Bloemendal, and by the French National Research Council (CNRS).

References

1. Robertson, J. D. (1963) The occurrence of a subunit pattern in the unit membrane of club endings in Mauthner cell synapses in goldfish brain. *J. Cell Biol.* **19,** 201–221.
2. Benedetti, E. L. and Emmelot, P. (1965) Electron microscopic observations on negatively stained plasma membranes isolated from rat liver. *J. Cell Biol.* **26,** 166–174.
3. Revel, J. P. and Karnovsky, M. J. (1967) Hexagonal array of subunits in intercellular junctions of the mouse heart and liver. *J. Cell Biol.* **33,** C7.
4. Kumar, N. M. and Gilula, N. B. (1996) The gap junction communication channel. *Cell* **84.** 381–388.
5. Bruzzone R., White, T. W., and Goodenough, D. A. (1996) The cellular internet: on-line with connexins. *BioEssays* **18,** 709–718.
6. Nicholson, S. and Bruzzone, R. (1997) Gap junctions: getting the message through. *Curr. Biol.* **7,** 340–344.
7. Jiang, J. X. and Goodenough, D. A. (1996) Heteromeric connexons in lens gap junction channels. *Proc. Natl. Acad. Sci. USA* **93,** 1287–1291.
8. Bennett, M. V. L. and Spray, D. C. (1985) *Gap Junctions,* Cold Spring Harbor Laboratory Press, Cold Spring Harbor, New York.
9. Staehelin, L. A. (1979) A simple guide to the evaluation of the quality of a freeze-fracture replica, in *Freeze-Fracture Methods, Artifacts, and Interpretations* (Rash, J. E. and Hudson, C. S., eds.), Raven Press, New York, pp. 11–17.

10. Peracchia, C. and Peracchia, L. L. (1988) Gap junction dynamics: reversible effect of hydrogen ions. *J. Cell Biol.* **87,** 719–727.

11. Pinto Da Silva, P., Parkison, C., and Dwyer, N. (1981) Fracture label: cytochemistry of freeze-fracture faces in the erythrocyte membrane. *Proc. Natl. Acad. Sci. USA* **78,** 343–347.

12. Pinto Da Silva, P. and Kan, F. W. K. (1984) Label fracture: a method for high resolution labeling of cell surfaces. *J. Cell Biol.,* **99,** 1156–1161.

13. Fujimoto, K. and Ogawa, K. (1991) Fracture-flip and fracture-flip cytochemistry to study macromolecular architecture of membrane surfaces: practical procedures, interpretation and application. *Acta Histochem. Cytochem.* **24,** 111–117.

14. Fujimoto, K. and Pinto Da Silva, P. (1992) Fracture-flip/Triton X-100 reveals the cytoplasmic surface of human erythrocyte membranes. *Acta Histochem. Cytochem.* **25,** 255–263.

15. Fujimoto, K. (1997) SDS-digested freeze–fracture replica labeling electron microscopy to study the two-dimensional distribution of integral membrane proteins and phospholipids in biomembranes: practical procedure, interpretation and application. *Histochem. Cell Biol.* **107,** 87–96.

15a. Rash, J. E. and Yasumura, T. (1999) Direct immunogold labeling of connexins and aquaporin-4 in freeze-fracture replicas of liver, brain, and spinal cord: factors limiting quantitative analysis. *Tissue Cell Res.* **296,** 307–321.

16. Hertzberg, E. L., Anderson, D. J., Friedlander, M., and Gilula, N. B. (1982) Comparative analysis of the major polypeptides from liver gap junctions and lens fiber junctions. *J. Cell Biol.* **92,** 53–59.

17. Dunia, I., Manenti, S., Rousselet, A., and Benedetti, E. L. (1987) Electron microscopic observations of reconstituted proteoliposomes with the purified major intrinsic membrane protein of eye lens fibers. *J. Cell Biol.* **105,** 1679–1689.

18. Gong, X., Klier, G., Huang, Q., Wu, Y., Lei, H., Kumar, N. M., Horwitz, J., and Gilula, N. B. (1997) Disruption of α3 connexin gene leads to proteolysis and cataractogenesis in mice. *Cell* **91,** 833–843.

19. White, T. W., Bruzzone, R., Goodenough, D. A., and Paul, D. L (1992) Mouse Cx50, a potential member of the connexin family of gap junction proteins, is the lens fiber protein MP70. *Mol. Biol. Cell.* **3,** 711–720.

20. Kistler, J., B. Kirkland, and S. Bullivant. (1985) Identification of a 70,000kD protein in lens membrane junctional domains. *J. Cell Biol.* **101,** 28–35.

21. Friedlander, M. (1980) Immunological approaches to the study of myogenesis and lens fiber junction formation. *Curr. Top. Dev. Biol.* **14,** 321–358.

22. Broekhuyse, R. M., Kuhlmann, E. D., and Stols, A. L. H. (1976) Lens membrane II. Isolation and characterization of the main intrinsic polypeptide (MIP) of bovine lens fiber membranes. *Exp. Eye Res.* **23,** 365–371.

23. Vallon , O., Dunia, I., Favard-Sereno, C., Hoebeke, J., and Benedetti, E. L. (1985) MP26 in the bovine lens: a post-embedding immunocytochemical study. *Biol. Cell* **53,** 85–88.

24. Milks, L. C., Kumar, N. M., Houghten, N., Unwin, N., and Gilula, N. B. (1988) Topology of the 32-kD liver gap junction protein determined by site-directed antibody localizations. *EMBO J.* **7,** 2967–2975.

25. Green, N., Alexander, H., Olson, A., Alexander, S., Shinnick, T. M., Sutcliffe, J. G., and Lerner, R. A. (1982) Immunogenic structure of the influenza virus hemagglutinin. *Cell* **28,** 477–487.

25. Paul, D. L., Ebihara, L., Takemoto, L. J., Swenson, K. I., and Goodenough, D. A. (1991) Connexin46, a novel lens gap junction protein, induces voltage-gated currents in nonjunctional plasma membrane of *Xenopus* oocytes. *J. Cell Biol.* **115,** 1077–1089.

26. Yeager, M. and Gilula, N. B. (1992) Membrane topology and quaternary structure of cardiac gap junction ion channels. *J. Mol. Biol.* **223,** 929–948.

27. Risek, B., Guthrie, S., Kumar, N., and Gilula, N. B. (1990) Modulation of gap junction transcript and protein expression during pregnancy in the rat. *J. Cell Biol.* **110,** 269–282.

28. Jarvis, L. J., Kumar, N. M., and Louis, C. F. (1993) The developmental expression of three mammalian lens fiber cell membrane proteins. *Invest. Ophthalmol. Vis. Sci.* **34,** 613–620.

29. Gruijters, W. T., Kistler, J., Bullivant, S., and Goodenough, D. A. (1987) Immunolocalization of MP70 in lens fiber 16–17–nm intercellular junctions. *J. Cell Biol.*, **104,** 565–572.

30. Kistler, J. and Bullivant, S. (1987) Protein processing in lens intercellular junctions: cleavage of MP70 to MP38. *Invest. Ophthalmol. Vis. Sci.* **28,** 1687–1692.

31. Li, J. S., Fitzgerald, S., Dong, Y., Knight, C., Donaldson, P., and Kistler, J. (1997) Processing of the gap junction connexin 50 in the ocular lens is accomplished by calpain. *Eur. J. Cell Biol.* **73,** 141–149.

32. Branton, D. (1966) Fracture faces of frozen membranes. *Proc. Natl. Acad. Sci. USA* **55,** 1048–1056.

33. Deamer, D. W. and Branton, D. (1967) Fracture planes in a ice-bilayer model membrane system. *Science* **158,** 655–657.

34. Pinto Da Silva, P. and Branton, D. (1970) Membrane splitting in freeze-teching: covalently bound ferritin as a membrane marker. *J. Cell Biol.* **45,** 598–604.

35. Moor, H. (1973) Cryotechnology for the structural analysis of biological material, in *Freeze-Etching, Techniques and Applications* (Benedetti, E. L. and Favard, P., eds.), French Society of Electron Microscopy, Paris, pp. 11–30.

36. Benedetti, E. L. and Favard, P. (1973) Freeze-etching, techniques and applications. French Society of Electron Microscopy, Paris, France.

37. Satir, B. H. and Satir, P. (1979) Partitioning of intramembrane particles during the freeze-fracture procedure, in *Freeze-Fracture Methods, Artifact, and Interpretations* (Rash, J. E. and Hudson, C. S., eds.), Raven Press, New York, pp. 43–49.

38. Torrisi, M. R. and Mancini, P. (1996) Freeze-fracture immunogold labeling. *Histochem. Cell Biol.*, **106,** 19–30.

39. Verkleij, J. (1984) Lipidic intramembranous particles. *Biochim. Biophys. Acta.* **779,** 43–63.

40. Hanna, R. B., Ornberg, R. L., and Reese, T. S. (1985) *Structural Details of Rapid Frozen Gap Junctions*, Cold Spring Harbor Laboratory Press, Cold Spring Harbor, New York, pp. 23–32.

41. Hirokawa, N. and Heuser, J. (1982) The inside and outside of gap-junction membranes visualized by deep etching. *Cell* **30,** 395–406.

42. Hoh, J. H., Sosinsky, G., Revel, J. P., and Hansena, P. K. (1993) Structure of the extracellular surface of the gap junction by atomic force microscopy. *Biophys. J.* **65,** 149–163.

43. Perkins, G., Goodenough, D., and Sosinsky, G. (1997) Tree-dimensional structure of the gap junction connexon. *Biophys. J.* **72,** 533–544.

44. Unger, V. M., Kumar, N. M., Gilula, N. B., and Yaeger, M. (1997) Proyection structure of a gap junction membrane channel at 7 Å resolution. *Nat. Struct. Biol.* **4,** 39–43.

45. Unger, V. M., Kumar, N. M., Gilula, N. B., and Yaeger, M. (1999) Three-dimensional structure of a recombinant gap junction membrane channel. *Science* **283,** 1176–1179.

46. Foote, C. I., Zan Zhou, and Nicholson, B. J. (1998) The pattern of disulfide linkages in the extracellular loop regions of Connexin 32 suggests a model for the docking interface of gap junctions. *J. Cell Biol.* **140,** 1187–1197.

47. White, T. W., Bruzzone, R., Wolfram, S., Paul, D. L., and Goodenough, D. A. (1994) Selective interactions among the multiple connexin proteins expressed in the vertebrate lens: the second extracellular domain is a determinant of compatibility between connexins. *J. Cell Biol.* **125,** 879–892.

48. Sosinsky, G. (1996) Molecular organization of gap junction membrane channels. *J. Bioenerg. Biomembr.* **28,** 297–309.

49. Ghroshroy, R., Goodenough, D. A., and Sosisnky, E. (1995) Preparation, characterization, and structure of half gap junctional layers split with urea and EGTA. *J. Membr. Biol.* **146,** 15–28.

50. Fujimoto, K., Nagafuchi, A., Tsukita, S., Kuraoka A., Ohokuma, A., and Shibata, Y. (1997) Dynamics of connexins, E-cadherin and α-catenin on cell membrane during junction formation. *J. Cell Sci.* **110,** 311–322.

51. Furuse, M., Fujimoto, K., Sato, N., Hirase, T., Tsukita, S., and Tsukita, S. (1996) Overexpression of occludin, a tight junction-associated integral membrane protein, induces formation of intracellular multilamellar bodies bearing tight junction-structures. *J. Cell Sci.* **109,** 429–435.

52. Hirase, T., Staddon, J. M., Saitou, M., Ando-Akatsuka, Y., Itoh, M., Furuse, M., Fujimoto, K., Tsukita, S., and Rubin, L. L. (1997) Occludin as a possible determinant of tight junction permeability in endothelial cells. *J. Cell Sci.* **110,** 1603–1613.

53. Hülser, D. F., Rehkopf, B., and Traub, O. (1997) Dispersed and aggregated gap junction channels identified by immunogold labeling of freeze-fractured membranes. *Exp. Cell Res.* **233,** 240–251.

54. Fujimoto, K., Umeda, M., and Fujimoto, T. (1996) Transmembrane phospholipid distribution revealed by freeze-fracture replica labeling. *J. Cell Sci.* **109,** 2453–2460.

55. Dunia, I., Recouvreur, M., Nicolas, P., Kumar, N., Bloemendal, H., and Benedetti, E. L. (1998) Assembly of connexins and MP26 in lens fiber plasma membranes studied by SDS-fracture immunolabeling. *J. Cell Sci.* **111,** 2109–2120.

56. Benedetti, E. L., Dunia, I., Dufier, J. L., Yit K-S., and Bloemendal, H. (1996) Plasma membrane-cytoskeleton complex in the normal and cataractous lens, in *The Cytoskeleton,* Vol. 3 (Hesketh, J. E. and Pryme, I. F., eds.), JAI Press, London, pp. 451–518.

57. Deen, P. M. T. and van Os, C. H. (1998) Epithelial aquaporins *Curr. Opin. Cell Biol.* **10**, 435–442.

58. Cullen, M. J., Walsh, J., Stevenson, S. A., Rothery, S. and Severs, N. J. (1998) Co-localization of dystrophin and β-dystroglycan demonstrated in "en face" view by double-immunogold labeling of freeze-fractured skeletal muscle. *J. Histochem. Cytochem.* **46**, 945–953.

59. Nomura, R. and Fujimoto, T. (1999) Tyrosine-phosphorylated caveolin-1: immunolocalization and molecular characterization. *Mol. Biol. Cell* **10**, 975–986.

60. Reynolds, J. A. and Tandford, Ch. (1970a) The gross conformation of protein-sodium dodecyl sulfate complexes. *J. Biol. Chem.* **254**, 5161–5165.

61. Reynolds, J. A. and Tandford, Ch. (1970b) Binding of dodecyl-sulfate to proteins at high binding ratios. Possible implications for the state of proteins in biological membranes. *Proc. Natl. Acad. Sci. USA* **66**, 1002–1007.

62. Helenius, A. and Simons K. (1975) Solubilization of membranes by detergents. *Biochem. Biophys. Acta* **415**, 29–79.

63. White, F. H. and Wright, G. (1984) Effect of structure-forming solutes on chicken eggwhite lysozyme after reductive cleavage of disulfide bonds. *Int. J. Pept. Prot. Res.* **23**, 256–270.

64. Rothe, G. M. and Maurer, W. D. (1986) One dimensional PAA-gel electrophoretic techniques to separate functional and denatured proteins, in *Gel Electrophoresis of Proteins* (Dunn, M. J., ed.), Wright Press, Bristol, UK, pp. 37–140.

65. Benedetti, E. L., and Emmelot, P. (1968) Hexagonal array of subunits in tight junctions separated from isolated rat liver plasma membranes. *J. Cell Biol.,* **38**, 15–24.

66. Dunia, I., Sen-Ghosh, K., Benedetti, E. L., Zweers, A., and Bloemendal, H. (1974) Isolation and protein pattern of eye lens fiber junctions. *FEBS Lett.* **45**, 139–144.

67. Robards, A. W. and Sleytr, V. B. (1985) Practical Methods of Electron Microscopy, in *Low Temperature Methods in Biological Electron Microscopy,* Vol. 10 (Glauert, A. M., ed.), Elsevier, Amsterdam.

68. Severs, N. J. and Shotton, D. M. (1995) Rapid freezing, freeze-fracture and deep etching, in *Techniques in Modern Biomedical Microscopy*, Wiley-Liss, New York.

69. Gross, H. (1979) Advances in ultrahigh vacuum freeze-fracturing at very low specimen temperature, in *Freeze-Fracture Methods, Artifacts and Interpretations* (Rash, J. E. and Hudson, C. S., eds.), Raven Press.,New York, pp. 127–139.

70. Gulik-Krzywicki, T. (1997) Freeze-fracture transmission electron microscopy. *Curr. Opin. Colloid Interphase Sci.* **2**, 137–144.

71. Raposo, G., Kleijmeer, M., Posthuma, G., Slot, J. and Geuze, H. (1997) Immunogold labeling of ultrathin cryosections:application in immunology, in *Weir's Handbook of Experimental Immunology,* 5th ed., Vol. 4. (Herzenberg, L. A., Weir, D. and Blackwell, C., eds.), Blackwell Science, Cambridge, MA, pp. 1–11.

72. Slot J. W., Posthuma, G., Chang, L. Y., Crapo, J. D., and Geuze, H. (1989) Quantitative aspects of immunogold labeling in embedded and non-embedded sections. *Am. J. Anat.* **185**, 271–281.

73. Griffiths, G. (1993) *Fine Structure Immunocytochemistry,* Springer-Verlag, New York, New York.

3

Purification of Gap Junctions

Gina E. Sosinsky and Guy A. Perkins

1. Introduction

   The descriptive term gap junction originally arose from the electron micrograph studies of Revel and Karnovsky in 1967 (*1*). This morphological name referred to a cell–cell contact area where two cell membranes were bridged by a specialized membrane protein complex. Gap junctions are found in almost all tissues where cells abut each other. The "gap" is a distinct morphological feature whereby heavy metal stains are able to intercalate into a space between the two joined cell membranes, as opposed to the situation in tight junctions, where the membranes fuse and the stain cannot penetrate in between the cells. Subsequent electron microscopy revealed that gap junctions consist of tens to thousands of membrane channels (also called intercellular channels) containing proteins generally called connexins. Each membrane channel was shown to contain two connexons, the hexamer of connexins, which dock at their extracellular surfaces. In essence, the gap junction membrane channel is a dimer of two hexamers joined together in the gap region. In this way, the membrane channel extends across both cell membranes. Both freeze–fracture and thin section electron micrographs showed that these gap junctional plaques contained quasicrystalline areas of the membrane channels. During the next decade, intensive efforts by many researchers were directed toward purifying the gap junction plaques first seen in these micrographs for both biochemical and structural analysis. Many man-years were spent developing and refining isolation protocols that reliably gave pure preparations of gap junctions, which were used for structure determination by electron crystallography (*2*) and X-ray diffraction (*3*). The specimens that are considered most desirable for electron crystallography are thin, uniform two-dimensional crystals. Early on, gap junctions were looked on as a tantalizing specimen, not only because of

From: *Methods in Molecular Biology, vol. 154: Connexin Methods and Protocols*
Edited by: R. Bruzzone and C. Giaume © Humana Press Inc., Totowa, NJ

their biological importance, but also because they form quasicrystals in the cell and with limited detergent treatments could be induced into well-ordered two-dimensional crystals *(4)*. It is clear from the structural work that has been done to date that the critical step in the structure determination is the isolation of crystalline material in sufficient and pure quantities (a problem that structural biologists also refer to as "no crystal, no grant"). In liver, the gap junction arrays are estimated to cover only ~0.1% of the surface area of a hepatocyte, therefore, one generally begins with large amounts of starting material to end with a small quantity of pure gap junctions *(5–7)*. Liver, lens and heart tissue *(8–10)* have classically been the organs of choice for isolating gap junctions because the majority of the tissue contains only one or two types of connexins and starting material can be obtained in large quantities. The situation has improved with the advent of molecular biology techniques for expression of the connexins in tissue culture systems such as the baby hamster kidney (BHK) cell line *(11)* and the insect cell line derived from *Spodoptera (12,13)*. In these cell lines, the gap junctions can be overexpressed, isolated and purified with larger yields than from native tissues. Recent advances in the three-dimensional structure determination at ~7Å have occurred because of the development of overexpression systems and generation of recombinant connexin material (*see* **ref. *14*** and Chapter 4 in *this volume*) as well as improved electron crystallographic structure determination methods. These EM crystallographic methods have been greatly enhanced by both improvements in electron microscopes as well as in computer reconstruction techniques (*see* Chapter 4 and also **ref. *15***).

The protocols used for isolating gap junctions can be divided into two classes based on the chemical treatment of the crude membrane fractions. One procedure uses two detergents. The first detergent solubilizes nonjunctional membranes and the second, sequential detergent both solubilizes nonjunctional membranes and extracts lipids within the gap junctional membranes that cause the channels to be more tightly packed, and therefore to be better ordered. The second procedure uses a high pH treatment of the crude membrane fraction to saponify the extrajunctional material. A detergent treatment may be subsequently used as well. It should be noted that the fraction obtained by the alkali incubation is typically less crystalline than with the detergent protocol. This may be used to an advantage if an investigator wants to solubilize gap junctional plaques into individual channels or connexons (as was done with insect cell gap junctions and liver gap junctions by Stauffer et al. *[12,13]*). Procedures for isolating soluble fractions containing connexons directly from plasma membranes (e.g., lens connexons, *[16,17]*) may also be found in Chapter 5 by Falk and in Chapter 6 by VanSlyke and Musil in *this volume*. The quality of the purified gap junctions has been assessed by both negative staining electron microscopy and sodium dodecylsulfate polyacrylamide gel electrophoresis (SDS-PAGE).

2. Materials

2.1. Detergent Extraction of Liver Gap Junctions

1. 1 mM NaHCO$_3$ buffer: 2.0 L of 1 mM NaHCO$_3$, 1 mM EGTA, 0.5 mM diisopropylfluorophosphate (DIFP) (0.168 g NaHCO$_3$, 20 mL of 100 mM EGTA stock, 184 µL DIFP, 1980 mL dH$_2$O) (*see* **Note 1**).

   Caution: DIFP is a neurotoxin, so it is essential to wear gloves when using it in its undiluted form. Place pipet tips used during dilution into a concentrated NaOH solution in a waste receptacle. When diluted in water, DIFP is much less harmful, although some care should be taken not to ingest any solutions containing DIFP.

2. 6 mM NaHCO$_3$ buffer: 3 L of 6 mM NaHCO$_3$, 1 mM EGTA, 0.5 mM DIFP (1.53 g of NaHCO$_3$, 30 mL of 100 mM EGTA stock, 276 µL of DIFP, 2970 mL of dH$_2$O).
3. Tris-HCl buffer: 3 L of 5 mM Tris-HCl pH 8.5, 1 mM EGTA, 0.5 mM of DIFP (150 mL of 100 mM Tris-HCl stock, 30 mL of 100 mM EGTA stock, 276 µL of DIFP, 2820 mL of dH$_2$O).
4. 2 L of 67% sucrose (1780.4 g of sucrose) in Tris-HCl buffer.
5. 1 L of 41% sucrose in Tris-HCl buffer (add 610 mL of 67% sucrose to 390 mL of Tris-HCl buffer).
6. 400 mL of 25% sucrose in Tris-HCl buffer (add 150 mL of 67% sucrose to 250 mL of Tris-HCl buffer).
7. 300 mL of 30% sucrose in Tris-HCl buffer (add 135 mL of 67% sucrose to 165 mL of Tris-HCl buffer).
8. 100 mL of 0.6% Sarkosyl detergent in Tris-HCl buffer (0.6 g/100 mL).
9. 100 mL of 0.6% Brij 58 detergent in Tris-HCl buffer (0.6 g/100 mL).

2.2. Alkali Extraction of Liver Gap Junctions

1. NaHCO$_3$ buffer: 20 L of 1 mM NaHCO$_3$, 0.5 mM DIFP.
2. 5 mL of 40 mM NaOH, 0.5 mM DIFP.
3. 25 mL of 20 mM NaOH, 0.5 mM DIFP.
4. 2 L of 67% sucrose (1780.4 g of sucrose) in 1 mM NaHCO$_3$ buffer.
5. 200 mL of 55% sucrose in 1 mM NaHCO$_3$ buffer (made from 67% sucrose solution).
6. 2 L of 37% sucrose in 1 mM NaHCO$_3$ buffer (made from 67% sucrose solution).
7. 5 mL of 30% sucrose in 1 mM NaHCO$_3$ buffer (made from 67% sucrose solution).
8. 2 mL of 41% sucrose in 1 mM NaHCO$_3$ buffer (made from 67% sucrose solution).

2.3. Detergent Extraction of Heart Gap Junctions

1. NaHCO$_3$ buffer: 1.0 L of 1 mM NaHCO$_3$, pH 8.2, 1 mM (phenylmethylsulfonyl fluoride (PMSF, added as a solid to solutions).
2. KI/Na$_2$S$_2$O$_3$/NaHCO$_3$ solution: 200 mL of 0.6 M KI, 6 mM Na$_2$S$_2$O$_3$, 1 mM NaHCO$_3$, pH 8.2, 1 mM PMSF.
3. KI/Na$_2$S$_2$O$_3$/Tris solution: 70 mL of 0.6 M KI, 6 mM Na$_2$S$_2$O$_3$, 5 mM Tris-HCl, pH 9.0, 1 mM PMSF.

4. Tris buffer: 400 mL of 5 mM Tris-HCl, pH 10.0, 1 mM PMSF.
5. 25 mL of 49% sucrose, 0.3% deoxycholate in 5 mM Tris-HCl, pH 10.0; solid PMSF added to a final concentration of 1 mM.
6. 35 mL of 35% sucrose, 0.3% deoxycholate in 5 mM Tris-HCl, pH 10.0; solid PMSF added to a final concentration of 1 mM.
7. 30 mL of 0.6% Sarkosyl in 5 mM Tris-HCl, pH 10.0, 1 mM PMSF (added as a solid to solutions).
8. 20 mL of 0.3% deoxycholate in 5 mM Tris-HCl, pH 10.0, 1 mM PMSF.

2.4. Detergent Extraction of Cx263T Gap Junctions from Stable Transfected BHK Cells

1. 4-(2-Hydroxyethyl)-1-piperazineethanesulfonic acid (HEPES) buffer: 180 mL of 10 mM HEPES, pH 7.4; 0.8% NaCl.
2. PMSF stock solution is 70 mg/mL in dimethyl sulfoxide (DMSO).
3. Buffered Tween-20: 10 mL of 10 mM HEPES, pH 7.5, 0.8% NaCl; solid PMSF added to a final concentration of 0.14 mg/mL, 200 mM KI, 2 mM sodium thiosulfate, Tween-20 at a final concentration of 2.8%.
4. Dialysis buffer: 1 L of HEPES buffer containing 2.5 mL of gentamicin.
5. 30 mL of 25% sucrose in 10 mM HEPES buffer.
6. 10 mL of Tween-20 in 10 mM HEPES buffer containing 200 mM potassium iodide, 2 mM sodium thiosulfate. The concentration of Tween-20 for this incubation is 2.7% per milligram of total membrane protein. After 12 h, 1,2-diheptanoyl-sn-phosphocholine (DHPC, Avanti Polar Lipids) at a concentration of 50× the critical micelle concentration is added to the detergent solution. The critical micelle concentration of Tween-20 is 0.06 mM. The Tween-20 was purified on a mixed ion exchange column (Bio-Rad AG-1X8 resin) to eliminate charged contaminants.

2.5. Alkali Extraction of Spodoptera Gap Junctions

1. NaHCO$_3$ buffer: 2.3 L of 1 mM NaHCO$_3$, pH 8.2, solid PMSF added to a final concentration of 1 mM.
2. 50 mL of 67% sucrose in NaHCO$_3$ buffer.
3. 30 mL of 30% sucrose in NaHCO$_3$ buffer.

2.6. Splitting of Liver Gap Junctions

1. Tris buffer: 10 mL of 5 mM Tris-HCl, pH 8.5, 1 mM EGTA, and 0.5 mM DIFP.
2. Urea solution: 10 mL of Tris buffer, 8 M urea, 5 mM EGTA, 10 mM dithiothreitol (DTT).

2.7. Electron Microscopy Reagents

1. 1–2% Uranyl acetate, 1–2% sodium phosphotungstate, or 2% potassium phosphotungstate (w/v in distilled water). Electron microscopy is necessary to check purity and crystallinity of gap junctions.
2. 300–400 copper mesh grids (available from microscope supply companies such as Pelco, PO Box 492477, Redding, CA 96049-2477, USA; Ernest F. Fullam,

900 Albany Shaker Rd., Latham, New York 12110, USA; and Structure Probe, 569 East Gay Street, West Chester, PA 19380, USA).

3. Mica sheets (available from microscope supply companies such as those listed in **step 2**).

4. 0.3% Formvar solution (w/v in ethylene dichloride).

2.8. Centrifuges and Rotors

All rotors are designations from Beckman-Coulter Instruments, Spinco Div., Palo Alto, CA, USA or Sorvall Instruments, Newtown, CT, USA except where noted. The web sites, http://www.beckmancoulter.com/beckman/biorsrch/prodinfo/cntrifug/rotrcalc.asp for Beckman-Coulter rotors or http://www.sorvall.com/support/rcf-calc.htm for Sorvall rotors are useful for converting run times between different rotors. *Please note:* most of the Relative Centrifugal Forces (RCF, expressed in the text here as numbers of g's) are converted from revolutions per minute (RPM) and are the *average* RCFs.

1. Specific rotors for the detergent extraction of Liver gap junctions: JA-10, JA-14, JA-20 (for use in Beckman J series Centrifuge) , SW-28 Ti (for use in Beckman Ultracentrifuge).

2. Specific rotors for the alkali extraction of liver gap junctions: JA-10, JA-20, SW-40 Ti (for use in Beckman Ultracentrifuge), JCF-Z continuous flow rotor, Ti-15 ultracentrifuge zonal rotor.

3. Specific rotors for the detergent extraction of heart gap junctions: Sorvall SS-34 (Sorvall Preparative Centrifuge), SW-28 Ti (for use in Beckman Ultracentrifuge).

4. Specific rotors for the detergent extraction of Cx263T gap junctions from BHK cells expressing Cx263T : Beckman JA-10 (for usc in Beckman J series Centrifuge), GH-3.8 (for use in a Beckman Allegra or GS6 Benchtop Centrifuge), SW-28 Ti (for use in Beckman Ultracentrifuge).

5. Specific rotors for the alkali extraction of *Spodoptera* gap junctions: JA-20, SW-40 Ti.

6. Specific rotors for the splitting of liver gap junctions: TLA-100 (for use in Beckman Table Tap Ultracentrifuge Model).

2.9. Other Small Apparatus

1. Probe sonicator with small probe, Branson Ultrasonics, 465 Borrego Ct No. C, San Dimas, CA, USA.

2. Reichert-Jung 10431 hand refractometer, from Reichert-Jung, Rheinlandstrasse 36, Frankfurt Am Main, 60529 Germany.

3. DotMetric protein assay, from Geno Technology, 3047 Bartold Ave., Maplewood, MO 63143, USA.

4. Two 4-L Cole Parmer buckets, Cole-Parmer Instrument, 625 East Bunker Court, Vernon Hills, IL 60061–1844, USA.

2.10. Other Reagents

1. All reagents are from Sigma–Aldrich (USA) except where noted.

3. Methods

3.1. Purification of Gap Junction Membranes from Liver

3.1.1. Detergent Extraction Protocol

Gap junctions are isolated as double membrane plaques by membrane fractionation and purification techniques (*see* **Fig. 1** for a schematic of the procedure). The isolation protocol, modified from Fallon and Goodenough *(5)*, produces exceptionally high yields of gap junctions (up to 1 mg/150 g of liver) with relatively high crystallinity and little amorphous material; collagen contamination is also low. The two detergents used are Sarkosyl and Brij 58. The following protocol is for livers obtained from 36 young rats (3–4 wk old). The total amount of starting liver tissue is ~150 g of liver.

1. In the cold room, cool two steel Cole-Parmer buckets and one swath of two layers and another swath of four layers of cheesecloth. Set out two dishpans of ice and body bags. Rinse Dounce and a pestle and 12 polyallomer JA-10 bottles. Cool 2 JA-10 rotors in separate centrifuges. All subsequent purification steps are performed on ice or in centrifuges cooled to 4°C, except where noted.
2. Pour 3 L of 6 mM NaHCO$_3$ buffer in a small Cole–Parmer steel bucket and place it on one pan of ice. Place the 2 -L container of 1 mM NaHCO$_3$ buffer on the other pan of ice.
3. Kill 36 rats, six rats at a time (*see* **Note 2**). Fill the Dounce homogenizer with cold 1 mM NaHCO$_3$ buffer (~50 mL). Put the livers in the Dounce and homogenize with 8–10 strokes. Homogenize with a Tissuemizer. Add this homogenate to the 3 L of 6 mM NaHCO$_3$ buffer.
4. When all the livers have been homogenized, let the total homogenate sit on ice for 10 min.
5. Filter this homogenate through two layers and then four layers of cheesecloth and funnel into the second Cole–Parmer steel bucket.
6. Distribute the filtrate into 12 ultraclear JA-10 bottles and centrifuge for 30 min at 10,976g; brake at max.
7. Aspirate the supernatant, taking special care to remove the white lipid layers. Resuspend the pellets in each of the 12 JA-10 bottles with 1 mM NaHCO3 buffer (~20 mL of buffer in each bottle). Mix and then distribute the resuspension into two ultraclear JA-10 bottles and fill these with 1 mM NaHCO3 buffer. Centrifuge for 30 min at ~11,000g.
8. Aspirate the supernatant and resuspend each pellet in ~100 mL of 1 mM NaHCO$_3$ buffer. Transfer into a 1-L flask. The volume of the sample should be 200–300 mL.
9. Place a large stirbar in the flask of sample, and while spinning, slowly add 67% buffered sucrose until the sample has a sucrose density of 50%. Approximately ~500 mL of 67% sucrose is needed. Use a hand refractometer to confirm this, such as a Reichert-Jung 10431 hand refractometer. The volume of the sucrose-laden sample will be approx 900 mL. Keep this cold.

10. Make six gradients in six ultraclear JA-10 bottles:
    a. Layer sample/50% sucrose on the bottom of six JA-10 bottles (~150 mL per bottle).
    b. Layer 150 mL of 41% sucrose on top of the sample (*see* **Note 3**).
    c. Layer ~37 mL of 25% sucrose on top of 41% sucrose (use automated pipet at slowest speed).
    d. Centrifuge overnight in the JA-10 at ~11,000*g*. Brake at 0.
11. Collect the 25%/41% interface using a 25-mL pipet into a 2 L flask. Usually this is 2×25 mL per bottle. Dilute the 300–400 mL of the interface material with 800 mL of 5 m*M* Tris buffer. Centrifuge in six JA-14 bottles at 18,879 g for 30 min.
12. While the sample is being centrifuged, prepare Sarkosyl detergent. Rinse a stirbar, six JA-20 tubes, and a glass syringe. Cool one JA-20 rotor to 4°C.
13. When the centrifugation in the JA-14 has been completed, aspirate the maximum amount of the supernatant and resuspend the pellet in a minimum amount of Tris buffer (~10 mL each tube). Collect these resuspended pellets into one JA-14 bottle. Rinse the centrifuge bottles in a minimum amount of buffer (~10 mL) and transfer the rinse into the JA-14 bottle. The total volume should be below 100 mL. Bring the volume to 100 mL with just enough Tris buffer.
14. While stirring, slowly pour 100 mL of Sarkosyl into the beaker. Stir at room temperature for 10 min. Briefly sonicate this solution with a probe sonicator. Decant the sample into six ultraclear JA-20 tubes and centrifuge for 20 min at ~31,400*g*.
15. Prepare the Brij 58 detergent. It takes 10–15 min for the Brij to dissolve. Use a circular shaker in a cold room.
16. Aspirate the supernatant from **step 14**. Resuspend the pellets with a TOTAL of 10 mL of Brij solution, that is, 1.5 mL/tube. Use plastic pipets to resuspend. Do not foam. Mix the resuspensions and briefly sonicate this solution with a probe sonicator.
17. Make four gradients in four ultraclear SW-28 tubes in the following way:
    a. Layer ~7 mL of 41% sucrose at the bottom.
    b. Layer ~26 mL of 30% on top of 41%.
    c. Layer the sample/Brij onto the 30% and top the centrifuge tubes up with Brij.
    d. Centrifuge the gradients in a SW-28 rotor for 2 h at ~82,700*g*.
18. To collect the sample at the 30%/41% interface, first clean the outsides of the gradient tubes with ethanol and water and dry them. Use an 18- or 25-gauge needle to puncture the tube at the bottom side of the tube. Use a fiber optics light to watch the interface come down (illuminate from the side); sometimes a black background helps to visualize the boundary. Then collect the interface, which appears white, into a clear SW-28 tube (~2 mL from each tube).
19. Top the SW-28 tube with Tris buffer and mix thoroughly. Spin in the SW-28 for 30 min at ~82,700*g*.
20. Resuspend the pellet with 0.5 mL of Tris buffer. Transfer to a 1.5 mL Eppendorf tube and add 1 μL of DIFP. Vortex-mix briefly.
21. (Optional) To further purify the gap junctions, **steps 14–20** may be repeated with the pellets from **step 19**, substituting 0.5% sodium deoxycholate for the 1.0%

Sarkosyl. The total volume should be no more than 20 mL. Hence, only one centrifuge tube is used. Continue to use one tube for **steps 16–19**.

3.1.2. Alkali Extraction Protocol

Initial enrichment of gap junctions is accomplished by the partial purification of liver plasma membranes, following the procedure described by Hertzberg *(7)*. Subsequently, nonjunctional membranes are removed by sonication in the presence of alkali. Differential centrifugation and density gradient centrifugation finally yields purified gap junction plaques containing Cx26 and Cx32. About the same amount of gap junction proteins can be obtained with this alkali procedure (about 1 mg/150 g of liver) compared to the detergent procedure. However, the crystalline quality of the junctional plaques and purity of preparations may not be as good as with detergent procedures.

1. Guillotine 40 rats (*see* **Note 4**). Excise the livers and place in 2 L of cold 1 mM NaHCO$_3$ buffer. For all steps, keep the sample on ice or at 4°C.
2. Mince liver into small pieces with scissors and rinse repeatedly with NaHCO$_3$ buffer until the supernatant is clear, approximately 4 L is typical for the repeated rinses to clear the tissue (*see* **Note 5**).
3. Homogenize the minced tissue with four to eight strokes (*see* **Note 6**) in 2 L of NaHCO$_3$ buffer in a loose-fitting large Dounce homogenizer that will hold at least 300 mL (*see* **Note 7**).
4. Dilute homogenate to 4 L with NaHCO$_3$ buffer and filter first through two layers and then four layers of cheesecloth.
5. Dilute homogenate to 8 L with NaHCO$_3$ buffer and pump at 250 mL/min through a JCF-Z rotor at ~3590 g for a total time of 32 min (*see* **Note 8**).
6. Resuspend the pellet with 300 mL of NaHCO$_3$ buffer and homogenize with no more than two gentle strokes with the same type of Dounce homogenizer used in **step 3**. Centrifuge at 16,000g for 10 min.
7. Resuspend the pellet in NaHCO$_3$ buffer to give a final volume of 300 mL. Add 600 mL of 67% sucrose solution and filter through four layers of cheesecloth.
8. To remove nuclei and obtain a fraction enriched in liver plasma membrane, use the zonal Ti-15 rotor. Place the sample in the four sectors of the rotor (*see* **Note 9**) and fill to the top with the 37% sucrose solution. Seal the rotor and fill the loading cap with the 55% sucrose solution using a 60-mL syringe through the outer set of holes in the vanes. Loading is finished when some of the 37% sucrose solution is forced out of the cap (*see* **Note 10**).
9. Spin in an ultracentrifuge by accelerating at the slowest acceleration setting of the ultracentrifuge to 84,000g and maintain at this speed for an additional 1.5 h. Stop without braking and collect at the sample/37% sucrose interface with a 50-mL syringe (~30 mL) (*see* **Note 11**).
10. Dilute 3× with NaHCO$_3$ buffer and centrifuge in a JA-10 rotor for 10 min at 16,000g.

11. Resuspend pellet (enriched plasma membranes) in a minimal volume of $NaHCO_3$ buffer (~20 mL); aliquot in 5-mL volumes, and store at –20°C until processed for purification of gap junctions (*see* **Note 12**).

12. To one of the 5-mL aliquots of enriched plasma membrane add 5 mL of 40 mM NaOH in polysulfone centrifuge tubes (*see* **Note 13**).

13. Sonicate with a probe sonicator at medium setting for 12 s, for example, Model W-225R, Heat System Ultrasonics with standard microtip using setting 6.

14. Add 25 mL of 20 mM NaOH and mix.

15. Sediment by centrifuging at 43,700g for 10 min in a JA-20 rotor.

16. Aspirate the supernatant and loose material on top of the stratified pellet.

17. Resuspend the pellet with 0.25 mL of $NaHCO_3$ buffer and disperse with gentle sonication. Dilute with additional $NaHCO_3$ buffer (~3 mL) and centrifuge at 43,700g for 10 min.

18. Disperse the pellet by sonication in ~3 mL of $NaHCO_3$ buffer and dilute further to 12 mL. Add 16.5 g of 67% sucrose in buffer and adjust volume to 25 mL. Confirm that the sample is 38% sucrose with a hand refractometer, for example, a Reichert-Jung 10431, and adjust if necessary.

19. Prepare six tubes (one sixth of the sample in each) of discontinuous sucrose gradients in ultraclear disposable tubes for the SW-40 Ti rotor by first adding 2 mL of 45% sucrose in buffer. Add 4 mL of the sample. Add 5 mL of 30% sucrose in buffer and finally top with $NaHCO_3$ buffer. Centrifuge at 270,000g for 1.5 h. Set brake on ultracentrifuge to off.

20. Use a 16- or 18-gauge needle on a 10-ml glass syringe to collect the material at all of the interfaces into one pool. Do not puncture, but instead collect by gently inserting the needle from the top. Use a fiber optics light to watch the interfaces (illuminate from the side); sometimes a black background helps to visualize the boundary, which appears white. When all of the interfaces have been collected (~4 mL/tube), combine the remaining sucrose as a "gradient residue" fraction.

21. Dilute both fractions 3× with $NaHCO_3$ buffer. Centrifuge in a JA-20 rotor at 43,700g for 20 min. Resuspend by sonication in ~2 mL of $NaHCO_3$ buffer. Dilute with ~2 mL of $NaHCO_3$ buffer and repeat the centrifugation.

22. Resuspend the pellets by sonication in ~1 mL of $NaHCO_3$ buffer. Purified gap junctions are in the "interface" pool. However, additional gap junctions, often highly pure, are in the "gradient residue" pool (*see* **Note 14**).

3.2. Purification of Gap Junctions from Heart

3.2.1. Detergent Extraction Protocol

The following procedure is that of Manjunath et al. *(18)* with modifications described by Manjunath and Page *(8)* and Yeager *(19)*. The yield is approx 0.4 mg of gap junction proteins per 1 g of heart. The two detergents used are Sarkosyl and deoxycholate. The protocol described here is for 4 rat hearts, but can be scaled up to 20 rat hearts *(19)*.

1. Anesthetize four rats (Sprague–Dawley, 300–350 g) with ether (*see* **Note 15**).
2. Excise hearts and place in 40 mL of ice-cold $NaHCO_3$ buffer. All subsequent steps are carried out on ice or at 4°C, unless otherwise noted.
3. Cut away the superficial fat and major vessels. Mince into small pieces with scissors.
4. Homogenize for 60 s with a Tissuemizer (for example, model SDT 100 EM, Tekmar Instruments set at maximum speed).
5. Dilute homogenate to 300 mL with $NaHCO_3$ buffer and filter through six layers of cheesecloth.
6. Centrifuge at ~21,400g for 15 min in a Sorvall SS-34 rotor.
7. Discard the supernatant and resuspend the pellet in 300 mL of $NaHCO_3$ buffer with a Dounce homogenizer with only three or four strokes.
8. Centrifuge the suspension at ~21,400g for 15 min in a Sorvall SS-34 rotor and discard the supernatant.
9. Suspend the pellet in 200 mL of $KI/Na_2S_2O_3/NaHCO_3$ solution and slowly stir overnight in a cold room. This extracts the contractile proteins.
10. Centrifuge the suspension at ~11,300g for 30 min in a Sorvall SS-34 rotor and discard the supernatant.
11. Add 70 mL of $KI/Na_2S_2O_3/Tris$ solution and resuspend. Centrifuge the suspension at ~11,300g for 30 min in a Sorvall SS-34 rotor and discard the supernatant.
12. Add 300 mL of Tris buffer and resuspend. Centrifuge the suspension at ~11,300g for 30 min in a Sorvall SS-34 rotor and discard the supernatant.
13. Resuspend in 30 mL of Tris buffer at room temperature.
14. Homogenize suspension with a 100–300 mL of Dounce homogenizer using three strokes.
15. Add 30 mL of Sarkosyl in buffer while stirring (*see* **Note 16**).
16. Stir for 10 min at room temperature.
17. Make three discontinuous gradients in three ultraclear SW-28 tubes:
    a. Layer 8 mL of 49% sucrose/deoxycholate on the bottom.
    b. Layer 10 mL of 35% sucrose/deoxycholate next.
    c. Layer 20 mL of sample on top.
    d. Centrifuge at ~64,100g for 60 min at 15°C in a SW-28 rotor. Brake at 0.
18. Collect band at the 35–49% sucrose interface and dilute with an equal volume of 0.3% deoxycholate in buffer.
19. Centrifuge at ~103,800g for 60 min at 15°C in a SW-28 rotor, which yields gap-junction enriched membranes in the pellet.
20. Suspend pellet in 2 mL of 0.3% deoxycholate and sonicate, for example, with a Sonifier Cell Disrupter, model W140, Heat Systems-Ultrasonics at setting 2, for 20 s in a cold room.
21. Make one discontinuous gradient
    a. Layer 3 mL of 49% sucrose/deoxycholate on the bottom (*see* **Note 17**).
    b. Layer 4.8 mL of 35% sucrose/deoxycholate.
    c. Layer sonicated suspension on top.
    d. Centrifuge at ~66,200 g for 60 min at 15°C in a SW-28.1 rotor.

22. Collect the 35/49% sucrose interface, which will contain the purified gap junction plaques (*see* **Note 18**).

3.3. Purification of Gap Junction Membranes from Cultured Cells

3.3.1. Truncated Form of Cx43 from Stable Transfected BHK Cells

The major cardiac gap junction protein is Cx43. Expression of a truncated form of Cx43, Cx-263T, has been achieved in a stable transfected BHK cell line *(14)*. The truncation occurs after Pro^{263} and removes most of the large C-terminal domain. The buffer used is HEPES, rather than the more commonly used bicarbonate and Tris buffers. The two detergents used are Tween-20 and DHPC. This protocol results in a much less pure fraction of gap junctions than can be obtained from either liver or heart tissue using the previously described protocols. The crystallinity of the purified gap junction plaques (~7Å), however, is the highest reported from all protocols and may be related to the deletion of a flexible cytoplasmic domain of Cx43. The following protocol is for confluent cell cultures grown on 20 T150 tissue culture flasks. *See* Chapter 4 in *this volume*.

1. Examine the cell confluency by light microscopy. When the cells appear nearly confluent, they are ready for induction. Aspirate the medium. Induce expression of the truncated form of Cx43, Cx-263T, by adding 100 μ*M* Zn-acetate to the culture medium of a confluent layer of stably transfected BHK cells that express Cx-263T under the control of the inducible mouse metallothionine promotor.
2. Collect cells 8 h after induction by using a cell scraper to dislodge the cells from the bottom of the plates. Place a 500-mL JA-10 centrifuge bottle into an ice bucket and insert a funnel into the bottle. Decant the media with the suspended cells into the funnel. Centrifuge at 2000*g* for 5 min at 4°C in a JA-10 rotor.
3. Decant the supernatant.
4. Carefully pipet 10 mL of ice-cold HEPES buffer into the centrifuge tube without disturbing the pellet.
5. Detach the pellet by swirling the bottle.
6. Pour the contents with the intact pellet into a 50-ml Falcon conical plastic tube.
7. Vortex-mix the tube to break up the pellet.
8. Add HEPES buffer until the volume is 50 mL.
9. Centrifuge at 162*g* for 5 min. at 4°C in a GH-3.8 rotor.
10. Decant the supernatant and add 5 mL of HEPES buffer containing 0.14 mg/mL of PMSF to the pellet.
11. Resuspend the pellet by vortex-mixing.
12. Add HEPES buffer with 0.14 mg/mL PMSF until the final volume is 15 mL.
13. Incubated in the Falcon tube for 5 min on ice.
14. Sonicate for 5 s 3× with a Branson sonicator with the output setting at 6–7, 60% duty cycle. Wait 15 s between each burst (*see* **Note 19**).

15. With a light microscope, examine a drop of the suspension to determine if the cell lysis is at least 90%. If the cell lysis is less than 90%, then repeat the sonication for 5 s.
16. Add HEPES buffer with 0.14 mg/mL of PMSF until the volume is 35 mL.
17. Centrifuge at ~82,700g for 45 min at 4°C in a SW-28 rotor.
18. Decant the supernatant and resuspend the pellet in 5 mL of cold HEPES buffer.
19. Add 49% sucrose in HEPES buffer to a SW-28 tube and pipet the 5-mL crude membrane homogenate on top.
20. Centrifuge at ~82,700g for 45 min. at 4°C in a SW-28 rotor.
21. Use a plastic disposable pipet to carefully remove the material floating on top.
22. Retrieve the 49% interface from the top of the tube with a different pipet and transfer to another SW-28 tube. The appearance of the homogenate should be milky white.
23. Add HEPES buffer until the volume is 35 mL.
24. Use a plastic pipet to mix the solution.
25. Centrifuge at ~82,700g for 45 min at 4°C in a SW-28 rotor.
26. Decant the supernatant.
27. Add 1 mL of cold HEPES buffer with 0.14 mg/mL of PMSF.
28. Disrupt the pellet by repeated passage through a plastic Eppendorf pipet.
29. Homogenize the sample by sonication for 5 s. The specimen is stable at 4°C for ~2 wk.
30. The following steps provide the procedure for *in situ* 2-dimensional crystallization of recombinant gap junction channels. Determine the protein concentration with an assay, for example, the DotMetric protein assay, Geno Technology.
31. Adjust the volume of the enriched gap junction suspension to a protein concentration of 1 mg/mL.
32. Extract lipids and dissolve nonjunctional membranes by adding to this solution 2.8% Tween-20, 200 mM KI, 2 mM sodium thiosulfate, 0.14 mg/mL of PMSF, 0.05 mg/mL of gentamicin in 10 mM HEPES buffer, pH 7.5, containing 0.8% NaCl. Because Tween-20 is contaminated with aldehydes, peroxides, and free acids, it must be purified immediately before use by ion-exchange chromatography (resin AG501-X8D; Bio-Rad, Richmond, CA, USA).
33. Add a magnetic mini-stirring bar to the tube and stir at 27° C for 12 h.
34. After the 12-h incubation, add solid DHPC at a final concentration of 13.5 mg/mL and continue extraction for 1 h at 27°C.
35. Make one discontinuous gradient in an ultraclear SW-28 tube:
    a. Layer on the bottom 30 mL of 25% sucrose in 10 mM HEPES buffer, 0.8% NaCl, pH 7.5.
    b. Layer sample on top (~5 mL).
    c. Centrifuge at ~82,700g for 60 min at 4°C in a SW-28 rotor. Brake at 0.
36. Collect the pellet, carefully removing the supernatant to avoid introducing any excess detergent in the sample. Wipe the tube clean with a Kimwipe to remove any residual sucrose or detergent.
37. Resuspend the pellet in 0.8 mL of HEPES buffer. Sonicate, if necessary, for a few seconds.

38. Dialyze against two changes of a 500-mL reservoir of HEPES buffer containing 2.5 mL of gentamicin for 1–2 d at 4°C to remove the residual detergent.
39. Store the suspension at 4°C in a 1.5-mL Eppendorf tube.

3.3.2. Gap Junctions Composed of Cx32 and Cx26 from Spodoptera Transfected Cultured Cells

Cx32 and Cx26 have been successfully expressed in the insect cell line Sf9 from *Spodoptera* (*12,13*). This procedure and subsequent extraction and purification of gap junction plaques is described in Stauffer (*13*). An alkali extraction procedure was used similar to the Hertzberg technique described previously (*7*), but with fewer centrifugation steps. Gap junctions consisting of Cx32, Cx26, or a combination of the two have been successfully expressed and purified. These are the connexins that are found in liver gap junctions. The recipe here is for 1 L of culture with a cell density of 2×10^6 cells/mL.

1. Harvest 1 L of connexin baculovirus *Spodoptera* insect cells (cell line Sf9) after a high level of expression of the coding sequence 60–90 h following infection.
2. Wash cells with 200 mL of 150 m*M* NaCl, 1 m*M* PMSF, 5 m*M* 4-morpholine-ethanesulfonic acid (MES), pH 6.2 buffer to remove medium (*see* **Note 20**).
3. Add 40 m*M* NaOH to an equal volume of cell extract so that the final concentration of NaOH is 20 m*M*. Break the cells by sonicating for 15 s, for example, with a Kontes probe sonicator operated at 5W using a 3-mm tip.
4. Incubate the broken cells on ice for 1 h. Afterwards, centrifuge in preparative rotor, such as a JA-20 rotor, at 35,000*g* for 20 min.
5. Resuspend pellets in 12 mL of NaHCO$_3$ buffer and place in a 100-mL flask. Place a large stirbar in the flask of sample, and, with spinning, slowly add 67% buffered sucrose until the sample has a sucrose density of 42%. Use a hand refractometer to confirm this, such as a Reichert–Jung 10431 hand refractometer.
6. Prepare six tubes (one sixth of the sample in each) of discontinuous sucrose gradients in ultraclear disposable tubes for the ultracentrifuge rotors, such as a SW-40 Ti or SW-28 rotor, by first adding 4–5 mL of the sample. Add 5 mL of 30% sucrose in buffer and finally top with NaHCO$_3$ buffer. Centrifuge at 100,000*g* for 100 min. Stop without braking.
7. Collect bands at both the 42%/30% and the 30%/NaHCO$_3$ interfaces. Also collect any material in the 30% sucrose bulk.
8. Pool the samples and dilute fivefold with NaHCO$_3$ buffer. Centrifuge in a preparative rotor, for example, JA-20 rotor, at 35,000*g* for 30 min.
9. Resuspend the pellet in 1 mL NaHCO$_3$ buffer.

3.4. Preparing Split Gap Junctions from Liver Gap Junctions

Gap junctions are isolated as double-membrane plaques of paired connexons. However, for certain biochemical and structural studies, it is advantageous to split the two membranes along their extracellular pairings to isolate single

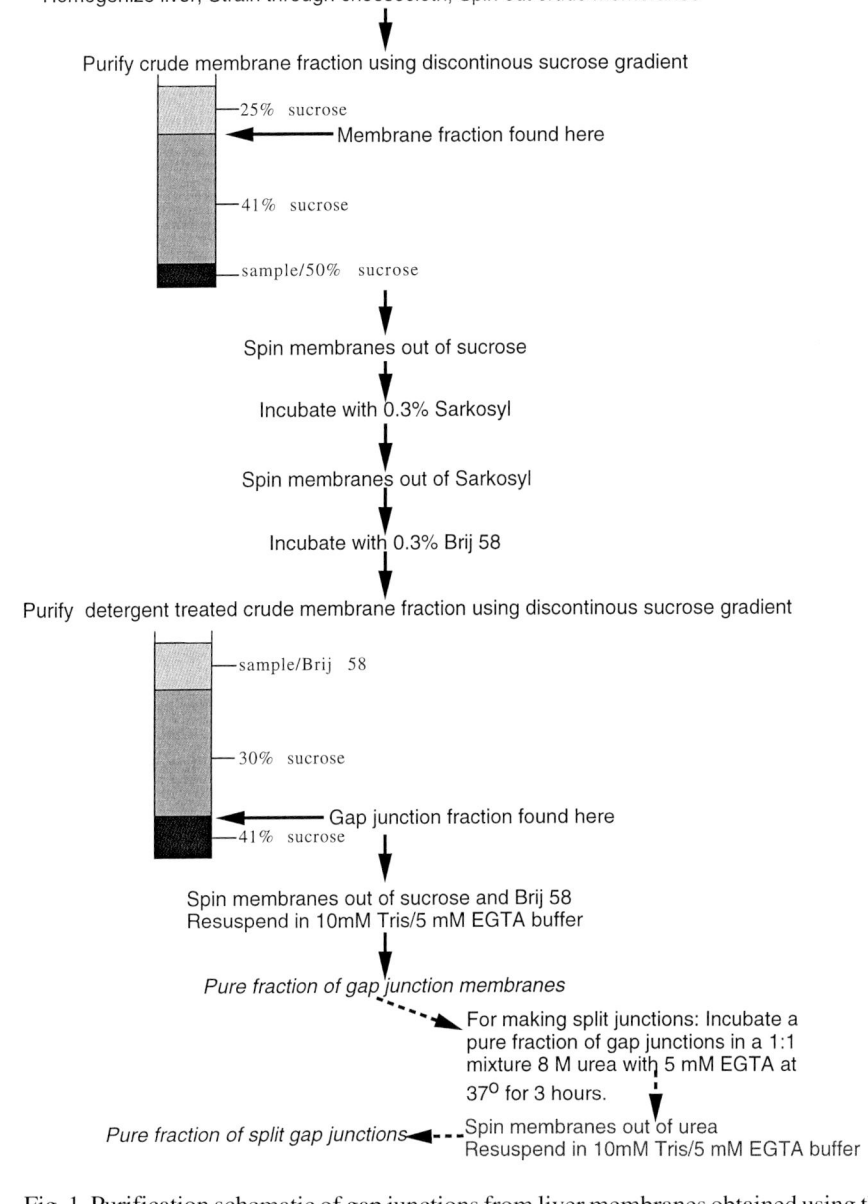

Fig. 1. Purification schematic of gap junctions from liver membranes obtained using the detergent protocol. Preparation of split junctions is shown at the bottom (end of protocol).

connexon layers. The splitting procedure is described by Ghoshroy et al. *(6)* (*see* **Fig. 1 bottom** for a schematic of the procedure). This procedure maintains the same level of crystallinity in the connexon layers that was present in the double-membrane plaques. The combination of urea, EGTA (a chelating agent), and elevated temperature consistently provides > 70% splitting. Alternative splitting procedures are described in Zimmer et al. *(20)* and Milks et al. *(21)*.

1. To 0.1 mL of gap junctions (~0.2– 0.5 µg/mL) purified according to **Subheading 3.1.1.**, add 0.1 mL of urea solution. Incubate at 37°C for 3 h (*see* **Note 21**).
2. Refrigerate at 4°C for 3–5 d.
3. Centrifuge sample at 280,000*g* for 45 min in a TL-100 centrifuge using a TLA100.2 rotor.
4. Resuspend sample in 0.1 mL of Tris buffer (*see* **Note 22**).

3.5. Assessing Purity of Gap Junction Preparations

The quality of gap junction preparations can be assessed using (1) negative stain electron microscopy, (2) (sodium dodecyl sulfate-SDS-PAGE), and (3) Western blotting. Negative staining provides information about the crystallinity and morphology of the preparation while SDS-PAGE and Western blotting indicate the biochemical purity. Thin-section electron microscopy can also be used to assay the morphological purity of the preparations (for representative examples *see* **refs. 6–10, 18, 21–22**). See **Fig. 2** for negative stain micrographs of liver gap junctions prepared by (1) the detergent protocol in **Subheading 3.1.1**, (2) the alkali protocol in **Subheading 3.1.2.**, and (3) split junctions as prepared in **Subheading 3.4.** (*see* **Note 23**). **Fig. 2D** shows an enlargement of a portion of a gap junction in **Fig. 2A** showing the crystalline arrangement of the connexons (magnification is ~6× that of **Fig. 2A**).

1. Float either a Formwar or carbon film onto water to make support films.
2. Place grids on the film and dry onto a filter paper.
3. Place a 3–5-µL drop of gap junction preparation on the grid and allow to adhere for 1–2 min.
4. Rinse with ~15 µL of negative stain, for example, a 2% aqueous solution of uranyl acetate, and place a 3–5 µL drop on top for 1–2 min.
5. Blot the solution with a piece of filter paper (at the side of the electron microscope grid) and allow to dry. Alternatively, a drop of gap junction preparation can be mixed with a drop of the negative stain and applied to the coated electron micropscope grid.
6. Allow material to adhere for 1–2 min and then dry with filter paper. Be sure to blot with filter paper at the side of the grid. Examine in an electron microscope at magnifications typically in the range of 20–60,000×.

4. Notes

1. In general, protease inhibitors are needed to prevent the proteolytic degradation of connexins. The two most frequently used for gap junctions are PMSF and DIFP. It is a good idea to add one of these protease inhibitors to all buffers until the sample is passed through density gradient centrifugation. A protease inhibitor or a bacte-

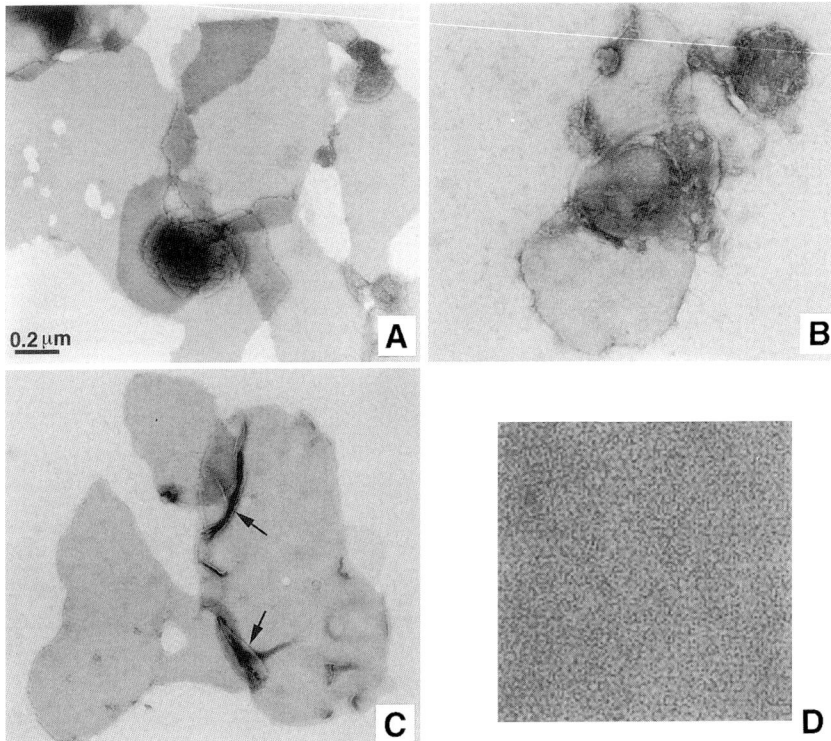

Fig. 2. Representative electron micrographs of gap junction preparations from three of the protocols in this chapter. The gap junctions were stained with uranyl acetate in all of these micrographs. (**A**) Liver gap junctions purified by the detergent protocol. (**B**) Micrograph of liver gap junctions purified by the alkali procedure. (**C**) Split gap junctions (single connexon layers) obtained by a urea treatment described in the text. The *arrows* point to the folds in the membrane indicating that these are single membranes. An enlargement of an area from one of the gap junctions in (A) is shown in (**D**). Note the "doughnut" appearance of the intercellular channels and the crystalline lattice. The magnification of (D) is about six times that of (A).

rial growth inhibitor is also generally added to the final solution used for sample storage. Keeping all solutions and samples at 4°C (except where noted in the protocols) is recommended to decrease any proteolysis and bacterial contamination.

2. Guillotining the rats is recommended to obtain the freshest liver tissue. When working with a large number of animals, it is best that two people work together, one to kill the rats, and the other to extract the livers. Rats that are 3–4 wk old (80–100 g) work best. Older rats can be used, but collagen contamination will increase and guillotining is also more difficult with older, larger animals.

3. When layering large volumes of sucrose solution on top of another layer, it is convenient to use an open syringe connected to small-bore tubing. An adjustable clamp is attached to the tubing to regulate flow rate. In this way, the sucrose solution can be dribbled down the inner side of the centrifuge tube to avoid mixing of layers. More sucrose solution can be added to the top of the syringe, supported by a standard stand and clamp, as needed. The amount of solution added can be monitored with the scale on the syringe.

4. To keep the cost down from using a large number of rats, it may be preferred to use retired breeders, which are relatively cheap and have larger livers than young rats.

5. An alternative to scissors mincing is a single passage through a Foley food mill. However, when the food mill is used, the minced tissue should not be rinsed because material would be lost from the more finely minced sample.

6. The cell breakage can be monitored with a phase-contrast optical microscope. About 90–95% breakage is optimal.

7. An alternative to the Dounce is a Polytron homogenizer operated at half-maximum power for 10–30 s.

8. An alternative to using the JCF-Z rotor is: Centrifuge in a JA-10 rotor for 10 min at ~4,420g with low brake. Aspirate the supernatant and resuspend the pellet in ~30 mL NaHCO$_3$ buffer with two gentle strokes with a Dounce homogenizer. Repeat centrifugation 2–4 times until the supernatant is relatively clear.

9. If a zonal rotor is not available, an angle rotor in a standard centrifuge can be used with a prolonged centrifuge time.

10. A zonal rotor can be loaded and unloaded at rest. A reorienting core can be used, but the procedure works adequately with a standard core.

11. An alternative to using the zonal rotor is: Divide the sample into six bottles for a JA-10 rotor and overlay with 37% sucrose solution to bottle capacity. Using slow acceleration, centrifuge at 16,000 overnight (15–17 h). Stop without the brake and collect the material at the 37% sucrose/sample interface.

12. If fewer rats are used *(7–10)* modifications to the procedure are simply made by using proportionately less buffer volume and rotors with smaller capacity, for example, JA-14 and JA-20.

13. Polysulfone bottles tolerate alkali better than polycarbonate bottles.

14. A similar alkali extraction procedure can be used for the purification of myocardial gap junctions.

15. Buffer volumes can be increased proportionately to allow for the isolation of gap junctions from a larger number of rats. Rotors with larger capacity may need to be employed.

16. The concentration of Sarkosyl is critical. With concentrations < 0.3%, partially purified preparations were obtained. With concentrations of 0.6% the purification was high, but the yield was lower.

17. By using a 49% sucrose cushion in the sucrose step gradient, contaminating collagen and other intercellular matrix proteins were pelleted and hence removed from the sample.

18. By retrieving the junctions at the 35%/49% sucrose interface and not pelleting them, aggregation of these junctional plaques was minimized.

19. Place the sonicator tip at the bottom of the tube to minimize foaming. The final suspension should appear milky and homogeneous.
20. Cells can be stored as a frozen pellet at $-70°C$ for several weeks. Resuspend pellet in NaHCO3 buffer (20 mL/500 mL) of culture volume.
21. The need for urea indicates that hydrophobic forces are important for holding two connexons together. Likewise, the need for EGTA suggests that Ca^{2+} ions are also important for holding the paired connexons together. Tables 1, 2, and 3 of Ghoshroy et al. (*6*) show the many conditions attempted to optimize the splitting procedure. One hundred percent splitting is possible, but only with severe membrane disruption.
22. Split junction preparations are best used as soon as possible after splitting protocol. While stable at $4°C$, the membranes may start to vesiculate after long storage (> 1 wk).
23. For characterization of the split gap junctions by SDS-PAGE, it is important to never boil the Cx32 connexons in SDS solution because severe aggregation of the connexin protein occurs. Instead of boiling, incubate samples just below the boiling point (~80–90°C). On the other hand, Cx43 connexons (heart) can withstand boiling without aggregation of connexin43 occurring.

Acknowledgments

We express our appreciation to Drs. Nalin Kumar and Vinzenz Unger for their critical reading of the manuscript and their enthusiasm for sharing with us the details of their protocols. This work is supported by a grant from the National Science Foundation (MCB-9728338 awarded to Gina Sosinsky). Some of the work included here was conducted at the National Center for Microscopy and Imaging Research at San Diego, which is supported by National Institutes of Health Grant 2P41 RR04050 awarded to Dr. Mark Ellisman.

References

1. Revel, J.-P. and Karnovsky, (1967) Hexagonal array of subunits in intercellular junctions of the mouse heart and liver. *J. Cell Biol.* **33,** C7–C12.
2. Amos, L. A., Henderson, R., and Unwin, P. N. (1982) Three-dimensional structure determination by electron microscopy of two-dimensional crystals. *Prog. Biophys. Mol. Biol.* **39,** 183–231.
3. Makowski, L., Caspar, D. L. D., Phillips, W. C., and Goodenough, D. A. (1977) Gap junction structure II. Analysis of the X-ray diffraction data. *J.Cell Biol.* **74,** 629–645.
4. Unwin, P. N. T. and Zampighi, G. (1980) Structure of the junction between communicating cells. *Nature* **283,** 545–549.
5. Fallon, R. F. and Goodenough, D. A. (1981) Five-hour half-life of mouse liver gap junction protein. *J. Cell Biol.* **90,** 521–526.
6. Ghoshroy, S., Goodenough, D. A., and Sosinsky, G. E. (1995) Preparation, characterization, and structure of half gap junctional layers split with urea and EGTA. *J. Membr. Biol.* **146,** 15–28.

7. Hertzberg, E. L. (1984) A detergent independent procedure for the isolation of gap junctions from rat liver. *J. Biol. Chem.* **259,** 9936–9943.

8. Manjunath, C. K. and Page, E. (1986) Rat heart gap junctions as disulfide-bonded connexon multimers: Their depolymerization and solubilization in deoxycholate. *J. Membr. Biol.* **90,** 43–57.

9. Yeager, M. and Gilula, N. B. (1992) Membrane topology and quaternary structure of cardiac gap junction ion channels. *J. Mol. Biol.* **223,** 929–948.

10. Yancey, S. B., John, S. A., Lal, R., Austin, B. J., and Revel, J.-P. (1989) The 43-kD polypeptide of heart gap junctions: immunolocalization, topology, and functional domains. *J. Cell Biol.* **108,** 2241–2254.

11. Kumar, N. M., Friend, D. S., and Gilula, N. B. (1995) Synthesis and assembly of human β1 gap junctions in BHK cells by DNA transfection with the human β1 cDNA. *J. Cell Sci.* **108,** 3725–3734.

12. Stauffer, K. A., Kumar, N. M., Gilula, N. B., and Unwin, N. (1991) Isolation and purification of gap junction channels. *J. Cell Biol.* **115,** 141–150.

13. Stauffer, K. A. (1995) The gap junction protein β_1-connexin (connexin-32) and β_2-connexin (connexin26) can form heteromeric hemichannels. *J. Biol. Chem.* **270,** 6768–6772.

14. Yeager, M., Unger, V. M., and Mitra, A. K. (1999) Three-dimensional structure of membrane proteins determined by two-dimensional crystallization: Electron cryo-microscopy and image analysis. *Methods Enzymol.* **294,** 135–180.

15. Special issue: Advances in Computational Image Processing for Microscopy, (1996) *J. Struct. Biol.* **116,** Carragher, B. and Smith, P. R., eds.

16. Jiang, J. X. and Goodenough, D. A. (1996) Heteromeric connexons in lens gap junction channels. *Proc. Natl. Acad. Sci. USA* **3,** 1287–1291.

17. Kistler, J., Goldie, K., Donaldson, P., and Engel, A. (1994) Reconstitution of native-type noncrystalline lens fiber gap junctions from isolated hemichannels. *J. Cell Biol.* **126,** 1047–1058.

18. Manjunath, D. K., Goings, G. E., and Page, E. (1985) Proteolysis of cardiac gap junctions during their isolation from rat hearts. *J. Membr. Biol.* **85,** 159–168.

19. Yeager, M. (1994) *In situ* two-dimensional crystallization of a polytopic membrane protein: the cardiac gap junction channel. *Acta Crystallogr.* D50, 632–638.

20. Zimmer, D. B., Green, C. R., Evans, W. H., and Gilula, N. B. (1987) Topological analysis of the major protein in isolated intact rat liver gap junctions and gap junction-derived single membrane structures. *J. Biol. Chem.* **262,** 7751–7763.

21. Milks, L. C., Kumar, N. M., Houghten, R., Unwin, N., and Gilula, N. B. (1988) Topology of the 32-kd liver gap junction protein determined by site-directed antibody localizations. *EMBO J.* **7,** 2967–2975.

4

Culturing of Mammalian Cells Expressing Recombinant Connexins and Two-Dimensional Crystallization of the Isolated Gap Junctions

Mark Yeager and Vinzenz M. Unger

1. Introduction

The overexpression of membrane proteins is not at all routine. The reader is referred to Grisshammer and Tate *(1)* for a recent review. Kumar et al *(2)* have generated a stably transfected baby hamster kidney (BHK) cell line that can be induced to express wild type and mutant forms of connexins. Our structural studies have been focused on a C-terminal truncation mutant of rat heart connexin43 (Cx43) connexin (designated α_1Cx-263T [Cx43 truncated after Pro^{263}]) that was expressed in BHK cells under control of the inducible mouse metallothionin promotor *(2–4)*. Freeze–fracture, thin-section, and negative stain electron microscopy of the BHK cell membranes demonstrated that the recombinant protein assembled with the characteristic septalaminar morphology of gap junctions *(2,5)*. In addition, dye transfer experiments *(2,6)* demonstrated that the recombinant gap junctions that assembled in BHK cells were functional. To our knowledge, our analysis of cardiac gap junction channels is the first example of a polytopic membrane protein that has been expressed in a heterologous system and examined by structural methods.

Two-dimensional (2D) crystallization of membrane proteins is usually accomplished by in vitro reconstitution in which detergent-solubilized lipids are mixed with the detergent-solubilized and purified protein, and the detergent is removed by dialysis *(7–9)*. An alternative to this approach is *in situ* 2D crystallization in which the protein is never removed from its native membrane *(10)*. However, a requirement of this approach is that the protein must already be fairly concentrated in the membrane or have a tendency to self-associate in

From: *Methods in Molecular Biology, vol. 154: Connexin Methods and Protocols*
Edited by: R. Bruzzone and C. Giaume © Humana Press Inc., Totowa, NJ

patches that can be isolated. 2D crystallization is induced using gentle conditions that extract membrane lipids without solubilizing the protein. Although the level of connexin expression in the BHK cells was quite low, microgram amounts of protein were sufficient to pursue *in situ* 2D crystallization experiments because hundreds of channels aggregate in plaques. Extraction with Tween-20 and 1,2-diheptanoyl-*sn*-phosphocholine (DHPC) not only improved the crystallinity of the naturally occurring 2D arrays but also enriched the preparation by solubilizing nonjunctional proteins.

Detailed here are the protocols for culturing of BHK cells expressing α_1Cx263T, isolation of membranes enriched for the recombinant channels, and *in situ* 2D crystallization of the gap junction plaques. The techniques that we used for electron-cryomicroscopy and image analysis of the 2D crystals have been previously described *(10)*.

2. Materials

2.1. Reagents

1. DMEM: Gibco BRL Dulbecco's modified eagle medium (high glucose with pyridoxine-HCl, with 110 mg/L of sodium pyruvate, without L-glutamine, stored at 2–8°C protected from light).
2. FBS: Gibco BRL fetal bovine serum (qualified, origin: Mexican), heat inactivated, performance tested, mycoplasma tested, virus tested, endotoxin tested (store at –20°C).
3. PSG: Gibco BRL penicillin–streptomycin–glutamine prepared with 10,000 U/mL of penicillin G sodium; 10,000 μg/mL of streptomycin sulfate; 29.2 mg/mL of L-glutamine; 10 mM of sodium citrate in 0.14% NaCl.
4. BHK medium: Add 50 mL of FBS and 5 mL of PSG solution to 500 mL of DMEM; store at 4°C.
5. MTX: 55 mM Methotrexate (Immunex) in water.
6. 75% Ethanol: Prepared from 95% ethanol by dilution with nanopure water.
7. Detach solution: 10 mM PBS, pH 7.4; Sigma packet (containing 138 mM NaCl, 2.7 mM KCl); 5 mM EDTA.
8. Dimethyl sulfoxide (DMSO): Fisher Molecular Biology Grade.
9. Stock PMSF solution: 70 mg/mL of phenylmethylsulfonyl fluoride (PMSF) (Calbiochem) in 100% DMSO.
10. Zinc acetate: 100 mM zinc acetate (Fisher) in nanopure water. In the tissue culture hood, filter the solution through a 0.2-μm filter into a sterile bottle.
11. 4-(2-Hydroxyethyl)-1-piperazineethanesulfonic acid (HEPES) buffer: 10 mM (Sigma, hemisodium salt), 0.8% NaCl, pH 7.5. PMSF (70 μg/mL) is added quickly and mixed by vigorous shaking. Because PMSF is hydrolyzable, it should be added just prior to use. If buffer aliquots are older than 30 min, they are discarded and fresh buffer is prepared.

12. 49% Sucrose: 49 g of sucrose (Fisher ultracentrifugation grade) dissolved in HEPES buffer; total volume 100 mL.
13. 28% Tween-20: Sigma (*see* **Note 1**).
14. DHPC: Avanti Polar Lipids (Birmingham, AL).
15. KI–sodium thiosulfate: 200 mM Potassium iodide (Sigma) + 2 mM sodium thiosulfate (Fisher Chemicals).
16. Gentamycin: 50 μg/mL of gentamycin (GIBCO) in 10 mM HEPES buffer, pH 7.5, containing 0.8% NaCl.
17. Bleach: Clorox.
18. 5% Sodium dodecyl sulfate (SDS): SDS, electrophoresis grade (Fisher Chemicals, Fair Lawn, NJ).
19. Ion-exchange resin: AG501-X8(D) (Bio-Rad, Richmond, CA).
20. DotMetric protein assay kit (Geno Technology) (*see* **Note 2**).
21. Water: Nanopure, deionized.

2.2. Equipment

1. Falcon INTEGRID 15 × 25 cm tissue culture dishes; polystyrene, nonpyrogenic (treated by vacuum gas plasma to optimize cell adherence).
2. Falcon Blue Max Polypropylene Conical Tubes, 15 and 50 mL (Becton Dickinson).
3. Sterile glass bottles, 250 mL (Pyrex).
4. Fisher disposable serological sterile pipets (2, 10, and 25 mL).
5. Fisher disposable Pasteur pipets, stored in Bellco glass cylinder and sterilized.
6. Plastic columns for chromatography of Tween-20 (Econo-Pac 1.5 × 12 cm polypropylene chromatography column, Bio-Rad).
7. Lab tissue (Kimwipes).
8. 0.2-μm Syringe filters (Acrodisc PF, Gelman Sciences).
9. Plastic funnel.
10. Fisher Disposable Cell Scraper.
11. 1-mL Polyethylene Disposable Transfer Pipets (Bio-Rad).
12. 2.0-mL Biofreeze Vials (Costar, Cambridge, MA).
13. Eppendorf microcentrifuge tubes, 0.5 and 1.5 mL.
14. Biotech dialysis devices for OMS 102; MW exclusion limit 100,000 Da.
15. Revco ultima CO_2 incubator.
16. Revco –20°C freezer.
17. Revco –70°C freezer.
18. Taylor–Wharton 18XT liquid nitrogen dewar.
19. SteriGARD II tissue culture hood (The Baker Company).
20. Zeiss axiovert 25 inverted light microscope.
21. PolyScience water bath.
22. Vortex mixer (MaxiMix Plus, Thermolyne).
23. Magnetic stirring bar (1 cm).
24. Plastic screw-capped bottle (Nalgene, 8 mL).
25. Beckman centrifuges: Refer to the following web site to convert rpms to maximum relative centrifugal force (maxRCF).

http://www.beckman-coulter.com/beckman/biorsrch/prodinfo/cntrifug/rotrcalc.asp).
    J-30I with JA10 rotor
    GS-6KR with GH-3.8 rotor
    LE-80K ultracentrifuge with SW28 rotor
26. Heat Systems Sonicator.
27. Kontes Micro-ultrasonic Cell Disrupter.
28. Barnstead Nanopure Infinity water purifier.
29. Drummond Pipet-Aid.
30. pH meter (ATI orion model 550).
31. Ice bucket and styrofoam boats.

3. Methods

3.1. General Tissue Culturing Techniques

1. Prior to working in the tissue culture hood, wash your hands and forearms with soap, and then spray with 75% ethanol.
2. Before using the hood, turn off the ultraviolet light.
3. Spray the surface of the tissue culture hood and any bottles or equipment that will be used in the hood with 75% ethanol.
4. Outside the tissue culture hood, pour the waste materials (e.g., used BHK medium) into a storage bottle (e.g., an empty DMEM bottle) labeled "tissue culture waste." When full, add bleach and dispose of the solution down the sink.
5. When you have completed your work, spray the tissue culture hood with 75% ethanol and wipe with lab tissues.
6. Turn on the ultraviolet light.

3.2. Thawing of Frozen BHK Cell Stocks

1. Warm about 20 mL of water to 37°C.
2. Transfer the cryovial from the liquid nitrogen storage dewar flask to the warm water for ~30 s.
3. Remove the cryovial from the water and roll it back and forth between your hands until the solution just starts to thaw.
4. Spray the cryovial with 75% ethanol and place the vial, a tissue culture plate, and the BHK medium (at 4°C) in the tissue culture hood.
5. Pipet 10 mL of BHK medium at 4°C into the tissue culture plate.
6. Use a 2-mL sterile pipet to transfer the 1.5 mL of thawed cells from the cryovial into the tissue culture plate containing the cold BHK medium. Swirl the plate to disperse the cells.
7. Transfer the tissue culture plate to the 37°C incubator for ~1 h.
8. Pipet 20 mL of BHK medium into a 50-mL conical plastic tube and warm to 37°C.
9. Examine the tissue culture plate by phase-contrast light microscopy to assess whether the cells have attached. Typically, cells that have not attached in 15–20 min are dead. Cells should never be left in DMSO for more than 1 h.

10. When the cells have attached, transfer the tissue culture plate to the hood and aspirate the medium (which contains some DMSO in which the cells were frozen).
11. Pour the 20 mL of fresh, warm BHK medium into the tissue culture plate, and place the plate back in the 37°C incubator.
12. On the day after thawing the cells, examine the tissue culture plate and check that the cells have recovered from the cold. The cells should have a scalloped rather than round appearance.

3.3. Methotrexate Selection of BHK Cells

1. The methotrexate (MTX) selection takes 4 d and should be done after thawing a new batch of cells or about once every month when continuously growing cells. Because the BHK cells are subcultured every 2 d, the MTX solution is added twice.
2. When the cells have grown to confluence, subculture as detailed in **Subheading 3.4.** but include 200 μL of MTX in each 20 mL of BHK medium for the final suspension. It is not necessary to add MTX in the 10-mL of BHK medium aliquotted for the centrifugation.
3. Incubate the plate for 2 d.
4. Repeat **steps 2** and **3**.

3.4. Subculturing of BHK Cells and Scaleup from 1–4 Plates (see *Note 3*)

1. Cells are grown at 37°C in an atmosphere of 100% relative humidity and 5% CO_2. If the cells are healthy, they are subcultured every other day.
2. Examine the cells by phase-contrast light microscopy and verify that they have grown to confluence. If so, transfer the plate to the tissue culture hood.
3. Pour 80 mL of BHK medium into two 50-mL conical plastic tubes, at least 10 mL of BHK medium into another 50-mL conical plastic tube, and 10 mL of detach solution into a 15-mL plastic conical tube.
4. Heat the solution aliquots to 37°C in a water bath.
5. Once the solutions have reach 37°C, spray the tops of the plastic conical tubes with 75% ethanol and transfer them into the tissue culture hood.
6. Attach a sterile glass pipet to the tygon tubing in the vacuum aspirator in the tissue culture hood.
7. Tilt the tissue culture plate to ~30°, lift the lid slightly, and aspirate the medium from a corner of the plate. Avoid touching the bottom of the plate to minimize disruption of the cells.
8. Clear the tygon tubing by aspirating ~5 mL of 75% ethanol that has been added to a 50-mL plastic conical tube.
9. Pour ~10 mL of detach solution into the tissue culture plate, and rotate and rock the plate to evenly disperse the solution over the cells lining the bottom of the plate. When the plate is tilted toward you, one can see that light scattering from the fully attached cells gives the bottom of the plate a semitranslucent gray appearance. After about a minute, the cells are presumably loosened from

the bottom of the plate, which increases the light scattering so that the gray appearance of the plate bottom is even more apparent. If the cells were confluent, clumps of cells will freely detach after approx 3–5 min.

10. At this point, detachment is facilitated by washing the plate with the detach solution. Attach a 10-mL sterile plastic pipet to the electronic Drummond Pipet-Aid and repeatedly aspirate and eject the detach solution over the bottom of the plate. Be sure to minimize bubbles as you pipet the cells. As the solution is sprayed over the bottom of the plate, one can see transparent streaks where the cells have completely detached from the plate. Continue to spray the plate until the bottom is completely transparent and all the cells are detached.

11. The cells suspended in detach solution are then added to the 50-mL plastic conical tube containing 40 mL of BHK medium.

12. Centrifugation 1 (228 maxRCF, 1000 rpm, 3 min, room temp, GH-3.8 rotor, Beckman GS-6KR centrifuge).

13. Decant the supernatant into a plastic conical tube that serves as a waste container in the sterile hood.

14. Resuspend the pellet with 10 mL of BHK medium by repeated aspiration into and out of a 10-mL pipet attached to the automatic pipettor.

15. Now transfer the resuspended cells to the remaining 70 mL of BHK medium, to give a total volume of 80 mL divided between two plastic conical tubes (*see* **Note 3**).

16. Place four new tissue culture plates in the hood; pipet 20 mL of the suspended cells into each plate; if large bubbles form, rotate the plate so the bubbles move to the perimeter of the medium. The individual cells and cell clumps settle very quickly. Therefore, when scaling up from a larger number of plates, swirl the larger volume of medium after every fifth plate (or aspirate a couple of times with a 20-mL pipet) to keep the cells suspended. If the cells are not uniformly mixed, the cell density will vary from plate to plate, making it difficult to select a good time point for induction.

17. Place the tissue culture plates in the 37°C incubator. Check that *(1)* the medium in each plate is level, *(2)* the water pan in the incubator contains water, and *(3)* the CO_2 tank is not empty.

3.5. Storage of BHK Stocks

1. Detach a confluent layer of tissue culture cells using 10 mL of detach solution (**steps 1–10** in **Subheading 3.4.**).

2. The cells suspended in detach solution are then added to the plastic conical tube containing at least ~10 mL of BHK medium.

3. Centrifugation 1 (228 maxRCF, 1000 rpm, 3 min, room temp, GH-3.8 rotor, Beckman GS-6KR centrifuge).

4. As DMSO is toxic, it should not be added full strength to the cells. Therefore, a 10% DMSO solution is prepared by dilution with BHK medium (e.g., 0.3 mL of DMSO and 2.7 mL of BHK medium). The cell pellet is then resuspended in this medium.

5. Immediately pipet 1.5 mL of the suspended cells into each cryovial, tighten the tops, and chill on ice.

6. For **steps 7–8**, transfer the vials to a well-insulated styrofoam container to reduce the cooling speed.
7. Transfer the styrofoam container with the vials to a –20°C freezer for 3–4 h.
8. Then transfer the container with vials to the –70°C freezer overnight.
9. The vials are then placed in a liquid nitrogen dewar flask for long-term storage. An optimal cooling rate of ~1°C/min is achieved by placing the vials into the top space of the flask.
10. Thaw one vial the following day to ensure that the freezing was successful.

3.6. Induction of Connexin Expression (see Chapter 3)

1. Examine the cells by light microscopy and assess the density. When the cells appear that they will be confluent in a few hours, they are ready for induction.
2. Add 100 µM zinc acetate to a sufficient volume of BHK medium (e.g., 540 mL of BHK medium for 27 plates will require 540 µL of the 100 mM zinc acetate stock solution) and heat to 37°C.
3. Transfer all the tissue culture plates and the BHK medium to the hood.
4. Aspirate the medium as in **steps 6–8** of **Subheading 3.4.**
5. Pipet 20 mL of fresh BHK medium containing 100 µM zinc acetate into each plate.
6. Transfer the plates back to the 37°C incubator for 8 h.

3.7. Isolation of Membranes Enriched for Gap Junctions

1. In our experience, gap junctions tend to adhere to glass, so plasticware should be used throughout the isolation procedure.
2. Place a 500-mL JA10 centrifuge bottle into an ice bucket and insert a funnel.
3. Use a plastic cell scraper to dislodge the cells from the bottom of the plates, and decant the media with the suspended cells into the JA10 bottle.
4. Centrifugation 1 (2000 maxRCF, 3,362 rpm, 5 min, 4°C, JA10 rotor, Beckman J-30I centrifuge).
5. Decant the supernatant.
6. Carefully pipet 10 mL of ice-cold HEPES buffer (10 mM HEPES, pH 7.5, 0.8% NaCl, 70 µg/mL of PMSF) into the JA10 centrifuge bottle so as not to disturb the pellet.
7. Detach the pellet by swirling the centrifuge bottle.
8. Pour the contents with the intact pellet into a 50-mL plastic conical tube.
9. Vortex-mix the plastic conical tube for ~15 s to break up the pellet.
10. Increase the volume to 50 mL with ice-cold HEPES buffer with 70 µg/mL of PMSF.
11. Centrifugation 2 (228 maxRCF, 1000 rpm, 5 min, 4°C, GH-3.8 rotor, Beckman GS-6KR centrifuge).
12. Decant the supernatant from the plastic conical tube, and add 5 mL of ice-cold HEPES buffer containing 140 µg/mL of PMSF to the pellet.
13. Suspend the pellet by vortex mixing.
14. Incubate the plastic conical tube for 5 min on ice.
15. Sonicate the sample 3× for 5 s (Heat Systems Sonicator, output setting 4.5), waiting 15 s between each burst. Keep the sonicator tip at the bottom of the

tube to minimize foaming. The final suspension should appear milky and homogeneous.

16. Examine a drop of the suspension by light microscopy to ensure that cell lysis is at least 90%. If there are still numerous intact cells, repeat the sonication for 5 s.

17. Transfer the solution to a centrifuge tube for the SW28 ultracentrifuge rotor. Add 35 mL of HEPES buffer with 140 μg/mL of PMSF and tare.

18. Centrifugation 3 (119,910 maxRCF, 25,000 rpm, 45 min, 4°C, SW28 rotor, Beckman LE-80K ultracentrifuge).

19. Decant the supernatant, add 5 mL of ice-cold HEPES buffer with 70 μg/mL of PMSF, and resuspend the pellet by repeated aspiration into a plastic pipet.

20. Disperse the pellet by sonication for ~5 s (Heat Systems Sonicator, output setting 4 or less if foaming occurs).

21. Add 30 mL of 49% sucrose in HEPES buffer with 70 μg/mL of PMSF chilled to 4°C to an SW28 ultracentrifuge tube. Carefully pipet the 5 mL of crude membrane homogenate onto the top of the sucrose cushion. Add additional HEPES buffer with 70 μg/mL of PMSF to fill the tube to within 2 mm of the top (total volume ~35 mL).

22. Centrifugation 4 (119,910 maxRCF, 25,000 rpm, 45 min, 4°C, SW28 rotor, Beckman LE-80K ultracentrifuge).

23. Use a plastic pipet to carefully remove material (e.g., solid PMSF) floating on the top of the tube. Then retrieve the 0/49% interface and transfer it to another SW28 tube. The homogenate should appear milky white.

24. Increase the volume to 35 mL with ice-cold, sucrose-free HEPES buffer with 70 μg/mL of PMSF. Use a plastic pipet to mix the solution to dilute the sucrose in the aspirated interface.

25. Centrifugation 5 (119,910 maxRCF, 25,000 rpm, 45 min, 4°C, SW28 rotor, Beckman LE-80K ultracentrifuge).

26. Decant the supernatant, add 1 mL of ice-cold HEPES buffer containing 140 μg/mL of PMSF, and disrupt the pellet by repeated passage through a plastic Eppendorf pipet.

27. Homogenize the sample by sonication for a few seconds (Kontes Micro-ultrasonic Cell Disrupter, output level 35).

3.8. Enrichment and In Situ 2D Crystallization of Recombinant Gap Junctions (*see* Note 4)

1. The protein concentration of the membrane sample is estimated using the DotMetric assay (Geno Technology) (*see* **Subheading 4.2.**).

2. The volume of the suspension is adjusted so that the protein concentration is 1 mg/mL, after addition of the following agents to the final concentrations that are indicated: 2.8% Tween-20, 200 mM KI, 2 mM sodium thiosulfate, 140 μg/mL of PMSF, 50 μg/mL of gentamycin in 10 mM HEPES buffer, pH 7.5, containing 0.8% NaCl. Because Tween-20 is contaminated with aldehydes, peroxides, and free acids, it must first be purified by ion-exchange chromatography immediately before use (*see* **Subheading 4.1.**).

3. The total sample volume is typically ~4 mL, which is placed in a screw-capped plastic bottle. Add a 1-cm magnetic stirring bar to the bottle and stir at 27°C for 12 h.

4. After a 12-h incubation, add solid DHPC to 13 mg/mL, and stir for an additional 1 h at 27°C.

5. Add 30 mL of 25% sucrose at room temperature (prepared in 10 mM HEPES, 0.8% NaCl, pH 7.5) to an SW28 centrifuge tube, and carefully add the entire contents on top of the sucrose cushion.

6. Centrifugation 1 (119,910 maxRCF, 25,000 rpm, 1 h, 4°C, SW28 rotor, Beckman LE-80K ultracentrifuge).

7. Carefully pipet off the detergent solution layer and sucrose so that the detergent meniscus does not touch the pellet, and wipe the tube clean with a lab tissue to remove any residual sucrose or detergent.

8. Resuspend the pellet in 600–800 μL of HEPES buffer, and sonicate for a few seconds as described previously so that the suspension is homogeneous.

9. Transfer the sample to a Bio-Tech International dialysis tube and dialyze for at least 24 h against 500 mL of the buffer (10 mM HEPES, containing 5 mL of gentamycin/1 L of buffer) with one change at ~12 h. The membranes tend to sediment to the bottom of the Bio-Tech tube and cover the dialysis membrane. With each buffer change, use a pipet to resuspend the membranes in the tube to facilitate dialysis.

10. Store the suspension at 4°C in a 2 mL screw-capped cyrovial. If, after a few days, a colored, soaplike film is observed when removing the cap, the membranes have residual detergent and the crystals will likely degrade within ~4 wk.

4. Notes

1. Purification of Tween-20. The Tween-20 is used at a final concentration of 2.8%. Prepare 28% Tween-20 by pouring 7 mL of Tween-20 into a plastic conical tube, bring the total volume to 25 mL with water, and mix well. Take ~10 mL of the ion-exchange resin and place in a beaker with water to create a slurry. Then set up a 15-mL column, and fill the column with ~5 mL of water. Pour the resin slurry into the column in aliquots of ~0.5 mL to create layers of mixed resin. The total bed volume is about 8–10 mL. Let the column settle, and flush with ~50 mL of water. Bring the water level to the top of the column. Put ~20 mL of the 28% Tween-20 on the column and let it flow freely by gravity (to displace free water from the column), then add the remaining 5 mL of Tween-20 to the column and control the flow rate at ~ one drop/5 s. Collect a minimum of 1–1.5 mL. The Tween is stable for about a week after purification; thereafter, it should be repurified, or a fresh batch should be purified. Alternatively, you can use protein grade 10% Tween-20 that is already purified by the manufacturer (Calbiochem). If you expect to reuse this stock solution, it is necessary to flush the tube with argon to remove air and prevent oxidation.

2. Dot Metric protein assay. Before starting the assay, prepare the "fixer" (9 mL water and 800 μL of fixing solution A) and "developer" (4 mL water and 800 μL of developer solution C). Remove a 10-μL aliqout of membrane suspension, add

10 μL of 5% SDS, and heat this mixture at 38°C for ~15 min. Now prepare a 5×
and a 10× dilution of the incubated sample in "dilution buffer 1" from the kit. For
each dilution, spot 1 μL twice on the strips, let it dry completely, and then place
in the fixer for 2 min. Now add 25 μL of the "sensitizing solution B" to the same
fixer vial, and let the sample incubate for another 2 min. Transfer the strip to
the developer tube, and let it incubate for ~2–5 min on ice in the dark. Blot the
strips gently on a lab tissue, let them dry sufficiently, and then use the scale in the
kit to estimate the protein concentration in the sample.

3. Comments on scaling up. If the cells are too diluted when they are plated, then
growth will be slow. We have found that for scaling up, a single plate of confluent
cells can usually be isolated and suspended in 80 mL to yield four plates (20 mL per
tissue culture plate). Hence, the scaleup is from 1 to 4 to 16 plates. When scaling up,
the increased volume of BHK medium can be contained in a 250-mL sterile glass
bottle. For subculturing >12 plates, the increased volume of medium can be contained in a 500-mL bottle, conveniently provided by an empty DMEM bottle.

4. An important factor for improving the resolution of the 2D crystals, the molecular
homogeneity of the protein. Liver (11) and cardiac tissue (12) are known to
contain multiple connexin subtypes, which may have contributed to a limiting
resolution of ~15 Å for gap junctions isolated from these tissues (13–15). A special advantage of the stably transfected BHK cells is that only a specific connexin
subtype is expressed. Although BHK cells naturally express full-length Cx43,
Western immunoblots showed that the full-length protein was detected only at
low levels in overexposed immunoblots. The C-tail in native cardiac gap junctions is sensitive to proteolysis (15,16). By removing the C-tail in the α_1Cx-263T
mutant, the protein was engineered to be more homogeneous. Resolution also
depends on the procedures adopted for 2D crystallization. The variables that were
tested to improve crystallinity included detergent screens (Tween-20, Tween-80,
Tween-85, Cymal), chaotropes (potassium iodide and sodium thiosulfate),
temperature (room temperature and 37°C), and incubation in glycerol. The Tween
detergents were tested on the basis of their success in growing 2D crystals of frog
rhodopsin (17). If the potassium iodide treatment was not included, then the
detergent extraction was not as successful. In addition, potassium iodide could
not be replaced with sodium chloride. Another factor for achieving higher resolution is the electron cryomicroscopy. The thermal and vibrational stability of
cryomicroscopes have certainly improved over the last decade. For the projection density map (3), a resolution of 7 Å was achieved using a conventional
Philips CM12 electron microscope operating at 100-kV, which was equipped
with a standard Gatan626 cryostage. For the 3D map (4), images of tilted crystals
were recorded using a Philips 200-kV electron microscope equipped with a field
emission gun. The improved stability and coherence of the 200-kV microscope
compared with the CM12 was important for recording high-resolution images
of tilted crystals. Nevertheless, high resolution images could not be recorded
from specimens tilted > 35°. Factors that limit the resolution of images of tilted
crystals include specimen flatness, beam-induced movement, and charging. The

same 200-kV microscope and cold stage were used to collect higher resolution images of 2D crystals of AQP1 tilted to ~50° *(18)*. Since the thickness of the AQP1 and gap junction crystals was ~60 Å and ~150 Å, respectively, we presume that the inability to record high-resolution images from highly tilted gap junction crystals was related to the increased thickness of the gap junction specimens. A final point that deserves consideration is the software used for image processing. The MRC image-processing package developed by R. Henderson and colleagues *(19–21)* was used to analyze the images and to correct for lattice distortions and effects of the contrast transfer function. Although lattice unbending has previously been used for the analysis of endogenous gap junctions *(14,15)*, the MRC software has undergone substantial improvements compared with the earlier program versions. In particular, the program MAKETRAN was used to calculate a theoretical reference area based on the initial 3D density map of the structure, which improved the accuracy of the cross correlation and unbending procedures.

Factors that probably did not significantly affect the improvement in resolution included the crystal size and specimen purity. The isolated gap junction plaques were formed by a mosaic of crystalline areas. This mosaicity as well as possible curvature of the plaques made it necessary to use areas with only a few hundred unit cells for image processing. The small size of the crystals was comparable to crystals derived from liver and heart tissue. Moreover, the gap junction preparations were substantially contaminated by nonjunctional membranes and were probably comparable in purity to crude plasma membrane preparations derived from liver *(22)*. In the early preparations of the recombinant membranes, the gap junctions were often obscured in the electron microscope by nonjunctional membranes deposited on the surface of the crystals. DHPC treatment was particularly useful for solubilizing nonjunctional membranes. Nevertheless, SDS gels of the enriched membrane preparations displayed a pattern of bands that was similar to crude preparations. Even in the enriched specimens, a Coomassie-stainable band that corresponded to the recombinant gap junction protein was not detectable. The yield of gap junctions was quite variable from preparation to preparation. Nevertheless, the crystals were quite reproducible with a variation in unit cell size of 76–79 Å. These observations suggest that the major limiting factors in producing crystals was the culturing of the BHK cells and expression of recombinant protein. In general, if negatively stained specimens showed at least five crystals per grid square, then frozen-hydrated specimens were prepared. Because specimens were rarely produced that had more than five crystals per grid square, the major limiting factors for the cryomicroscopy were the number of crystals on the grid as well as the difficulty of recording images from tilted crystals.

Acknowledgments

Portions of this chapter have been adapted from **refs. 5** and *10* with the permission of Academic Press. We acknowledge contributions to this protocol by Nalin Kumar, Daniel Entrikin, and Shannon Lewis and especially thank

Anchi Cheng for a critical reading. The writing of this chapter was supported by the NIH National Heart Lung and Blood Institute (M. Y.). Mark Yeager is a recipient of a Clinical Scientist Award in Translational Research from the Burroughs Wellcome Fund.

References

1. Grisshammer, R. and Tate, C. G. (1995) Overexpression of integral membrane proteins for structural studies. *Q. Rev. Biophys.* **28,** 315–422
2. Kumar, N. M., Friend, D. S. and Gilula, N. B. (1995) Synthesis and assembly of human α_1 gap junctions in BHK cells by DNA transfection with the human α_1 cDNA. *J. Cell Sci.* **108,** 3725–3734.
3. Unger, V. M., Kumar, N. M., Gilula, N. B., and Yeager, M. (1997) Projection structure of a gap junction membrane channel at 7Å resolution. *Nat. Struct. Biol.* **4,** 39–43.
4. Unger, V. M., Kumar, N. M., Gilula, N. B., and Yeager, M. (1999) Three-dimensional structure of a recombinant gap junction membrane channel. *Science* **283,** 1176–1180.
5. Unger, V. M., Kumar, N. M., Gilula, N. B., and Yeager, M. (1999) Expression, 2D crystallization, and electron cryo-crystallography of recombinant gap junction membrane channels. *J. Struct. Biol.* **128,** 98–105.
6. Guan, X., Cravatt, B. F., Ehring, G. R., Hall, J. E., Boger, D. L., Lerner, R. A., and Gilula, N. B. (1997) The sleep-inducing lipid oleamide deconvolutes gap junction communication and calcium wave transmission in glial cells. *J. Cell Biol.* **139,** 1785–1792
7. Kühlbrandt, W. (1992) Two-dimensional crystallization of membrane proteins. *Q. Rev. Biophys.* **25,** 1–49.
8. Jap, B. K., Zulauf, M., Scheybani, T., Hefti, A., Baumeister, W., Aebi, U., and Engel, A. (1992) 2D crystallization: from art to science. *Ultramicroscopy* **46,** 45–84.
9. Engel, A., Hoenger, A., Hefti, A., Henn, C., Ford, R. C., Kistler, J., and Zulauf, M. (1992) Assembly of 2-D membrane protein crystals: dynamics, crystal order, and fidelity of structure analysis by electron microscopy. *J. Struct. Biol.* **109,** 219–234.
10. Yeager, M., Unger, V. M., and Mitra, A. K. (1999) Three-dimensional structure of membrane proteins determined by two-dimensional crystallization, electron cryomicroscopy, and image analysis. *Methods Enzymol.* **294,** 135–180.
11. Nicholson, B., Dermietzel, R., Teplow, D., Traub, O., Willecke, K., and Revel, J.-P. (1987) Two homologous protein components of hepatic gap junctions. *Nature* **329,** 732–734.
12. Kanter, H. L. Saffitz, J. E., and Beyer, E. C. (1992) Cardiac myocytes express multiple gap junction proteins. *Circ. Res.* **70,** 438–444.
13. Unwin, P. N. T. and Ennis, P. D. (1984) Two configurations of a channel-forming membrane protein. *Nature* **307,** 609–613.
14. Gogol, E. and Unwin, N. (1988) Organization of connexons in isolated rat liver gap junctions. *Biophys. J.* **54,** 105–112.

15. Yeager, M. and Gilula, N. B. (1992) Membrane topology and quaternary structure of cardiac gap junction ion channels. *J. Mol. Biol.* **223,** 929–948.
16. Manjunath, C. K., Goings, G. E., and Page, E. (1985) Proteolysis of cardiac gap junctions during their isolation from rat hearts. *J. Membr. Biol.* **85,** 159–168.
17. Schertler, G. F. and Hargrave, P. A. (1995) Projection structure of frog rhodopsin in two crystal forms. *Proc. Natl. Acad. Sci. USA* **92,** 11,578–11,582.
18. Cheng, A., van Hoek, A. N., Yeager, M., Verkman, A. S., and Mitra, A. K. (1997) Three-dimensional organization in a human water channel. *Nature* **387,** 627–630.
19. Henderson, R., Baldwin, J. M., Downing, K. H., Lepault, J., and Zemlin, F. (1986) Structure of purple membrane from *Halobacterium halobium:* recording, measurement and evaluation of electron micrographs at 3.5 resolution. *Ultramicros.* **19,** 147–178.
20. Henderson, R., Baldwin, J. M., Ceska, T. A., Zemlin, F., Beckmann, E., and Downing, K. H. (1990) Model for the structure of bacteriorhodopsin based on high-resolution electron cryo-microscopy. *J. Mol. Biol.* **213,** 899–929.
21. Crowther, R. A., Henderson, R., and Smith, J. M. (1996) MRC image processing programs. *J. Struct. Biol.* **116,** 9–16.
22. Sikerwar, S. S. and Unwin, N. (1988) Three-dimensional structure of gap junctions in fragmented plasma membranes from rat liver. *Biophys. J.* **54,** 113–119.

5

Connexins/Connexons

Cell-Free Expression

Matthias M. Falk

1. Introduction

With a few exceptions, all secretory and plasma membrane proteins studied to date are synthesized in the endoplasmic reticulum (ER) membrane. Then, they are transported by successive vesicle budding and fusion from the ER through the Golgi stacks to the plasma membrane following the general intracellular transport route referred to as secretory pathway (originally reviewed in *1*). Gap junction connexins have been shown to follow this pathway.

A large portion of the current knowledge on the synthesis and translocation of secretory and transmembrane proteins (reviewed in *2–4*) has been obtained by synthesizing these proteins in a cell-free translation system. In principle, the system consists of a translation competent cell-lysate supplemented with ER-derived membrane vesicles referred to as microsomes *(5)*, and a synthetic protein-encoding RNA. The cell-free translation system has been found to accurately reproduce the steps involved in translation, translocation, as well as co- and posttranslational protein modifications that occur in vivo and therefore has become a standard assay system to study those processes. Because secretory as well as transmembrane proteins were found to use the same translocation machinery in the ER membrane *(3,6,7)*, the lysate system proved suitable to study both types of proteins.

Expressing a protein in a cell-free system appears attractive, as protein biosynthesis can be studied independently from the complex mechanisms occurring in a living cell. In addition, the system is readily accessible to scientific manipulations. Other advantages of this method include speed, a relative ease to interpret results because only one or a small number of known RNA species

From: *Methods in Molecular Biology, vol. 154: Connexin Methods and Protocols*
Edited by: R. Bruzzone and C. Giaume © Humana Press Inc., Totowa, NJ

are present and will be translated, simple detection of the synthesized radiolabeled proteins on sodium dodecyl sulfate (SDS)-polyacrylamide gels, no detectable endogenous protease activity, and a wide range of co- and posttranslational protein modifications, as the system is derived from eukaryotic cells. The system, however, is not suitable for the production of large quantities of protein. Yields of protein obtained with this method will be only in the picomole range. In addition, the standard cell-free translation system will not allow the study of processes occurring in intracellular compartments downstream of the ER, although a modification of the method has been described in which the microsomal vesicles are replaced by detergent-permeabilized whole cells *(8)*. Transport through the Golgi apparatus, as well as insertion into the plasma membrane, was observed in this modified system.

In this chapter, the synthesis and membrane integration of proteins into microsomes is described first, with special emphasis on connexins. Second, posttranslational protein assays are described that allow differentiation between cytoplasmic, transmembrane, and secretory proteins, and allow detection of posttranslational modifications such as signal peptide processing and core glycosylation. Third, methods are described to analyze potential oligomerization of the newly synthesized proteins, and finally protein detection methods are discussed.

Another description of these methods, with special emphasis on the analysis of homo- as well as hetero-oligomerization, and the characterization of connexin-specific assembly signals, has been published recently *(9)*.

2. Materials

2.1. Reagents

1. Reagents required for molecular biology, including restriction endonucleases, modifying enzymes, plasmids, connexin cDNAs, and in vitro transcription vectors.

2.1.1. In Vitro Transcription

1. Transcription kit containing 5× transcription buffer, 0.1 M DL-dithiothreitol (DTT), ribonucleotide trisphosphates (rNTPs), acetylated bovine serum albumin (BSA), and control DNA (Promega).
2. Sterile, high-quality deionized water.
3. RNase inhibitor such as RNasin (Promega).
4. SP6, T3, and/or T7 RNA polymerase (Promega).
5. DNA grade agarose for agarose gel electrophoresis.
6. Ethidium bromide.

2.1.2. In Vitro Translation/Membrane Integration

1. Reticulocyte lysate or wheat germ extract, including amino acid mixture minus methionine, or cysteine, and control RNA (Promega).

2. Canine pancreatic rough microsomes (Promega).
3. Isotopes: This example uses high-quality [^{35}S]methionine (SJ1515, Amersham).

2.1.3. Reagents for Posttranslational Assays

1. Buffer 1 for microsomal membrane pelleting: 150 m*M* K-acetate (KOAc), 50 m*M* triethanolamine (TEA)-acetic acid, pH 7.0, and 2.5 m*M* Mg-acetate (MgOAc).
2. 1 *M* NaOH for alkali extraction.
3. Sucrose for gradients.
4. 1 *M* CaCl$_2$ to stabilize microsomes.
5. Protease K or trypsin (Boehringer Mannheim).
6. Appropriate protease inhibitors such as diisopropyl fluorophosphate (DFP) or phenylmethylsulfonyl fluoride (PMSF) (both from Sigma). **Caution:** DFP is highly toxic and should be used only under a chemical hood. However, it is approx 10 times more efficient than PMSF and therefore preferable.
7. Detergents such as Triton X-100 (TX-100), Nonidet P-40 (NP-40), deoxycholate (DOC), and SDS.
8. Buffer 2 for membrane lysis: 150 m*M* NaCl, 1% NP-40, 0.5% DOC, 50 m*M* Tris, pH 7.5.
9. Endoglycosidase F or H (Boehringer Mannheim).
10. 1 × PBS.
11. Linear 5–20% sucrose gradients containing 150 m*M* NaCl, 50 m*M* Tris, pH 7.6; and the respective detergent used for solubilization.
12. Scintillation fluid for aqueous solutions.
13. Standard proteins with known sedimentation coefficient (S values) such as myoglobin, 2S; ovalbumin, 3.5S; BSA, 4.3S; catalase, 11.5S.
14. Connexin-specific antibodies (Zymed Laboratories).
15. Immunoprecipitation buffer containing 150 m*M* NaCl, 50 m*M* Tris, pH 7.6, and either no detergent, nonionic detergent such as TX-100, or ionic detergent (SDS), respectively.
16. Protein A–sepharose beads (Pharmacia).
17. 10% and 12.5% standard SDS Laemmli acrylamide gels. **Caution:** Acrylamide is a cumulative neurotoxin. It is recommended that gloves be worn whenever it is handled.
18. Protein sample buffer containing 3% SDS, 0.5% β-mercaptoethanol.
19. Mixture of prestained SDS gel marker proteins (Bio-Rad).
20. 1 *M* sodium salicylate (Sigma).

2.2. Equipment

1. Adjustable heating block or bath.
2. Agarose gel apparatus.
3. Micropipets and 0.5- and 1.5-mL microcentrifuge tubes. Tips and tubes must be autoclaved.
4. Airfuge or tabletop ultracentrifuge (both from Beckman Instruments).

5. Ultracentrifuge and SW55Ti rotor, or equivalent.
6. Refractometer.
7. Liquid scintillation counter.
8. Laboratory rocker or shaker.
9. SDS-PAGE apparatus such as Bio-Rad Mini Protean II.
10. Equipment to dry SDS-polyacrylamide gel electrophoresis (PAGE) gels.
11. X-ray film, X-ray cassettes including intensifying screens.
12. Densitometer.

3. Methods

3.1. Connexin Protein Synthesis and Membrane Integration

Cell-free translation systems consist of a translation competent cell lysate, containing ribosomes, precursor tRNAs, an energy generating system, factors involved in the translation process, a mixture of unlabeled amino acids except in general methionine, and radioactively labeled methionine for protein detection. On addition of a synthetic RNA that encodes a protein, the RNA will be translated into protein. If the RNA encodes a secretory or transmembrane protein and the lysate is supplemented with microsomes (*see* **Subheading 3.1.3.**) the protein will cotranslationally translocate into the microsomes. The lysate can then be electrophoresed on an SDS polyacrylamide gel and the translated proteins can be visualized by autoradiography using X-ray film or a phosphor imager system (*see* **Fig. 1**).

3.1.1. In Vitro Transcription

Before a desired protein can be translated a synthetic RNA transcript has to be synthesized. Several vectors are commercially available that have an SP6, T7, or T3 bacteriophage promoter cloned upstream of a multiple cloning site suitable for inserting the cDNA encoding the desired protein. These bacteriophage promoters are highly specific for their RNA polymerases. Therefore, even cDNAs encoding proteins that are toxic for *E. coli* can be cloned and amplified in these vectors. All three bacteriophage RNA polymerases are commercially available in very good qualities (also as complete transcription kits) allowing efficient synthesis of transcripts of up to 5–10 kb in length *(10–12)* (*see* **Note 1**).

Synthetic RNAs can be synthesized as capped, as well as uncapped RNAs. While capping is required for transfection of mammalian cells *(13)*, uncapped RNAs will be translated efficiently in the lysate system. A somewhat higher translation efficiency of capped RNAs is in general compensated by the more efficient synthesis of uncapped RNAs. Poly(A) tails are also not required for efficient translation in lysate systems.

1. Although the vectors contain a transcription termination signal downstream of the cloning site, linearization of the plasmid downstream of the inserted cDNA is

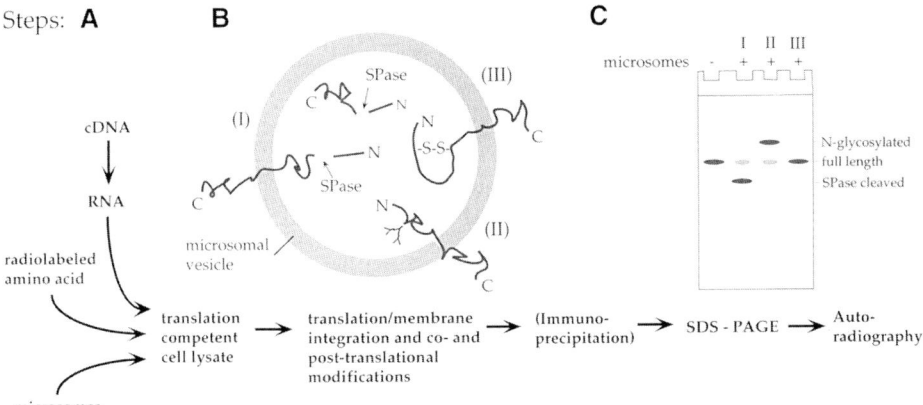

Fig. 1. Schematic representation of the components of a cell-free translation system and of the steps involved in a cell-free translation reaction. Secretory and transmembrane proteins can be expressed and cotranslationally translocated across and into membranes when the lysate system is supplemented with endoplasmic reticulum derived membrane vesicles (microsomes) (**A**). Secretory and transmembrane proteins will be co- and posttranslationally modified inside the microsomes (**B**). Modifications include signal peptidase (SPase)-mediated cleavage of N-terminal signal peptides (I), core glycosylation (II), and disulfide bridge formation (III). Polypeptide modifications I and II, leading to an increased (I), or decreased (II) mobility on SDS protein gels, can readily be detected by SDS-PAGE and autoradiography (**C**).

desirable and substantially increases the yield of RNA copies. Standard procedures for synthetic RNA synthesis are supplied in Promega Biotech's Technical Manual "Transcription in vitro Systems" (part no. TM016).

2. A typical transcription reaction used for the synthesis of connexin cRNAs *(14–16)* is as follows (*see* **Note 2**):

    10 µL of 5× transcription buffer
    5 µL of 0.1 M DTT
    10 µL of 2.5 mM rNTPs
    1 µL of 1 mg/mL of acetylated BSA
    1 U/µL of transcription volume RNasin
    0.3 U/µL of transcription volume SP6-polymerase
    0.1–1 µg of linearized plasmid DNA
    Water to 50 µL

3. Mix at room temperature and incubate at least 1 h at 40°C.

4. If the synthetic RNA is to be synthesized with a 5' CAP structure, only the concentration of rGTP is decreased fivefold, and 1 µL of 1 m*M* CAP analog (m⁷G[5']ppp[5']G) (Pharmacia) is added to the transcription protocol given previously.

5. Synthesized RNAs can be used without further purification in following translation reactions. However, if additional background bands appear on the gels, phenol–

chloroform purification, followed by ethanol precipitation *(17)* is recommended (*see* **Note 3**).

6. Efficiency of RNA synthesis, and quality of the RNA transcripts can be checked by electrophoresing a small aliquot (1–2 µL) on a freshly prepared standard agarose gel prepared with TAE or Tris/Borate/EDTA (TBE) buffer. To recognize the newly synthesized RNA on the gel and to avoid confusion with the linearized DNA that was used as a template and is also present in the transcription reaction, a lane containing linearized plasmid alone should be electrophoresed as a control in parallel as well.

7. RNA can be visualized by standard ethidium bromide staining *(17)* (*see* **Note 4**).

3.1.2. In Vitro Translation

Two types of translation-competent cell lysates are commercially available. Both lysates are prepared from cells highly active in protein biosynthesis. Reticulocyte lysate is prepared from the red blood cells of rabbits that have been injected with phenylhydrazine to destroy their mature red blood cells, and wheat germ extract is prepared from sprouting wheat seeds. Both extracts are depleted of endogenous RNAs and produce only minimal amounts of endogenous proteins. However, it is always advisable to run a control translation reaction without adding any synthetic RNA with each new batch of lysate. Translation in wheat germ lysates is sensitive to the concentration of potassium and magnesium ions and their concentration has to be optimized for each RNA. For most RNAs optimal potassium ion concentrations range from 120–160 mM, and optimal magnesium ion concentrations range from 1.5 to 4 mM. Although both lysates can be prepared in the laboratory *(18,19)*, best connexin translation results have been obtained with Promega Biotech's nuclease-treated Rabbit Reticulocyte Lysate System. Protocols for standard translation reactions are provided in the Promega Biotech Technical Manuals "Rabbit Reticulocyte Lysate System" (part no. TM232), and "Wheat Germ Extract" (part no. TM230). A standard protocol for connexin translation is given in **Subsection 3.1.3.**

3.1.3. Membrane Protein Translocation

Secretory as well as transmembrane proteins in general are cotranslationally translocated into the membrane of the ER. Several protocols have been developed to prepare ER membrane-derived vesicles (microsomes) from cells highly active in protein secretion. Most common, and the source of commercially available microsomes (e.g., Promega Biotech), are microsomes prepared from the acinar cells of canine pancreas (*see* **Note 5**). This organ is low in RNase concentration, soft in texture, and the secretory acinar cells are very rich in rough ER membranes. The microsomes are depleted of endogenous RNAs and give only few background bands. However, it is always advisable to run a con-

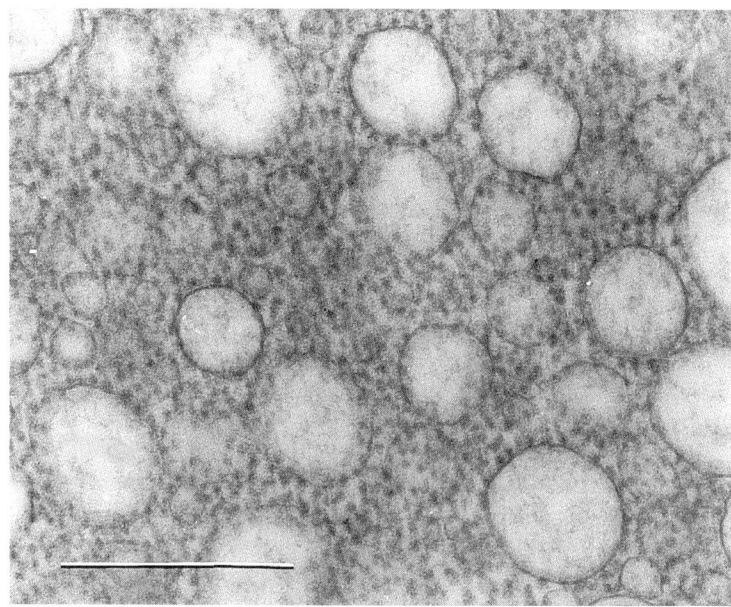

Fig. 2. Electron micrograph of negatively stained microsomal vesicles prepared from the endoplasmic reticulum of canine pancreatic acinar cells. Such microsomes are competent to cotranslationally translocate secretory and membrane proteins. Note that each microsomal vesicle has many ribosomes bound to its surface, indicating that each vesicle has multiple protein insertion sites. This is a prerequisite for the assembly of an oligomeric connexon within microsomes. Scale bar = 500 nm.

trol translocation reaction without adding any synthetic RNA with each new batch of microsomes (*see* **Note 6**).

The microsomes are patches of rough ER membranes that vesiculate upon their isolation (*see* **Fig. 2**). The microsomes have been described to contain all the components required for cotranslational protein translocation, including signal recognition particle (SRP), ribosomes, energy supplying molecules, and other factors (*see* **Note 7**). Furthermore, the microsomes have signal peptidase *(20)*, and core glycosylation activity *(21)*. N-terminal signal peptides will be cleaved from secretory and certain transmembrane proteins, and certain asparagine residues located in the lumen of the microsomes can be glycosylated. In **Fig. 3**, the expected translation products of a secretory (**A**), a transmembrane protein with an N-terminal cleavable signal sequence (**B**), a transmembrane protein without cleavable signal sequence (**C**), an N-glycosylated membrane protein without cleavable signal sequence (**D**), and a cytoplasmic protein (**E**) are schematically shown. A translation reaction in the presence of microsomes

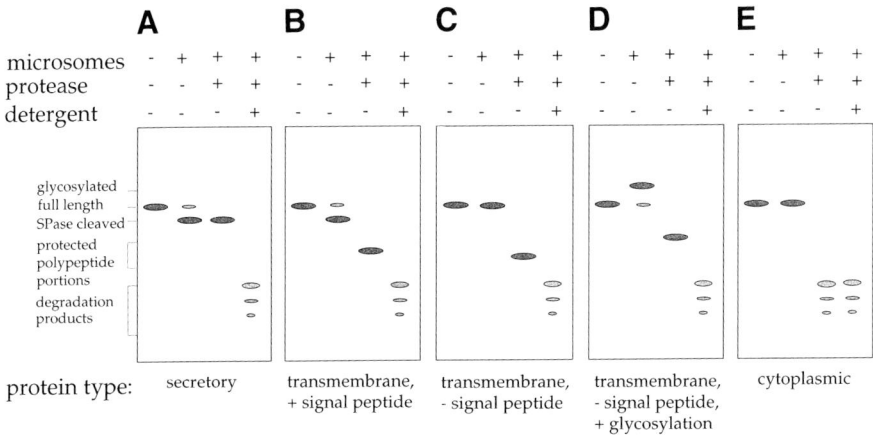

Fig. 3. Schematic representation of the protein band patterns expected for different types of proteins after translating these proteins in translation competent cell lysates supplemented with microsomes, SDS-PAGE, and autoradiography. The expected translation products of a secretory protein (**A**), a transmembrane protein with an *N*-terminal cleavable signal peptide (**B**), a transmembrane protein without cleavable signal peptide (**C**), a core glycosylated membrane protein without cleavable signal peptide (**D**), and a cytoplasmic protein (**E**) are shown. For each protein type the translation products generated in the absence (–) (*lanes 1*) and in the presence of (+) microsomes (*lanes 2*) are shown on each gel. Furthermore, polypeptide domains located inside the microsomes and protected from proteolytic degradation are shown after addition of protease to the lysate when translation is completed (*lanes 3*), and addition of protease in combination with detergent (*lanes 4*). Full-length, unmodified polypeptides are generated in the absence of microsomes (*lanes 1*). Secretory and transmembrane proteins encoding a signal sequence can be recognized by their increased electrophoretic mobility on SDS protein gels when translated in the presence of microsomes (*lanes 2, gels A and B*). Proteins not encoding a cleavable signal sequence migrate unaltered (*gel C*). Core glycosylated proteins can be recognized by their decreased mobility (*gel D*). Note that in a core glycosylated protein encoding a signal peptide, increase and decrease in electrophoretic mobility will interfere. In addition, deglycosylation is necessary to correctly recognize this protein type (*see* **Fig. 7**). Membrane-translocated proteins (*gels A–D*) can be distinguished from non-translocated proteins (*gel E*) by the generation of protease-protected polypeptide portions (*lanes 3, gels A–D*). Protected portions are degraded when the microsomal membrane seal is resolved by the addition of detergent (*lanes 4, gels A–D*).

(shown in **Fig. 3**, *lanes 2, 6, 10, 14,* and *18*) should always be accompanied by a parallel translation reaction in the absence of microsomes (shown in **Fig. 3**, *lanes 1, 5, 9, 13,* and *17*) to indicate the electrophoretic mobility of unmodified, full-length polypeptides.

1. Standard connexin translation reactions are 10–25 µL in volume and mixed as follows:

   > 25 µL of reticulocyte lysate
   > 2.5 µL of amino acid mixture minus methionine
   > 2.5 µL of [^{35}S]methionine (15 µCi/µL) (*see* **Note 8**)
   > 0.5–2 µg of synthetic RNA
   > water to 47.5 µL

2. Divide into two aliquots of 25 and 22.5 µL volume.
3. Add 2.5 µL (corresponding to 1 Eq/10 µL reaction volume) of microsomes to the smaller sample.
4. Incubate at 30–37°C for 30–60 min (*see* **Note 9**).

3.1.4. Increasing Translation Efficiency

A problem sometimes encountered with soluble as well as membrane proteins is that the synthetic RNA encoding the desired protein is not translated efficiently in translation-competent cell lysates. This was also found for connexin polypeptides, and may be related to the length of the 5' untranslated region, the structure of the translation initiation sequence adjacent to the translation Start-AUG *(22)*, or the absence of required factors. Optimization of the translation initiation sequence by site directed mutagenesis as well as adding subcellular fractions can improve translation efficiency. However, connexin translation efficiency was not substantially increased by these approaches (M. Falk, unpublished data).

Another approach is based on cloning the 5' untranslated region of an efficiently translated protein, such as globin, upstream of the cloned cDNA. Cloning the connexin cDNAs into the transcription vector pSP64T *(23)* that encodes the 5' noncoding region of *Xenopus* β-globin *(24)* immediately downstream of an SP6 promoter was found to dramatically increase the translation efficiency of the connexins as well as other cDNAs *(14–16,25)* (**Fig. 4**). An advanced version of pSP64T, pSPUTK is commercially available from Stratagene. Reducing the length of the 5' untranslated portion of the desired cDNA may further increase translation efficiency to prevent scanning ribosomes from unspecifically falling off the template cDNA.

Another approach to enhance translation efficiency is based on using the CAP independent translation initiation sequences (IRES = Internal Ribosomal Entry Site elements; or CITE = CAP-Independent Translation Enhancer) from picornaviruses (*see* **ref. 26** for review). Several vectors using these sequences, such as Novagen's pCITE-1, are commercially available. In general it is advisable to ensure that no AUG codon in any of the three possible reading frames is encoded in the 5' untranslated region upstream of the Start-AUG to prevent ribosomes scanning along the cDNA from initiating at wrong, upstream AUG codons.

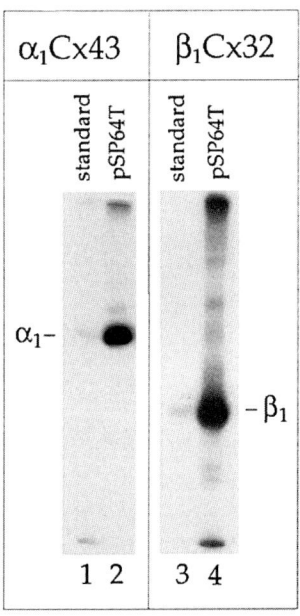

Fig. 4. Increasing translation efficiency of connexins in cell-free translation systems. α_1Cx43 and β_1Cx32 cDNAs were cloned into standard transcription vectors (*lanes 1, 3*), and into the transcription vector pSP64T (*lanes 2, 4*) that contains the 5' noncoding region of *Xenopus* β-globin as a translation enhancer upstream of the cDNA. cDNAs were transcribed and translated in rabbit reticulocyte lysates in the absence of microsomes. In each lane 0.5 µL of translation reaction was analyzed on SDS-polyacrylamide gels and translation products were visualized by autoradiography. Note the dramatically increased translation efficiency of transcripts containing the translation enhancer sequence (*lanes 2, 4* vs *1, 3*). Additional bands detectable on the gels represent aggregates of translation products, and translation products initiated at AUG codons located downstream from the connexin authentic initiation codons, and were generated mostly due to nonoptimized conditions. Gels were purposely overexposed to visualize these products. (Compare **Fig. 5** for connexin-expression profiles obtained after optimizing translation conditions.) Substantially increasing the translation efficiency of connexins allowed the subsequent synthesis and assembly of oligomeric connexons in cell-free translation systems, and to analyze the assembly behavior of different connexin isotypes.

3.1.5. Synthesis of Full-Length, Membrane Integrated Connexins in Cell-Free Translation Systems

We found that the translation of connexin polypeptides in standard cell-free translation systems supplemented with ER-derived microsomes resulted in a complete but inappropriate proteolytic processing that affected all connexin

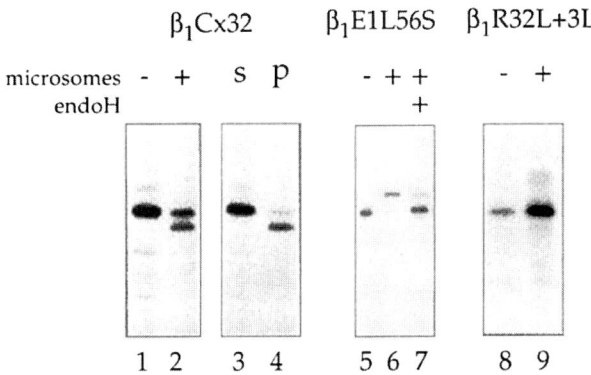

Fig. 5. Synthesis of cleaved and full-length membrane-integrated connexins. Translation of connexins in cell-free translation systems supplemented with pancreatic microsomes results in a complete, aberrant processing by the ER resident protease signal peptidase that removes an N-terminal portion including the N-terminal domain, and the first transmembrane spanning domain of connexins (*lane 2*). In the absence of microsomes no cleavage occurs (*lane 1*). The cleaved connexins pellet (p) together with the microsomes (*lane 4*), while the full-length connexins stay in the soluble lysate fraction (s) (*lane 3*). This result indicates that all membrane-integrated connexin polypeptides are cleaved; and that the full-length connexins also synthesized in the presence of microsomes (*lane 2*) are synthesized on nonmembrane bound ribosomes, and have failed to insert cotranslationally into the microsomal membranes. Full-length, glycosylated, and membrane integrated connexins were synthesized when an N-glycosylation site was introduced into the first extracellular loop (L56S amino acid exchange) (*lane 6*). Endoglycosidase H (endoH) removes the carbohydrate sidechain. The electrophoretic mobility of the deglycosylated polypeptides corresponds to the mobility of full-length connexins (*lane 7*). Full-length, membrane-integrated connexins are also synthesized when the length and hydrophobicity of the first transmembrane spanning domain is increased (R32L + 3L amino acid exchange) (*lane 9*).

polypeptides upon their membrane integration (reported in *14,16*) (*see* **Fig. 5**). A careful analysis of the cleavage reaction revealed that the relatively weak hydrophobic character of the first transmembrane spanning domain, which acts as internal signal anchor sequence, is responsible for this inappropriate cleavage. Our results indicate that the connexin signal anchor sequence is aberrantly recognized and positioned as a cleavable signal peptide within the endoplasmic reticulum translocon, and that this mispositioning enabled signal peptidase to access the cleavage sites *(16)*. These observations provide direct evidence for the involvement of yet uncharacterized cellular factors in the membrane integration process of connexins that are absent or inactive in the standard cell-free translation system *(14,16)*.

Two different methods have been found that prevent the cleavage and allow the efficient synthesis of full-length, correctly integrated connexin polypeptides into microsomes. Single amino acid exchanges, introduced by site-directed mutagenesis into the first extracellular loops of α_1Cx43, β_1Cx32, as well as β_2Cx26 (Glu[57] to Ser in α_1Cx43, Leu[56] to Ser in β_1Cx32, and Leu[54] to Ser in β_2Cx26), completely inhibited the cleavage reaction, most likely because of a steric hindrance between oligosaccharyl transferase (OST), the enzyme that recognizes the core glycosylation sequence and transfers core glycosyl groups from dolichol onto asparagine (*see* **Subheading 3.2.3.**), and signal peptidase (*15,16*, M. Falk, unpublished results). The point mutations result in the creation of core glycosylation sites within the connexin sequences and the binding of OST to the nascent connexin polypeptides (*see* **Subheading 3.2.3.**). However, core glycosyl groups were added only to Cx32, and Cx26 polypeptides during translation, and not to the Cx43 sequence (*15,16*, M. Falk, unpublished results).

The other method is based on increasing the hydrophobic character of the signal anchor sequence (the first transmembrane spanning domain) of connexins. Increasing the length of the hydrophobic core of the first transmembrane spanning domain in Cx32 from 18 amino acids (Val[23] to Ala[40]) in the wild-type protein to 21 amino acids and exchanging the central Arg[32] with an uncharged amino acid (leucine) completely abolished the cleavage as well (**Fig. 7** in **ref.** *16*).

3.1.6. Combined Transcription/Translation Systems

A few years ago combined transcription/translation systems were introduced that synthesize protein from an added cDNA in one step. These systems (e.g., TNT lysate from Promega) were primarily designed to test if a cloned cDNA or open reading frame produced a protein with the expected molecular weight. Such systems are probably better not used for the integration of connexins into microsomes because the combination of different reactions into one vial may complicate the interpretation of results, and may generate less clean translation reactions.

3.2. Assaying for Membrane Integration and Transmembrane Orientation

After translation of a membrane protein is completed its membrane integration can be analyzed. Great care should be taken in the interpretation of results that indicate potential membrane integration. This is especially true for membrane proteins that are normally not modified during or after their membrane integration by signal peptide cleavage, or core glycosylation. Electrophoretic mobility of such proteins will remain unaltered (*see* **Fig. 3**). Three methods (*see* **Subheadings 3.2.1.–3.2.3.**) are commonly used to determine membrane translocation that produce more or less convincing results. For polytopic membrane proteins with extended hydrophobic regions such as the connexins, a combination of different methods is generally required to obtain convincing results. We have used all three methods with connexins *(14,16)*.

3.2.1. Alkali Extraction

This method is based on the perforation of membranes at alkaline pH. When microsomes are exposed to alkaline pH after a completed translation reaction, a secretory protein, translocated into the lumen of the microsomes, would be extracted from the microsomes, while a transmembrane protein would still remain with the microsomal membranes. In control experiments, microsomes treated at neutral pH are assayed. Furthermore, it is important to assay well-characterized transmembrane, secretory, and cytoplasmic control proteins in parallel for a valid interpretation of results (*see* ref. *14*). This is especially important for proteins with extended hydrophobic regions, such as the connexins, as they tend to adhere strongly to hydrophobic surfaces. Therefore, it can be difficult to determine if a protein is indeed translocated into the membranes, or only adhered to the microsomal membrane surface (*see* **Note 10**).

We have used a modification of the original method described by Gilmore and Blobel *(27)* with connexins. Alkali-treated and control microsomes were pelleted after treatment with an Airfuge ultracentrifuge (Beckman Instruments, Palo Alto, CA). This centrifuge is very well suited for the small volume translation reactions. The standard A-100/18 rotor holds six tubes with a maximum capacity of 175 µL each. Alternatively, a tabletop microcentrifuge for Eppendorf tubes or a Beckman TL-100 tabletop ultracentrifuge can be used, although centrifugation times have to be extended substantially.

1. Following translation, 15 µL translocation reaction aliquots are adjusted to 50 µL volume with buffer 1. For alkali extractions, aliquots are adjusted to pH 11.5 with 1 *M* NaOH, and adjusted to 50 µL volume.
2. Samples are incubated on ice for 10 min, overlaid onto a 100 µL cushion of 0.5 *M* sucrose in buffer 1 or a 100 µL cushion of 0.2 *M* sucrose in 30 m*M* HEPES adjusted to pH 11.5, 150 m*M* KOAc, 2.5 m*M* MgOAc, and fractionated into a supernatant and pellet fraction.
3. Proteins in the supernatant fractions are then precipitated with trichloroacetic acid (TCA) and neutralized with saturated Tris base, before processing for SDS-PAGE.

At neutral as well as alkaline pH, membrane-integrated connexin polypeptides remain predominantly associated with the microsomal membranes and are detected in the pellet fraction. A similar result is obtained with the transmembrane control protein acetylcholine receptor subunit α7, while the secretory control proteins prolactin and yeast α-factor were largely extracted from the

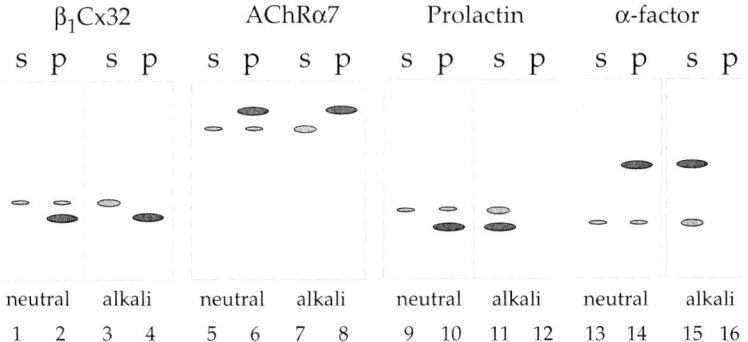

Fig. 6. Assaying for protein membrane integration. Schematic representation of the protein band patterns expected for transmembrane, and secretory proteins after incubating the microsomes at neutral and alkaline pH. At neutral as well as alkaline pH, membrane-integrated connexin polypeptides remained associated with the microsomal membranes and were detected in the pellet fraction (p) (*lanes 2* and *4*). Full-length, not membrane integrated connexins were detected in the soluble supernatant fraction (s) (*lanes 1* and *3*). A similar result was obtained with the glycosylated, transmembrane control protein acetylcholine receptor subunit α7 (AchRα7) (*lanes 5–8*), while the secretory control proteins prolactin and yeast α-factor were extracted from the microsomal membranes at alkaline pH and were detected in the supernatant fractions (*lanes 10, 14,* and *12, 16*).

microsomal membranes at alkaline pH and were detected in the supernatant fractions (*see* **Fig. 4** in **ref. *14***, and **Fig. 6**).

3.2.2. Protease Protection Assays

Translocation of many transmembrane proteins can be assayed better using a protease protection assay. Microsomes are incubated with externally added protease after translation is completed. Domains of a membrane integrated polypeptide protruding from the outside surface of the microsomes will be digested, while transmembrane domains, and domains located in the lumen of the microsomes are protected from proteolytic degradation (*see* **Fig. 3**, *lanes 3, 7, 11,* and *15*). In control, detergent is added to an aliquot of the translocation reaction prior to addition of protease that resolves the vesicle barrier (*see* **Fig. 3**, *lanes 4, 8, 12,* and *16*). After the digest is completed protected polypeptide domains can be characterized by immunoprecipitation analysis using domain-specific antipeptide antibodies (**ref. *14***, and **Subheading 3.3.2.**).

The following method has been used to determine membrane integration and transmembrane orientation of connexin polypeptides integrated into micorosomal vesicles:

1. 10 µL Aliquots of the connexin translocation reactions are diluted to 50 µL in buffer 1 (*see* **Subheading 3.2.1.**).
2. To increase the seal of the vesicles, microsomes are stabilized by the addition of CaCl$_2$ to a final concentration of 10 m*M* and incubated for 10 min on ice.
3. Either water or proteinase K (predigested for 30 min at 30°C) is added to final concentrations of 0.1 mg/mL and 0.5 mg/mL, or trypsin (predigested in the same way) is added to final concentrations of 0.5 mg/mL and 1 mg/mL from 10× stock solutions in water, respectively. In addition, where indicated, NP-40 is added to a final concentration of 1% from a 10% (w/v) stock in water. All digests are incubated for 1 h at 37°C (*see* **Note 11**).
4. Protease activity is blocked by the addition of 5 m*M* DFP, 1% SDS (final concentrations) and boiling for 5 min, following the method described by Chavez and Hall *(28)* (*see* **Note 12**).
5. Samples are then diluted 10× with buffer, chilled on ice, and processed for immunoprecipitation analysis, as described in **Subheading 3.3.2.**
6. Immunoprecipitated polypeptide fragments (*see* **Note 13**) are analyzed on special 20% SDS-gels allowing the resolution of small polypeptide fragments *(29)* (*see* **Fig. 3**).

3.2.3. N-*Glycosylation Tagging*

Probably the most convincing method to determine the integration of a protein into the membrane bilayer, which in addition allows determination of the overall transmembrane topology of a membrane protein, is to introduce core glycosylation sites into the polypeptide sequence. This approach is based on the activity of oligosaccharyl transferase (OST) that is present in the lumen of the ER and within microsomes. The enzyme transfers mannose-rich carbohydrate moieties onto asparagine residues that are part of *N*-glycosylation consensus sequences (Asp/X/Ser or Thr) *(30)*. Since glycosylation can occur only at sites located at the lumenal face of the ER (or the microsomes), glycosylation at any given site can be taken as proof of the ER lumenal localization (which coincides with the extracellular face) of that polypeptide domain. Each core glycosylation group added to the polypeptide chain increases its molecular mass by ~2.5 kDa. The accompanying reduction in electrophoretic mobility on SDS-polyacrylamide gels is generally large enough to be readily visible with proteins up to ~70 kDa in size.

We have used this method to demonstrate a cotranslational membrane integration of α_1Cx43, β_1Cx32, and β_2Cx26 into microsomes, and that the transmembrane topology of connexin polypeptides integrated into microsomes is identical to their topology in the plasma membrane *(16*, M. Falk, unpublished results).

N-Glycosylation consensus sequences were introduced into the extracellular loops E1 and E2, and into the intracellular loop of α_1Cx43, β_1Cx32, and

β_2Cx26 by substituting specific amino acid residues using site directed mutagenesis (*15,16*, M. Falk, unpublished results). Whereas the mutants with the consensus sites in the extracellular loop E1 or E2 were efficiently glycosylated after cell-free translation in the presence of microsomes resulting in a reduced mobility on SDS-polyacrylamide gels, the mutants with the consensus sites in the intracellular loop were not glycosylated. In control aliquots, endogly-cosidase H (endoH) digestion was used to remove the carbohydrate side chains. Electrophoretic mobility was shifted back to the electrophoretic mobility of wild-type connexin proteins (*see* **Fig. 1** in **ref.** *15*, **Fig. 2B** in **ref.** *16*, and **Fig. 5**, *lanes 5–7*).

3.3. Synthesis of Oligomeric Connexons in Microsomes

It was commonly thought that the oligomerization of membrane proteins does not occur in cell-free translation/membrane integration systems owing to the low probability for multiple polypeptide insertions into a microsomal vesicle *(31)*. However, micrographs of thin-sectioned microsomes showed that each microsomal vesicle has many ribosomes bound to its membrane surface (**Fig. 2**), indicating that each vesicle has multiple protein insertion sites, and as such would potentially allow the integration of several polypeptides into the same vesicle. Later, functional expression and assembly of a Shaker type K$^+$-channel (Shaker H4, an oligomeric structure consisting of four identical copies of a protein traversing the membrane bilayer six times) *(32)*, assembly of a human HLA–DR histocompatibility molecule (an $\alpha/\beta/\gamma$-heterotrimer) *(33,34)*, and assembly of the asialoglycoprotein receptor (an α/β hetero-oligomer) *(35)*, although probably not into a functional receptor molecule *(36)*, were reported to occur during cell-free expression in microsomes. These observations, combined with the expression of gap junction connexons reported in Falk et al. *(15)*, provide compelling evidence that the assembly of functional membrane structures consisting of several subunit proteins can take place in microsomes, and that the cell-free translation system can be used to study protein oligomerization processes.

Substantially increasing the translation efficiency of connexins in lysates (*see* **Subsection 3.1.4.**) was a crucial prerequisite to achieve multiple connexin polypeptide insertions into individual microsomes and the successful assembly into connexons within microsomes. Furthermore, the increased translation efficiency allowed subsequent analysis of the assembly behavior of connexin isotypes, and determination of assembly signals within the connexin polypeptide sequences (*15,37,38*, M. Falk, unpublished results).

Translation and oligomeric assembly analysis conditions are as follows (*see* also **ref.** *15*):

1. Translation reactions are generally 50–100 µL in volume for subsequent hydrodynamic oligomerization analysis. Volumes of 10–25 µL were used for the assembly analysis of connexins by immunoprecipitation.

2. To maximize translation efficiency, rabbit reticulocyte lysates are programmed with large amounts of the appropriate cRNA (typically 2–4 µg of RNA, as estimated from an ethidium bromide stained agarose gel, per 100 µL reaction volume).

3. To maximize connexin polypeptide membrane integration, low amounts of microsomes are used in the translation reactions. Typical concentrations are 5 Eq/100 µL reaction volume.

4. Translation reactions are incubated at 30°C for 1–3 h at 30°C to allow complete posttranslational folding and association of the newly synthesized polypeptides before subsequent oligomerization analysis (*see* **Note 14**).

3.3.1. Assembly Assayed by Hydrodynamic Analysis

Connexon assembly was assayed by hydrodynamic analysis using linear 5–20% sucrose gradients. This technique is generally used for assaying assembly of protein subunits into oligomeric structures and was used for connexins by others before *(39–41)*.

1. Nonmembrane integrated connexin polypeptides that are synthesized to a certain extent in the cell-free translation reactions as a byproduct on nonmembrane bound, free ribosomes (*see* **Subheadings 3.1.3.** and **3.2.2.**, and **Fig. 5**) are separated from membrane integrated connexins prior to gradient analysis by pelleting the microsomes through a 0.5 *M* sucrose cushion made in 1× PBS, using an Airfuge ultracentrifuge (Beckman Instruments, Palo Alto, CA) as described in **Subheading 3.2.1.**

2. To remove unincorporated radioactive label, microsomes are washed twice in 0.25 *M* sucrose, 1× PBS and pelleted as before.

3. Microsomes are then solubilized in 1% TX-100 (or other detergents) for 30 min at 4°C.

4. Detergent-insoluble material is precipitated by a high-speed centrifugation (15 min, 30 psi) using the Airfuge ultracentrifuge.

5. Supernatants are loaded on top of 5-mL linear 5–20% (w/v) sucrose gradients containing 150 m*M* NaCl; 50 m*M* Tris, pH 7.6; and the respective detergent used for solubilization.

6. After centrifugation for 16 h at 4°C in a SW55Ti rotor (Beckman Instruments Inc.), at 43,000 rpm (g_{av} = 160,000; g_{max} = 210,000), gradients are fractionated by puncturing the bottom of the tube with a 26-gauge needle, and approx 20 0.25-mL fractions are collected.

7. Aliquots of 25 µL of the fractions are analyzed by liquid scintillation counting and SDS-PAGE. In general, between 2000 and 15,000 counts per minute (cpm) are obtained per 5S peak fraction.

8. The cpm recorded in each fraction are corrected by the background activity (fraction with the lowest cpm), and plotted in percent activity per fraction (**Fig. 7**).

9. Aliquots of all fractions are also analyzed by immunoprecipitation (*see* **Subheading 3.3.2.**) using connexin-specific antibodies.

10. To verify assembly of connexins into connexons, control aliquots of the translocation reactions are solubilized in 0.1% SDS to resolve the oligomeric assem-

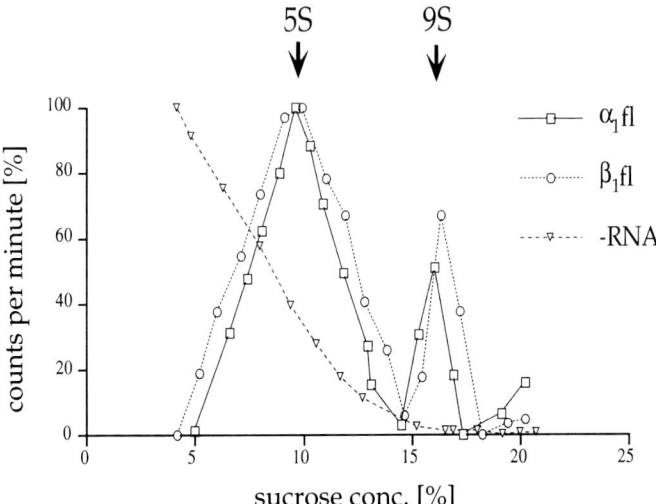

Fig. 7. Assembly of homo-oligomeric connexons in a cell-free translation system. Reticulocyte lysates were supplemented with microsomes and programmed with α_1Cx43 RNA, β_1Cx32 RNA, and without (–) RNA in control. After translation was full-length (fl) completed microsomes were harvested, lysed in nonionic detergent, and assembly of connexins was analyzed by hydrodynamic analysis on linear sucrose gradients. Gradients were fractionated from the bottom, and sucrose concentration, radioactivity, and connexin protein content was determined in each fraction. Radioactivity recovered from each fraction was plotted vs the sucrose concentration after subtracting the lowest counts from each fraction. 9S particles represent assembled connexons, while 5S particles represent unassembled connexin polypeptides. More than 30% of the connexin polypeptides were recovered as assembled hexameric gap junction connexons. (Reproduced with kind permission from **ref. 15**).

blies and analyzed in parallel as described previously, except that the gradients are prepared with 0.1% SDS, and run at 20°C to prevent precipitation of the SDS in the gradients.

11. The refractive index of each fraction is measured using a refractometer and converted into the corresponding sucrose concentrations using a standard conversion table.

12. Standard proteins with known sedimentation coefficients (myoglobin, 2S; ovalbumin, 3.5S; BSA, 4.3S; catalase, 11.5S) and gap junction connexons consisting of α_1Cx43, β_1Cx32, or β_2Cx26 expressed and purified from baculovirus infected insect cells *(42)* are analyzed on separate gradients to compare the connexin specific S-values with corresponding sucrose concentrations.

13. Dependent on the detergent used for the solubilization of microsomes, up to 32% of the membrane-integrated connexin polypeptides were recovered as hexameric connexons (**ref. 15**, and **Fig. 7**).

3.3.2. Assembly Assayed by Coimmunoprecipitation

Connexin specific monoclonal and antipeptide antibodies directed against different regions of α_1Cx43, β_1Cx32, and β_2Cx26 that display no detectable cross-reactivity with other connexin isotypes are used for the immunoprecipitation of connexin polypeptides from in vitro translation reactions (*see* **Note 15**). Oligomerization of connexin polypeptides into homo- and hetero-oligomeric complexes can be analyzed by immunoprecipitation following general methods described by Harlow and Lane *(43)*. Connexins are translated in different combinations together with other connexin isotypes, connexin mutants, or nonconnexin transmembrane proteins. Connexin polypeptides are then immunoprecipitated either from complete translocation reactions or from microsomes that had been pelleted through sucrose cushions as described in **Subheading 3.2.1.**, followed by their resuspension in 1× PBS, 0.25 *M* sucrose (*see* **Fig. 8**).

1. Microsomes are solubilized for 30 min on ice in immunoprecipitation buffer containing 150 m*M* NaCl, 50 m*M* Tris, pH 7.6; 1% TX-100, which lysed the microsomal vesicles without disrupting the oligomeric connexin complexes.
2. Insoluble material is precipitated by high-speed centrifugation using the Airfuge ultracentrifuge.
3. Aliquots of the supernatant corresponding to 10–25 μL of translocation reaction, a connexin-specific antibody against α_1Cx43, β_1Cx32, or β_2Cx26 (*see* above), and preswollen Protein A–Sepharose (where required) are incubated together in 1 mL of immunoprecipitation buffer for 2 h at room temperature, or at 4°C overnight with shaking (*see* **Note 16**).
4. Beads are sedimented by centrifugation and washed 2× with immunoprecipitation buffer prior to the addition of SDS protein sample buffer.
5. Precipitated antigens and associated polypeptides are detected by SDS-PAGE and autoradiography (*see* **Note 17**).
6. As controls, aliquots of the translocation reactions or resuspended microsomes are solubilized in immunoprecipitation buffer containing 0.1% SDS, or no detergent (*see* **Fig. 8** for a schematic representation).

3.4. Connexin Protein Detection Methods

1. Connexin translation products were analyzed qualitatively on 10% and 12.5% SDS Laemmli Bio-Rad minigels (acrylamide/*bis*-acrylamide ratio 29:1). Samples are solubilized in SDS sample buffer containing 3% SDS and 5% β-mercaptoethanol and analyzed without heating to prevent unspecific aggregation of connexin polypeptides, a phenomenon often observed with polytopic membrane proteins.
2. Following electrophoresis, gels are soaked for 10 min in 1 *M* sodium salicylate (Sigma) to enhance ^{35}S and ^{3}H autoradiography, dried, and exposed to Kodak X-AR film at –70°C using an intensifying screen.
3. To quantitatively analyze connexin expression and compare amounts of different translation products with each other, autoradiographs are scanned densitome-

trically using an AlphaImager 2000 Digital Imaging and Analysis System (Alpha Innotech Corporation, San Leandro, CA) as described in Falk and Gilula *(16)*.

4. Because the specific radioactivity of a radiolabeled amino acid incorporated during cell-free translation is provided by the manufacturer, and the frequency of this amino acid in the connexin polypeptide sequence is known, the amount of connexin polypeptides synthesized in the cell-free system can be determined quantitatively using a liquid scintillation counter. We have used this method to quantify the number of polypeptides adsorbed to antibody coated Protein A–Sepharose beads. Technical details are given in Kahn et al. *(25)*.

4. Notes

1. Bacteriophage RNA polymerases are active for several hours and yield of RNA transcripts increases with longer incubation times.
2. If several different transcripts will be synthesized, it is desirable to mix all components except the individual cDNAs to reduce pipetting steps, then aliquot the mix, and finally add cDNAs.
3. Shorter RNA transcripts (1–2 kb in length) can be frozen and thawed several times without significant degradation. Longer transcripts (5–10 kb) are best used fresh to avoid degradation by the freeze–thaw cycle.
4. Great care has to be taken to avoid any possible contamination with RNases. Only autoclaved materials should be used, and tubes and pipet tips should not be touched with bare hands. All electrophoresis equipment needs to be cleaned thoroughly immediately before use, and buffers should be prepared fresh to reduce cRNA degradation.
5. When purchased, microsomes are relatively costly. I have prepared microsomes from fresh canine pancreas following the protocol of Walter and Blobel *(21)* with great success. The prepared microsomes had at least a similar translocation efficiency, and low backgrounds comparable to those of purchased microsomes, and were very efficient in connexin protein integration. Inside-out microsomes were not detected in the preparations (*see* **Fig. 2**).
6. Microsomes are quite stable when stored at –70°C (2–3 yr). However, they should not be thawed and refrozen more than once or twice. Therefore, when first used, aliquoting them out into smaller volumes is highly recommended.
7. To investigate if the membrane translocation of a protein is dependent on signal recognition particle (SRP) wheat germ extract has to be used as a lysate, in combination with SRP-depleted microsomes and purified SRP. SRPs present in the wheat germ lysate do not interact with mammalian microsomes. Although we have used this methodology to prove that the membrane integration of connexins into microsomes is SRP-mediated *(14)*, details of this specific application are not described in this more general chapter. However, protocols for the high-salt extraction of microsomes and the preparation of SRP are described in several publications by Peter Walter and co-workers (University of California, San Francisco).

8. To reduce background, a highly purified methionine, such as SJ1515 (Amersham Biotech) should be used in the translation reactions. Translabel (ICN) was found to produce additional bands, and therefore should be avoided for this application.

9. To avoid nonspecific aggregation of connexin polypeptides that failed to integrate into microsomes during translation and are detected as a protein band on top of the separating gel, shorter incubation times of only 30–45 min are recommended.

10. The posttranslational membrane integration of α_1Cx43 and β_2Cx26 recently reported by two laboratories *(27,28)* is solely based on the results obtained by pelleting microsomes through sucrose cushions, and a protease protection assay using an inadequate protease *(Staphylococcus aureus* V8 protease; *see* below). Therefore, the observation may reflect a strong adhesion of these connexin isotypes to the outer microsomal membrane surface owing to their strong overall hydrophobicity, rather than real posttranslational membrane translocation. Further experiments using adequate technology (*see* **Subheadings 3.2.1.–3.2.3.**) and control proteins are required to convincingly demonstrate a posttranslational ER membrane integration of connexin polypeptides.

11. Proteases with a very low specificity, such as proteinase K or trypsin, are recommended for this assay because exposed polypeptide domains should be degraded completely. More specific proteases, such as *Staphylococcus aureus* V8 protease, which cleaves only peptide bonds C-terminally at glutamic acid and with a 3000-fold lower rate at aspartic acid *(44)* that have been used in connexin protease protection assays previously *(44)* are not suitable. The large polypeptide fragments generally produced by V8 protease do not make it possible to determine if a detected polypeptide fragment was indeed protected from proteolytic degradation and located inside the microsomal vesicle, or simply did not contain a protease V8 specific cleavage site (*see* **ref. 47** for further reading).

12. Protease activity has to be blocked completely before the microsomes are resolved in protein sample buffer; otherwise the protected polypeptide fragments will be degraded. This is especially important when the protected fragments will be characterized subsequently by immunoprecipitation. Proteinase K is extremely active, exhibiting residual activity even in protein sample buffer. A combination of either DFP or PMSF, together with SDS, and heating to at least 80°C as described above was found to efficiently block proteinase K, as well as trypsin activity.

13. The main disadvantage of protease protection assays is that sometimes the protease concentration cannot be increased high enough to obtain a complete digest of exposed polypeptide domains, before encountering leak of protease activity into the lumen of the microsomal vesicles. As a result, exposed but tightly membrane-associated domains may remain undigested. This was observed, for example, with a large portion of the C-terminal domain of the coronavirus glycoprotein E1 *(13)*. Although potentially accessible to proteolytic degradation the intracellular loop region of connexins was found to be relatively resistant to protease digestion *(48,49)*, which resulted in the precipitation of connexin polypep-

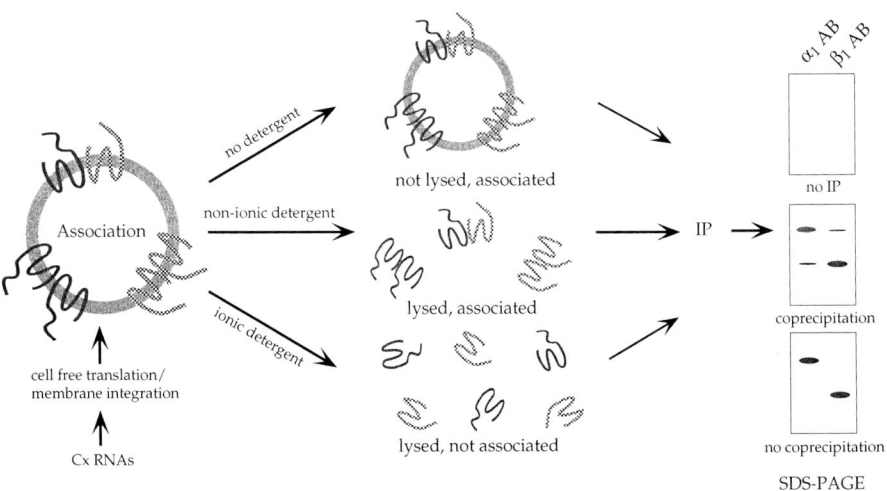

Fig. 8. Schematic representation of a connexin assembly assay. Oligomerization of connexins was analyzed by co-immunoprecipitation. Different connexin isotypes were translated simultaneously in a cell-free translation reaction in the presence of microsomes. Connexin assembly into homo-and hetero-oligomeric complexes was allowed when several connexin polypeptides inserted into the same microsomal vesicle. Microsomes were lysed in nonionic detergent that kept the oligomeric complexes intact. Co-immunoprecipitation using connexin specific antibodies indicated co-assembly of the different connexin isotypes. In control aliquots microsomes were not lysed in detergent. No immunoprecipitation occurred under theses conditions due to the relatively large size of the microsomes. Or microsomes were lysed in SDS that also disrupted the oligomeric connexin complexes. No co-immunoprecipitation occurred under these conditions.

tide fragments that included the intracellular loop domain in such protease protection assays (*see* **Fig. 7** in **ref.** *14*). This result intially misled us to believe that the transmembrane topology of connexins integrated into microsomes may be partially inverted *(50)*. The assumption was revised *(14)*, and the correct transmembrane topology of connexins integrated into microsomes was proven *(16)*.

14. Previously, 3–5 mM (final conc.) oxidized gluthatione (GSSG; Sigma, St. Louis, MO) added to the translation reactions was described to promote proper folding of membrane proteins by maintaining oxidized conditions in the translation reactions *(34,51)*. However, addition of GSSG was not found to be necessary for connexon assembly.

15. Several researchers have used chemical crosslinking to detect connexin–connexin interactions *(39,41,45)*. However, because this technique requires a chemical alteration of the polypeptides that may produce misleading results, especially if the concentration of the connexin polypeptides is low compared to the amount of total protein present in the microsomes, this technique was not further developed for the assembly analysis of cell-free expressed connexins.

16. Antibodies are used in combination with Protein A–Sepharose beads (Sigma). However, the cleanest immunoprecipitations are obtained with connexin-specific antibodies covalently bound to protein G–Sepharose beads (Pharmacia, Piscataway, NJ). To link the antibodies to the beads, 1 mL of protein G bead slurry is washed and incubated in 10 mM sodium phosphate buffer, pH 7.3, with 15 mg of monoclonal antibodies for 1 h at room temperature with continuous rocking. Antibody–bead complexes are washed twice in 0.1 M sodium borate, pH 9.0, and bound antibodies are covalently crosslinked to the Protein A with 20 mM dimethyl pimelimidate · 2HCl (DMP, Pierce, IL), in 0.1 M sodium borate, pH 9.0, for 30 min at room temperature with continuous rocking. The reaction was quenched by washing twice with 0.2 M ethanolamine, pH 8.0, for 1 h. Antibody–bead complexes were resuspended in 1× PBS, 0.01% Thimerosal and stored at 4°C until used.

17. Cell-free expressed connexons were found to be functional by means of channel activity. Single-channel activities were characterized by electrophysiological analysis of channels obtained after fusion of microsomes containing cell-free expressed connexins with planar lipid membranes. This method requires special experimental setup and is not described in this chapter. However, the experimental conditions are described in detail in Buehler et al. *(52)* and Falk et al. *(15)*.

Acknowledgments

The author is grateful to Norton B. Gilula, Ewald Beck, and Heiner Niemann in whose laboratories much of the experimental work leading to the development of the methodology described in this chapter has been obtained. Work in the author's laboratory is supported by the National Institute of Health (Grant R01 GM55725). Matthias M. Falk is a former recipient of a Deutsche Forschungsgemeinschaft fellowship Fa 261/1–1.

References

1. Pfeffer, S. R. and Rothman, J. E. (1987) Biosynthetic protein transport and sorting by the endoplasmic reticulum and Golgi. *Annu. Rev. Biochem.* **56,** 829–852.
2. Rapoport, T. A., Jungnickel, B., and Kutay, U. (1996) Protein transport across the eukaryotic endoplasmic reticulum and bacterial inner membranes. *Annu. Rev. Biochem.* **65,** 271–303.
3. High, S. and Laird, V. (1997) Membrane protein biosynthesis—all sewn up? *Trends Cell Biol.* **7,** 206–210.
4. Johnson, A. E. (1997) Protein translocation at the ER membrane: a complex process becomes more so. *Trends Cell Biol.* **7,** 90–95.
5. Blobel, G. and Dobberstein, B. (1975) Transfer of proteins across membranes. I. Presence of proteolytically processed and unprocessed nascent immunoglobulin light chains on membrane-bound ribosomes of murine myeloma. *J. Cell Biol.* **67,** 852–862.
6. Borel, A. C. and Simon, S. M. (1996) Biogenesis of polytopic membrane proteins: Membrane segments of *p*-glycoprotein sequentially translocate to span the ER membrane. *Biochemistry* **35,** 10,587–10,594.

7. Mothes, W., Heinrich, S. U., Graf, R., Nilsson, I. M., von Heijne, G., Brunner, J., and Rapoport. T. A. (1997) Molecular mechanism of membrane protein integration into the endoplasmic reticulum. *Cell* **89,** 523–533.
8. Jadot, M., Hofmann, M. W., Graf, R., Quader, H., and Martoglio, B. (1995) Protein insertion into the endoplasmic reticulum of permeabilized cells. *FEBS Lett.* **371,** 145–148.
9. Falk, M. M. (2000) Cell-free synthesis for analyzing the membrane integration, oligomerization, and assembly characteristics of gap junction connexins. *Methods* **20,** 165–179.
10. Falk, M. M., Grigera, P. R., Bergmann, I. E., Zibert, A., Multhaup, G., and Beck, E. (1990) Foot-and-mouth disease virus protease 3C induces Specific proteolytic cleavage of host cell histone H3. *J. Virol.* **64,** 748–756.
11. Falk, M. M., Sobrino, F., and Beck, E. (1992) VPg gene amplification correlates with infective particle formation in foot-and-mouth disease virus. *J. Virol.* **66,** 2251–2260.
12. Zibert, A., Maass, G., Strebel, K., Falk, M. M., and Beck, E. (1990) Infectious foot-and-mouth disease virus derived from a cloned full-length cDNA. *J. Virol.* **64,** 2467–2473.
13. Mayer, T., Tamura, T., Falk, M. M., and Niemann, H. (1988) Membrane integration and cellular transport of the coronavirus glycoprotein E1, a class III membrane glycoprotein. *J. Biol. Chem.* **263,** 14,956–14,963.
14. Falk, M. M., Kumar, N. M., and Gilula, N. B. (1994) Membrane insertion of gap junction connexins: polytopic channel forming membrane proteins. *J. Cell Biol.* **127,** 343–355.
15. Falk, M. M., Buehler, L. K., Kumar, N. M., and Gilula, N. B. (1997) Cell-free synthesis and assembly of connexins into functional gap junction membrane channels. *EMBO J.* **10,** 2703–2716.
16. Falk, M. M. and Gilula, N. B. (1998) Connexin membrane protein biosynthesis is influenced by polypeptide positioning within the translocon and signal peptidase access. *J. Biol. Chem.* **273,** 7856–7864.
17. Sambrook, J., Fritsch, E. F., and Maniatis, T. (eds.) (1989) *Molecular Cloning. A Laboratory Manual.* Cold Spring Harbor Laboratory Press, Cold Spring Harbor, NY.
18. Jackson, R. J. and Hunt, T. (1983) Preparation and use of nuclease-treated rabbit reticulocyte lysates for the translation of eukaryotic messenger RNA. *Methods Enzymol.* **96,** 50–74.
19. Erickson, A. and Blobel, G. (1983) Cell-free translation of messenger RNA in a wheat germ system. *Methods Enzymol.* **96,** 38–50.
20. Evans, E. A., Gilmore, R., and Blobel, G. (1986) Purification of microsomal signal peptidase as a complex. *Proc. Natl. Acad. Sci. USA* **83,** 581–585.
21. Walter, P. and Blobel, G. (1983) Preparation of microsomal membranes for cotranslational protein translocation. *Methods Enzymol.* **96,** 84–93.
22. Kozak, M. (1989) The scanning model for translation: an update. *J. Cell Biol.* **108,** 229–241.

23. Krieg, P. A. and Melton, D. A. (1984) Functional messenger RNAs are produced by SP6 *in vitro* transcription of cloned cDNAs. *Nucleic Acids Res.* **12,** 7057–7070.
24. Williams, J. G., Kay, M. and Patient, R. K. (1980) The nucleotide sequence of the major β-globin mRNA from *Xenopus laevis*. *Nucleic Acids Res.* **8,** 4247–4258.
25. Kahn, T. W., Beachy, R. N., and Falk, M. M. (1997) Cell-free expression of a GFP-fusion protein allows protein quantitation *in vitro* and *in vivo*. *Curr. Biol.* **7,** R207–R208.
26. Jackson, R. J., Howell, M. T., and Kaminski, A. (1990) The novel mechanism of initiation of picornavirus RNA translation. *Trends Biochem. Sci.* **15,** 477–483.
27. Gilmore, R. and Blobel, G. (1985) Translocation of secretory proteins across the microsomal membrane occurs through an environment accessible to aqueous perturbants. *Cell* **42,** 497–505.
28. Chavez, R. A. and Hall, Z. W. (1992) Expression of fusion proteins of the nicotinic acetylcholine receptor from mammalian muscle identifies the membrane-spanning regions in the α and γ subunits. *J. Cell Biol.* **116,** 385–393.
29. Thomas, J. O. and Kornberg, D. (1975) An octamer of histones in chromatin and free in solution. *Proc. Natl. Acad. Sci. USA* **72,** 2626–2630.
30. Hart, G. W., Brew, K., Grant, G. A., Bradshaw, R. A., and Lennarz, W. J. (1978) Primary structural requirements for the enzymatic formation of the N-glycosidic bond in glycoproteins. *J. Biol. Chem.* **254,** 9747–9753.
31. Anderson, D. J. and Blobel, G. (1981) *In vitro* synthesis, glycosylation, and membrane insertion of the four subunits of *Torpedo* acetylcholine receptor. *Proc. Natl. Acad. Sci. USA* **78,** 5598–5602.
32. Rosenberg, R. L. and East, J. E. (1992) Cell-free expression of functional *Shaker* potassium channels. *Nature* **360,** 166–169.
33. Kvist, S., Wiman, K., Claesson, L., Peterson, P. A., and Dobberstein, B. (1982) Membrane insertion and oligomeric assembly of HLA-DR histocompatibility antigens. *Cell* **29,** 61–69.
34. Qu, D. and Green, M. (1995) Folding and assembly of a human MHC class II molecule in a cell-free system. *DNA Cell Biol.* **14,** 741–751.
35. Sayer, J. T. and Doyle, D. (1990) Assembly of a heterooligomeric asialoglycoprotein receptor complex during cell-free translation. *Proc. Natl. Acad. Sci. USA* **87,** 4854–4858.
36. Yilla, M., Doyle, D., and Sawyer, J. T. (1992) Early disulfide bond formation prevents heterotypic aggregation of membrane proteins in a cell-free translation system. *J. Cell Biol.* **188,** 245–252.
37. Falk, M. M., Buehler, L. K., Kumar, N. M., and Gilula, N. B. (1998) Cell-free expression of functional gap junction channnels, in *Gap Junctions* (Werner, R., ed.), IOS Press, Amsterdam, Berlin, Oxford, Tokyo, Washington DC, pp. 135–139.
38. Yeager, M., Unger, V., and Falk, M. M. (1998) Synthesis, assembly, and structure of gap junction intercellular channels. *Curr. Opin. Struct. Biol.* **8,** 517–524.
39. Musil, L. S. and Goodenough, D. A. (1993) Multisubunit assembly of an integral plasma membrane channel protein, gap junction connexin43, occurs after exit from the ER. *Cell* **74,** 1065–1077.

40. Kistler, J., Goldie, K., Donaldson, P., and Engel, A. (1994) Reconstitution of native-type noncrystalline lens fiber gap junctions from isolated hemichannels. *J. Cell Biol.* **126,** 1047–1058.

41. Cascio, M., Kumar, N. M., Safarik, R., and Gilula, N. B. (1995) Physical characterization of gap junction membrane connexons (hemi-channels) isolated from rat liver. *J. Biol. Chem.* **270,** 18,643–18,648.

42. Stauffer, K. A., Kumar, N. M., Gilula, N. B., and Unwin, N. (1991) Isolation and purification of gap junction channels. *J. Cell Biol.* **115,** 141–150.

43. Harlow, E. and Lane, D. (ed.) (1988) *Antibodies. A Laboratory Manual.* Cold Spring Harbor Laboratory Press, Cold Spring Harbor, NY.

44. Zhang, J.-T., Chen, M., Foote, C. I., and Nicholson, B. J. (1996) Membrane integration of *in vitro*-translated gap junctional proteins: co- and posttranslational mechanisms. *Mol. Biol. Cell* **7,** 471–482.

45. Ahmad, S., Diez, J. A., George, H., and Evans, W. H. (1999) Synthesis and assembly of connexins *in vitro* into homomeric and heteromeric functional gap junction hemichannels. *Biochem. J.* **339,** 247–253.

46. Sørensen, S. B., Sørensen, T. L., and Breddam, K. (1991) Fragmentation of proteins by *S. aureus* strain V8 protease. *FEBS Lett.* **294,** 195–197.

47. Morimoto, T., Aprin, M., and Gaetani, S. (1983) Use of proteases for the study of membrane insertion. *Methods Enzymol.* **96,** 121–150.

48. Zimmer, D. B., Green, C. R., Evans, W. H., and Gilula, N. B. (1987) Topological analysis of the major protein in isolated intact rat liver gap junctions and gap junction-derived single membrane structures. *J. Biol. Chem.* **262,** 7751–7763.

49. Milks, L. C., Kumar, N. M., Houghten, R., Unwin, N., and Gilula, N. B. (1988) Topology of the 32-kd liver gap junction protein determined by site-directed antibody localizations. *EMBO J.* **7,** 2967–2975.

50. Falk, M. M., Kumar, N. M., and Gilula, N. B. (1995) Biosynthetic membrane integration of connexin proteins. *Prog. Cell Res.* **4,** 319–322.

51. Marquardt, T., Hebert, D. N., and Helenius, A. (1993) Posttranslational folding of influenza hemagglutinin in isolated endoplasmic reticulum-derived microsomes. *J. Biol. Chem.* **268,** 19,618–19,625.

52. Buehler, L. K., Stauffer, K. A., Gilula, N. B., and Kumar, N. M. (1995) Single channel behavior of recombinant β_2 gap junction connexons reconstituted into planar lipid bilayers. *Biophys. J.* **68,** 1767–1775.

6

Biochemical Analysis of Connexon Assembly

Judy K. VanSlyke and Linda S. Musil

1. Introduction

The formation of gap junctions is a multistep process that begins with synthesis in the membrane of the endoplasmic reticulum (ER) of connexins (Cx), members of a highly homologous family of polytopic integral membrane proteins (*see* **Fig. 1**, *no. 1*) *(1)*. The first stage in gap junction assembly is the noncovalent oligomerization of six connexin monomers into an annular structure known as a connexon or hemichannel (*see* **Fig. 1**, *no. 2*). In many *(2,3)* but apparently not all *(4)* situations, this event takes place after exit from the ER within an intracellular compartment that is most likely the trans-Golgi network. The connexon complex is then transported to the cell surface (*see* **Fig. 1**, *no. 3*), where it associates head-to-head with a connexon on the surface of an adjacent cell to form an intercellular channel (*see* **Fig. 1**, *no. 4*). Such channels then cluster at up to $10,000/\mu m^2$ to form gap junctional "plaques," the endpoint in gap junction assembly (*see* **Fig. 1**, *no. 5*). An emerging concept is that connexon assembly is not simply a spontaneous default process but instead shows a high degree of selectivity, as evidenced by the finding that only certain types of connexins can co-oligomerize to form heteromeric connexons *(5)* and the ability of osteoblastic cells to assemble endogenously expressed Cx43, but not Cx46 *(3)*. Determining the mechanism and regulation of connexon formation is therefore a fundamental part of understanding gap junction biosynthesis and function.

Because gap junctions are present in almost all types of animal cells, a large variety of tissues as well as cultured cell lines can be used as starting material for the purification of connexons. Members of the connexin family exhibit tissue-specific expression, a factor that must be taken into account when selecting a system for study. For instance, rat liver is a good source of Cx32 but not of

From: *Methods in Molecular Biology, vol. 154: Connexin Methods and Protocols*
Edited by: R. Bruzzone and C. Giaume © Humana Press Inc., Totowa, NJ

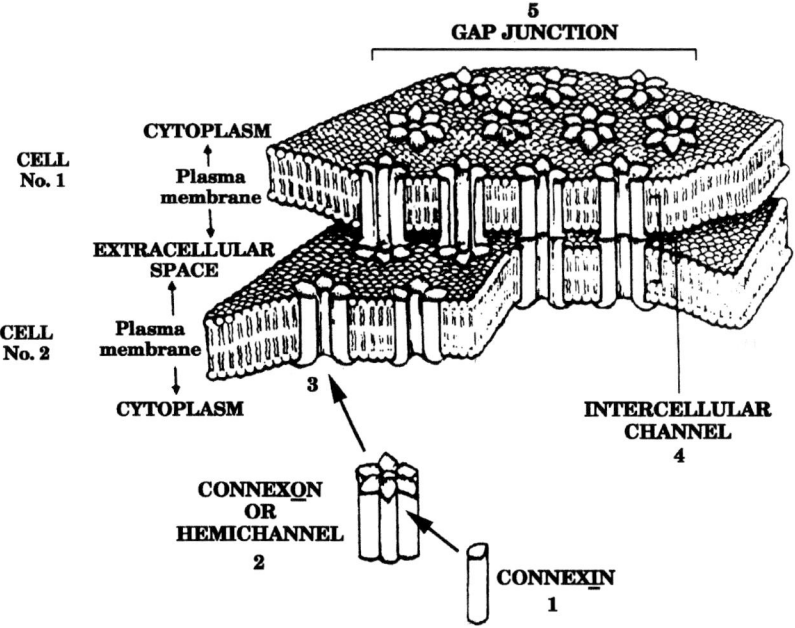

Fig. 1. Schematic of multistep assembly of gap junctional plaques. Assembly begins with the biosynthesis of integral membrane proteins of the connexin family (*1*). Multiple connexins oligomerize intracellularly to form a connexon (also known as hemichannel) (*2*) prior to delivery to the plasma membrane (*3*). Two connexons on apposing cell surfaces associate to form an intercellular channel (*4*) and channels then cluster into tightly packed arrays classified as gap junctional plaques (*5*).

Cx43. Conversely, cardiac muscle tissue and pregnant rat myometrium (day of labor) are rich in Cx43, but not Cx32.

The first described and still most widely used methods for isolating and analyzing connexons are chemical crosslinking and sucrose gradient velocity sedimentation. In this chapter, we present protocols that we have developed to purify connexons composed of either Cx43 or Cx32 (*2*). We will also refer to modifications of these techniques developed by other investigators, with the caveat that we do not have extensive personal experience with these variations. The tissues that we have successfully isolated connexons from include rat liver (Cx32), chicken lens epithelium, and rat myometrium (both Cx43). These techniques have also been used with a wide variety of tissue culture cells that either endogenously express connexins or that have been transfected with connexin-encoding cDNAs, as well as with *Xenopus* oocytes programmed with connexin cRNAs (*2*). Continuous cell lines that we routinely use that express measurable amounts of endogenous connexons include normal rat kidney (NRK), Chinese hamster ovary (CHO), and mouse sarcoma (S180) cells for Cx43 and rat hepatoma cells (MH_1C_1) for Cx32.

2. Materials

2.1. Reagents

Unless otherwise noted, all reagents are available from Sigma.

2.1.1. Labeling Medium for Low CO_2 Incubators

1. Sterile tissue culture stock solutions: 200 mM L-glutamine (100×), 79 mM L-leucine (200×), 80 mM L-lysine HCl (200×), 5000 U/mL of penicillin–streptomycin (50×), 7.5% sodium bicarbonate, 1 M HEPES 4-(2-hydroxyethyl)-1-piperazifneethanesulfonic acid buffer solution, dialyzed fetal calf serum, 1 M NaOH.
2. ^{35}S-Express label: Express™ Methionine/Cysteine Protein Labeling Mix, ^{35}S, >1000 Ci (37.0 TBq)/mmol (NEN cat. no. NEG072).
3. 2× Matlin's medium stock: Dissolve the powder for 1 L of deficient Eagle minimum essential medium (MEM) tissue culture medium (–methionine, –lysine, –leucine) in 455 mL of double-distilled H_2O (ddH_2O). Add 5 mL of 200× leucine stock, 5 mL of 200× lysine stock, 15 mL of 1 M HEPES buffer solution, and 20 mL of 50× penicillin–streptomycin. Filter sterilize and store at 4°C.
4. 1× Matlin's medium (–methionine) stock: To make 500 mL of medium, mix 250 mL of 2× Matlin's medium stock, 250 mL of sterile ddH_2O, 2.35 mL of 7.5% sodium bicarbonate, and 1 mL of 1 M NaOH. The final pH should be 7.2–7.3.
5. Matlin's starve/labeling medium: Supplement 1× Matlin's medium with 5% dialyzed fetal calf serum and 2 mM glutamine.

2.1.2. Isolation of Connexons from Tissue or Tissue Culture Cells

1. 20% TX-100: Dilute 2 mL of Triton X-100 (TX-100) in 8 mL of ddH_2O by rotating at room temperature until it is a homogeneous mixture. Filter through a Nalgene SFCA 0.2-µm filter (syringe type with 25-mm membrane). Prepare fresh weekly.
2. 20× NEM: 200 mM N-Ethylmaleimide in ddH_2O. Heat to 65°C for 15 s (no longer). Prepare and add to buffers within an hour of use.
3. PMSF: 100 mM phenylmethylsulfonyl fluoride in ethanol. Can be stored at –20°C for up to a week. Add to buffers at 200 µM (or 1 mM where specified) within an hour of use.
4. 50× Leupeptin–soybean trypsin inhibitor (SBTI): Dissolve 10 mg of leupeptin in 1 mL of ddH_2O and 25 mg of SBTI (type II-S) in 1 mL of PBS. Combine and divide into 100-µL aliquots. Store at –20°C.
5. Protein assay kit: BCA Protein Assay Reagent Kit (Pierce cat. no. 23225).
6. PBS stock: Dulbecco's phosphate-buffered saline, without $CaCl_2$ or $MgCl_2$ (Gibco-BRL cat. no. 14190-151).

7. hHB buffer: 5 mM HEPES, 1 mM EDTA, 1 mM EGTA, adjusted to pH 8.2. Supplement with 10 mM NEM and 200 μM PMSF immediately before use.
8. Incubation buffer: 136.8 mM NaCl, 5.36 mM KCl, 0.336 mM Na$_2$HPO$_4$, 0.345 mM KH$_2$PO$_4$, 0.8 mM MgSO$_4$, 2.7 mM CaCl$_2$, 20 mM HEPES, adjusted to pH 7.5, filter sterilized, and stored at 4°C.
9. Incubation buffer*: Supplement incubation buffer with 10 mM NEM and 200 μM PMSF.
10. Incubation buffer**: Supplement incubation buffer with NEM, PMSF, and leupeptin–SBTI.

2.1.3. Chemical Crosslinking of Connexons

1. Crosslinking reagents: Dissolve 7.5 mg of DSP (dithio-*bis*[succinimidyl propionate]) (Pierce) in 600 μL of dimethyl sulfoxide (DMSO) or 50 mg of EGS (ethylene glycol-*bis*[succinimdylsuccinate]) (Pierce) in 1 mL of DMSO by rigorous vortex-mixing plus a 1-min incubation at 37°C. A few particles may remain insoluble.
2. 1 M Glycine, pH 9.2, and 1 M glycine, pH 7.2: Prepare solutions fresh each time or filter sterilize and store at room temperature. Remove small aliquots (sterilely) as needed.

2.1.4. Velocity Sedimentation of Connexons in Sucrose Gradients

1. Incubation buffer and 20% TX-100: Described in **Subheading 2.1.2.**
2. Incubation buffer + TX-100: Incubation buffer supplemented with 20% TX-100 to a final concentration of 0.1%.
3. 20% (w/w) sucrose solution: Dissolve 20 g of sucrose in incubation buffer + TX-100 and bring the final volume to 100 mL. Aliquot and store at –20°C.
4. 5% (w/w) sucrose solution: Dilute 20% sucrose solution with incubation buffer + TX-100 to a final concentration of 5% sucrose.

2.1.5. Immunoprecipitation Reactions

1. 20× NEM, PMSF, and 20% TX-100: Described in **Subheading 2.1.2.**
2. 10% SDS: Dissolve 0.5 g of sodium dodecyl sulfate (purified, BDH Laboratory Supplies) completely in 4.5 mL of ddH$_2$O and filter through a Nalgene SFCA 0.2-μm filter (syringe type with 25-mm membrane). Prepare fresh weekly.
3. 10% BSA: Dissolve 10 g of bovine serum albumin, fraction V, in 90 mL of ddH$_2$O, and bring up to 100 mL final volume. Aliquot into 5 mL fractions and store at –20°C.
4. Immunoprecipitation buffer: 100 mM NaCl, 20 mM sodium borate, 15 mM EDTA, 15 mM EGTA, 0.02% NaN$_3$, adjusted to pH 8.5, filter-sterilized, and stored at 4°C.
5. Immunoprecipitation dilution buffer: Supplement immunoprecipitation buffer with 0.7% BSA, 1.2% TX-100, 10 mM NEM, and 200 μM PMSF within an hour of use.
6. Wash buffer 1: Supplement immunoprecipitation buffer with 0.5% BSA, 0.5% TX-100, 0.1% SDS, 10 mM NEM, and 200 μM PMSF.

7. Wash Buffer 2: Supplement immunoprecipitation buffer with 0.05% TX-100, 0.1% SDS, 10 mM NEM, and 200 µM PMSF.
8. 50% Slurry of Protein A conjugated to Sepharose beads: Protein A-Sepharose 4B® (Zymed cat. no. 10-1041).

2.1.6. SDS-PAGE and Immunoblot Analysis

1. 10× Sample loading buffer: Mix 2.3 mL of glycerol; 800 µL of ddH$_2$O; 124 µL of 0.5 M Tris-HCl, pH 6.8; and 0.8 g of electrophoresis-grade SDS. Incubate at 60°C until homogeneous. Add bromophenol blue (until dark blue), mix, aliquot in small fractions, and store at –20°C.
2. 1× Sample loading buffer: Warm the 10× stock to 100°C until melted and dilute 10-fold in ddH$_2$O. For reducing conditions, add 20 µL of β-mercaptoethanol per milliliter of 1× solution immediately before using.
3. 10% SDS: Dissolve 10 g of electrophoresis-grade SDS in 100 mL of ddH$_2$O.
4. Electrophoresis gel components: 30% Acrylamide/Bis solution, 37.5:1 (2.6% C), Bio-Rad cat. no. 161–0158; ammonium persulfate and TEMED (Bio-Rad); resolving gel buffer, 1.5 M Tris-HCl, pH 8.8; stacking gel buffer, 0.5 M Tris-HCl, pH 6.8.
5. Running Gel Buffer: 24.8 mM Tris base, 192 mM glycine, 0.1% SDS.
6. Fix/destain solution: 10% Glacial acetic acid, 20% methanol.
7. Transfer buffer: 24.8 mM Tris base, 192 mM glycine, 20% methanol. Make a 10× stock of Tris and glycine and store at room temperature. Dilute to 1× and add methanol just before use.

2.2. Apparatus

1. Small household refrigerator.
2. N-Con Bod-Cubator (N-Con Systems, Larchmont, NY).
3. Barnant 100 Thermocouple Thermometer, model no. 600–2820 (JKT) (Barnant, Barrington, IL).
4. Tabletop centrifuge with swinging bucket rotor (Beckman GPR).
5. Dounce tissue grinder with type A pestle (Fisher), polypropylene pellet pestle (Kontes Glass cat. no. 749521-1500).
6. Tabletop ultracentrifuge Optima TLX and TLA100.3 rotor (Beckman), microcentrifuge polyallomer tubes (Beckman cat. no. 357448).
7. L8M preparative ultracentrifuge and SW60 Ti rotor (Beckman), ultraclear centrifuge tubes 7/16 × 2 3/8 in. (11 × 60 mm) (Beckman cat. no. 344062).
8. Small gradient maker SG5 (Hoefer).
9. Hand-held refractometer, Leica cat. no. 10431 (VWR).
10. Mini-Protean II Electrophoresis Cell (Bio-Rad).
11. Trans-Blot Electrophoretic Transfer Cell with plate electrodes (Bio-Rad cat. no. 170-3946).
12. X-OMAT film (Kodak) or PhosphorImaging system (PhosphorImager™ 445 SI, Molecular Dynamics).

3. Methods

3.1. Triton X-100 Solubilization of Connexons

The first step in the biochemical detection of connexons by either chemical crosslinking or sucrose gradient velocity sedimentation is to solubilize these species from membranes in a manner that faithfully preserves their oligomeric state. In principle, this could be achieved by either dissociating fully assembled gap junctional plaques (*see* **Fig. 1**, *no. 5*) into individual connexons or by capturing nascent connexons (*see* **Fig. 1**, *no. 2* and *no. 3*) prior to incorporation into junctional plaques. In the cell types we have examined, the first approach is of limited value. This is because most gap junctional plaques are exceptionally resistant to solubilization in common nondenaturing detergents, including Triton X-100 (TX-100), Triton X-114 (TX-114), octylpolyoxyethylene (POE), and octylglucoside (OG) when used in physiological salt solutions at 4°C *(6)*. Harsher detergent treatments cause extensive breakdown of gap junctional plaques to individual connexin subunits rather than into connexon hexamers. Two situations in which connexons have been generated from fully assembled gap junctional plaques have been described in the literature. First, Kistler and colleagues have demonstrated that gap junctional plaques purified from mature lens fiber cells can be dissociated into individual connexons and connexon pairs with TX-100 and certain other mild detergents (POE, OG, maltopyranoside) *(7)*. Although the basis for the anomalous detergent sensitivity of fiber gap junctions is unknown, a likely possibility is that the unique lipid composition of lens fiber cells changes the extractability properties of the gap junctional channels embedded in them. Second, Stauffer et al. have reported that gap junctional plaques purified from rat liver could be dissociated into hexameric connexons in a mixture of 2 M NaCl, 10 mM EDTA, 100 mM dithiothreitol (DTT), 5% dodecyl maltoside, and 100 mM glycine, pH 10, at 4°C *(8)*. In our hands, however, this procedure is not universally applicable in that it results in a very low yield of connexons from Cx43-expressing tissue culture cells.

Because of the difficulties we have encountered in generating connexons from fully assembled gap junctions, we usually analyze connexons prior to their incorporation into plaques. The overall strategy is to lyse cells in the presence of TX-100 at 4°C, conditions under which gap junctional plaques remain insoluble and can be removed by high-speed centrifugation *(6)*. The supernatant fraction contains both newly synthesized connexin monomers and connexons in a state suitable for subsequent analysis by chemical crosslinking or sucrose gradient velocity sedimentation. We describe techniques optimized for the solubilization of connexons from monolayer tissue culture cells as well as from two types of animal tissue (Cx32 from rat liver and Cx43 from pregnant rat myometrium). Similar results were obtained with material processed immediately after harvesting or after snap-freezing in liquid nitrogen and storage at −80°C. All manipulations should be conducted at 4°C and solutions should be prechilled. Buffers and tubes should be sterilized as added insurance against artifactual proteolytic degradation.

3.1.1. TX-100 Solubilization of Tissue Culture Cells

Monolayer tissue culture cells that have been radioactively labeled at either 37°C or 20°C or unlabeled cells can be used (*see* **Note 1**).

1. Place the tissue culture dishes on a metal plate on ice (never allow the plates to warm >20°C).
2. Wash the monolayers 3× with incubation buffer* and then scrape the cells from the dish with a rubber policeman in more incubation buffer*. Rinse the dish once more with Incubation Buffer and add this wash to the cell suspension.
3. Centrifuge at 180*g* in a tabletop centrifuge for 10 min at 4°C.
4. Resuspend the cell pellet in incubation buffer** (1 mL/confluent 60-mm dish; scale volume up or down proportionately if a dish of a different size is used) and passage through a 25-gauge needle attached to a 1-mL syringe multiple times to completely homogenize the cell suspension.
5. Add TX-100 from a 20% stock solution to bring the homogenate to a final concentration of 1% TX-100 and vortex-mix well. Incubate on ice 30 to 40 min.
6. Centrifuge the extract at 100,000*g* in an ultracentrifuge for 50 min at 4°C. This is equivalent to 43,000 rpm in a TLA100.3 rotor in a tabletop Beckman ultracentrifuge. Collect the TX-100-solubilized extract without disturbing the pellet and store on ice until further analysis.

3.1.2. TX-100 Solubilization of Tissue

Although conceptually identical to solubilization of connexons from tissue culture cells, quantitative solubilization of connexons from animal tissue requires consideration of three additional factors. First, an initial tissue homogenate must be made. If the tissue is easily macerated (e.g., liver) or if small amounts of material are to be utilized, polypropylene pestles that fit into 1.5-mL microcentrifuge tubes can be used. Alternatively, a Dounce-type homogenizer with a type A pestle can be employed. Second, the concentration of connexin protein in the TX-100-solubilized extract must be high enough to be readily detectable over any nonspecific background by the antibodies used for immunoblotting. In the case of liver, this was achieved by removing cytosolic components prior to treatment of the membrane fraction with TX-100. With pregnant rat myometrium, the levels of Cx43 expression and the specificity of the anti-Cx43 antibodies were sufficient to allow addition of TX-100 directly to the whole tissue homogenate. Lastly, the amount of TX-100 added to the tissue must be in excess of that required for complete solubilization of membranes.

3.1.2.1. RAT LIVER TISSUE

1. Remove the liver from a 35-d-old rat and use either immediately or after snap-freezing in liquid nitrogen and storage at –80°C. Weigh 1.5–2 g of the fresh or frozen liver tissue (about one third of the entire organ). Rinse the tissue with cold PBS and then cold hHB buffer to remove as much blood as possible and place in a Dounce homogenizer with 9 mL of hHB.
2. Homogenize with 20 strokes of a type A pestle on ice and let sit on ice for 10 min. Filter homogenate through 16 layers of cheesecloth.
3. Centrifuge the filtered material at 100,000g at 4°C for 30 min. This is most easily achieved using the TLA100.3 rotor in a Beckman tabletop ultracentrifuge. Remove and discard the supernatant and resuspend the membrane pellet in 5 mL of chilled incubation buffer*.
4. Remove a small aliquot of the membrane fraction and perform a protein assay as follows. To 160 µL of the membrane sample, add 600 µL of ddH$_2$O and 40 µL of 10% SDS. Vortex-mix and incubate at room temperature for 20 min. Centrifuge the solution at 15,000 rpm (~16,000g) in a microcentrifuge for 10 min. Measure the protein concentration of the supernatant using a SDS-compatible assay, such as the Pierce BCA kit. A concentration of ~15 mg of protein/mL is expected.
5. Dilute the membrane fraction with additional incubation buffer* to a final concentration of 1.5 mg of protein/mL of buffer.
6. Bring the sample to 1.2% TX-100 (final concentration) with 20% TX-100 stock solution and vortex-mix well, taking care not to allow the solution to warm. Incubate on ice for 40 min. Extraction of membrane proteins is optimal at this TX-100/protein ratio.
7. Centrifuge the detergent-treated material at 100,000g for 50 minutes at 4°C (43,000 rpm in the TLA100.3 rotor). Collect the TX-100-solubilized supernatant and save on ice.
8. Perform another protein assay on the TX-100-solubilized extract using the Pierce BCA kit. Dilute 100 µL of extract with 70 µL of ddH$_2$O to make the TX-100 concentration 0.7%, the maximum compatible with this protein assay. Use between 10 and 75 µL of the diluted protein sample in the assay and determine the protein concentration against a standard of BSA in 0.7% TX-100. One rat liver should yield between 30 and 150 mg of TX-100-solubilized protein.
9. Dilute the TX-solubilized extract with enough incubation buffer*, supplemented with TX-100, to bring the final concentration to 0.5 mg protein/mL in 1% TX-100.

3.1.2.2. RAT MYOMETRIAL TISSUE (DAY 23 OF PREGNANCY)

1. After removal of endometrium, homogenize as thoroughly as possible ~ 100 mg of uterine tissue in 1 mL of incubation buffer* (with 1 mM PMSF) in a 1.5 mL microcentrifuge tube using a polypropylene pestle on ice.
2. Bring the homogenate to 1.2% (final concentration) TX-100 and vortex-mix well, taking care not to allow the sample to reach room temperature. Incubate on ice for 40 min. This amount of Triton is sufficient for optimal extraction of membrane proteins.

3. Centrifuge the sample at 100,000g for 50 min at 4°C (43,000 rpm in a TLA100.3 rotor in a Beckman tabletop ultracentrifuge). Harvest the TX-100-solubilized fraction and save on ice.

4. Perform a protein assay on the TX-100-solubilized extract as described previously for liver samples. Triton solubilization of 100 mg of tissue should yield about 2.5–3 mg of protein. Dilute the sample to a final concentration of 1.5 mg of protein/mL in incubation buffer supplemented with enough TX-100 to bring the final detergent concentration to 1%.

3.2. Chemical Crosslinking of Connexons

The oligomeric state of connexins in TX-100-solubilized extracts generated as described in **Subheading 3.1.** can be analyzed using chemical crosslinkers *(2)*. We have successfully used the homobifunctional, amine-reactive reagents EGS (ethylene glycol-*bis*[succinimidylsuccinate]), DSP (dithio-*bis*[succinimidyl propionate]), DTSSP (3,3'-dithio-*bis*[sulfosuccinimidyl propionate]), and BS^3 (*bis*[sulfosuccinimidyl]suberate) to covalently link connexin monomers within a connexon. The use of DMS (dimethylsuberimidate · HCl) has been reported by Cascio et al. *(9)*. Crosslinks formed by DSP and its water-soluble analog DTSSP have thiol bridges that are easily cleaved by reducing agents (*see* **Note 2**), allowing identification of the individual connexin components within an oligomeric complex. Crosslinks generated by EGS are reversible by prolonged incubation in 1 M hydroxylamine, pH 8.5, whereas BS^3 is uncleavable. The concentration of crosslinker, as well as the length and temperature of the crosslinking reaction, are critical parameters and should not be varied. Too little crosslinking results in a ladder of partially crosslinked products whereas over crosslinking can cause virtually all of the connexin in the sample to become incorporated into heterogeneous, nonspecific high molecular mass (>200 kDa) complexes that do not reflect assembly into connexons.

Completely crosslinked, fully assembled Cx43 connexons migrate on a 4–11% polyacrylamide gradient gel (*see* **Note 8**) slightly faster than a 200-kDa molecular weight marker (*see* **Fig. 2**), well resolved from any remaining unassembled monomer at ~40 kDa. The M_r of Cx32-containing connexons is ~150 kDa and of the monomers, is ~27 kDa. We have observed that in contrast to Cx43, Cx32 is difficult to completely crosslink in a connexon and instead tends to yield a ladder of crosslinked intermediates (*see* **Fig. 2**). We suspect that Cx32 hemichannels are not completely stable in TX-100 solutions and that partial dissociation of connexons and/or the fact that Cx32 is prone to aggregation may account for the crosslinking results we as well as other investigators *(9)* have obtained (*see* **Note 3**).

The protocol described next is for optimal crosslinking of fully assembled connexons. We have empirically found that whereas either DSP or EGS is suitable for Cx43, Cx32 connexons are most efficiently crosslinked with EGS.

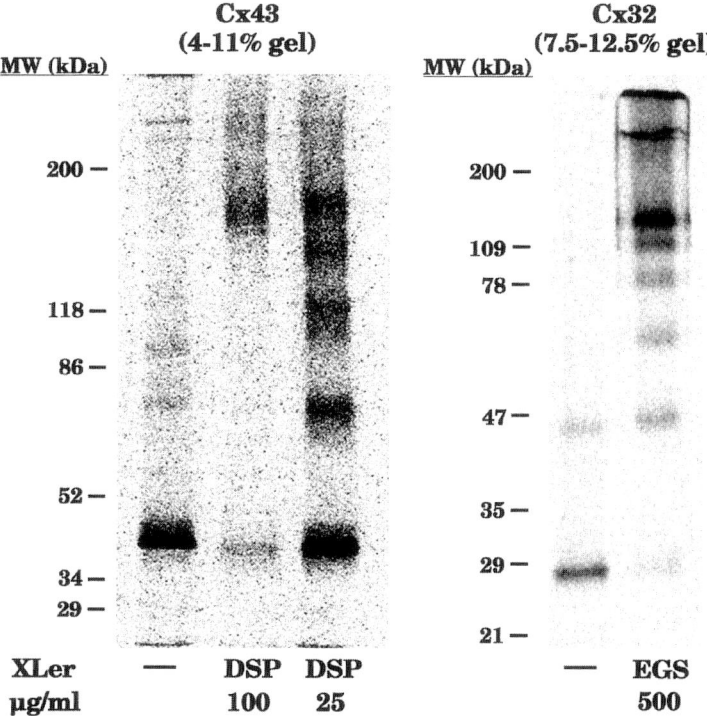

Fig. 2. Crosslinking analysis of connexons from tissue culture cells. NRK cells expressing either Cx43 endogenously or Cx32 exogenously were metabolically labeled at 20°C. Connexons in TX-100-solubilized cell extracts were chemically crosslinked with either DSP or EGS and visualized by SDS-PAGE on gradient gels after immuno-precipitation with affinity purified anti-connexin antibodies. The final concentration of crosslinker in µg/mL (in DMSO) is given below each gel lane; mock crosslinking (–) was acheived with DMSO alone. Molecular mass markers are indicated on the left of the gels in kilodaltons (kDa).

1. Dilute one volume of TX-100-solubilized extract, prepared as described in **Subheading 3.1.**, with two volumes of cold incubation buffer (solution is now 0.3% TX-100), mix, and then divide the sample into two equal portions. The amount of soluble extract required for recovery of easily detectable quantities of connexons is addressed in **Note 4**.

2. Prepare a fresh stock in DMSO of either DSP (12.5 mg/mL, 125× stock) or EGS (50 mg/mL, 100× stock). Add crosslinker to one of the duplicate samples (final concentration = 100 µg/mL for DSP; 500 µg/mL for EGS), vortex-mix well immediately, and place on ice. Add an identical volume of DMSO to the other duplicate for a mock crosslinked control. Incubate the samples on ice for 30 min with occasional vortex-mixing.

3. Stop the reaction by adding to all samples 20 µL/mL of 1 M glycine, pH 9.2, vortex-mix, and incubate on ice for 30 min. The glycine provides excess amino groups to competitively inhibit protein–protein crosslinking and the higher pH increases the rate of reagent hydrolysis.

4. Add 10 µL/mL of 1 M glycine, pH 7.2, to each sample to restore the pH and vortex-mix. Store samples on ice until proceeding with either immunoprecipitation or chloroform–methanol protein precipitation, as described in **Subheading 3.4.**

3.3. Velocity Sedimentation of Connexons in Sucrose Gradients

This procedure was adopted from studies of the assembly of the nicotinic acetylcholine receptor *(10)*. It takes advantage of the fact that a 5–20% (w/w) linear sucrose gradient is essentially isokinetic; that is, a particle moves in the gradient at a constant velocity such that the distance through which it sediments is directly proportional to its sedimentation coefficient. As expected for a monomer that has a predicted molecular mass of 43 kDa, unassembled Cx43 is recovered in the 5S region (containing ~9–11% sucrose) after centrifugation on a 5–20% sucrose gradient. Cx43 connexons (either with or without prior crosslinking) migrate at the 9S position (~14–16% sucrose), similar to the comparably sized pentameric nicotinic acetylcholine receptor *(2)*. No Cx43 is recovered in the 9S position from cells in which connexon assembly was blocked, demonstrating that velocity sedimentation does not induce artifactual post-lysis aggregation of connexins.

While the gradient profile for Cx43 *(2)* or Cx46 *(3)* shows a clean separation of monomers from connexons, this is not true for Cx32-containing extracts analyzed identically. Cx32 is recovered from the gradient in a single peak that spans from the 5S to the 9S regions. If Cx32 is crosslinked before sucrose gradient velocity sedimentation, the completely crosslinked hexamers are recovered in the fractions containing ~13–14.5% sucrose and the monomers in ~8–10% sucrose, in keeping with the expected molecular weights of these species. Partial complexes containing two to five Cx32 molecules are recovered in the intermediate fractions. The lack of resolution of Cx32 into two distinct peaks after sucrose gradient sedimentation therefore appears to be due to the same instability and/or aggregation phenomena that are responsible for the inability to completely crosslink multimeric forms of Cx32 to fully assembled, hexameric connexons (*see* **Subheading 3.2.**).

1. Prepare stock solutions containing either 5% or 20% sucrose in incubation buffer and 0.1% TX-100. With both of the stopcocks closed, add 2 mL of the 20% sucrose solution to the chamber of a 5-mL gradient maker closest to the outflow tube. Add 2 mL of the 5% sucrose solution to the other chamber. Open the stopcock to the outflow tube and then the stopcock separating the two chambers and pour a linear sucrose gradient into an 11 × 60 mm ultra-clear tube (Beckman cat.

no. 344062) designed for an SW60 rotor. The sucrose solution is continuously mixed during gradient preparation on a stirplate by means of a stirbar inserted into the 20% sucrose solution. Chill the tubes to 4°C, being careful not to disturb the sucrose gradient.

2 Carefully overlay each sucrose gradient with approximately 0.5–0.75 mL of TX-100-solubilized extract, prepared as described in **Subheading 3.1.** Solubilized protein from one confluent 60-mm dish of tissue culture cells is roughly the maximum amount of sample that can be loaded onto a 4-mL gradient without affecting the resolution of connexin species, whereas ~ 1 mg of TX-100-solubilized extract from either liver or myometrium has been successfully subjected to gradient analysis (*see* **Note 4**).

3. Centrifuge the gradients in an SW60 swinging bucket rotor in a Beckman ultracentrifuge at 49,000 rpm for 12 h at 4°C (i.e., 250,000g; 1.15×10^{12} radians^2/s). Allow acceleration to proceed slowly over 3 min and deceleration to occur without braking.

4. Puncture the bottom of the centrifuge tube with an 18-gauge needle just slightly to the side and above the pellet. Collect the sucrose solution dropwise into microcentrifuge tubes on ice, about 400 μL of sample per tube.

5. Measure the percent sucrose in each fraction using a hand-held refractometer (Leica cat. no. 10431). Store the samples on ice until further analysis by immunoprecipitation or chloroform–methanol protein precipitation as described in **Subheading 3.4.2.** If the collected fractions are to be crosslinked, read **Note 5**.

3.4. SDS-PAGE Analysis of Connexons

After chemical crosslinking or sucrose gradient velocity sedimentation, monomeric and assembled connexin species must be immunoprecipitated and/or analyzed by Western blotting using connexin-specific antibodies (*11*). Immunoprecipitants of connexins metabolically labeled with [^{35}S] methionine are detected after SDS-polyacrylamide gel electrophoresis by autoradiography or with a PhosphorImager. Unlabeled connexins are analyzed by immunoblotting either with or without prior immunoprecipitation (*see* **Note 9**). In the latter case, the presence of substantial amounts of TX-100 in the samples and their large volumes hinder analysis on minigels. We therefore precipitate the proteins in such samples using the method of Wessel and Flugge (*12*) prior to SDS-PAGE. This simple procedure yields near quantitative recovery of proteins and removes salts and detergents.

3.4.1. Immunoprecipitation

1. Denature samples for immunoprecipitation by adding SDS to a final concentration of 0.6%, vortex-mix, and heat to 100°C for 3 min or (depending on the type of connexin to be analyzed) incubate at room temperature for 30 min (*see* **Note 6**). Dilute the denatured samples with 2.5 volumes of immunoprecipitation dilution buffer, centrifuge at maximum speed in a microcentrifuge (15,000 rpm, ~16,000g)

or (for larger samples) in a tabletop low-speed centrifuge in a swinging bucket rotor for 10 min and transfer the supernatant to a new tube.

2. If necessary, preclear samples by incubating for 2 h with 30 μL of a 50% slurry of Protein A conjugated to Sepharose 4B while rotating at 4°C. Centrifuge the samples as described in step 1 and transfer the supernatant to a new tube. This step reduces the nonspecific background of immunoprecipitants and is used when samples are prepared from whole tissue or when Cx32 is immunoprecipitated from some cell types.

3. Add anti-connexin antibodies (either affinity purified or as crude serum) capable of specifically immunoprecipitating SDS-denatured protein and incubate while rotating overnight at 4°C. For further information concerning antibodies, *see* **Note 7**.

4. Add 30 μL of Protein A–Sepharose beads (50% slurry) to each immunoprecipitation reaction. Incubate while rotating at 4°C for at least 2 h to allow antibody–antigen complexes to bind to the Protein A. Pellet the beads by centrifugation for 1 min in a microcentrifuge and wash the beads 4× (1 mL each) with wash buffer 1.

5. After removal of the fourth wash, add 1 mL of wash buffer 2 and transfer the samples to a new microcentrifuge tube. Pellet the beads as described previously and completely remove all of the buffer using a 27-gauge needle attached to a syringe or an aspirator.

6. Resuspend the Protein A–Sepharose beads in 1× sample loading buffer containing 2% β-mercaptoethanol (β-ME). If the sample has been crosslinked with DSP or DTSSP, omit β-ME unless cleavage of the crosslinks is desired (*see* **Note 2**). Heat the samples to 100°C for 5 min or (depending on the connexin species) incubate at room temperature for 30 min (*see* **Note 6**).

7. Load the samples onto an SDS-polyacrylamide minigel. Specifications of the optimal percentages of acrylamide for the separation of different connexin species are provided in **Note 8**. Run the gel until the dye front reaches the bottom of the resolving gel.

8. If the gel contains radioactive immunoprecipitants, fix the gel with fix/destain solution for at least 30 min, dry, and then expose to X-ray film or a Phosphor-Imaging screen. If the proteins in the gel are to be detected by immunoblot analysis, prepare the gel for Western transfer as described in **Note 9**.

3.4.2. Chloroform/Methanol Precipitation of Proteins for Immunoblotting Analysis

1. Dilute samples with four volumes of methanol, vortex-mix; add one volume of chloroform, vortex-mix; and finally add three volumes of ddH$_2$O and vortex-mix. Centrifuge at 9000*g* for 5 min at room temperature (11,000 rpm in a microcentrifuge) and discard the upper layer (solution will be in two phases), avoiding the interface. Add three volumes of methanol, vortex-mix, and again centrifuge at 9000*g* for 5 min. Remove the supernatant and allow the pellet to dry completely. At this point, the samples can be stored indefinitely. For analysis, resuspend the pellets in 2× sample loading buffer ± β-ME (*see* **Note 2**) containing 5% SDS.

2. Proceed with SDS-PAGE and immunoblot analysis as described in **Subheading 3.4.1**, **step 7**, and **Notes 8** and **9**.

4. Notes

1. Under steady-state conditions, the amount of connexon available for analysis (**Fig 1**; *2* and *3*) will be limited by the rate of assembly of monomeric connexins into connexons as well as by the incorporation of connexons into Triton-insoluble and therefore unassayable gap junctional plaques (**Fig 1**; *5*). If connexons are to be extracted from tissue, the fraction of the total connexin population in the form of TX-100-soluble connexons has been determined in vivo. However, if primary or continuous cell cultures are to be used, temperature manipulations that promote accumulation of TX-100-soluble connexons can be employed. Incubation of cells at 20°C inhibits trafficking through the secretory pathway between the trans Golgi network (TGN) and the plasma membrane *(13)*. Because connexons in most cell types assemble within the TGN, metabolic radiolabeling at 20°C can be used to inhibit the incorporation of connexons into insoluble cell surface gap junctional plaques *(2)*. Extended (5–18 h) incubation of unlabeled tissue culture cells at the same temperature can also be used to accumulate connexons. Although other systems (e.g., refrigerated water baths) can be used, we conduct 20°C incubations in a cold room in a small household refrigerator (with the cooling element turned off) containing an N-Con Bod-Cubator (from N-Con Systems) heating unit. A thermometer probe (placed in a small container of water inside the refrigerator near where the cells are to be incubated) linked to an external monitor (such as a JKT Thermocouple Thermometer, Barnant) is essential for setting and continuously monitoring the temperature (must be $\geq 19°C$ and $< 21°C$).

    Most tissue culture media are formulated for use in 5–10% CO_2 environments and will therefore become unacceptably alkaline when used at low (ambient) CO_2 levels. We therefore use a modification of a HEPES-buffered minimal essential medium developed by Matlin and Simons for low CO_2 incubations *(13)*. The recipe given in **Subheading 2.** is for a methionine-deficient version of this medium and is designed for radiolabeling with [^{35}S]methionine. For other uses, the medium can be supplemented with 200 μM unlabeled L-methionine.

    Our standard procedure for metabolically radiolabeling connexons at 20°C is as follows:

    a. Wash a 60-mm dish of confluent adherent tissue culture cells twice with methionine-free Matlin's medium.

    b. Overlay with 2.5 mL of Matlin's medium supplemented with 5% dialyzed fetal calf serum (dFCS). Incubate at 37°C for 30 min under ambient CO_2 conditions in a tissue culture incubator (either a standard unit with the compressed CO_2 source turned off or a non-CO_2 unit). This "starve" depletes the cell's endogenous pools of methionine and enhances the subsequent incorporation of [^{35}S]methionine into newly synthesized proteins.

    c. Remove the medium and add 2.5 mL of fresh Matlin's medium + 5% dFCS containing 0.3 mCi of [^{35}S]methionine (37.8 μL of fresh ^{35}S-Express label from NEN, cat. no. NEG 072).

    d. Incubate the cells at 37°C in the same incubator for 20 min, chill the tissue culture dish for 1–2 min on a metal plate on ice, and then place the monolayer in a 20°C low-CO_2 incubator (described previously) for 4 h and 40 min (total labeling time is 5 h). The initial 20 min of labeling at 37°C increases the amount of radiolabeled connexin synthesized but is too short to permit detectable assembly of the labeled molecules into gap junctional plaques. After labeling, harvest the cells as described in **Subheading 3.1.1.** If metabolic labeling of connexons at 37°C is desired, a minimum labeling period of approx 45 min is required to accumulate sufficient quantities of [^{35}S]methionine-labeled connexin in the TGN to allow detectable connexon assembly.

2. Crosslinked bonds generated by thiol-sensitive reagents such as DSP and DTSSP can be broken by incubating samples for either 30 min in 10–50 mM DTT at 37°C or by heating to 100°C for 5 min in sample loading buffer containing 2 to 5% β-ME. We have noticed that connexons that have been exposed to SDS are sometimes more resistant to cleavage of crosslinks than those in TX-100. Crosslinks can be reduced prior to SDS addition by incubation for 1 h at room temperature with 20 mM DTT. The DTT must be inactivated with 50 mM NEM prior to addition of immunoprecipitating antibodies.

3. Increasing the amount of EGS to > 500 µg/mL does not improve the yield of fully crosslinked Cx32-containing connexons but instead results in the concentration-dependent conversion of Cx32 (along with other proteins in the extract) into very large, unidentifiable, and most likely nonspecific complexes. Examination of ladders of partially crosslinked Cx32 connexons reveals six distinct bands spaced at ~27-kDa intervals, consistent with the dogma that connexons are hexameric. In contrast, partial crosslinking of Cx43 connexons achieved by decreasing the concentration of crosslinking reagent by 75% (e.g., 25 µg/mL DSP or 125 µg/mL EGS) generates a ladder with only five "rungs" with electrophoretic mobilities expected for species containing one to five Cx43 molecules (*see* **Fig. 2**). The most likely explanation for this phenomenon is that fully crosslinked Cx43 hexamers for some reason migrate anomalously fast on SDS-PAGE in a position indistinguishable from that of a pentameric partial complex. Alternatively, it may be possible that the stochiometry of Cx32-containing and full-length Cx43-containing connexons differs.

4. The amount of TX-100-solubilized extract required for the generation of easily detectable levels of crosslinked connexons on a minigel depends on the level of expression and the assembly efficiency of connexins in the cell type of interest. For good Cx43 expressers such as NRK or CHO cells, the TX-100-solubilized extract from one third of a 60-mm dish of metabolically-labeled tissue culture cells is adequate. If crosslinked connexin species are to be further analyzed by sucrose gradient velocity sedimentation, a half of a 60-mm dish of cells is necessary because of unexplained loss of connexin signal during the latter procedure. To accomplish this, a more concentrated TX-100-solubilized extract can be prepared by solubilizing the cells from a 60-mm dish in 500 µL (instead of 1 mL) of incubation buffer and 1% TX-100. After dilution with two volumes of incubation buffer and the crosslinking reaction, approximately half of the cell extract will be in a volume (750 µL) that can be loaded on a 4-mL sucrose gradient.

If the connexons are from tissues and are to be analyzed by Western blotting, a limiting factor is the amount of protein that can be loaded onto a minigel (about 100–150 µg/lane of a 10-well comb). Fifty to 100 µg of protein prepared from liver or myometrium as described in the **Subheading 3.1.2.** is sufficient to detect connexons after crosslinking. The protein concentration of samples in SDS-PAGE sample loading buffer can be determined using the Pierce BCA assay. Make sure the final concentration of SDS in the protein assay is <0.5% and that no reducing agent is present; β-ME can be added after the quantitation. Western blotting can also be conducted after immunoprecipitation of TX-100-solubilized extracts with anti-connexin antibodies. In this case the volume of extract can be increased at least 7- to 10-fold because the total amount of protein in the immunoprecipitant is much lower than the limit for electrophoretic separation.

5. Optimal crosslinking of Cx43 connexons recovered from the 9S peak of a sucrose gradient requires twice the concentration of crosslinking reagent as is used to modify connexons in an unfractionated TX-100-solubilized extract.

6. Some connexins, including Cx32 and Cx26, are prone to aggregation, especially when heated above 37°C in SDS. Such connexins can be denatured by incubating TX-100-solubilized extracts with 0.6% SDS for 30 min at room temperature prior to dilution with immunoprecipitation dilution buffer. Immediately before loading samples onto a SDS-PAGE gel, noncrosslinked samples are incubated in sample loading buffer for 30 min at room temperature whereas crosslinked material is heated at 37°C for 5 min. TX-100-solubilized extracts containing connexins that do not aggregate in heat plus SDS (e.g., Cx43) are treated with 0.6% SDS at 100°C for 3 min before immunoprecipitation. Optimal resolution of Cx43-containing species requires that the samples be heated to 100°C for 5 min in sample loading buffer shortly before SDS-PAGE.

7. The volume of anti-connexin antibody to be used in immunoprecipitation reactions depends on the concentration and specificity of the reagent. If possible, the antibody should be titered to ensure quantitative recoveries and to minimize the level of nonspecific background. Interestingly, some antibody stocks that exhibit an initial linear relationship between the amount of immunoglobulin added and the amount of connexin protein immunoprecipitated show the opposite behavior when supersaturating levels of antibody are added. Because one half of a single antibody molecule can potentially immunoprecipitate an entire connexon complex, less antibody is required to immunoprecipitate the same number of connexin molecules if they are oligomerized than if they are monomeric. Immunoprecipitation of crosslinked connexons with subsaturating amounts of antibody yields the cleanest results with the least nonspecific background.

When immunoprecipitating connexin species for subsequent immunoblot analysis, covalently cross-linking the immunoprecipitating antibody to Protein A–Sepharose beads reduces the amount of immunoglobulin that is loaded onto the gel and that can react with the secondary antibodies used to detect connexin on the blot. A procedure adapted from the Harlow and Lane antibody manual *(14)* is outlined as follows:

a. Wash Protein A–Sepharose beads 3× with PBS using a microcentrifuge to pellet the beads.
b. To 400 µL of washed, packed beads, add 200 µL of crude serum (or monoclonal supernatant if the antibodies bind Protein A) and 800 µL of PBS. Mix by rotating at 4°C overnight.
c. Wash the beads 2× with PBS at room temperature to remove the unbound serum components. Wash 2× with 4 mL of 0.2 M sodium borate, pH 9.0.
d. Resuspend the beads in 4 mL of sodium borate buffer and add the covalent crosslinker dimethyl pimelimidate (as a solid) to a final concentration of 20 mM.
e. Rotate at room temperature for 30 min.
f. To stop the cross-linking reaction, pellet the beads and resuspend in 4 mL of 0.2 M monoethanolamine, pH 8.0. Pellet again and resuspend in 4 mL of the same.
g. Incubate while rotating at room temperature for 2 h.
h. Wash the beads 3× with PBS and then resuspend in 2 mL of freshly prepared 100 mM glycine, pH 3.0. Pellet the beads immediately and resuspend in another 2 mL of 100 mM glycine. Incubate at room temperature for 3 min (not longer).
i. Immediately wash the beads 4× with PBS and once with PBS supplemented with 0.1% TX-100 and 0.005% thimerosol (a preservative). Resuspend the beads in enough PBS + TX-100 + thimerosol to bring the final volume to 800 µL. Use 30 µL of this 50% suspension (after washing once or twice in PBS–TX-100 to remove the thimerosol) per immunoprecipitation reaction. Can be stored at 4°C for up to a year.

8. For visualization of monomeric connexins, 10% acrylamide gels are routinely used for Cx43 and 11% gels for Cx32. When crosslinked forms are to be analyzed, 4–11% acrylamide (for Cx43) and 7.5%–12.5% (for Cx32) gradient gels are used for maximum resolution of connexin-containing species. Gradient minigels are prepared using the same apparatus and essentially the same procedure as is employed for pouring linear sucrose gradients (*see* **Subheading 3.3.**), substituting 1.9 mL of the higher percentage acrylamide solution for the 20% sucrose solution and 2 mL of the lower percentage acrylamide solution for the 5% sucrose solution. Immediately after pouring, overlay the top of the gel mix with water-saturated butanol. After polymerization at room temperature for an hour, overlay with a 3% acrylamide stacking gel *(15)*.

9. SDS-PAGE gels are transferred to polyvinylidene fluoride (PVDF) membranes (Immobilon-P, Millipore) using a standard Tris–glycine–methanol procedure *(16)*. If only monomeric connexins are to be detected, transfer for 45–60 minutes using a transfer apparatus with plate electrodes. If crosslinked oligomeric forms are also present, the transfer time is increased to 2 h.

After transfer, the PVDF membrane is blocked and then probed with anti-connexin antibodies. Chemiluminescent detection protocols are more sensitive than colorimetric methods, but either will work. The Tropix Western-Light Detection system (Tropix, Bedford, MA) utilizes secondary antibodies conjugated with alkaline phosphatase and is compatible with chemiluminescence-sensitive phosphorimaging screens.

References

1. Goodenough, D. A., Goliger, J. A., and Paul, D. L. (1996) Connexins, connexons, and intercellular communication. *Annu. Rev. Biochem.* **65,** 475–502.
2. Musil, L., S. and Goodenough, D. A. (1993) Multisubunit assembly of an integral plasma membrane channel protein, gap junction connexin43, occurs after exit from the ER. *Cell* **74,** 1065–1077.
3. Koval, M., Harley, J. E., Hick, E., and Steinberg, T. H. (1997) Connexin46 is retained as monomers in a trans-Golgi compartment of osteoblastic cells. *J. Cell Biol.* **137,** 847–857.
4. Kumar, N.M, Friend, D. S., and Gilula, N. B. (1995) Synthesis and assembly of human B1 gap junctions in BHK cells by DNA transfection with the human B1 cDNA. *J. Cell Sci.* **108,** 3725–3734.
5. Jiang, J. X. and Goodenough, D. A. (1996) Heteromeric connexons in lens gap junction channels. *Proc. Natl. Acad. Sci. USA*, **93,** 1287–1291.
6. Musil, L. S. and Goodenough, D. A. (1991) Biochemical analysis of connexin43 intracellular transport, phosphorylation, and assembly into gap junctional plaques. *J. Cell Biol.* **115,** 1357–1374.
7. Kistler, J., Goldie, K., Donaldson, P., and Engel, A. (1994) Reconstitution of native-type noncrystalline lens fiber gap junctions from isolated hemichannels. *J. Cell Biol.* **126,** 1047–1058.
8. Stauffer, K. A., Kumar, N. M., Gilula, N. B., and Unwin, N. (1991) Isolation and purification of gap junction channels. *J. Cell Biol.* **115,** 141–150.
9. Cascio, M., Kumar, N. M., Safarik, R., and Gilula, N. B. (1995) Physical characterization of gap junction membrane connexons (hemi-channels) isolated from rat liver. *J. Biol. Chem.* **270,** 18,643–18,648.
10. Blount, P. and Merlie, J. P. (1988) Native folding of the acetylcholine receptor a subunit expressed in the absence of other receptor subunits. *J. Biol. Chem.* **263,** 1072–1080.
11. Musil, L. S., Cunningham, B. A., Edelman, G. M., and Goodenough, D. A. (1990) Differential phosphorylation of the gap junction protein connexin43 in junctional communication-competent and -deficient cell lines. *J. Cell Biol.* **111,** 2077–2088.
12. Wessel, D. and Flugge, U. I. (1984) A method for the quantitative recovery of protein in dilute solution in the presence of detergents and lipids. *Anal. Biochem.* **138,** 141–143.
13. Matlin, K. S. and Simons, K. (1983) Reduced temperature prevents transfer of a membrane glycoprotein to the cell surface but does not prevent terminal glycosylation. *Cell* **34,** 233–243.
14. Harlow, E. and Lane, D. (1988) *Antibodies: A Laboratory Manual*, Cold Spring Harbor Laboratory Press, Cold Spring Harbor, NY.
15. Hames, B. D. and Rickwood, D., ed. (1990) *Gel Electrophoresis of Proteins. A Practical Approach,* IRL Press, Oxford, England.
16. Towbin, H., Staehelin, T., and Gordon, J. (1979) Electrophoretic transfer of proteins from polyacrylamide gels to nitrocellulose sheets: procedure and some applications. *Proc. Natl. Acad. Sci. USA* **76,** 4350–4354.

7

Expression and Imaging of Connexin-GFP Chimeras in Live Mammalian Cells

Dale W. Laird, Karen Jordan, and Qing Shao

1. Introduction

Gap junction (GJ) proteins, connexins (Cx), possess many properties that are atypical of other well-characterized integral membrane proteins (1). Oligomerization of Cxs into hemichannels (connexons) has been shown to occur after the protein exits the endoplasmic reticulum (2). Once delivered to the cell surface, connexons from one cell dock with connexons from a neighboring cell and cluster into GJ plaques, a process that is facilitated by cadherins (3). Connexins have short half-lives and GJs are turned over rapidly (3). To study GJ formation, removal, and connexin intracellular trafficking in real time, we recently used the green fluorescent protein (GFP) as a fusion partner to construct Cx-GFP chimeras (see **Fig. 1**).

GFP is a 238 amino acid protein originally cloned from the jellyfish *Aequorea victoria* (4). GFP is rapidly becoming an important reporter molecule for monitoring gene expression and protein localization in vivo and in vitro in live cells (5). Fusion constructs that join GFP to Cxs (Cx-GFP chimeras) can provide important insights into localization of Cxs, mechanisms of Cx trafficking, and pathways for GJ disassembly, internalization, and degradation. We have now fused GFP to a number of Cxs, without apparent disruption of Cx trafficking to the cell surface or inhibition of gap junction plaque formation (Laird, Jordan, and Shao, unpublished data). In the case of Cx43-GFP, this fusion protein assembles into GJ plaques (see **Fig. 2**) and functional channels that exhibit wild-type characteristics (6).

The coding sequence of GFP can be directly fused in-frame to either the amino (N)- or carboxyl (C)-terminus of the target Cx. Alternatively, polypeptide linkers consisting of 4–22 amino acid residues can be added in-frame to

From: *Methods in Molecular Biology, vol. 154: Connexin Methods and Protocols*
Edited by: R. Bruzzone and C. Giaume © Humana Press Inc., Totowa, NJ

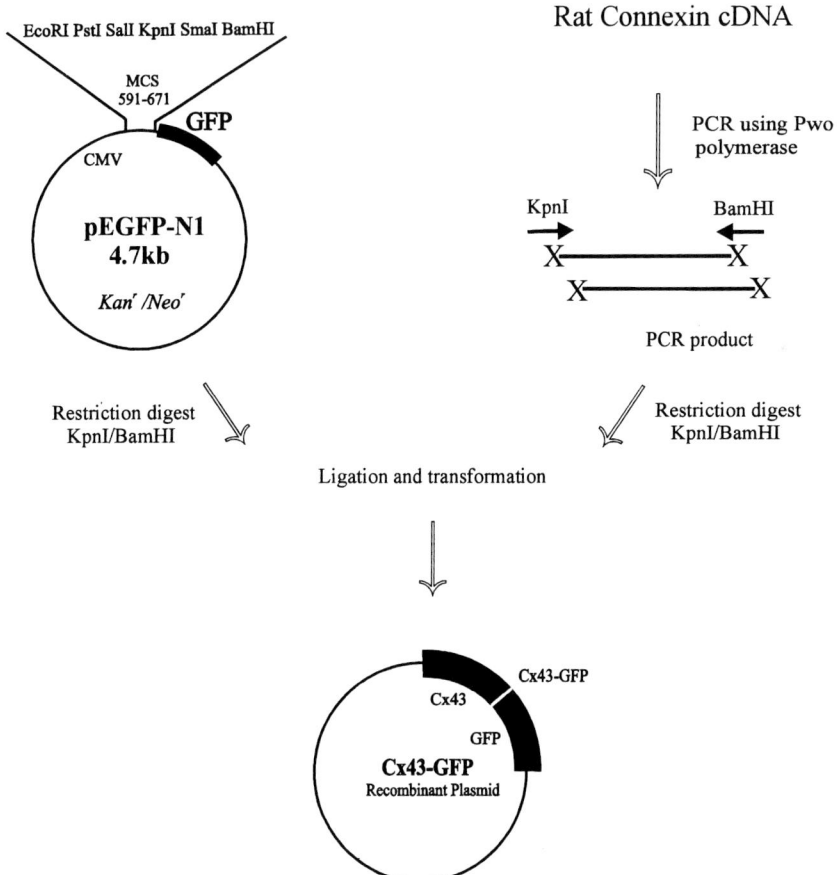

Fig 1. Schematic diagram of the pEGFP vector containing the Cx43 insert. The Cx43 is fused in-frame to the N-terminal domain of GFP.

the N- or C-terminals of the Cx to tether the GFP and enhance the likelihood that the Cx portion of the fusion protein will be functional. The cDNA encoding the Cx-GFP fusion protein can be transiently or stably transfected into mammalian cells. Because Cx-GFP fusion proteins have inherent fluorescence, they can be localized directly in living cells without further manipulation. In addition, time-lapse imaging can be performed and the resulting series of images can be animated in movie sequences.

We are currently using Cx-GFP chimeras expressed in living cells to examine the trafficking and mechanisms involved in gap junction assembly and turnover. Using Cx-GFP fusion proteins, we can: (1) examine the vesicle populations and/or tubular extensions that are involved in delivering Cx-GFPs to the

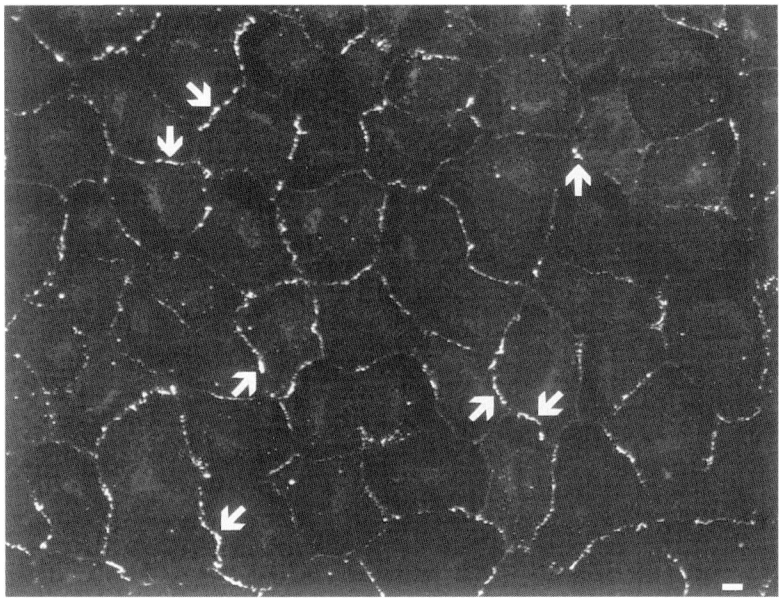

Fig 2. Expression and localization of Cx43-GFP in normal rat kidney cells. Normal rat kidney cells stably expressing Cx43-GFP were cultured on glass coverslips for 2 d prior to fixing in 80% methanol–20% acetone for 10 min at –20°C. The cells were imaged on a Zeiss LSM confocal microscope using a scan speed of 32 s and a pinhole setting of 20 which yields an optical slice of ~1 μm. Note that Cx43-GFP assembles into gap junctions that resemble the classical punctate distribution pattern (*arrows*). Bar = 10 μm.

cell surface, (2) investigate the sites of vesicle docking at the cell surface, (3) track the mobility of Cx-GFP hemichannels within the plasma membrane, (4) examine the mobility and clustering of cell surface gap junction plaques, (5) determine the mechanism of GJ disassembly and internalization, and (6) follow the fate and degradation of internalized GJs. These trafficking studies are aided by the availability of several well-characterized and reversible inhibitors of protein trafficking (i.e., brefeldin A, monensin, etc.), dominant-negative effectors for key trafficking molecules (i.e., rabs, dynamin, etc.) and inhibitors of cytoskeletal elements (i.e., nocodazole, cytochalasin D, etc.).

2. Materials

1. pEGFP-N or pEGFP-C (CLONTECH Laboratories) (*see* **Notes 1** and **2**).
2. Connexin cDNAs.
3. Design and order polymerase chain reaction (PCR) primers (*see* **Note 3**).

4. *Pwo* DNA polymerase and buffer (Boehringer Mannheim) (*see* **Note 4**).
5. Restriction enzymes and buffers: Available from many manufacturers (*see* **Note 5**).
6. T4 DNA ligase and 10× ligation buffer (Boehringer Mannheim).
7. A suitable *E. coli* stain such as JM109, JM101, or XL1-Blue.
8. Plasmid purification kit that yields transfection quality plasmid DNA (QIAGEN).
9. LipofectAMINE or LipofectAMINE Plus for transfection (Life Technologies).
10. Custom-made glass coverslip bottomed 35-mm tissue culture dishes.
11. Opti-MEM1 medium containing 10 m*M* 4-6(2-hydroxyethy)-1-piperazine-ethanesufonic acid (HEPES), pH 7.2.
12. Heated stage for confocal microscope (20/20 Technology, Bionomic controller, Model no. BC-100).
13. Zeiss LSM 410 inverted confocal microscope.
14. Zeiss time series imaging software.
15. Jaz disk storage.
16. Quicktime and Adobe Image Premiere software.
17. VCR.
18. High-resolution color dye sublimation printer.

3. Methods

3.1. Engineering and Expressing Cx-GFP Chimeras

1. Generate Cx cDNA with appropriate restriction sites for subcloning by using 10–100 ng of existing plasmid containing the Cx of interest and appropriate PCR primers. Obviously, anyone who chooses to make appropriate fusion proteins will have to customize their primers with respect to the connexin being used and the 4–22 amino acid linker sequence that we note in the introduction (*6*).
2. Digest the pEGFP vector and PCR products with restriction enzymes (*see* **Note 6**).
3. Purify the PCR products (*see* **Note 7**) and digested pEGFP from a 1% agarose gel.
4. Ligate the linearized pEGFP vector and insert Cx DNA at a molar ratio ranging from 1:1 to 1:3 in 1 µL of 10× ligation buffer, 1 U of T4 DNA ligase, and distilled water (dH$_2$O) up to 10 µL at 15°C overnight.
5. Transform *E. coli* (JM109) with a fraction of the ligation reaction and spread on LB plates containing 30 µg/mL kanamycin (*see* **Note 8**).
6. Pick individual bacterial colonies and extract the DNA using a standard plasmid miniprep protocol.
7. Verify insertion of the Cx cDNA by digestion with appropriate restriction enzymes and gel electrophoresis.
8. Sequence the Cx-GFP plasmid (*see* **Note 9**).
9. Prepare a large-scale plasmid preparation of the Cx-GFP construct.
10. Transfect mammalian cells grown to 50–70% confluency in 60 mm dishes with 3 µL/mL of LipofectAMINE and 1 µg/mL of construct DNA (premix LipofectAMINE and DNA) in a total volume of 2 mL of serum-free medium for 2–5 h (*see* **Notes 10** and **11**).

3.2. Time-Lapse Imaging and Animation of Cx-GFP Chimeras in Live Cells

1. Culture cells that stably or transiently express Cx-GFP on sterile 12-mm round glass coverslips (Fisher) in 60-mm culture dishes until cells have grown to required confluency (usually it takes 2–4 d for most cell lines to reach 70% confluency).
2. Remove coverslips from 60-mm culture dishes and place them face down in a glass coverslip bottomed 35-mm dish (*see* **Note 12**) containing 2 mL of Opti-MEM1 (Life Technologies) supplemented with 10 m*M* HEPES, pH 7.2 (*see* **Note 13**).
3. Place the 35-mm dish containing the glass coverslip on a Zeiss LSM 410 confocal microscope equipped with a 20/20 Technology temperature controlled stage preset to 37°C.
4. View and image Cx-GFP in living cells using a 1.4 numerical aperture 63× plan-apochromat oil lens (*see* **Note 14**). Cx-GFP is excited with a 488-nm laser line emitted from an argon/krypton mixed gas laser (*see* **Note 15**). Optical scans of approx 1 μm in thickness are collected every 16–32 s for periods up to 60 min (*see* **Note 16**).
5. Store individual images and time-lapse image series on a Jaz disk (*see* **Note 17**).
6. Animate time-lapse image series using Adobe Image Premiere or Quicktime software to examine Cx-GFP trafficking (*see* **Note 18**).

4. Notes

1. Vectors of pEGFP-C or -N are used for the expression of C-terminal or N-terminal EGFP fusion proteins. The EGFP gene encodes a red-shifted variant (enhanced variant) of the *Aequorea victoria* GFP (*7*) optimized for brighter fluorescence and higher expression in mammalian cells (Ex$_{max}$ 488 nm; Em$_{max}$ 507 nm). For pEGFP-C, connexin cDNA is inserted into the multiple cloning site (MCS) downstream of the EGFP coding sequence and the resulting protein is expressed fused to the C-terminus of EGFP. For pEGFP-N, connexin cDNA cloned into the MCS is expressed fused to the N-terminus of EGFP. Care must be taken to ensure that the inserted connexin cDNA and EGFP remains in the same open reading frame.
2. At present, we do not highly recommend using the pEBFP (blue) vector to construct Cx-BFP chimeras as the fluorescence emitted from BFP (Ex$_{max}$ 380 nm; Em$_{max}$ 440 nm) is at least an order of magnitude less, making imaging of these fusion proteins difficult. In addition, if Cx-BFP fusion proteins are to be imaged on a confocal microscope, the microscope must be equipped with a UV laser. Although we have not tested the pEYFP (yellow) vector in our laboratory, the wavelength needed to excite YFP (Ex$_{max}$ 513 nm) is available on most conventional epifluorescent or confocal microscopes and the fluorescence (Em$_{max}$ 527 nm) should be similar in intensity to GFP.
3. Constructing primers that introduce restriction sites into the coding region of GFP for in-frame fusion to connexin coding sequences is straightforward. Primers

designed to include appropriate restriction sites and connexin coding sequences work well for PCR amplification. The typical length of primers used for connexin application is 18–22 basepairs. The products of PCR allow for the generation of sticky ends suitable for cloning into pEGFP vectors. When fusing to the C-terminus of a connexin, the stop codon must be deleted and is replaced by one of the restriction enzyme sites indicated in **Fig. 1**, which shows a schematic outline of GFP fusion to the C-terminal of Cx43. To fuse GFP to the connexin N-terminus, the start codon is replaced by a restriction site in the 5' sense primer without losing the connexin methionine.

4. When using *Taq* DNA polymerase (error rate of 2×10^{-4} errors/base) about 56% of a 200-bp amplification product will contain at least a single error after 1-million-fold amplification. In contrast, when using *Pwo* DNA polymerase for amplification only 10% of the products will contain an error under the same conditions.

5. The selected restriction sites should have at least two additional nucleotide bases 5' to the recognition sequence (this depends on the restriction enzyme used) to ensure that the enzymes will in fact recognize and cleave the sequence (*see* the New England Biolabs catalog for the details of individual enzymes).

6. If you use one restriction enzyme to digest the vector, the vector needs to be dephosphorylated with calf intestinal alkaline phosphatase and proper orientation will need to be confirmed.

7. Sometimes the efficiency of digesting PCR products by restriction enzymes is poor. To compensate for this, we first subclone the PCR products into a T-vector (Promega) and then use restriction enzymes to digest the restriction sites introduced by the primer.

8. For transformation, 5 µL of ligation reaction is added to 50 µL of competent *E. coli* cells and the cells are heat shocked for 45 s in a water bath at 42°C.

9. DNA sequence analysis is used to confirm in-frame fusion and to rule out PCR errors, point mutations and deletions.

10. We recommend using half the amount of LipofectAMINE (or LipofectAMINE Plus) as suggested by the manufacturers (Life Technologies) for transfecting cells to reduce the amount of autofluorescence generated by LipofectAMINE in transient transfections. Transient expression of Cx-GFP is observed 48 h after the transfection.

11. Stable transfectants can be selected by resistance to G418 (neomycin). Generation of cell lines that stably express Cx-GFP fusion proteins can be difficult. In several instances clonal cells revert and lose Cx-GFP expression after several passages. This can be overcome by periodic subcloning of stable cell lines that express Cx-GFP, maintaining transfected cells under G418 selection pressure and by liquid nitrogen storage of a large number of vials that contain early passage cells.

12. Coverslips with attached mammalian cells are placed face down on the customized cover slip bottomed tissue culture dishes to facilitate examination with high-resolution, high numerical aperture lenses. Focusing on cells through two no. 1 coverslips, as would be the case if the cells were placed face up, is less than

optimal. The inversion of the cover slip onto glass bottom plates appears to have no detrimental effect on the viability or morphology of the cells. Furthermore, the cells remain accessible to drug treatments (i.e., Brefeldin A, nocodazole, ammonium chloride) during the experiment.

13. We typically maintain the pH at 7.2 by using 10 mM HEPES. However, use of a bicarbonate buffer system would be advantageous providing the stage is equipped with a CO_2 chamber. To avoid high levels of background fluorescence, we use phenol red-reduced OptiMEM-1 medium. Riboflavins also contribute to background fluorescence, however, we find that the background fluorescence is sufficiently low without using riboflavin-deficient medium.

14. A number of optical lenses can be used to view Cx-GFP fluorescence; however, we have found that one of the best and most versatile lenses is the 63× plan-apochromat oil lens. To date, this is the highest numerical aperture lens sold that provides for excellent capturing of the GFP fluorescence. Other lenses with high numerical apertures (i.e., 40× plan-neofluar, 1.3 numerical aperture) will also provide high-resolution images.

15. Typically, we collect Cx-GFP fluorescence from live cells after passage through a series of dichroic and barrier filters. The fluorescence emitted from GFP is allowed to first pass through a 510-nm dichroic followed by a long pass 515-nm barrier filter. If the cells that are being imaged are transiently transfected, there may be autofluorescence contributed from the LipofectAMINE. To reduce image contamination with background fluorescence, a 515–565 nm barrier filter is used to replace the long pass 515-nm barrier filter. This approach is also necessary if the cells to be imaged are fixed and immunolabeled with an antibody against another cellular component. Fluorescence from an immunolabeled antigen must have spectral characteristics distinct from GFP and is usually collected on a second photomultiplier.

16. The quality of the images obtained is directly correlated with the scan speed used to acquire the images and the optical slice thickness. In general, the slower the scan speed, the less pixel noise and the better the image quality. Likewise, 1-μm optical sections tend to yield higher quality images than optical sections that are thicker. A potential disadvantage of using 1-μm optical sections rests in the fact that Cx-GFP vesicles or plaques may move out of the Z plane being imaged and some data will be lost from the time-lapse series. This problem can be overcome by increasing the pinhole size and, consequently, the optical slice thickness. We typically use scan speeds of 16–32 s for image fields of 768 × 576 pixels. The disadvantage of using these rather slow scan speeds is the possibility that Cx-GFP may spatially redistributed at a faster rate than the images are being acquired. While this potential problem cannot be readily overcome on a confocal microscope, new generation CCD-cooled cameras are now becoming available that can acquire high resolution and high signal-to-noise images in less than 2 s. At present, we are setting up this technology to further examine Cx-GFP trafficking.

17. Image data (~250 kB/image) can be stored on any number of devices (hard disk, Zip, Jaz, CDs, etc.). We chose to use an Iomega Jaz drive and the Jaz disk storage

system mainly because these disks are economical (1 GB storage/disk) and the drives have access speeds comparable to hard disks. The recent availability of CD Rom recorders allows us to back-up our Jaz disk data on CDs.

18. In the past, one of the factors that dampened many investigators' enthusiasm for performing GFP studies in living cells was the difficulty in creating, presenting and publishing the time-lapse movie sequences. Now with the ease in creating world wide web sites this problem has been eliminated and several cell biology journals (i.e., *Journal of Cell Biology, Molecular Biology of the Cell*) readily accept and promote publication of movies in Quicktime format. We generate time-lapse movies for depositing on the world wide web and for accessing with Quicktime using Adobe Image Premiere software. In generating the movie sequences, it is important to have a high-end computer with no less that 64 MB RAM.

Acknowledgments

The authors would like to thank Dr. Suzanne Bernier for her critical review of this manuscript. This work was supported by the Medical Research Council of Canada (Grant MT 12241).

References

1. Bruzzone, R., White, T. W., and Paul, D. L. (1996) Connections with connexins: the molecular basis of direct intercellular signaling. *Eur. J. Biochem.* **238,** 1–27.
2. Musil, L. S. and Goodenough, D. A. (1993) Multisubunit assembly of an integral plasma membrane channel protein, gap junction connexin43, occurs after exit from the ER. *Cell* **74,** 1065–1077.
3. Laird, D. W. (1996) The life cycle of a connexin: gap junction formation, removal, and degradation. *J. Bioenerg. Biomembr.* **28,** 311–318.
4. Chalfie, M., Tu, Y., Euskirchen, G., Ward, W. W., and Prasher, D. C. (1994) Green fluorescent protein as a marker for gene expression. *Science* **263,** 802–805.
5. Sciaky, N., Presley, J., Smith, C., Zaal, K. J., Cole, N., Moreira, J. E., Terasaki, M., Siggia, E., and Lippincott-Schwartz, J. (1997) Golgi tubule traffic and the effects of brefeldin A visualized in living cells. *J. Cell Biol.* **139,** 1137–1155.
6. Jordan, K., Solan, J. L., Dominguez, M., Sia, M., Hand, A., Lampe, P. D., and Laird D. W. (1999) Trafficking, assembly and function of a connexin43–green fluorescent protein chimera in live mammalian cells. *Mol. Biol. Cell* **10,** 2033–2050.
7. Yang, T. T., Sinai, P., Green, G., Kitts, P. A., Chen, Y. T., Lybarger, L., Chervenak, R., Patterson, G. H., Piston, D. W., and Kain, S. R. (1998) Improved fluorescence and dual color detection with enhanced blue and green variants of the green fluorescent protein. *J. Biol. Chem.* **273,** 8212–8216.

8

Analysis of Connexin Expression in Brain Slices by Single-Cell Reverse Transcriptase Polymerase Chain Reaction

Laurent Venance

1. Introduction

The molecular characterization of gap junction channels has revealed the existence of a large number of the subunits forming these channels, the connexins (Cx) (for reviews *see* **refs.** *1,2*). As a consequence of their differential expression and their ability to form homo- and heterotypic channels, the molecular and functional characterization of native gap junction channels in defined cell subpopulations has been a difficult enterprise so far.

Using the *in situ* hybridization technique, it was possible to study the differential expression patterns of Cxs in the brain. However, the identification of Cx expression in certain cells may not be possible when expression levels are low. Furthermore, the technique permits the concomitant colocalization of two or three Cxs at the most.

Hence, the use of the single-cell reverse a polymerase chain reaction (RT-PCR) technique in acute brain slices offers an extremely powerful approach to study the differential expression of Cxs in identified neuronal and glial cells.

The single-cell RT-PCR technique combined with electrophysiological studies has been successfully used in a number of studies over the last 10 yr (*3–7*); (for reviews *see* **refs.** *8,9*). The whole-cell patch-clamp configuration is used to characterize functional properties of single neurons and permits the subsequent harvesting of the cell content for the molecular analysis. Single-cell PCR studies include studies in which the analysis of gene families with specific primers has been employed and studies in which the aim was the analysis of different genes using the "multiplex PCR" approach (for examples *see* **refs.** *7,10–12*).

From: *Methods in Molecular Biology, vol. 154: Connexin Methods and Protocols*
Edited by: R. Bruzzone and C. Giaume © Humana Press Inc., Totowa, NJ

Important technical aspects that need to be considered when single-cell PCR experiments in the acute brain slice preparation are performed have been described elsewhere *(8)*. Here I will outline the exact procedures that have been used for the analysis of Cx expression in identified neurons recorded from brain slices.

2. Materials
2.1. Reagents
2.1.1. cDNA Synthesis

1. 5× primer–dNTP mix contains 25 µ*M* hexamer random primers (Boehringer Mannheim, Mannheim, Germany) and 2.5 m*M* of each deoxyribonucleotide (Pharmacia, Upsala, Sweden) in 10 m*M* Tris-HCl, pH 8.0.
2. 200 m*M* Dithiothreitol (DTT) (Biomol, Plymouth Meeting, PA, USA) dissolved in autoclaved water and filtered (0.2 µm pore size).
3. 40 U/µL of ribonuclease inhibitor (Promega, Mannheim, Germany).
4. 200 U/µL of superScript II reverse transcriptase (Gibco-BRL Life Technologies GmbH, Eggenstein, Germany).

2.1.2. PCR Amplifications

1. Taq DNA polymerase (Gibco-BRL).
2. 10× buffer (Gibco-BRL): 200 m*M* Tris-HCl, pH 8.4, 500 m*M* KCl.
3. 50 m*M* MgCl$_2$ (Gibco-BRL).
4. dNTPs, 5 m*M* each (Pharmacia).
5. Mineral oil (Sigma Chemical, St Louis, MO, USA).

2.1.3. Subcloning of the PCR Product into the M13 RF-DNA Vector

1. 10× ligation buffer: 500 m*M* Tris-HCl, pH 8.1, 100 m*M* MgCl$_2$, 100 m*M* DDT, 1 m*M* EDTA.
2. 10 m*M* ATP (Sigma).
3. T4 DNA ligase (1 U/µL, Boerhinger Mannheim).
4. Cloning vector: MP13mpRF (Stratagene, La Jolla, CA, USA).
5. Competent cells: XL2-Blue cell (Stratagene).
6. Columns P100 Chroma Spin + TE-100 (Clontech Lab., Palo Alto, CA, USA).
7. Gel extraction kit, Qiae II (Qiagen, Valencia, CA, USA).
8. Nitrocellulose membranes : Protan BA 85 (Schleicher & Schuel, Dassel, Germany).

2.1.4. Oligonucleotide Radiolabeling

1. 10× buffer (same buffer as for the ligation).
2. [γ-^{32}P]ATP (Amersham, Little Chalfont, UK).
3. T4 polynucleotide kinase (New England Biolabs, Beverly, MA, USA).
4. Bio-spin 6 column (Bio-Rad, Hercules, CA, USA).

2.1.5. Southern Blot

1. Nylon membrane (Porablot NY amp, Macherey-Nagel, Düren, Germany).
2. Denaturation buffer: 1.5 M NaCl and 0.5 M NaOH in milliQ water.
3. Hybridization buffer (2×): 0.05% sodiumpyrophosphate, 10× saline sodium citrate (SSC) (Amresco, Solon, OH, USA), 10× Denhart's (Sigma), 100 μg/mL yeast RNA.
4. Formamide (Sigma).

2.2. Solutions

The compositions of the patch-clamp recording solutions used in these experiments were the following:

1. Internal solution: 140 mM KCl, 3 mM MgCl$_2$, 5 mM EGTA, and 5 mM 4-(2-hydroxyethyl-1-piperazifneethanesulfonic acid (HEPES), pH 7.3 adjusted with KOH (*see* **Notes 1** and **2**).
2. External solution: 125 mM NaCl, 2.5 mM KCl, 1 mM MgCl$_2$, 2 mM CaCl$_2$, 1.25 mM NaH$_2$PO$_4$, 25 mM glucose, and 25 mM NaHCO$_3$, with a pH 7.35 after bubbling with a gas mix containing 95% O$_2$–5% CO$_2$.

2.3. Apparatus

The following equipment was used in our experiments but an adequate alternative is of course possible.

1. Patch-clamp puller pipets: Flaming-Brown P-97 puller (Sutter Instruments, Novato, CA, USA). The glass used to pull the patch-clamp pipets is made of borosilicate (glaskapillaren from Hilgenberg, Masfeld, Germany). The outer diameter is 2 mm, the thickness 0.5 mm, and the length 75 mm.
2. Antivibration table: Physik Instrumente (Waldbronn, Germany).
3. Upright microscope: Axioskop 2 with DIC optics and objectives (2.5×, 40×, and 60×) from Carl Zeiss (Göttingen, Germany).
4. Camera and camera controller used for the infrared difference interference contrast (IR-DIC) videomicroscopy: Hammamastu C2400 (Hammamatsu, Japan). Video monitor WV-5410 from Panasonic.
5. Patch-clamp amplifiers EPC-7, EPC-9, and software for acquisition and analysis Pulse and Pulse Fit 8.11 from HEKA Elektronik (Lambrecht, Germany). Analysis was done also with Igor Pro from WaveMetrics (Lake Oswego, OR USA).
6. Micromanipulator system: LN combi 25 and controller SM I from Luigs & Neumann (Ratingen, Germany).
7. Microslicer: DTK-1000 from Dosaka Co. Ltd. (Kyoto, Japan).
8. For expelling procedure: Picopritzer II (General Valve, Fairfield, USA). Binocular from Nikon (Japan).
9. PCR amplifications: DNA Thermal Cycler 480 (Perkin Elmer, Norwalk, CT, USA).
10. Sequencer: 377 DNA sequencer, ABI prism™ (Perkin Elmer).

11. Crosslinker: UV Stratalinker 2400 (Stratagene).
12. X-ray film developing machine: Hyper Processor, Amersham Life Science.

3. Methods
General Cautions

Precautions to avoid contamination should be taken, as the aim of this technique is to harvest and handle minute amounts of mRNA.

A room should be dedicated solely to these electrophysiological and molecular biology experiments. During experiments, powder-free gloves should be worn and draughts (open windows, doors, etc.) should be avoided. Whenever possible, instruments (water-bath, centrifuge, PCR machine, etc.) should be used for these experiments only.

3.1. Electrophysiological Part

3.1.1. Patch-Clamp Recording on Slice

Rat brain slices of 300–400 μm thickness are stored at 32°C for 1 h after slicing and then at room temperature. For these experiments, large-tip-diameter pipets (1.5–2.5 MΩ) (Hilgenberg, Masfeld, Germany) are used to harvest as much as possible of the cell content in a short time.

1. To reduce the risk of contaminations the following recommendations should be considered:
   a. The glass tubing for the patch-clamp pipette is sterilized before use by heating from 4–6 h at 200°C.
   b. For the internal and external solutions autoclaved milliQ water should be used.
   c. The internal solution is autoclaved before every experiment and filtered (0.2-μm filters) before filling the patch-clamp pipet. The pipet is filled with 8 μL of internal solution.
   d. Gloves should be worn and for every manipulations a sterile forceps should be used.
   e. The microscope, the electrode holder, and the stage supporting the perfusion chamber should be carefully cleaned with 70% ethanol before each experiment.
2. In a first step, a cell in the slice is selected based on morphological criteria using the infrared difference interference contrast (IR-DIC) videomicroscopy. The cell is then functionally characterized using the patch-clamp technique in the acute slice preparation *(13)* (*see* **Notes 3** and **4**). To avoid an obstruction of the tip of the pipet, positive pressure is applied before lowering the pipet into the bath solution and the slice.
3. Once the seal configuration is established and stabilized ($R_m > 2$ GΩ), the whole-cell configuration is achieved by gently breaking the patch membrane using a negative pressure pulse. Then the cell may be precisely identified using biophysi-

cal parameters under current-clamp conditions. According to these parameters, it is possible to discriminate between glial cells and neurons and between subpopulations of neurons.

3.1.2. Harvesting of the Cell Content

1. The harvesting step is performed immediately after the electrophysiological identification of the cell. To harvest the cytoplasmic content of the cell, negative pressure (approx 50 mbars) is applied to the patch-clamp pipet. This step is performed under IR-DIC videomicroscopy visual control and concomitant control of the seal resistance. To avoid contamination, a constant giga-ohm seal during this procedure is an absolute requirement. The negative pressure that is applied at the beginning of the harvesting procedure critically determines the amount of cytoplasmic material that is collected into the pipet. This pressure needs to be relatively low (50 mbars) and constant for the first 1–2 min and then higher (100 mbars) toward the end of the harvesting procedure. If the pressure is too high at the beginning, the nucleus of the targeted cell may obstruct the tip of the pipet (this event typically increases the access resistance). Because the expression level of most neuronal Cxs is low, it is necessary to harvest as much as possible of the cell content.
2. After harvesting, the pipet is removed very carefully to avoid contamination from the surrounding tissue. To this end, it is ideal to obtain an outside-out patch configuration. Furthermore, several passages (five or six) the pipet through the air–external solution interface may be required to remove some debris attached to the surface of the pipet. The condition of the slice is also crucial. Unhealthy or old slices (more than 5 h after slicing) will lead to a high increase in contaminations risks. After each harvesting, the patch-clamp electrode wire should be electrochemically rechlorided in a 30 mM KCl solution.
3. Two kinds of controls have to be done for each experiment: slice controls and an intracellular solution control. For the former, a patch pipet is lowered into the slice in a place close to a neuron for 1–2 min but without patching it. For the latter, 8 µL of internal solution is used. In both cases the material is then processed in an identical fashion as the harvested cell content.

3.1.3. Expelling

The content of the patch pipet is expelled into a PCR tube (Perkin Elmer, Gene amp PCR tubes 0.5 mL) using positive pressure. For this purpose, we use a home-made expeller permitting the precise positioning of the pipet into the PCR tube and avoiding uncontrolled movement of the pipet when high positive pressure is applied for expelling *(8)*. The pressure of the nitrogen, which is passed through a gas filter, is controlled by a Picopritzer.

1. As soon as the harvesting is achieved, 2 µL of primer/dNTP mix and 0.5 µL of DTT are added to the PCR tube. These solutions must be kept on ice during the

experiment and new aliquots used for each experiment. Then the PCR tube and the harvesting pipet are placed in the expeller. The bottom of the tip of the pipet is positioned under visual control (10× binocular microscope) on the surface of the solution in the PCR tube.

2. Positive pressure (4 bars) is applied for 30–60 s to expell as much as possible of the harvested material contained in the tip of the pipet as this material represents the majority of the harvested cell content.
3. The tip of the pipet is then gently broken by touching the bottom of the PCR tube. The remaining volume present in the pipette (8 μL) can be expelled under a lower pressure (approx 0.2 bar) within 10 s.

3.2. Molecular Biology Part: Analysis of the mRNA Expression

3.2.1. cDNA Synthesis

After the expelling procedure the reverse transcription (RT) step is carried out immediately. The enzymes that are kept at –20°C are rapidly added: 0.5 μL of RNasin and 0.5 μL of superScript II reverse transcriptase. New aliquots of these enzymes are used for each experiment. The tube is gently flicked and then centrifuged for several seconds using a bench-picofuge (Stratagene). The reaction is then incubated at 40°C for 60 min. The samples are stored at –20°C until the PCR reactions are performed.

3.2.2. PCR Amplifications

All 10 μL of the RT reaction are used for the subsequent PCR reaction, as many Cx mRNAs are present in low amounts in neurons (*see* **Note 3**). Also, two PCR reactions (PCR1 and PCR2) are required to amplify enough material from a single-cell for the subsequent analysis. PCR1 is performed directly in the PCR tube containing the cDNA reaction.

3.2.2.1. FIRST PCR AMPLICATION

Mix A:
>   7 μL of 10× buffer (Gibco-BRL)
>   3 μL of 50 mM MgCl$_2$
>   47 μL of H$_2$O
>   10 μL of of the reverse transcription reaction

The MgCl$_2$ concentration is variable. For each primers pair different MgCl$_2$ concentrations (from 0.5–4 mM) should be tested (using plasmids and/or total brain cDNA as templates) to optimize the PCR reaction *(14)*. The water volume is adjusted to obtain a final volume of 70 μL per PCR tube for mix A. Mix A is directly pipetted into the tube containing 10 μL of the RT reaction and two drops of mineral oil are added to the tube.

The PCR tubes are then placed in the PCR machine for a hot start, 94°C for 5 min. This step is followed by the addition of 30 μL of mix B per tube.

Mix B:

  3 μL of 10× with buffer (Gibco-BRL)
  2 μL of 5' primer (10 pmol/μL)
  2 μL of 3' primer (10 pmol/μL)
  22.5 μL of H_2O
  0.5 μL of *Taq* DNA polymerase (Gibco-BRL)

Primer dilutions are made in 10 μ*M* Tris-HCl, pH 8. The optimal primer concentration is tested using plasmids and total brain cDNA as templates. The water volume is adjusted to a final volume of 30 μL. Mix B should not be vortex-mixed but just flicked and then centrifuged to avoid damaging of the Taq polymerase. The PCR cycling program consists of 35 cycles using the following cycling program:

  a. Denaturation step, 94°C for 30 s.
  b. Annealing step, 56°C for 30 s, except for the degenerate primers (up primer PCR1, lo primer PCR1/2, and up nested primer PCR2); the annealing step is 53°C for 30 seconds.
  c. Extension step, 72°C for 40 s.

After an extension step at 72°C for 10 min, the PCR reaction is stopped by cooling the samples to 4°C. The PCR tubes are stored at –20°C until the second PCR amplification (PCR2) is performed.

3.2.2.2. SECOND PCR AMPLIFICATION

For the second amplification, 1–4 μL of the PCR1 reaction are used as a template. This volume has to be optimized so as to have a clear signal without nonspecific amplication. Of the primers used for the second amplification, at least one primer should be a nested primer. Both primer concentration and $MgCl_2$ concentration have to be optimized for this reaction as well. The different volumes for Mix A and B are the same than in PCR1, except that 3 μL of dNTPs, 5 m*M* each (Pharmacia), have to be added in Mix A and the volume of H_2O should be adjusted. The PCR cycling program for the PCR2 is similar to the one used for the PCR1.

3.2.3. Strategies for Cx Identification and Analysis of the PCR2 Products

3.2.3.1. IDENTIFICATION OF THE α- AND β-FAMILY CXS WITH DEGENERATE PRIMERS

Degenerate primers are used to first detect which Cxs are expressed in the harvested cells (*see* **Fig. 1**, *3.1.–3.5.*). These primers are located on highly conserved regions (extracellular loops, E1 and E2) of exon 2 (*see* **Note 5**). Nested degenerate primers are used for the PCR2 (*see* **Table 1**). Visualization of the PCR product on an agarose gel followed by southern-blot analysis with specific probes for different Cxs are then performed.

1. Electrophysiological identification and characterization

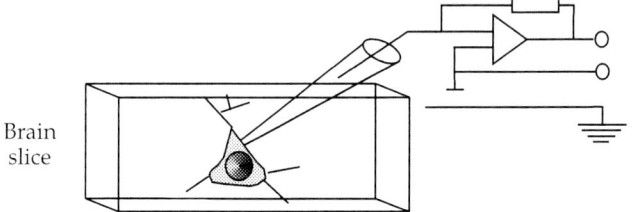

Brain
slice

2. Harvesting of the cell content and cDNA synthesis

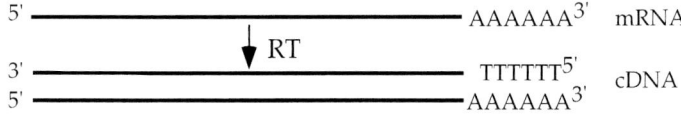

3. 1. Identification of the α- and β-family Cxs PCR1 with degenerated primers

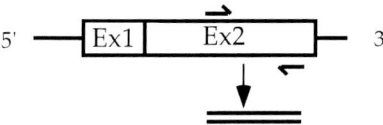

3. 2. Nested PCR2 with degenerate primers and southern-blot analysis
with specific probes

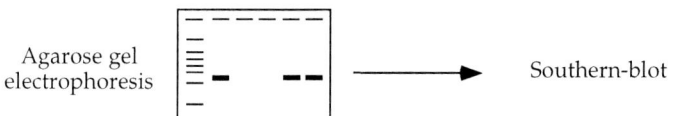

3. 3. Cloning of the PCR2 product and transformation in E-coli

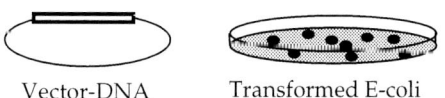

3. 4. Hybridization with specific probes

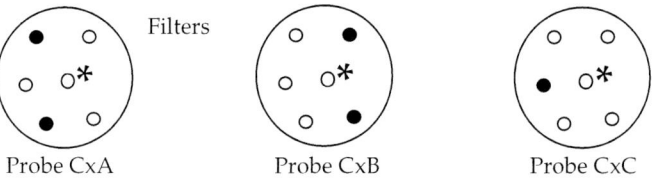

3. 5. Direct sequencing

4. 1. Confirmation of the identity of Cxs with specific primers

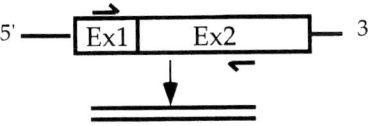

4. 2. PCR2 multiplex with specific primers for the different identified Cxs

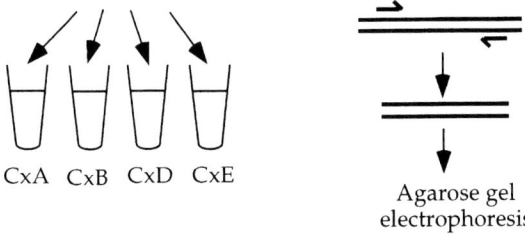

CxA CxB CxD CxE

Agarose gel
electrophoresis

4. 3. Confirmation of the identity of the PCR2 product

- Agarose gel electrophoresis (length)

- Southern-blot with specific probes

- Direct sequencing (eventually)

Fig. 1. Overview of the different steps involved in the single-cett RT-PCR analysis for connexins.

This step is aimed at the identification of the Cx expression pattern in particular neuronal subpopulation. The drawback of this approach consists in the fact that it is impossible to discriminate between amplification from genomic material or mRNA, as the primers are located on the same exon. This discrimination can be achieved by subsequently carrying out experiments in which subunit-specific primers spanning an intron are used (*see* **Subheading 3.2.3.2.** and **Fig. 1**, *no. 4.1.–4.3.*).

Table 1
Sequences of PCR Primers and Southern Blot
Specific Probes, for Connexins Subunit Analysis

Identification of the α- and β-family connexins
Lo primer PCR1: 5'-TGGG(CAG)C(TG)GGA(CAG)A(CTG)GAAGCAGT-3'
Lo nested primer PCR2: 5'-TTCCCCATCTC(CTG)CA(CT)(GA)T(CTG)CG-3'
Up primer PCR1/2: 5'-GGCTGT(GA)A(CAG)AA(TC)GTCTGCTA(TC)GAC-3'

Confirmation of the identity of the connexins and exclusion of the genomic amplifi
 cation multiplex PCR for Cx26 and Cx32
Lo primer PCR1: 5'-CGGA(CT)GTG(GA)GAGATGGGGAA-3'
Up primer Cx32 PCR1/2: 5'-CCCTACACAGACATGAGACC-3'
Up primer Cx26 PCR1/2: 5'-ACTCCGGACCTGCTCCTTAC-3'
Lo nested primer Cx26 PCR2: 5'-GTCGTAGCACACATTCTTACAGCC-3'
Lo nested primer CCx32 PCR2: 5-GTCATAGCAGACGCTGTTACAGCC-3'

Amplification of Cx36
Lo primer PCR1: 5'-GACAGTCGAGTACCGGCGTTCTC-3'
Lo nested primer PCR2: 5'-AACAGAGACTGGGGGTGCACACC-3'
Up primer PCR1/2: 5'-ATGGGGGAATGGACCATCTTG-3'

Southern blot specific probes
Cx26[a]: 5'-CATGATGTAAAAGACATACATGAAGACAGCTTCGAA
      GATGACCCG-3'
Cx32[a]: 5'-CTTTACCTCTTCCAGGTGAAGGGGGTCCCCATGCCC-3'
Cx26: 5'-GTTTGTTGACACCCCCGAGGATGCTCTGGAGTGTGCC-3'
Cx32: 5'-TGCCGATTCACGCCACTGAGCAAGGTGTATAGACC-3'
Cx36: 5'-ATCGTACACCGTCTCCCCTACAATGGCCAC-3'

[a]These specific probes hybridize wih the PCR2 product of the degenerated primers used for
the identification of the α- and β-family CXs.

3.2.3.1.1. Cloning of the PCR2 cDNA product. An agarose gel (1.5%) electrophoresis is performed by loading 10 μL of the PCR2 product. The primers,
used for the PCR2, contain restriction sites allowing directional cloning into a
vector (e.g., M13 or BS). After isolation of the PCR2 product from the gel
using standard methods (gel extraction kit Qiagen) and cleavage with adequate
restriction enzymes, the PCR fragment is cloned into MP13mp18RF (Stratagene) *(15)*. The digestion product should be cleaned by phenol–chloroform
extraction followed by elution of the digest on columns P100. For the ligation
a three- to tenfold molar excess of the PCR fragment vs the vector is used. The
transformation is performed using XL2-Blue competent cells (Stratagene) *(16)*.

3.2.3.1.2. Plaque identification. Plaque identification is achieved by hybridization of the nitrocellulose filters (Scleicher & Schuel) with specific radiolabeled probes. Cx subunit-specific oligonucleotide probes that are 30–45 nucleotides long are used. 5'-End radiolabeling of the oligonucleotide probes is performed in a standard fashion:

2 μL of 10X ligation buffer

4 pmol of oligonucleotide

8 pmol [γ-^{32}P] ATP (>5000 Ci/mmol, Amersham)

water to a final volume of 19 μL

All these steps are performed on ice. After the addition of 1 μL of T4 polynucleotide kinase (10 U, New England Biolabs), the reaction is incubated at 37°C for 30 min and then stopped by adding 30 μL of a 50 mM EDTA solution. The 50 μL samples are loaded on chromatography columns (Bio-spin 6, Bio-Rad) to elute only the radiolabeled probe.

The membranes are incubated overnight at 34°C with the radiolabeled probes in the following hybridization solution: 0.5 L 2× hybridization buffer, 0.3 L formamide (Sigma) and 0.2 L water.

The stringency of the washing conditions depends on the probe that is used. For our experiments the conditions were as follows: 2× washes with 0.5× saline sodium citrate (SSC) (Amresco), the first one at room temperature and the second at 56°C. The signal is detected by placing the dried membranes into an autoradiographic cassette against an X-ray film (Kodak, X-OMAT AR) for variable time depending of the strength of the signal and then developed (Hyper Processor, Amersham Life Science).

One radiolabeled oligonucleotide should be designed to recognize all members of the Cx family. In this way, a negative result obtained with the specific probes and a positive one obtained with the nonspecific Cx probe indicates that a new Cx might be expressed.

3.2.3.1.3. Sequencing. Southern blot results are confirmed either by direct sequencing of the PCR product or by sequencing the DNA obtained after plaque or colony picks. The negative plaques should mean that a new Cx is expressed and then must be sequenced.

3.2.3.2. CONFIRMATION OF THE IDENTITY OF THE CXS

3.2.3.2.1. PCRs performed with specific primers. As indicated previously, the possible repertoire of Cx subunit expression in identified neurons requires confirmation by PCR experiments in which subunit-specific primers that span an intron are used. Thus, if PCR experiments with the degenerate primers indicate, for example, that Cx26 and Cx32 are coexpressed in a particular cell type, this result needs to be verified by performing additional experiments in which

PCR experiments with Cx26- and Cx32-specific primers are used (*see* **Fig. 1**, *no. 4.1., 4.2.* and **Table 1**).

3.2.3.2.2. Multiplex PCRs. The aim of this step is to identify which Cxs are coexpressed in the studied cells to confirm the PCR results obtained with degenerate primers (*see* **Note 6**). For the PCR1, a 3'-degenerate primer located on exon 2 is used with 5'-specific primers located on exon 1 that are specific for each identified Cx (*see* **Fig. 1**, *no. 4.1.* and **Table 1**). These different primers are used at the same concentration. Two microliters of the PCR1 reaction are used for the second PCR. The PCRs2 are performed with different sets of specific primers for each identified Cx (*see* **Fig. 1**, *no. 4.2.* and **Table 1**). Also these primers should be intron-spanning. The PCR protocols are identical to those already described in **Subheading 3.2.2.**

3.2.3.2.3. Southern blot and sequencing. Diagnostic agarose gel electrophoresis of the PCR2 are performed and directly process for Southern blotting. Southern blots with radiolabeled oligonucleotides specific for each identified Cx are carried out systematically (**Table 1**). Agarose gels (as many as specific radiolabeled probes) are incubated twice for 10 min in a denaturation buffer and then blotted against a nylon membrane (Porablot). Covalent links between the cDNA and the membrane are established with UV crosslinking. The membranes are incubated overnight (shaking, 34°C) with the different radiolabeled oligonucleotides, then washed and exposed against X-ray films. Direct sequencing of the PCR products obtained after multiplex PCR should be carried out to confirm the specificity of the PCRs amplifications.

4. Notes

1. It is not possible to use ATP compounds (e.g., ATP-Mg) in the RT-PCR experimental conditions because the internal solution has to be autoclaved to be sterile. These compounds are degrated during the autoclaving procedure.
2. Biocytin or Lucifer Yellow filling of the cells with subsequent RT-PCR analysis is possible. Biocytin (5%) or Lucifer yellow (1%) do not alter cDNA synthesis and the PCR reactions. However, the autoclaving procedure of the internal solution significantly diminishes the fluorescence of Lucifer Yellow. Another restriction of these molecules is their long diffusion time of (at least 30 min for biocytin to study properly dye-coupling between neurons) which increases the degradation of the Cx mRNAs. In conclusion, these molecules, used in the RT-PCR conditions, are better suited for the morphological characterization and localization of the injected neurons rather than for the study of interneuronal coupling via Gjs.
3. In some cases, a neuronal subpopulation cannot be identified using electrophysiological parameters only, but additional analysis of the expression of different markers may be required. For example, three subpopulations among the medium spiny neurons of the striatum cannot be revealed by an electrophysiological study but only by analysis of the combination of the expression of two peptides mark-

ers, substance P and enkephalin (*10,11*). It will be necessary in these cases to couple the analysis of Cxs to that of different markers by single-cell RT-PCR. This means that the RT product from one cell should be split into different samples: one for the analysis of Cx expression and the others for the different markers. As the Cx expression is quite low in neurons, this procedure will lead to an important decrease of the efficiency of the Cx detection.

A solution may be to add a step after the RT: the reamplification of the total cDNA (*4,19*). This procedure consists of:

a. Performing the RT with oligo(dT) that are extended at the 5' end with the T7 RNA polymerase promotor.
b. Once the cDNA synthesis is achieved, using the T7 RNA polymerase and synthesizing amplified RNA (aRNA).
c. With random primers and Klenow enzyme, synthesizing the aRNA-cDNA hybrid.
d. Double-stranded cDNA synthesis.

In this way, larger amounts of total cDNA should permit to split the RT products into different samples without losing the efficiency for the Cx expression analysis.

4. The single-cell RT-PCR technique has been used to study distribution of mRNAs in neurites (*20*). Because the coupling sites between neurons seem to be localized on the neurites, this approach could be a good alternative to study distribution of Cx mRNAs.
5. The entire coding sequence of the Cxs are located on exon 2 except for the recently cloned Cx36 (*17,18*). This Cx belongs to the d-family and is expressed specifically in the neurones. The degenerate primers proposed in this chapter are unable to permit the amplification of the Cx36. For this reason, specific primers designed for the amplification of this neuronal Cx are proposed as an alternative.
6. For members of the same family, such as Cxs, it is possible to perform multiplex PCRs. Controls need to be performed that ascertain the amplification of all transcripts with the primers that are chosen. Often, the optimal PCR conditions for one template do not coincide with those of another one.

Acknowledgments

I thank Prof. H. Monyer for scientific supervision, helpful discussions, and critically reading the manuscript. I thank also Prof. P. Seeburg for his constant support and U. Amtmann for technical assistance.

References

1. Bruzzone, R., White, T. W., and Paul, D. L. (1996) Connections with connexins: the molecular basis of direct signalling. *Eur. J. Biochem.* **238**, 1–27.
2. Kumar, N. M. and Gilula, N. B. (1996) The gap junction communication channel. *Cell* **84**, 381–388.
3. Van Gelder, R. N., Von Zastrov, M. E., Yool, A., Dement, W. C., Barchas, J. D., and Eberwine, J. H. (1990) Amplified RNA synthesized from limited quantities of heterogenous cDNA. *Proc. Natl. Acad. Sci. USA* **87**, 1663–1667.

4. Eberwine, J., Yeh, H., Miyashiro, K., Cao, Y., Nair, S., Finnell, R., Zettel, M., and Coleman, P. (1992) Analysis of gene expression in single live neurons. *Proc. Natl. Acad. Sci. USA* **89,** 3010–3014.

5. Surmeier, D. J., Eberwine, J. H., Wilson, C. J., Cao, Y., Stefani, A., and Kitai, S. T. (1992) Dopamine receptor subtypes colocalize in rat striatonigral neurons. *Proc. Natl. Acad. Sci. USA* **89,** 10,178–10,182.

6. Lambolez, B., Audinat, E., Bochet, P., Crepel, F., and Rossier, J. (1992) AMPA receptor subunits expressed by single Purkinje cells. *Neuron* **9,** 247–258.

7. Jonas, P., Racca, C., Sakmann, B., Seeburg, P., and Monyer, H. (1994) Differences in Ca^{2+} permeability of AMPA-type glutamate receptor channels in neocortical neurons caused by differential GluR-B subunit expression. *Neuron* **12,** 1281–1289.

8. Monyer, H. and Jonas, P. (1995) Polymerase chain reaction analysis of ion channel expression in single neurons of brain slices, in *Single Channel Recording*, 2nd edit. (Sakmann, B. and Neher, E., eds.), Plenum Press, New York, pp. 357–373.

9. Monyer, H. and Lambolez, B. (1995) Molecular biology and physiology at the single-cell level. *Curr. Opin. Neurobiol.* **5,** 382–387.

10. Surmeier, D. J., Song, W.-J., and Yan, Z. (1996) Coordinated expression of dopamine receptors in neostriatal medium spiny neurons. *J. Neurosci.* **16,** 6579–6591.

11. Mermelstein, P. G., Song, W.-J., Tkatch, T., Yan, Z., and Surmeier, D. J. (1998) Inwardly rectifying potassium (IRK) current are correlated with IRK subunit expression in rat nucleus accumbens medium spiny neurons. *J. Neurosci.* **18,** 6650–6661.

12. Cauli, B., Audinat, E., Lambolez, B., Angulo, M. C., Ropert, N., Tsuzuki, K., Hestrin, S., and Rossier, J. (1997) Molecular and physiological diversity of cortical nonpyramidal cells. *J. Neurosci.* **10,** 3894–3906.

13. Sakmann, B. and Stuart, G. (1995) Patch-pipette recordings from the soma, dendrites, and axon of neurons in brain slices, in *Single Channel Recording*, 2nd edit. (Sakmann, B. and Neher, E., eds.), Plenum Press, New York.

14. Roux, K. H. (1995) Optimization and troubleshooting in PCR, in *PCR Primer. A Laboratory Manual* (Dieffenbach, C. and Dveksler, G., eds.), Cold Spring Harbor Laboratory Press, Cold Spring Harbor, NY, pp. 53–62.

15. Yanish-Perron, C., Vieira, J., and Messing, J. (1985) Improved M13 phage cloning vectors and host strains: nucleotide sequences of M13mp18 and pUC19 vectors. *Gene* **33,** 103–119.

16. Sambrook, J., Fritsh, E. F., and Maniatis, T. (1989) Cloning into bacteriophage M13 vectors and transformation of competent bacteria, in *Molecular Cloning*. 2nd edit., Cold Spring Harbor Laboratory Press, Cold Spring Harbor, NY, pp. 4.33–4.38.

17. Condorelli, D. F., Parenti, R., Spinella, F., Salinaro, A. T., Belluardo, N., Cardile, V., and Cicirata, F. (1998) Cloning of a new gap junction gene (Cx36) highly expressed in a mammalian brain neurons. *Eur. J. Neurosci.* **10,** 1202–1208.

18. Söhl, G., Degen, J., Teubner, B., and Willecke, K. (1998) The murine gap junction gene connexin36 is highly expressed in mouse retina and regulated during brain development. *FEBS Lett.* **428,** 27–31.

19. Chiang, L. W. (1998) Detection of gene expression in single neurons by patch-clamp and single-cell reverse transcriptase polymerase chain reaction. *J. Chromatogr. A* **806,** 209–218.
20. Miyashiro, K., Dichter, M., and Eberwine, J. H. (1994) On the nature and differential distribution of mRNAs in hippocampal neurites: implications for neuronal functioning. *Proc. Natl. Acad. Sci. USA* **91,** 10,800–10,804.

9

Use of Retroviruses to Express Connexins

Jean X. Jiang

1. Introduction

Gap junction-mediated intercellular communications are known to be important for tissue growth, differentiation, and signaling (1). To experimentally probe the roles played by connexins in tissue development and function, we describe here a retroviral approach that has been successfully used to express nonviral exogenous connexins in embryonic chick lens in vivo (2). The expression of the retrovirally transduced connexin genes is found to be colocalized with their connexin counterparts in vivo.

This avian retroviral helper-independent vector was originally developed by Hughes and colleagues (3). These vectors utilize the same efficient and precise integration machinery of naturally occurring retroviruses to produce a single copy of the viral genome stably integrated into the host chromosome. These vectors, derived from the Rous sarcoma virus (RSV), are transformation defective and replication competent strains, suggesting that their amplification is directly associated with host cell DNA replication and cell division. They are capable of continuously spreading into dividing cells until cell proliferation ceases. Thus, if administered properly, virtually all of the virus-encountering cells will be infected. Compared to avian replication competent retroviruses, most murine retroviruses are of limited use for many in vivo studies, as they induce effective gene transfer in only a small number of cells (4).

There are two potential applications of retroviral approaches in the study of gap junction function. First, the retroviral vector provides a unique tool to transduce full-length connexin or mutant genes into target organs *in situ* and to study the effect of connexins on the development of various organs and tissues during avian embryogenesis. Retroviral constructs containing viral proteins (env, pol, and gag) conjugated with different connexins or their mutants can be

From: *Methods in Molecular Biology, vol. 154: Connexin Methods and Protocols*
Edited by: R. Bruzzone and C. Giaume © Humana Press Inc., Totowa, NJ

microinjected into target organs or tissues and the developmental and functional impacts of those genes can be analyzed directly. Second, it is known that infection of DNA constructs into primary cell culture is often more difficult and less efficient. The use of replication competent retroviral constructs eliminates these problems and the efficiency of infection can be close to 100%.

After identification of specific connexin gene(s) and target organ(s) or tissue(s) to study, the overall experimental approach should proceed in the following order (details described in **Subheading 2.**): (1) Clone exogenous connexin gene(s) into retroviral vector in a two step process: first into an adapter vector and then into a viral vector. (2) Prepare high-titer viruses from packaging cell cultures. (3) Microinject retroviral vectors containing connexin genes into target organs or tissues *in situ* or infect primary cell cultures. (4) Analyze the consequences of exogenous gene expression by various biochemical and histochemical means.

2. Materials

1. CEF cells, retroviral vector, RCAS(A) were generously provided by Dr. Constance Cepko's lab (Harvard Medical School, Boston, MA). For RCAS(A), RC stands for replication competent; A for avian leukemia virus; S for a splice acceptor; (A) for a subgroup of retroviruses. Avian viruses have multiple envelope (env) types, called subgroups A–E. Subgroup A envelope allows infection of most commercially available chicken strains.
2. Anti-FLAG epitope monoclonal antiserum (M2) from mouse is from Sigma (St. Louis, MO). The FLAG marker peptide is: NH_3-Asp-Tyr-Lys-Asp-Asp-Asp-Asp-Lys-COOH.
3. QT-6, a chemically transformed quail fibroblast cell line, is from ATCC (Bethesda, MD).
4. Restriction enzymes, DNA polymerases, and T4 ligase are from New England Biolabs (Revere, MA).
5. Klentaq and 10X Klentaq buffer are from Sigma (St. Louis, MO).
6. Deoxynucleotides triphosphates (10 mM) are from Sigma (St. Louis, MO). A stock solution of 1.25 mM is prepared with distilled water and stored at –20°C.
7. Lipofectamine, Dulbecco's modified Eagle medium (DMEM), and OPTI-MEM are from Gibco (Grand Island, NY).
8. Fetal bovine serum (FBS) is from Hyclone (Logan, UT).
9. Chick serum is from Sigma (St. Louis, MO).
10. Gel isolation and large quantity DNA purification kits are from Qiagen (Santa Clarita, CA). The experimental procedures are exactly based on the user's manual provided by the manufacturer.
11. P27 (antiviral protein gag) polyclonal antibody raised from rabbit is obtained from SPAFAS (Norwich, CT).
12. Blocking solution (1% bovine serum albumin [BSA], 2% normal goat serum, 2% fish skin gelatin, and 0.25% Triton in phosphate-buffered saline [PBS]).

13. Peroxide stock solution (30%).
14. 3,3'-Diaminobenzidine tetrahydrochloride (DAB) from Sigma (St. Louis, MO).
15. Fertilized eggs from Specific Pathogen Free (SPF-standard) white Leghorn chickens from SPAFAS (Norwich, CT). Prior to incubation at 37°C, fertilized eggs can be stored at 14°C for no longer than 2 wk. The fertilized eggs are incubated for the desired times in a humidified 37°C rotating wooden incubator from Petersime (Gettysburg, OH) specially designed for hatching chick or bird eggs.
16. 70% Ethanol.
17. Small curved scissors and fine forceps (Storz, St. Louis, MO), previously soaked in ethanol.
18. 18-gauge needle attached to a 5-mL syringe.
19. Scotch tape (1.2 cm width) and clear packaging tape (4.5 cm width).
20. Egg holder, homemade styrofoam rings with 2-inch diameter and one-half inch height.
21. Dissection microscope (Olympus SZ-4045 Zoom Stereomicroscope).
22. Fiberoptic dual light source (Olympus, Japan).
23. Microinjector (PLI-100 Picoinjector, Medical System, Greenvale, NY).
24. Micromanipulator (MM-33A, Sutter, Novato, CA).
25. Micropipet grinder (EG-40, Narishige, Japan) for beveling tips for microinjection pipets.
26. Vertical micropipet puller (P-30, Sutter, Novato, CA).
27. Approximately 10 glass micropipets (outer diameter, 1 mm; inner diameter, 0.75 mm, Sutter, Novato, CA); pull the pipets by a micropipet puller, break the tip with forceps, and smooth the tip with a micropipet grinder.
28. Fast Green (0.25% in water), filter sterilized and stored frozen as aliquots.
29. Concentrated virus, store at –80°C, thaw, and place on ice.
30. Benchtop nonrotating incubator.
31. Tissue-Tek compound used to mount tissues for frozen sections is from Miles Scientific (Naperville, IL).
32. Formaldehyde (16% stock solution), for cell fixation, is obtained from Electron Microscopy Science (Ft. Washington, PA) and stored in the dark at 4°C.
33. Rhodamine-conjugated goat antimouse IgG or goat antirabbit IgG are from Cappel (West Chester, PA) and are stored in the dark at 4°C.
34. Fluorescein isothiocyanate (FITC)-conjugated goat antirabbit IgG is from Boehringer Mannheim (Indianapolis, IN) and is stored in the dark at 4°C.
35. Anti-rabbit or mouse alkaline phosphatase conjugated Ig is from Promega (Madison, WI).
36. Pap pen for making wax rings is from Binding Site (San Diego, CA).
37. Vectashield® fluorescent mounting medium containing anti-quenching reagents and ABC staining kit for immunohistochemical staining are from Vector Laboratories (Burlingame, CA). The ABC kit contains two solutions, avidin-DH and biotinylated enzyme.
38. Superfrost® glass slides are from Fisher Scientific (Pittsburgh, PA).
39. HM505 cryostat from Microm (Walldorf, Germany).
40. Tmax 400 film is from Kodak (Rochester, NY).

41. Lysis buffer (5 mM each of Tris-HCl, EDTA, and EGTA plus 10 mM N-ethylamleimide, 2 mM phenylmethyllsulfonyl fluoride [PMSF] and 0.2 mM leupeptin, pH 8.0).

42. Sample loading buffer 2× (0.05 M Tris-HCl, pH 6.8, 1% sodium dodecyl sulfate [SDS], 2% β-mercaptoethanol, 35% glycerol).

43. Nitrocellulose membrane for Western blot is from Schleicher & Schuell (Keene, NH).

44. Electrophoresis gel components: 30% *bis*-acrylamide solution, ammonium persulfate, TEMED, Mini-Protean II Electrophoresis Cell, and Trans-Blot Electrophoretic Transfer Cell with plate electrodes are from Bio-Rad (Hercules, CA).

45. Resolving gel buffer, 1.5 M Tris-HCl, pH 8.8, stacking gel buffer, 0.5 M Tris-HCl, pH 6.8.

46. Running gel buffer: 24.8 mM Tris base, 192 mM glycine, 0.1% SDS.

47. Fix/destain solution (10% glacial acetic acid, 20% methanol).

48. Transfer buffer: 24.8 mM Tris base, 192 mM glycine, 20% methanol. Make a 10× stock of Tris and glycine and store at room temperature. Dilute to 1× and add methanol just before use.

49. Blotto (10% nonfat dried milk and 0.02% sodium azide in PBS).

50. Secondary antibody washing buffer: 150 mM NaCl, 50 mM Tris-HCl, pH 7, 0.5% Tween-20.

51. Nitroblue tetrazolium (NBT) (100 mg/mL in 70% dimethylformamide) and 5-bromo-4-chloro-3-indolyl phosphate (BCIP) (50 mg/mL in 100% dimethylformamide) from Sigma (St. Louis, MO).

52. Alkaline phosphatase reaction buffer: 100 mM Tris-HCl, 100 mM NaCl, 5 mM MgCl$_2$, pH 9.5.

53. All other chemicals and reagents are obtained from either Sigma (St. Louis, MO) or Fisher Scientific (Pittsburgh, PA).

3. Methods

3.1. Making Retroviral DNA Constructs

Cloning of full-length connexin cDNA or connexin gene fragments into viral DNA vectors involves two steps. First the DNA insert is cloned into an adaptor plasmid and then into a viral vector. The adaptor plasmid called CLA12NCO was originally constructed by Hughes et al. *(3)*. This miniplasmid can convert virtually any DNA segment into a *Cla*I fragment suitable for insertion into the retroviral vector (RCAS). The CLA12NCO plasmid consists of a polylinker region with multiple restriction enzyme sites (**Fig. 1A**) and start codon (ATG), which is encoded within the *Nco*I site. There are two unique *Cla*I sites on both sides of the polylinker regions.

3.1.1. Polymerase Chain Reaction (PCR) to Make Connexin DNA Fragments

1. Design primers for PCR. The 5' end sense strand PCR primer designed contains the ATG starting codon encoded within an *Nco*I site and 15–18 additional coding nucleotides (nt) downstream of the start codon sequence. The 3' end antisense

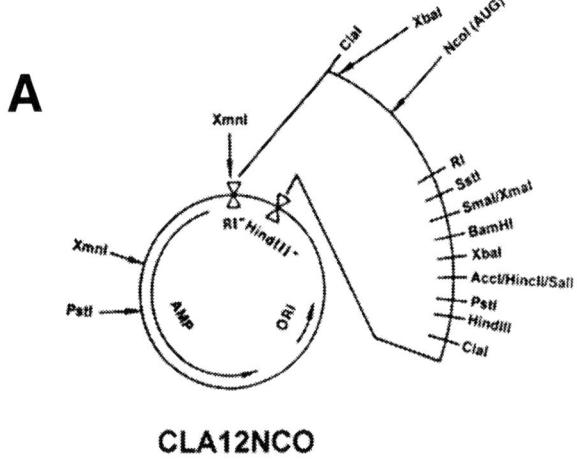

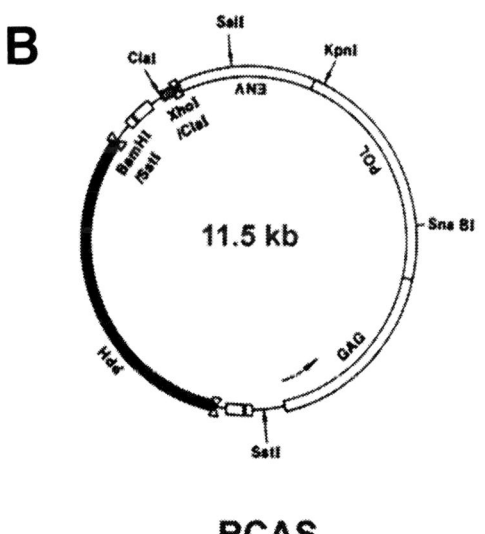

Fig. 1. Maps of adaptor plasmid, CLA12NCO (**A**) and retroviral vector, RCAS (**B**). PPH, *E.coli* replicon. The *arrow* points to the direction of gene transcription.

strand primer includes the last 15–18 nt of the coding sequence, the FLAG epitope tag sequence (5'-GACTACAAGGACGACGATGACAAG-3'), a stop codon, and one of the restriction enzyme sites (i.e., *Eco*RI) present within the polylinker region of CLA12NCO (**Fig. 1A**). Total length of the antisense primer is approx 50–60 nt (*see* **Note 1**).

2. PCR reactions. The typical PCR reaction in our experiments contains: 0.5 μg of each connexin template, 100 picomol of sense and antisense primers, 200 μM dNTP (8 μL of 1.25 mM stock), 5 μL of 10× Klentaq buffer (Sigma), and 0.5 μL of Klentaq DNA polymerase (4.5 U) (Sigma) in 50 μL. Thirty cycles of the following program are performed: 94°C for 1 min, 58°C for 1 min (depending on the T_m derived from primer sequences. For the way to estimate T_m, *see* **Note 2**), and 72°C for 1 min. The reaction is completed with a final extension for 10 min at 72°C.

3. Separate PCR products on a 1% agarose gel. Excise the DNA band from the gel and purify it using a gel extraction kit (Qiagen) to elute DNA from the gel.

4. Digest purified PCR products with restriction enzymes in separation reactions, *Nco*I for the 5' end and *Eco*RI for the 3' end. The choice of the restriction enzymes is flexible for 3' end primers, as there are multiple restriction sites in the polylinker regions, whereas the 5' end has to be a *Nco*I site encoding the start codon. Digestion by certain restriction enzymes requires additional nucleotides flanking the consensus sequences for optimum performance (*see* **Note 1**).

3.1.2. Making Adaptor Plasmid Containing DNA Inserts

The aforementioned restriction enzyme treated PCR products are subcloned into the adaptor plasmid CLA12NCO. The CLA12NCO vector (shown in **Fig. 1A**) is linearized by separate digestion with *Nco*I and *Eco*RI, and then is gel isolated. Restriction enzyme treated PCR fragment and CLA12NCO are mixed at molar ratio of 2–5:1 and the sticky end-ligation reaction is performed at 16°C with T4-ligase for 16 h, and followed by transformation using competent cells (i.e., DH5α). The colonies on the culture plates are collected, inoculated overnight at 37°C, and minipreps are obtained. The constructs with DNA inserts of the correct size are selected and a large quantity of DNA is purified using a DNA isolation kit (Qiagen) according to the manufacturer's instructions. To ensure the accuracy of these DNA sequences, the constructs are sequenced using primers derived from CLA12NCO sequences.

3.1.3. Cloning into Viral Vector, RCAS(A)

The CLA12NCO vector containing the correct connexin gene is then digested with *Cla*I and the resulting fragments are gel isolated. The linearized viral vector, RCAS(A), is prepared by digestion with *Cla*I and pretreated with alkaline phosphatase to eliminate the 5' end phosphate group, thus preventing self-ligation of the vector. The DNA fragments with *Cla*I sites on both ends are then ligated into *Cla*I linearized RCAS(A) and transformed into competent cells (i.e., DH5α). The constructs with the right size inserts are selected. Because *Cla*I creates the same sticky ends on both 5' and 3' ends of the inserts, DNA fragments can be inserted into RCAS(A) by both sense and antisense orientations. The restriction enzyme *Sal*I is used to distinguish these two possibilities. There are two *Sal*I sites: one at approx 900 bp downstream of the *Cla*I site on RCAS(A) plasmid (**Fig. 1B**), and another site is in the polylinker region of CLA12NCO, downstream from the *Nco*I and *Eco*RI sites (**Fig. 1A**). The arrow points toward the direction (counterclockwise) of the gene transcription. If the DNA insert is in the correct orientation, *Sal*I digestion should release a fragment with 900 bp plus the size of the full-length inserts, whereas in the wrong orientation the size of the excised fragment is approx 900 bp. Finally, large amounts of RCAS(A) containing the subcloned connexins are prepared using a Maxi-Qiagen kit. The DNA concentration is determined by spectro-photometric reading at 260 nm ($A_{260} \times 50 \times$ dilution factor = μg/mL) and is confirmed by DNA band intensity comparison with a known concentration of DNA molecular weight standard on agarose gels.

3.2. Preparation of High-Titer Viral Stocks and Titering

3.2.1. Preparation of High-Titer Viral Stocks from Primary Chick Embryonic Fibroblast (CEF) Packaging Cells

Preparation of high-titer retrovirus from avian packaging cells is modified from the published procedure *(5)*. To obtain chick RCAS(A) retroviruses, CEF, a virus packaging cell line, is commonly employed. CEF cells should be used with the least possible passages to achieve high titer viral stocks.

The medium used for culturing CEF cells is DMEM supplemented with 10% FBS and 2% chick serum.

1. Plate CEF cells at a density of 7.5×10^5 cells on a 60-mm culture plate the day before transfection to reach 50–75% cell confluence for transfection (the cell density is critical for transfection; *see* **Note 3**).
2. In 600 μL of OPTI-MEM, mix 4.5 μg of DNA with 36 μg of lipofectamine and incubate for 20 min at room temperature. After incubation, dilute samples with 2.4 mL of OPTI-MEM and add it to CEF cells that have been prewashed once with OPTI-MEM. Then, incubate cell cultures at 37°C for 6 h before adding 3 mL of medium (DMEM supplemented with 20% FBS and 4% chick serum).
3. Approx 2 d after transfection without changing medium, when cells are confluent, split cells to a 100-mm plate. After cultures in 100-mm plates reach confluence, split cells onto a 150-mm plate.
4. When confluent, split onto four 150-mm plates for preparation of viral stocks and two 35-mm plates for analysis of gene expression. One 35-mm plate should con-tain a 22×22 mm glass coverslip (*see* **Subheading 3.4.1.**).
5. Feed cultures daily with 10 mL of fresh medium (DMEM plus 10% FBS and 2% chick serum) for each 150-mm plate until cells reach confluence. Begin collect-ing the medium 2 d after cultures have reached confluence by filtering through a 45-μm (not 22-μm) pore-size filter to eliminate cell debris. Each day, collect, pool, and filter the medium collected from 4×150 mm culture plates. Continue this process for 3 consecutive days. Store the medium containing virions in a –80° freezer for an unlimited period of time.

To avoid repeated passages of the virion containing cells and decrease the chance of cell contamination, an alternative, short protocol for preparation of viral stocks is described including initial transfection followed by later infection procedures. The yield is compatible with the aforementioned multiple-step preparation.

1. The day before transfection, plate CEF cells at a density of 2.5×10^5 cells on a 35-mm culture plate, instead of on a 60-mm plate.
2. In 200 µL of OPTI-MEM, mix 1.5 µg of DNA with 12 µg of lipofectamine and incubate for 20 min at room temperature. After incubation, dilute samples with 0.8 mL of OPTI-MEM and add it to CEF cells that have been prewashed once with OPTI-MEM. Then, incubate cells at 37°C for 6 h before adding 1 mL of growth medium.
3. Two days after transfection, when the cells are confluent, feed cultures with 2 mL of fresh medium. The next day, collect the medium and store at –80°C. Repeat medium change and collection for the following 3 consecutive days. These aliquots contain low titers of viruses.
4. Prepare CEF cells, in parallel, on four 150-mm plates the evening before infection. This will result in a 20–30% confluence by the next day and will give better infection results (*see* **Note 3**).
5. For infection, dilute 2 mL of medium collected in **step 3** with 18 mL of culture medium. Add 5 mL of this 1:10 dilution to each 150-mm plate and incubate for 2 h at 37°C. Then add additional 10 mL of fresh culture medium to each plate. Change medium daily until cells are confluent. Change with 10 mL of fresh growth medium, collect the next day, and pool the medium from four 150-mm plates. Filter the collected medium through a 0.45-µm filter and store at –80°C. Keep feeding cells and collecting medium daily for 3–4 d.

3.2.2. Concentrate Viral Stocks

1. Prior to centrifugation to concentrate viral stock, rinse Beckman SW28 rotor buckets and tubes with 70% ethanol and dry them under UV in the culture hood overnight (maintain sterile conditions for preparation of viral stocks; *see* **Note 4**).
2. Thaw collected culture medium containing virions, stored at -80°C, on ice. Divide 40 mL of medium collected from four 150-mm plates into two centrifuge tubes and centrifuge them at 72,000*g* (Beckman SW28 rotor, 20,000 rpm) for 2 h at 4°C.
3. Carefully decant supernatant and aspirate the residual liquid attached to the inside wall of the centrifuge tubes starting 0.5 cm from the bottom of the tube. Resuspend the viral pellets with a residual amount of medium (~50 µL) and shake vigorously on a platform shaker for 2 h at 4°C. Then aliquot resuspended viruses (10 µL each into 0.5-mL Eppendorf tubes) and store at –80°C for an unlimited period of time. This step concentrates the viruses up to 500-fold.

3.2.3. Titering Viral Stock

The concentrated viruses are titered according to a modified procedure (*6*), using quail QT-6 cells. This cell line can reach only < 50% spreading even with prolonged cell passages, thus providing an ideal culture system for the titering assay.

1. The day before titering, plate QT-6 cells (approx 2×10^5 per well) into six-well culture plates in the same culture medium as for CEF cells.
2. Make serial dilutions (10^4, 10^5, 10^6, 10^7, and 10^8) of concentrated viral stocks with culture medium and add separately 0.5 mL of each dilution to five wells of a six-well culture plate. The sixth well is a nonviral infection control. Incubate plates for 3 h at 37°C and then add an additional 0.75 mL of culture medium.
3. After approx 48 h of incubation, when cells reach about 100% confluence (the confluence is very critical for the success of titering assay; *see* **Note 5**), fix cells with 4% paraformaldehyde (diluted in PBS from a 16% stock solution) (*see* **Note 6**) for 10 min and wash 3× with PBS, 5 min each.
4. Incubate cells with 1 mL of blocking solution containing 2% normal goat serum, 2% fish skin gelatin, 0.5% Triton X-100, and 1% bovine serum albumin (BSA) in PBS for 30 min.
5. To determine the expression of gag, one of the viral proteins, use 1 mL of rabbit polyclonal antiserum P27, at a 1:150 dilution (in blocking solution) for 1 h at room temperature.
6. Detect antigenic signals using the ABC kit exactly as indicated in the manufacturer's instructions. In brief, first incubate cultures with 1 mL of biotinylated anti-rabbit antibody (1:200 dilution) in blocking solution for 1 h and then wash 3× with PBS. ABC staining steps include mixing four drops of solution A and B each into 10 mL of 1% normal goat serum in PBS (this solution needs to be prepared at least 30 min earlier) and adding 1 mL of ABC solution to the cells for 30 min.
7. Wash the cells 3× with PBS and incubate with 80 μL of 1% peroxide (prepared by diluting a 30% stock solution with PBS) plus 8 mL of 3,3'-diaminobenzidine tetrahydrochloride (DAB) (160 μL of 10 mg/mL DAB in 8 mL of PBS). The staining process lasts for approx 5 min until the development of a brownish color. Stop the reaction by washing 3× in PBS. Identify positive colonies as clusters of brownish-colored cells and count the clusters to obtain the titer, defined as colony-forming units per milliliter (cfu/mL). For example, if we count 15 colonies in 10^7 dilution wells, the viral titer should be 15×2 (add 0.5 mL instead of 1 mL) $\times 10^7 = 3 \times 10^8$ cfu/mL. The good virus titers we have obtained are approx $1-5 \times 10^8$ cfu/mL after concentration.

3.3. Microinjection and Expression of RCAS(A) Connexin Constructs into Chick Embryos In Situ

To examine the roles played by connexins in chicken embryonic development, the retroviral transduced connexin genes can be delivered into certain regions of chick embryos at various developmental stages. By varying the site and time of infection, targeted gene transfer can be confined to certain selected populations of cells without the need for tissue-specific promoters. For example, during early development of the eye lens, formation of the lens vesicle with a central lumen provides an ideal site for microinjection of retroviral con-

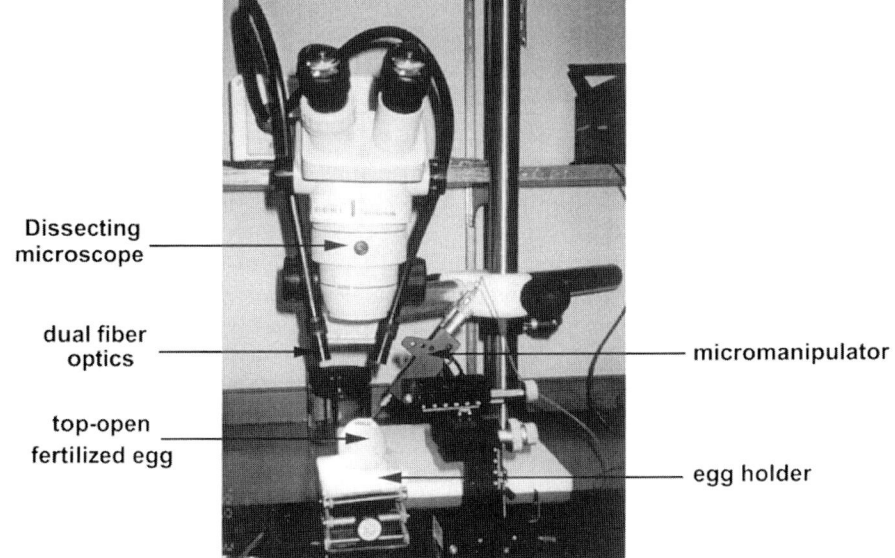

Fig. 2. Setup for microinjection of retroviruses into chick embryos. An egg with an open shell exposing the embryo to be injected is shown.

structs to infect surrounding lens cells *(2)*. After injection, the embryos can resume their development, thus permitting analysis of the result of forced connexin expression and of the manipulation of cell–cell communication *in situ* within the living organism. The microinjection of virus into chick embryos is partially based on a published procedure and the setup is shown in **Fig. 2 (6)**.

3.3.1. Procedures for Microinjection

1. Warm fertilized eggs to room temperature and then incubate them in a humidified, rocking incubator approx 37.5°C for the designated time. The incubation time depends on the developmental stage required for injection. Detailed criteria for determination of the developmental stage are described by Hamburger and Hamilton *(7)*.
2. Pull glass micropipets with a micropipet puller, break the tips using a pair of forceps, and grind them to make smooth tips. The diameter of broken tips is approx 0.5 μm.
3. Remove a batch of eggs from the incubator, drench with 70% ethanol, and air-dry.
4. Lay an egg on its side in an egg holder and use an 18-gauge needle attached to a 5-mL syringe to produce a small hole at the larger end of the eggs. Insert the needle about 0.5 cm into the albumin at a steep angle to avoid hitting the yolk and withdraw 2–3 mL of albumin and discard. (If some of the yolks are withdrawn, discard the egg). Seal the holes with Scotch tape.

5. Locate the embryo by turning off the main light sources in the room and directly illuminate the egg with a fiberoptic light source. Circle the embryo shallow with a pencil on the shell of the egg and attach a piece of transparent packaging tape on the top. Use small curved scissors to cut through the transparent tape to prevent small pieces of shell from falling into the egg. Open and enlarge the hole to approx 2 cm in diameter following the pencil marks to expose the embryo for injection.

6. Illuminate the embryo with fiberoptics light source under a dissection microscope. Embryos younger than 2 d can usually be injected directly. Embryos older than 2 d require additional preparation because of a thickening and increased vascularization of the embryonic membranes. With fine dissecting forceps and scissors, cut off the amnionic membrane directly overlying on top of the embryos. Try not to cut too widely, otherwise the embryos might sink into the yolk mass, and avoid touching blood vessels if possible. Cover the opening with the lid of a 60-mm Petri dish and set aside.

7. Dilute 1 µL of 10× Fast green into one vial (10 µL) of concentrated viral stocks. Centrifuge for approx 10 s in a benchtop microcentrifuge (10,000*g*) at 4°C to pellet off big chunks of undissolved material, which might clog the micropipet tip. Keep the virus stock on ice during microinjection process. One virus stock can be used to inject a batch of embryos (usually about a few dozen per experiment).

8. Insert a micropipet into the microinjector tubing attached to a micromanipulator. Place 1 µL of Fast Green containing viral stock on a piece of parafilm, and lower the tip of the micropipet into the viral stock. Fill the micropipet with viral solutions. Place the tip of the filled micropipet in sterile saline whenever there is a lag in the procedure.

9. Lower the micropipet into the target region of the embryo using a micromanipulator. For injection into the embryonic eye lens, adjust light source until the outline of lens vesicle is visible.

10. Inject about 0.05–0.5 µL of viral stocks. The electrode is not calibrated and it is recommended that as much as possible be injected. After injection, reseal the hole with the packaging tape (4.5 cm width) and return to a benchtop incubator without rotating motion at 37°C.

3.3.2. Preparation of Tissue Sections

After incubation of virus-injected embryos in a benchtop incubator for the designated time, dissect out the organs or whole embryos where viruses have been injected. By now, the overall changes (i.e., size, shape, developmental defects, etc.) can be examined and recorded. To study the results in more detail, tissue sections are prepared as follows:

1. Wash organs or whole embryos with PBS and fix with 2% paraformaldehyde (*see* **Note 6**) in PBS for 30 min.

2. After fixation, immerse the specimen into 1 *M* sucrose and saturate cells with sucrose molecules. Sucrose can prevent ice particle formation inside the organ

during the freezing process, thus minimizing any tissue damage. Tissues will first float on the surface because of the high density of the sucrose, but they will gradually sink to the bottom of the container as sucrose molecules permeate the specimens. The bigger the specimen is, the longer it takes to sink to the bottom The time varies from a few hours to overnight.

3. Make cone-shape freezing tissue holders by wrapping aluminum foil around the bottom end of an Eppendorf tube used as the mold. Put one or two drops of Tissue-Tek (Miles Scientific) OCT compound into the holder. Immerse tissues into OCT compound under the microscope and adjust them with forceps to ensure a proper orientation. Quickly freeze them by immersion in liquid nitrogen and store at –80°C for an unlimited period of time.

4. Prepare tissue sections using an HM505 cryostat (Microm, Walldorf, Germany). Unwrap the tissues from aluminum foil holders, mount them on disc holders fitted with a cryostat using a drop of OCT, and fix them by freezing them inside the cytostat until the OCT hardens and turns whitish. Collect 10–15-μm thick sections onto Superfrost® glass slides (Fisher Scientific) and store those slides at –80°C.

3.4. Connexin Expression

3.4.1. Detection of Connexin Expression During Viral Stock Preparation Process

During the amplification steps for viral stock preparation, expression of viral and inserted connexin genes can be examined. As described in **Subheading 3.2.1.**, **step 4**, cultures are split from one 150-mm plate to four 150-mm plates and two additional 35-mm plates, one of which contains a 22×22 mm glass coverslip. Once cultures in these two 35-mm plates are confluent, they can be used for detection of viral and inserted gene expression either by immunofluorescence or by Western blot.

3.4.1.1. IMMUNOFLUORESCENCE STAINING

1. After an initial wash (3× with PBS) of the 35-mm cultures with a glass coverslip inside, cells are fixed with 1 mL of 2% paraformaldehyde in PBS (diluted from a 16% stock solution) (*see* **Note 6**) for 30 min. After fixation, rinse 3× with PBS and then incubate in blocking solution (*see* **Subheading 3.2.3.**, **step 4** for its composition) for 30 min.

2. To detect expression of the viral protein gag, incubate cells with 1 mL of anti-gag antibody (P27, 1:150 dilution) for 1 h and wash 3× with PBS. To detect connexin insert expression, incubate cells with corresponding a 0.5 mL of anti-connexin antibody or anti-epitope tag FLAG antibody (10 μg/mL) for 1 h. Incubation at 4°C overnight can enhance the specificity of the signal.

3. Wash cells 3× with PBS and incubate for 1 h at room temperature with 0.5 mL of rodamine or fluorescein FITC conjugated anti-rabbit or anti-mouse secondary antibodies, at a 1:500 dilution (*see* **Note 7**).

4. After secondary antibody incubation, wash the cell with PBS 3×. Pick up the glass coverslip from a 35-mm plate with cultures on top, flip it over, and place it

on a drop of mounting medium (Vector Laboratories) on the surface of a glass slide. Immobilize the glass coverslip with nail polish.

5. Examine immunostaining using a fluorescence microscope and record images with a cool-headed digital camera or on film (Tmax 400, Kodak, Rochester, NY).

3.4.1.2. PREPARATION OF A CRUDE MEMBRANE AND WESTERN BLOTTING

Because connexins are transmembrane proteins, separation of a crude membrane fraction from soluble components prior to Western blotting helps to concentrate the samples and to decrease the possible interference of signals from soluble proteins.

1. Wash the cells 3× with PBS.
2. Scrape the virus-infected cultures from a 35-mm culture plate with a rubber policeman with 1 mL of lysis buffer. Disrupt cells by repeated passages (20×) through an 18-gauge needle in lysis buffer.
3. Centrifuge whole cell lysates in a Beckman SW60 Ti rotor, 100,000g (28,000 rpm) for 30 min. Discard the supernatant and resuspend the pellet in 25 µL of PBS. Add 25 µL of 2× concentrated sample loading buffer (0.05 M Tris-HCl, pH 6.8, 1% SDS, 2% β-mercaptoethanol, 35% glycerol) and boil for 5 min. Separate proteins on a 10% SDS-polyacrylamide gel (SDS-PAGE), using a minigel apparatus.
4. Transfer proteins to a nitrocellulose membrane at 75V for 1 h with ice-cold gel transfer buffer. Transferred proteins are visualized by immersing the nitrocellulose membrane in 0.2% Ponceau S (in 1% acetic acid) until the pink protein bands are visible. Mark the positions of molecular weight standards with a ballpoint pen and rinse off the red Ponceau S staining with repeated washes in distilled water.
5. Block membranes with 5 mL of Blotto (10% nonfat dried milk and 0.02% sodium azide in PBS) for 1 h with constant shaking. Wash twice with PBS and add 5 mL of primary antiserum dissolved in PBS plus 0.5% Tween-20 for 1 h. To detect FLAG tagged connexin insert expression, dilute anti-FLAG antibody (Sigma) to 10 µg/mL. When using connexin antibodies, dilute them according to the recommendations of the supplying laboratory or manufacturer (*see* **Note 8**).
6. Wash 3×, 5 min each with secondary antibody washing buffer: 150 mM NaCl, 50 mM Tris-HCL, pH 7, 0.5% Tween-20. Add 5 mL of secondary antibody, antirabbit or antimouse alkaline phosphatase conjugated Ig (1:5000) (Promega) to the aforementioned secondary antibody washing solution for 1 h. Visualize the results by alkaline phosphatase substrates, 33 µL of NBT (100 mg/mL in 70% dimethylformamide) and 33 µL of BCIP (50 mg/mL in 100% dimethylformamide) dissolved in 10 mL of alkaline phosphatase reaction buffer (100 mM Tris-HCl, 100 mM NaCl, 5 mM MgCl$_2$, pH 9.5). Stop the reaction when the color is developed.

3.4.2. Immunostaining of Tissue Sections Infected with RCAS(A) Constructs

1. Prior to staining, leave slides (prepared as described in **Subheading 3.3.3.**) on the bench top for a few minutes to let the condensation accumulated during the

freezing process dry up. Fix the slides in ice-cold acetone for 3 min and dry at room temperature.

2. Make wax circles, using a pap pen (Binding Site, San Diego, CA), surrounding the tissue sections (try to enclose the least empty space possible), and let the wax dry.

3. Place PBS into the circled area (make sure that the liquid does not leak), incubate for 5 min, and aspirate. Add blocking solution (*see* **Subheading 3.2.3.**, **step 4**, for its composition) for 30 min.

4. Wash and incubate with primary and secondary antibodies exactly as described in **Subheading 3.4.1.1.** For tissues containing the same connexins as retroviral transduced ones, anti-FLAG antibody can be used to detect expression of the virally delivered exogenous connexins.

5. After immunostaining, add a drop of mounting medium, carefully cover the section with a glass coverslip (slowly and avoiding trapping any air bubbles), and immobilize it with nail polish.

6. Observe under a fluorescence microscope and store images with a digital camera or on film using a regular camera.

4. Notes

1. For amplification of PCR fragments for cloning into CLA12NCO, the primer for the region containing restriction sites needs to be carefully designed. Certain restriction enzymes require some additional nucleotides surrounding the restriction consensus sequences; otherwise the activities can be dramatically lower (for reference see the 1998/99 Catalog of New England Biolab, Appendix "cleavage close to the end of DNA fragments," pp. 258–259). It is also a good idea to perform the restriction digestion separately by using two different enzymes unless both enzymes use exactly the same buffer at the most optimal conditions. Second, make sure that the coding sequence does not contain additional sites recognized by these restriction enzymes. In one instance, there are two additional *Nco*I sites within the connexin56 (Cx56) sequence, necessitating triple ligation reactions with three fragments.

   The following is an example of a pair of PCR primers designed to make full-length Cx45.6 fragment for cloning into CLA12NCO vector:
   Sense primer: 5'-CGGAATTCCATGGGTGACTGGAG 3'
   Antisense primer (Flag sequence underlined): 5'-GCGAATTC<u>TTACT TGTCATCGTCGTCCTTGTAGTCT</u>ACAGTCAGATCGTC-3'.

2. Estimation of approximate T_m is based on the following **equation (10)**:

$$\text{No. } [G + C] \times 4°C + \text{No. } [A + T] \times 2°C = T_m$$

3. For initial transfection by DNA constructs and infection by viral stocks, the cell culture density is very critical. For transfection, the optimal density is approx 50–75% confluence, whereas best results for infection are obtained when cells are approx 20–30% confluence.

4. The entire procedure of preparation of viral stocks takes about 2 wk to accomplish. All of the steps have to be performed under sterile conditions, as bacterial

contamination of the injected material can cause lethality to chick embryos. Each batch of viruses can be tested for contamination by incubating 1 μL of concentrated viral stock in 1 mL of culture medium at 37°C for a few days. If viral stocks get contaminated, yeast or fungus will grow in those mediums. In addition, care must be taken to prevent cross-contamination of different viral producing cells because RCAS-based viruses can spread! One might want to handle one type of cells at any given time in the tissue culture hood, and avoid repeated pipetting to minimize chances of carrying over viruses.

5. For titering viruses, cultures of infected QT-6 cells need to be 100% confluent. The empty spaces between cells may cause problems by creating artificial signals, as peroxidase staining tends to give very strong false-positive signals for cells along the edges of empty spaces.

6. For fixation of the tissues or cell cultures, a final concentration of 1–4% formaldehyde is adequate for most purposes. It should be noted that a higher concentration of formaldehyde might disrupt the antigenicity of certain epitopes, especially of those recognized by monoclonal antibodies.

7. For immunofluorescence staining of cell cultures or tissue sections, red-fluorescent rhodamine is more stable and less light sensitive than green-fluorescent FITC. However, fluorescein staining is likely to give lower background.

8. Anti-FLAG monoclonal antibody (Sigma) works well for both immunoblots and immunostaining of frozen tissue sections and cell cultures with low background labeling. For connexin antibodies, there are two resources, one prepared from individual research laboratories, and another from commercial sources. Commercially available antibodies include anti-Cx26, 32, 40, 43, and 50 from Zymed Laboratories (South San Francisco, CA) and Chemicon (Temecula, CA). The working concentrations are based on the manufacturer's recommendations. The antibodies we used, anti-Cx45.6 and Cx56, are prepared from fusion proteins, and glutathione-*S*-transferase (GST) plus C-terminal portions of connexins, and polyclonal antibodies are raised from rabbit (Pocono Rabbit Farm, Canadensis, PA). The detailed procedures are described in Jiang et al *(8)*. Because rabbit serum will contain anti-GST antibodies and other nonspecific antibodies in addition to anti-connexin antibodies, further purification steps are needed, following the procedures detailed by Harlow and Lane *(9)*. After affinity purification, we normally used 1:500 dilution of antibodies for immunoblots and immunofluorescence staining.

References

1. Goodenough, D. A., Goliger, J. A., and Paul, D. L. (1996) Connexins, connexons, and intercellular communication. *Annu. Rev. Biochem.* **65,** 475–502.
2. Jiang, J. X. and Goodenough, D. A. (1998) Retroviral expression of connexins in embryonic chick lens. *Invest. Ophthamol. Vis. Sci.* **39,** 537–543.
3. Hughes, S. H., Greenhouse, J. J., Fetropoulos, C. J., and Sutrave, P. (1987) Adaptor plasmids simplify the insertion of foreign DNA into helper-independent retroviral vectors. *J. Virol.* **61,** 3004–3012.

4. Cepko, C. (1988) Retrovirus vectors and their application in neurobiology. *Neuron* **1,** 345–353.

5. Brasier, A. R. and Fortin J. J. (1993) Introduction of DNA into mammalian cells, in *Current Protocols in Molecular Biology* (Ausubel, F. M., Brent, R., Kingston, R. E., Moore, D. D., Seidman, J. G., Smith, J. A., Struhl, K., eds.), John Wiley & Sons, New York, pp. 901–917.

6. Fekete, D. M. and Cepko, C. L. (1993) Replication-competent retroviral vectors encoding alkaline phosphatase reveal spatial restriction of viral gene expression/transduction in the chick embryo. *Mol. Cell. Biol.* **13,** 2604–2613.

7. Hamburger, V. and Hamilton, H. L. (1951) A series of normal stages in the development of the chick embryo. *J. Morphol.* **88,** 49–92.

8. Jiang, J. X., White, T. W., Goodenough, D. A., and Paul, D. L. (1994) Molecular cloning and functional characterization of chick lens fiber connexin 45.6. *Mol. Biol. Cell* **5,** 363–373.

9. Harlow, E., and Lane, D. (1988) Antibodies. *A Laboratory Manual,* Cold Spring Harbor Laboratory Press, Cold Spring Harbor, NY, pp. 313–315.

10. Itakura, K., Rossi, J. J., and Wallace, R. B. (1984) Synthesis and use of synthetic oligonucleotides. *Annu. Rev. Biochem.* **53,** 323–356.

10

Spatiotemporal Depletion of Connexins Using Antisense Oligonucleotides

Colin R. Green, Lee yong Law, Jun Sheng Lin, and David L. Becker

1. Introduction

Antisense oligodeoxynucleotides have considerable potential as agents for the manipulation of specific gene expression *(1–3)*. They function by entering the cell and binding to the messenger RNA (mRNA), thereby blocking protein translation (*see* **Note 1**). The short half-life of oligonucleotides can be overcome, sustained delivery can be achieved, and specific tissue targeting can be obtained using a Pluronic gel delivery system. It is therefore possible to gain an understanding of the roles of connexins, which are expressed in dynamic and spatially controlled patterns, through transient reduction in the expression of specific connexin proteins. This leads to dramatic but precisely defined aberrations in embryonic development, for example *(3)*. The major advantage of the approach is the ability to achieve spatiotemporal regulation of connexin gene expression while reducing some of the problems associated with gene knockout approaches (*see* **Tables 1** and **2**).

In this chapter we describe the use of antisense oligodeoxynucleotides and a Pluronic F-127 gel delivery system to regulate specific connexin expression. Pluronic gel is liquid at low temperature (4°C) but sets as it warms to physiological temperatures. It is a mild surfactant that aids antisense penetration into cells (it is used as a cell loading reagent at lower concentrations *[4]*), and acts as a reservoir providing sustained release of the antisense oligonucleotide being applied. This helps reduce the effect of intracellular nuclease breakdown. The gel also allows the use of oligonucleotides at relatively low concentrations, reducing artifacts that can occur at higher dose levels, including nonspecific imperfectly matched mRNA binding *(5,6)* and the accumulation of

From: *Methods in Molecular Biology, vol. 154: Connexin Methods and Protocols*
Edited by: R. Bruzzone and C. Giaume © Humana Press Inc., Totowa, NJ

Table 1
Advantages of an Antisense Oligodeoxynucleotide Knockdown Approach
Compared with the Gene Knockout Approach
for Regulating Gene Expression

Antisense oligodeoxynucleotides	Gene knockout
Ease of use	Specialist equipment required
Species optional	Suitable for a limited number of species
Dose controllable	On or off
Low cost	Expensive to set up, colony expensive to maintain
Transient (temporal specificity)	Early developmental events may occlude later events
Site directable (spatial specificity)	Perturbations in one area may mask others, internal controls are not possible
Compensation by related gene family members less likely *(3,15)*	Compensation by other members of a gene family *(16)*
Incomplete (knockdown, not knockout) allowing analysis where knockout is lethal	Phenotypes may be lethal
Naturally occurring	
Clinical application possible	

nucleotide and nucleoside breakdown products which can effect cell proliferation and differentiation *(2)*. Finally, the gel can be precisely placed onto the target tissue. In a further modification of the technique, plugs of antisense oligonucleotide containing gel are injected directly into tissues. With these methods it is possible to achieve connexin knockdown for 24–48 h using unmodified antisense oligonucleotides providing precise spatiotemporal control of connexin gene expression.

2. Materials

Materials and equipment required will vary depending on the application envisaged. Those listed in the following subheadings cover the specific examples given in this chapter; suppliers listed are those that we have used.

2.1. Equipment

1. Vortex mixer (e.g., VF2–IKA Labortechnik).
2. Rocking agitator.

Table 2
Some Disadvantages of the Antisense Oligodeoxynucleotide Knockdown Approach Compared with the Gene Knockout Approach for Regulating Gene Expression

Antisense oligodeoxynucleotides	Gene knockout
Short half-life (breaks down)	Stable
Potential nonspecificity	Highly specific
Multiple mechanisms may operate (mRNA binding, DNA triplet formation)	Single mechanism
Transient (less useful for low turnover proteins and extended events)	Total
Incomplete (knockdown, not knockout so interpretation may be more difficult)	Knockout is complete
Each animal is affected to a different degree	A stable population may be obtained
Permeability (Access and delivery may be variable)	

3. 200-μL Pipetman (Gilson).
4. Fine curved scissors (Dupont) and fine forceps (Dupont Inox no. 5).
5. Confocal laser scanning microscope (Leica TCS 4D) or epifluorescence microscope (to check penetration of fluorescein isothiocyanate [FITC]-tagged oligodexoynucleotides).
6. Egg Incubator (Domex, Dominion Industries, Hamilton, New Zealand).
7. Cell culture incubator (carbon dioxide).
8. Stereotactic table and micrometer stage (to hold animals for accurate lesioning and antisense application in brain lesion studies).
9. Dental drill and 0.5-mm diameter grinding tip (Emesco Dental) or engraving tool (Arlec Engraver).
10. Vibratome (Oxford Laboratories, USA).

2.2. Chemicals

1. Antisense oligodeoxynucleotides (Oligos Etc., Eugene, OR).
2. Pluronic F-127 gel (BASF; also available through Sigma).
3. Dimethyl sulfoxide (DMSO) and hydrogen peroxide (Sigma).
4. Culture medium (depending upon cell type being studied) and Opti-MEM reduced serum medium (Gibco-BRL).
5. Phosphate buffered saline (PBS) (Oxoid, Basingstoke, UK).
6. Rotring ink (Rotring Gmbh, Germany).
7. 70% Ethanol, Ringer's solution.

8. Disposable scalpel blades (nos. 11, 15, 22); 5-mL and 1-mL syringes; 21G (0.8 × 38-mm) and (19G (1.1 × 38–mm) syringe needles (Beckton Dickinson).
9. 50-mL Disposable screw capped tubes (Sarstedt or Falcon).
10. 1500-mL Eppendorf tubes (Eppendorf) and 200-mL Pipetman yellow tips (Costar).
11. Sellotape (regular, not Magic tape) T3 silk sutures (Deknatel).

3. Methods

3.1. Oligodeoxynucleotides

1. Antisense oligodeoxynucleotides (ODNs) need to be approx 18–30 bases in length; longer ODNs are less able to penetrate to the cell, shorter ones may be less specific.
2. In designing suitable ODNs you must identify sequences that will not form stem loop structures or stable secondary structures (homodimerization or hetero-duplexs), or at least have loop melting temperatures below that at which the experiments will be carried out *(7)*. A number of computer programs are available to provide this information, or use one supplied by the ODN supplier (e.g., OligoTech from Oligos Etc., Eugene, OR, USA). Avoid more than three guanines in a row. ODNs directed against mRNA hairpin stem regions should be avoided even though the predicted folding pattern may not necessarily be that occurring in the cell. Hairpin stems in target mRNA can be predicted using computer programs such as GENEWORK (Oxford Molecular Group). An increasing number of computer programs are in fact available to assist in ODN design (MacVector and GENEWORK from Oxford Molecular Group, Oligo from National Biosciences). Oligo, for example, is able to predict all possible target positions on the mRNA of interest for ODNs of any given length; the complementary sequences then become candidate ODNs. ODNs must be specific to the connexin(s) of interest and ODN sequences should be aligned carefully with genes such as the other members of connexin family to ensure there are no homologous regions of more than eight continuous bases matching. Finally the ODN should also be checked against databases (GenBank/EMBL/DDBJ/PDB) to eliminate obvious homologies with unrelated but potentially perturbable genes.
3. The ODN must also be capable of binding to your mRNA at physiological temperatures. The computer programs listed previously will also predict the melting temperature (T_m) of an antisense for an unstructured RNA. Your antisense needs a higher than average T_m for the target RNA. For example, two of the antisense ODNs we use against Cx43 mRNA have primer to target T_ms of 78.1°C and 69.1°C.
4. We have found 0.5–1.0 μ*M* oligonucleotide concentrations most effective and have observed nonspecific toxicity effects with some applications above 10 μ*M* concentration. Preliminary screens for characterizing the efficacy is essential and should be carried out at concentrations between 0.05 μ*M* and 50 μ*M*.
5. One advantage of the Pluronic gel delivery system is that unmodified phospho-diester oligomers can be used, the gel reservoir replacing oligonucleotides as they are degraded by nucleases. Modified ODNs are not recommended as they

may cause artifacts (*see* **Note 2**, however).

6. For some ODNs addition of 1% DMSO to the gel may increase efficacy of the antisense, presumably by aiding entry into the tissue of interest. This should be avoided if possible however, it adds another variable to be controlled for.

7. Some researchers have noted that oligonucleotides from some suppliers are better than others. It may be necessary to design several probes before finding one that works well in blocking translation of the protein of interest (*see* **Note 3**).

3.2. Pluronic Gel

1. Prepare a 25–30% w/w solution of Pluronic F-127 gel (BASF) in PBS (molecular grade water). The percentage concentration determines the softness and persistence of the gel. Higher concentrations may set too quickly, but stay in place longer before dissolving away. The gel may take 24–48 h to dissolve and is best done on a rocker at 4°C. It will not dissolve at room temperature (which is above the setting temperature) and shaking or stirring the solution vigorously simply causes frothing.

2. Add the ODNs (antisense or controls) to the desired concentration, and thoroughly vortex-mix several times, keeping on ice between mixes. The mixture can then be aliquoted in 1.5-mL Eppendorf tubes and stored at –80°C. Alternatively, aliquot the ODNs in stock solutions of 50–100 μM stored frozen and prepare fresh for each experiment from a stock of gel kept at 4°C.

3. Typically 5–10 μL of gel should be applied, depending on the area of tissue to be covered, by using a chilled 0.5–10-μL pipet tip. Because the gel sets rapidly it is necessary to work swiftly during the application process and the gel must be kept on ice. A fresh tip should be used for every application.

4. Although the antisense is essentially single-stranded DNA it is recommended that RNAse free conditions be used where possible.

3.3. Controls

Controls are essential and specific inhibition must be demonstrated. To check for tissue sensitivity to the loading technique use gel-only controls. To check for specificity of the antisense of interest use nonspecific oligomers which must be designed under the same conditions (check for the probability of homodimerization and hairpin looping; check that they do not recognize any other protein sequences in the species of interest which might be effected). The random control ODNs should have a similar base composition to the antisense ODN.

1. Sense controls can be useful. These can be used independently but beware that stable DNA triplets can be formed, in some instances giving a similar result to the antisense ODN of interest *(8)*. Sense controls can also be premixed with the antisense at equal concentrations in a competition control.

2. Check for depth of penetration and the time course of entry of antisense into a tissue by using ODNs with a label tag. A number of options are available from an

ODN supplier including Biotin and fluorescent tags. We use a fluorescein tag in conjunction with confocal laser scanning microscopy which provides direct imaging of penetration without the need to section the sample.

3. It is important to check that the protein of interest is being knocked down and that any biological effects seen can therefore be attributed to this. Use immunohistochemistry to follow connexin expression in experimental and control tissues to confirm that your ODN is specific to the connexin of interest (and you are not knocking down other connexin types in a nonspecific manner). We have not observed compensation through up-regulation of other connexin types but this should also be checked in any experimental model used. Immunohistochemistry will also provide the time course for knockdown and recovery. It has distinct advantages over Western blotting in providing spatial distribution of protein knockdown in targeted tissues, and with confocal microscopy enables rapid quantification of protein expression *(9,10)*. Northern blots and RNase protection assays are of little use in determining antisense ODN efficiency.

4. It should be possible to replicate the same biological effect using another antisense to the gene of interest. We have, for example, used two different ODNs to knock down Cx43 in the chick embryo, both giving similar effects which are quite distinct from toxicity effects.

5. Addition of Rotring ink (Rotring GmbH, Germany) to the gel can help in checking placement on the tissue, and in determining how long it remains in position. In tissues undergoing much movement the gel will dissipate more rapidly but in most cases it will remain in place for at least 12 h.

3.4. Chicken Embryos

The Pluronic gel antisense ODN approach is ideal for studying connexin roles during development of the chick embryo (*see* **Note 4**).

1. Fertilized eggs (*see* Chapter 9 in this volume for more details on the manipulation of fertilized chick eggs) should be incubated at 38°C and staged according to Hamburger and Hamilton stages *(11)*. On removal from the incubator eggs are rotated 10× to ensure that the embryo is floating free at the top of the egg.

2. Clean eggs with 70% alcohol and then make a small hole with sterile forceps into the blunt end of the egg where the air sac is found. Insert a 21G (0.8 × 38-mm) hypodermic syringe with the needle vertically through the hole to the bottom of the egg and withdraw a small amount (< 1 mL) of albumin.

3. Place Sellotape over the top of the egg and make a small hole through the tape and shell using forceps. A larger hole or window can now be cut from this start point in the top of the shell using small fine scissors *(12)*.

4. Once the egg has been windowed the vitelline and amniotic membranes over the area to be treated are gently opened using fine forceps; it is necessary to remove these membranes to achieve antisense penetration of the tissue of interest. Take care not to damage the embryo or associated blood vessels. At early

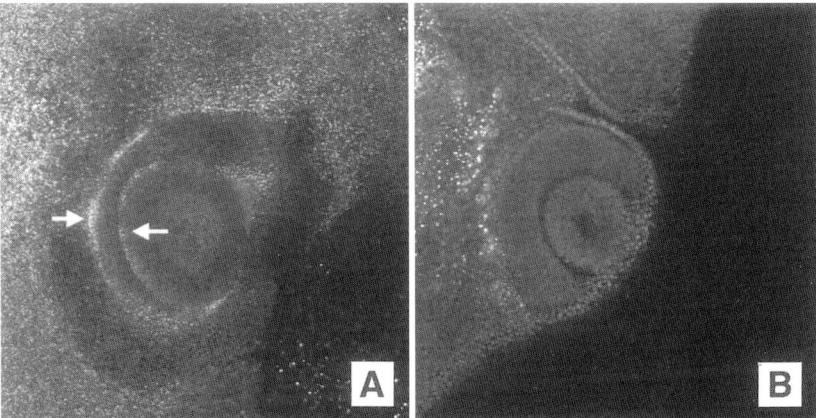

Fig. 1. Immunohistochemical labeling of CX43 protein in the developing eyes of stage 13/14 chick embryos. The embryos were treated with random control (**A**) or Cx43 specific antisense (**B**) ODNs in Pluronic gel at stage 11. In (**A**) Cx43 is at normal levels for this stage embryo, but in (**B**) the Cx43 levels are much lower in the developing sensory and pigmented layers of the eye (*arrows*).

   stages of development a 30% rotring or India ink–PBS solution can be injected
   under the amniotic sac to reveal the outline of the embryo more clearly.
5. Pluronic gel (5–10 μL) is applied to the target site. The drops of gel can be placed
   with high accuracy to parts of an embryo, tissue, or organ culture, and will mold
   around the site of application and set in place as they do so. The volume of the
   drops can be varied as required for the experiment and applications can be repeated
   at later time points if required.
6. Following Pluronic gel application eggs are sealed with Sellotape and replaced in
   the incubator where they develop until ready for analysis of protein knockdown
   or developmental perturbations (*see* **Fig. 1**).

3.5. Rat Brain Lesions

   The Pluronic gel plug-antisense ODN method has been used to study the
effect of Cx43 knockdown during astrocytosis which occurs following
lesioning of the cerebral cortex of the mammalian brain.

1. Anesthetize animals and hold the head in a stereotaxic clamping device. Shave
   the region around the lesion site, and slit the skin with a scalpel and pull it back to
   leave the skull plates clear. Drill a 1.5-mm diameter hole through the skull plate
   using a hand drill or dentist's drill and make a lesion using a 19G (1.1 × 38-mm)-
   gauge syringe needle attached to a micrometer stage. The stage allows accurate
   directional control and precise penetration depth (in our experiments, 2 mm).

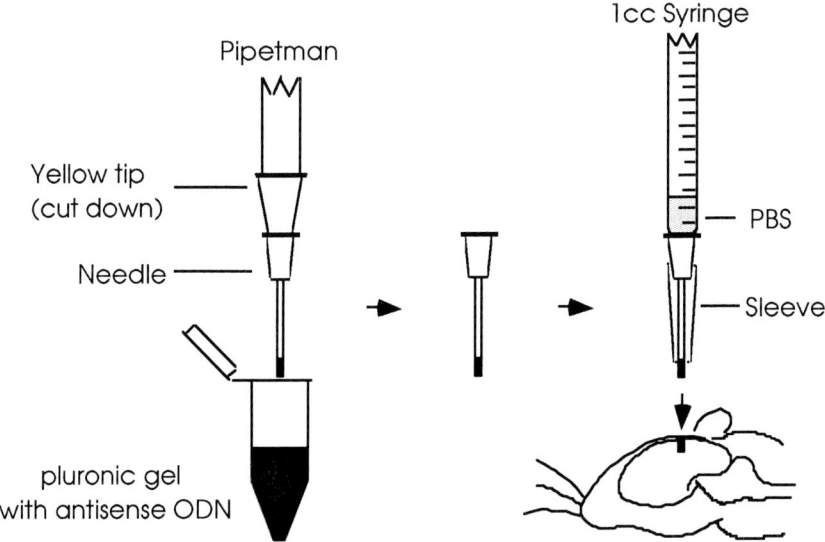

Fig. 2. A schematic representation of the method for applying Pluronic gel plugs into tissues. The gel containing the ODN of interest is sucked into an ice-cold syringe needle filed off so as to have a flattened tip. The needle is attached to a pipettor using a cut down yellow tip. The gel is allowed to set in the needle tip which is then transferred to a 1-mL syringe containing a small amount of PBS. Pressure is gently applied to the plunger and the gel will "pop" from the needle into a premade lesion in the target tissue. When injecting into a brain lesion the needle is covered with a sleeve (a cut down yellow tip) so that the needle tip can only penetrate the skull to a set depth, the sleeve coming against the skull preventing overpenetration.

2. With the animal prepared, suck ice-cold Pluronic gel containing the ODN of interest (or a control ODN) into a precooled 19G (1 · 1 × 38-mm) gauge syringe needle filed off so as to have a flat tip (*see* **Fig. 2**). The syringe needle can be attached to a volumetric pipet via a cut down yellow pipet tip. The gel will then set as the needle warms to room temperature. In most cases 10 µL is sufficient.
3. Transfer the needle with the gel plug at its tip to a 1-mL syringe containing PBS, and place a sleeve (a cut down yellow pipet tip) over the needle shaft so that the needle tip can be lowered into the lesion with the sleeve (coming up against the skull) preventing overpenetration (*see* **Fig. 2**). Gentle pressure on the syringe plunger will "pop" the gel plug out of the needle into the lesion. This can be felt, though not seen, by the operator. There is insufficient pressure to expel the gel using the volumetric pipet.
4. Treat the wound with hydrogen peroxide to stop bleeding and suture the skin back into place. Animals should be carefully monitored and left until ready for sacrifice.

3.6. Brain Slices or Dissociated Neurons in Culture (*see* **Notes 5** and **6**)

1. Following humane killing by cervical dislocation or decapitation the brain should be dissected free as quickly as possible and placed in chilled oxygenated ringer.
2. Slices should be taken rapidly using either a vibratome or wire slice and transferred to appropriate tissue culture medium and incubation conditions for the tissue and allowed to acclimatize in an incubator at 37°C for 1 h. Alternatively the tissue should be dissociated and plated out at appropriate cell densities required for tissue culture and allowed to adhere to the dish or attain desired confluence.
3. Medium is removed and the slices or cells rinsed with serum-free medium. Serum-free medium is used to reduce the effects of nuclease activity in the medium and thereby extend the life-span of the ODNs. The serum-free medium is removed and Pluronic gel can be applied to the desired part of the slice or area of the cultured cells using the application methods described for chick embryos. Serum-free medium is replaced and the preparation returned to the incubator for 24–48 h before analysis.

4. Notes

1. Antisense oligodeoxynucleotides have a complementary base sequence to the targeted mRNA which it then recognizes and binds to, halting protein production. The bound mRNA is then usually cleaved by RNase H and the resulting fragments, recognized by the cell as foreign, are broken down.
2. The typical-half life of an unmodified phosphodiester oligonucleotide is only 20 min *(2)* and phosphorothioate or methylphosphonate ODNs are often used as an alternative. In general, however, these have a weaker affinity for the message of interest, can be less efficient at entering cells, and can cause nonspecific inhibition by binding to essential proteins *(2,13,14)*. In drug trials they have caused side effects *(14)*. Second-generation ODNs may provide alternatives, for example, three chimeric antisense ODNs (patented to Oligos Etc.) or end modified ODNs may survive longer in the cell without the ill effects of fully modified backbone ODNs. For further information check the Oligo Etc. web site http://www.oligosetc.com or http://www.gene-tools.com.
3. RNA can adopt a variety of secondary structures in the cell and is highly protein bound, both of which may limit antisense access. Therefore, when multiple antisense probes are screened against an RNA target, some are more effective than others. In the past, antisense probes have been designed to target the translation initiation codon but recent studies indicate that most regions of the mRNA are accessible to ODNs. We have achieved effective targeting of carboxyl-terminal regions.
4. Antisense oligonucleotides are effective only when applied during periods of connexin gene transcription. They will have no effect on existing connexin channels whose removal from the membrane will be dependent upon protein turnover kinetics. The Pluronic gel delivery system provides a sustained release vehicle

for approx 24 h. Knockdown effects in the developing chicken embryo are observed from 2 h through to 48 h after ODN delivery. In developmental experiments the efficacy of the ODN is improved if they are applied to tissues just prior to expression of the target protein.

5. An antisense ODN approach has advantages and disadvantages when compared with gene knockout studies (**Tables 1** and **2**).

6. In general antisense ODNs have been less effective on cell cultures than in tissues. The Pluronic gel delivery system appears to overcome some of the limitations.

References

1. Stein, C. A. (1992) Anti-sense oligodeoxynucleotides—promises and pitfalls. *Leukemia* **6,** 967–974.
2. Wagner, R. W. (1994) Gene inhibition using antisense oligodeoxynucleotides. *Nature* **372,** 333–335.
3. Becker, D. L., McGonnell, I., Makarenkova, H. P., Patel, K., Tickle, C., Lorimer, J., and Green, C. R. (1999) Roles for $\alpha 1$ connexin in morphogenesis of chick embryos revealed using a novel antisense approach. *Dev. Genet.* **24,** 33–42.
4. Haughland, R. P. (1996) *Handbook of Fluorescent Probes and Research Chemicals,* 6th Edit. Molecular Probes, Eugene, OR, USA.
5. Woolf, T. M., Melton, D. A., and Jennings, C. G. B. (1992) Specificity of antisense oligonucleotides in vivo. *Proc. Natl. Acad. Sci. USA* **89,** 7305–7309.
6. Boiziau. C., Moreau, S., and Toulme, J-J. (1994) A phosphorothioate oligonucleotide blocks reverse transcription via an antisense mechanism. *FEBS Lett.* **340,** 236–240.
7. Spector, D. L., Goldman, R. D., and Leinwand, L. A. (1997) Cells: *A Laboratory Manual,* Vol. 2. Cold Spring Harbor Laboratory Press, NY, pp. 89.1–89.19.
8. Neckers, L. and Whitesell, L. (1993) Antisense technology: biological utility and practical considerations. *Am. J. Physiol.* **265,** L1–L12.
9. Green, C. R., Peters, N. S., Gourdie, R. G., Rothery, S., and Severs, N. J. (1993) Validation of immunohistochemical quantification in confocal scanning laser microscopy: a comparative assessment of gap junction size using confocal and ultrastructural techniques. *J. Histochem. Cytochem.* **41,** 1339–1349.
10. Becker, D. L., Evans, W. H., Green, C. R., and Warner, A. (1995) Functional analysis of amino acid sequences in connexin 43 involved in intercellular communication through gap junctions. *J. Cell Sci.* **108,** 1455–1467.
11. Hamburger, V. and Hamilton, H. (1951) A series of normal stages in the development of the chick embryo. *J. Morphol.* **88,** 49–92.
12. Tickle, T. (1993) Chick limb buds, in Essential Developmental biology (Stern, C. D., and Holland, P. W. H., eds.), Oxford University Press, Oxford, pp. 119–125.
13. Milligan, J. F., Matteucci, M. D., and Martin, J. C. (1993) Current concepts in antisense drug design. *J. Medic. Chem.* **36,** 1923–1937.
14. Shimmings, A. (1998) Future rosy for antisense agents. *SCRIP* **2380,** 24–25.
15. Edelman, E. R., Simons, M., Sirois, M. G., and Rosenberg, R. D. (1995) c-*myc* in vasculoproliferative disease. *Circ. Res.* **76,** 176–182.

16. De Zeeuw, C. I., Hansel, C., Bian, F., Koekkoek, S. K., van Alphen, A. M., Linden, D. J., and Oberdick, J. (1998) Expression of a protein kinase C inhibitor in Purkinje cells blocks cerebellar LTD and adaption of the vestibulo-ocular reflex. *Neuron* **20,** 495–508.

11

Transfection and Expression
of Exogenous Connexins in Mammalian Cells

Dieter Manthey and Klaus Willecke

1. Introduction

Currently, 15 different connexin genes have been described in the murine genome *(1–4)*. Different connexins are expressed in a cell specific fashion, and in most cell types two or more connexins have been demonstrated. To characterize defined connexin channels independently of other connexins that are coexpressed in most mammalian cells, exogenous connexin DNAs can be transfected and expressed in cultured cells that show a low level of endogenous gap junction channels *(5)*.

By using appropriate experimental techniques, one can analyze the function of intercellular gap junction channels composed of hemichannels of the same connexin type (homotypic) or of different connexin types (heterotypic) *(1)*. Furthermore, the expression of modified connexins can help to evaluate the function of connexin domains or amino acid residues for regulation of gap junction mediated intercellular communication (e.g., docking and gating). Finally, one can try to study the interaction of connexins by using inducible expression systems *(6)*.

Expression of connexin proteins in recipient cells can be accomplished by RNA or DNA transfer. Injection of cRNA into paired *Xenopus* oocytes *(7)* or of cDNA in human HeLa cells has been reported *(8)*. Transfer of connexin genes by DNA in cultured cells can be carried out to achieve transient and stable conditions of expression. Both expression systems have advantages but also certain limitations when compared to the much more complex situation of gap junction mediated communication in the animal. The expression of exogenous connexins in paired *Xenopus* oocytes is described in Chapter 13 by Skerrett et al. in this volume. It is a fast procedure that allows expression of the

From: *Methods in Molecular Biology, vol. 154: Connexin Methods and Protocols*
Edited by: R. Bruzzone and C. Giaume © Humana Press Inc., Totowa, NJ

injected connexin cRNA at high efficiency. Electrophysiological measurements and tracer transfer between pairs of relatively large oocytes are easier to perform than with small mammalian cells. On the other hand, expression of the endogenous *Xenopus* connexin38 must be experimentally suppressed in oocytes *(9)*, and some differences have been reported in the communication behavior of connexin channels expressed in *Xenopus* oocytes and in mammalian cells *(5)*.

For unequivocal expression of exogenous connexins in mammalian cells, recipient cell lines should show no or little expression of endogenous connexins. Cultured primary cells usually express one or more endogenous connexins. Several tumorigenic cell lines, however, show decreased or absent expression of connexins. These findings led to the suggestion that loss or reduction of gap junction mediated communication may correspond to loss of growth control during tumorigenesis *(10)*. The partial reversal of the transformed phenotype after expression of exogenous connexins via transfection in tumorigenic cell lines, such as human HeLa cervix carcinoma cells *(11)*, human SK HEP-1 hepatoma cells *(12)* and rat C6 glioma cells *(13)* supports this notion. In addition, mouse N2A neuroblastoma cells *(14)* and ROS rat osteogenic sarcoma cells *(15)* have all been exploited to achieve expression of different connexins.

The methods described in this chapter have been used in our laboratory to express exogenous connexins mainly in HeLa cells but, with slight modifications, they can be adjusted also to other transformed cell lines. First, the exogenous connexin gene, cDNA, or coding region from genomic DNA needs to be cloned in an expression vector containing a cloning site downstream of a strong (mostly viral) promoter, a polyadenylation site, and resistance marker genes for selection in pro- and eukaryotic cells. To introduce the DNA into mammalian cells, several gene transfer techniques have been described, for example, transfection with calcium phosphate *(16)* or DEAE-Dextran *(17)*, and transfer by electroporation *(18)* or lipofection *(19,20)*. After transfer of connexin DNA, selection, isolation, and functional characterization of the transfected cells can be carried out. The expected connexin transcript needs to be analyzed by Northern blot hybridization to a specific connexin probe. Protein mass and location of the expressed exogenous connexin in contacting plasma membranes can be checked by immunoblot and immunofluorescence analysis, respectively. For functional analyses, measurements of tracer transfer or electrical conductance between transfected cells should be carried out.

2. Materials

2.1. Reagents

All reagents were used in p.a. quality from Merck (Darmstadt, Germany), except where noted.

1. DMEM cell culture medium (standard medium): 10 g/L OF Dulbecco's modified Eagles medium (Gibco-BRL, Eggenstein, Germany) supplemented with 3.7 g/L NaHCO$_3$, 10% fetal calf serum (Gibco-BRL), 100 µg/mL OF streptomycin (Sigma, St. Louis, MO, USA), 100 U/mL OF penicillin (Sigma), and 2 mM L-glutamine. HeLa transfectants were maintained in standard medium containing puromycin (1 µg/mL; Sigma, cat. no. P-8833).

2. Human cervix carcinoma HeLa cells were obtained from the American Type Culture Collection (Rockville, MD, USA), Catalogue No. CCL2. We use HeLa cells with low endogenous background of gap junctional communication (based on measurements of tracer transfer and electrical conductance) *(21,22)*. Very low levels of endogenous connexin (Cx) mRNAs were detected for Cx26, Cx31, and Cx45 in these HeLa cells. No transcripts were found for Cx30, Cx37, Cx49, Cx43, Cx46, and Cx50 *(23)*.

3. Expression vector pBEHpac18 *(25)*. This vector contains a pUC18 multiple cloning site (Stratagene, La Jolla, CA, USA), flanked by a 5× repeated SV40E promoter sequence and a polyadenylation motif. For selection in prokaryotic cells, the vector contains a gene conferring resistance to ampicillin and, for selection in eukaryotic cells, the *pac* gene that codes for puromycin N-acetyltransferase mediated resistance against puromycin.

4. Primary antibodies, specific and sensitive for the connexins to be examined.

5. Secondary antibodies, specific for the F$_c$ part of the primary antibodies, conjugated with a fluorophore.

6. Phosphate buffered saline (PBS): 8 g/L of NaCl, 0.2 g/L of KCl, 1.15 g/L of Na$_2$HPO$_4$, 0.2 g/L of KH$_2$PO$_4$, adjust pH to 7.2 with HCl.

7. PBS*: 0.1 g/L of CaCl$_2$ and 0.1 g/L of MgCl$_2$ in PBS.

8. PBT: PBS and 0.1% (v/v) Tween-20 (polyoxyethylenesorbitan monolaurate, Sigma).

9. FCS/PBT–PBT and 10% fetal calf serum (FCS).

10. Paraphenylendiamine solution: 10% (v/v) in PBS, 90% (v/v) glycerol (Sigma), 0.1% (w/v) para-phenylenediamine, adjust pH to 8.0 with carbonate–bicarbonate buffer. Store solution in the dark at –20°C. Caution: paraphenylendiamine is a toxic compound.

11. Rapid mounting medium for microscopy, for example, Entellan (Merck) or nail polish.

12. Trypsin solution: 8 g of NaCl, 0.4 g of KCl, 1 g of glucose, 0.2 g of EDTA, 16 mL of solved trypsin (2.5% trypsin in PBS), add bidistilled H$_2$O to 1 L, adjust pH to 7.8.

13. CaCl$_2$ solution (2.5 M): 183.7 g of CaCl$_2$ add bidistilled H$_2$O to 500 mL, filter sterilize through a nitrocellulose filter (0.2 µm pore size, Schleicher & Schuell, Dassel, Germany). Store the solution at –20°C in 50 mL aliquots.

14. 2× 4-(2-hydroxyethyl)-1-piperazineethanesulfonic acid (HEPES)-buffered saline (HeBS) solution: 16.4 g of NaCl, 11.9 g of HEPES acid, 0.21 g of Na$_2$HPO$_4$, 800 mL of bidistilled H$_2$O. Titrate to pH 7.05 with 5 N NaOH; add bidistilled H$_2$O to 1 L. Filter sterilize through a nitrocellulose filter (0.2 µm pore size, Schleicher & Schuell). Store at –20°C in 50-mL aliquots. Exact pH adjustment is very important for efficient transfection. The optimal pH range is 7.05–7.12.

15. Silicone grease: Rotisilon C/D (Roth, Karlsruhe, Germany), autoclaved.

2.2. Cell Culture Equipment

1. Incubator (37°C, 10% CO_2), cell culture hood, tissue culture plates (25 or 75 cm^2, Falcon, Becton Dickinson, Plymouth, UK), tissue culture flasks (3.5- and 10-cm diameters, Falcon), multiwell tissue culture plates (12 and 24 wells, Falcon), conical tubes (13- and 50-mL, Sarstaedt, Nuembrecht, Germany), and cryotubes (1 mL, Nunc, Roskilde, Denmark).
2. For cell cloning: mechanical pipetting aid (Accu-jet™, Brand, Wertheim, Germany), stainless steel rings (inner diameter: 5 mm, outer diameter: 8 mm).
3. For immunofluorescence analysis: Microscope (phase contrast and UV light) and UV filter, appropriate for the fluorescent tag of the secondary antibodies used.

3. Methods

3.1. Cell Culture and Vector Construction

Before introduction and expression of DNA into mammalian cells, a suitable cell line must be chosen that shows no or low endogenous connexin expression. Several tumorigenic murine and human cell lines have been used for this purpose. In our laboratory we have most experience with human HeLa cervix carcinoma cells *(4,22,24)* (*see* **Note 1**). For transfection of other cell lines, it may be necessary to determine empirically optimal culture and transfection conditions. HeLa cells grow in standard medium in a 37°C incubator with a moist atmosphere of 10% CO_2. Under optimal conditions, the cells double within 18–24 h. Cells are plated at 20% confluence on sterile culture dishes or flasks. Medium is replaced every 3 d. The cells are passaged before reaching 90% confluence.

Electrophysiological characterization of human HeLa cells yielded very low electrical conductance *(21)*. Usually, this low level of endogenous conductance did not influence the relatively high level of conductance measured in HeLa–connexin transfectants. For transfection of HeLa cells, we used the vector pBEHpac18 *(25)*. The cDNA sequence or coding region of a connexin was inserted into the multiple cloning site of the vector (*see* **Note 2**). Detailed protocols of this procedure are described in method books for molecular cloning (*see*, e.g., *26,27*).

3.2. Cell Transfection

Here we describe two protocols for HeLa cell transfections: first, coprecipitation of DNA and calcium phosphate, and second, lipofection of DNA using a commercial agent. Both methods can be used to obtain transient and stable transfections. The protocol of HEPES-buffered calcium phosphate is based on a method described by Graham and van der Eb *(16)*. The plasmid DNA is incorporated into adherent cultured cells via a precipitate deposited on the cell surface. Transfection of HeLa cells by lipofection is based on the ability of the lipofection reagent to form vesicles that can fuse with cell membranes, whereby plasmid DNA encapsulated in these vesicles can be delivered to the cytoplasm of recipient cells.

3.2.1. HeBS Calcium Phosphate Protocol *(see* **Note 3***)*

1. On the day before transfection, 10^6 HeLa cells are plated on a tissue culture plate (10 cm diameter) to get a 30% confluent cell layer for transfection. The 30% starting confluency should provide a near confluent cell layer at the second day after transfection when the cells are diluted into selective medium. To maintain the cells in exponential growth, they need to be fed with 9 mL of standard medium 4–8 h before transfection.
2. The plasmid DNA to be transfected is precipitated with ethanol and air-dried after preparation. Detailed protocols for plasmid preparation can be found in books describing molecular cloning *(26,27)*. The plasmid DNA should yield an extinction ratio of >1.8 at 260–280 nm, which indicates little or no protein contamination.
3. Plasmid DNA is resuspended at a concentration of 1 µg/µL in sterile water. To 20 µL (20 µg) of nonlinearized plasmid DNA *(see* **Note 5***)*, a solution of 430 µL of sterile water and 50 µL of 2.5 *M* CaCl$_2$ is added.
4. 2× HeBS (500 µL) are placed in a conical tube (13 mL). With a plugged 1-mL plastic pipet, attached to a mechanical pipettor, air is continuously bubbled through the HeBS solution. The DNA–CaCl$_2$ solution is added dropwise with a plugged Pasteur pipet. Then the solution is immediately vortex-mixed for 5 s.
5. Incubate the solution for 20 min at room temperature, while the precipitate is forming *(see* **Note 6***)*.
6. Distribute the precipitate solution dropwise with a Pasteur pipet into the medium on the culture dish, then mix gently by swirling the culture dish.
7. Incubate the cells for 16 h in an incubator. During this time, the precipitate settles onto the bottom of the culture dish and can be seen under the microscope as little crystals on the harvested cells.

3.2.2. Lipofection Procedure

For transfection of HeLa cells, we use the lipofection reagent Tfx-20™ (Promega, Madison, WI, USA) *(see* **Note 7***)*.

1. HeLa cells and plasmid DNA are prepared as described previously for the calcium phosphate protocol *(see* **Subheading 3.2.1.**, **steps 1–2***)*.
2. To 6 mL of standard medium in a sterile conical tube (13 mL), add 10 µg of nonlinearized plasmid-DNA (1 µg/µL in sterile H$_2$O) *(see* **Note 5***)* and vortex-mix the solution for 20 s. Afterwards, add the Tfx-20 suspension at a ratio of 4:1 (v/v) to DNA solution and vortex-mix the solution for 20 s.
3. Incubate the mixture for 10–15 min at room temperature.

4. Remove medium from cells and pipet the 6 mL of DNA–Tfx-20 medium onto the cultured cells. Place cells in incubator for 45–60 min.
5. At the end of the incubation time, add 6 mL of prewarmed (37°C) standard medium to the cells and incubate for 16–24 h in the incubator.

3.3. Selection of Stably Transfected Cells (Common to Both Transfection Procedures)

1. After 16 h of incubation with the DNA calcium phosphate precipitate or DNA–lipofectin vesicles, the medium is removed. Wash the cells twice with prewarmed (37°C) PBS, then incubate the cells for further 24 h with standard medium.
2. The next day, the cells are distributed into culture dishes (10 cm diameter) and incubated with 10 mL of selection medium. Dilution of the cells at 1:5–1:20 is necessary for selection. Nontransfected cells die within 3 d after selection; transiently transfected cells survive for more than 7 d under selective conditions in the presence of puromycin. A lower dilution may lead to confluent cell layers before the selection becomes effective. Then transfected cell clones cannot be separated from each other after selection.
3. Change the selection medium every 4 d for 3–4 wk. At the first time, wash the dishes twice with prewarmed PBS in order to remove dead nontransfected cells.
4. After 2 wk, stable transfected cells become visible as small clones likely to be derived from single cells. The cell clones should be removed when they reach a diameter of 2–3 mm. For cell cloning, sterile steel or glass rings ("cloning rings") are recommended (*see* **Note 8**). First, the position of the cell clones is marked on the bottom of the dish with a permanent marker.
5. Remove medium and wash the cells twice with PBS. In the next 2–3 m the cloning rings must be placed on the marked cell clones. Before that, the cloning ring was dipped with its bottom into silicone grease, and then placed on top of the clone, so that the inner part of the ring forms a sealed chamber enclosing the cell clone.
6. Apply one drop of prewarmed trypsin solution with a Pasteur pipet into the ring and incubate for 5–10 min in the incubator for detachment of the cells to be harvested.
7. Place four or five drops of selection medium into the ring and pipet the suspension up and down 5×. Transfer the cell suspension into a well of a 24-well tissue culture plate (Falcon), add 0.5 mL of selection medium and incubate the plate for 1 d in the incubator.
8. The next day remove the medium and incubate the cells with 0.5 mL of selection medium. It is usually not necessary to remove trypsin, as it is inactivated by trypsin inhibitor present in the serum of the culture medium.
9. Dilute the cells into a well of a 12 well tissue culture plate (Falcon), when the cell layer reaches confluency.
10. Feed the cells every 3 d. At 90% confluency, dilute the cells at a ratio of 1:2. After trypsination and centrifugation (800 rpm, room temperature, Beckman GPKR Centrifuge), the cell pellet is resuspended in 2 mL of selection medium.

11. An aliquot (1.1 mL) is used for plating in a 25-cm^2 tissue culture flask or 3.5-cm culture dish, 0.9 mL of the cell suspension is placed into a cyrotube (Nunc), and 0.1 mL of dimethylsulfoxide (DMSO) is added.

12. Seal the tube and shake 5–6× by inverting, then place the tube in a special freezing tray of a liquid nitrogen tank. Alternatively, put it into a closed polystyrene box and place it for 24 h in a –80°C refrigerator to allow for slow cooling of the cells. It is important that freezing starts immediately after addition of DMSO and continues at a temperature decrease of 1°C/min to –80°C for efficient survival of the cells.

13. The next day, the tube can be stored in liquid nitrogen (*see* **Note 9**).

14. When the cells reach confluency in the 3.5 cm dish, analysis of the cell clone can start. **Caution**: Freeze enough cells for further investigations.

3.4. Screening of Cell Clones

For identification and characterization of the exogenous connexin in transfected cell clones, it is recommended to confirm transcription of the exogenous connexin by Northern blot hybridization (*see* **Note 10**). Furthermore, it is necessary to demonstrate by immunofluorescence analysis that the transfected cells express the exogenous connexin protein in the plasma membranes of contacting neighboring cells (*see* **Note 11**).

3.4.1. Northern Blot Hybridization

For preparation of total RNA, cells are plated in a cell culture dish (6 cm, 4×10^5 cells/dish) or 10 cm diameter (1.2×10^6 cells/dish) and grown to 80–90% confluency. For optimal yields of RNA, the cells should reach this confluency 2 or 3 d after plating. The cells should be fed with fresh culture medium 6 h before preparation of RNA. For isolation of total RNA, we follow the procedure described by Chomczynski and Sacchi *(28)*. This commonly used method is detailed in molecular biology manuals *(26,27)*. Alternatively, commercially available, modified isolation methods, for example, TRIzol (Total RNA Isolation Reagent, Gibco-BRL) could be followed.

After isolation of total RNA, it is usually not necessary to purify mRNA further, since expression of transfected connexin genes is regulated by viral promoter and enhancer elements, resulting in high levels of mRNA of the exogenous connexin. For subsequent radioactive or nonradioactive Northern blot hybridization (described in standard laboratory manuals) only 2–3 µg per lane (width: 1 cm) of total RNA is needed.

By Northern blot hybridization, it is possible to identify transfected cell clones that express the expected connexin mRNA at the highest relative level. In some cases, no hybridization signal occurs, or the hybridization signal does not show the expected size, or on the same lane more than one signal is detected. This may be caused by partial loss or multiple recombinations of the transfected construct into the cell genome. These clones should obviously be discarded.

3.4.2. Immunofluorescence

The analysis of indirect immunofluorescence is a fast and sensitive assay that helps to identify transfected cell clones that express the exogenous connexin in regions of the plasma membrane contacting neighboring cells. It is a simple and efficient technique that requires specific antibodies to connexin proteins. Polyclonal antibodies should be affinity purified and should not show any cross-reactivity with other connexins or proteins. The following protocol is inexpensive and reliable.

1. Place sterile glass coverslips in a cell culture dish (3.5- or 6-cm diameter). Add prewarmed culture medium and plate freshly diluted, transfected cells into the dish so that the cells reach about 70% confluency 2 d after plating (for a 6-cm 6×10^5 exponentially growing cells are needed).
2. After 2 d, the culture dish is washed twice with ice cold PBS*. Place each coverslip, using a tweezer, in one well of a 12- or 24-well culture plate, with the cell side facing up. Then the cells on the coverslips are fixed by incubation with methanol (precooled at –20°C) for 10–20 min at –20°C. The methanol is removed and cells are washed twice for 5 min with PBT at room temperature.
3. To block unspecific binding of antibodies, the cells on the coverslips are incubated for at least 30 min with FCS–PBT at room temperature.
4. Before incubation, the first antibody must be diluted (1:20–1:2000 usually works well for most antibodies) in FCS–PBS. Then the FCS–PBT solution is removed from the culture dish and replaced by the antibody–FCS–PBS solution. We recommend 50 µL of antibody–FCS–PBS solution per cm² of coverslip. To prevent evaporation of the antibody–FCS–PBS solution, the entire culture dish is covered with parafilm and stored in a moisture-saturated box to prevent drying. Incubate for 2 h at room temperature or overnight at 4°C.
5. Remove excess antibody by washing the coverslip 3× with PBT for 5 min at room temperature.
6. During this time, centrifuge secondary antibodies at 12,000 rpm for 10 s in a table centrifuge (e.g., Beckman Microfuge). Then dilute secondary, fluorescence tagged antibodies that bind specifically to the F_c domain of the first antibodies, in an adequate volume of FCS–PBT (1:200–1:10000; see instructions of manufacturer). Incubate antibody solution (*see* **step 4**) on the coverslip for 30–60 min at room temperature. (Place the box with the coverslip in the dark to prevent bleaching of the fluorescent tag).
7. Remove excess of secondary antibodies by washing the coverslip 3× with PBT (5 min at room temperature).
8. Place a drop of para-pheneylenediamine solution on an object slide and place the coverslip with the cell side onto this drop. Seal the coverslip with Entellan or nail polish and examine cells under a fluorescence microscope.

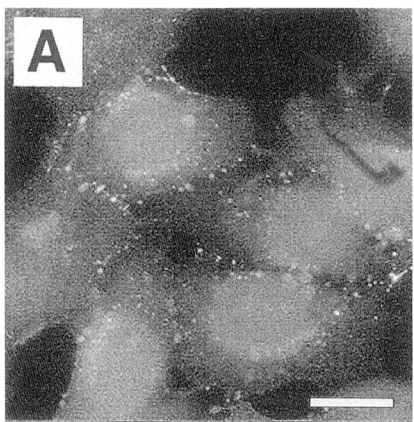 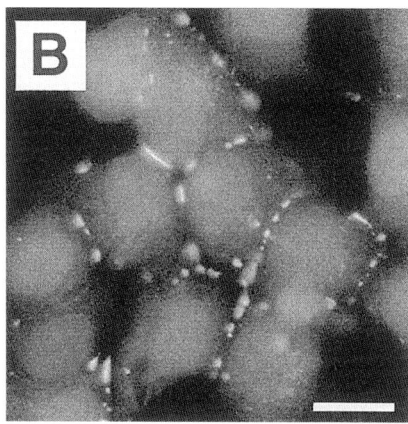

Fig. 1. Immunofluorescence analysis of connexin (Cx) transfected Hela cells using rabbit affinity purified anti-Cx26 antibodies *(31)* and goat antirabbit IgG–fluorescein isothiocyanate (Dianova, Hamburg, Germany). **(A)** HeLa cells with moderate expression level of mouse Cx26 protein. **(B)** HeLa cells with a high expression level of mouse Cx26. Both clones exhibit in contacting plasma membranes the characteristic punctate pattern of fluorescent signals, whose intensity depends on the expression level of the exogenous connexin DNA. Bar = 25 µm.

9. The fluorescent label is rather stable for 2 d, but it is advisable to inspect cells on the same day. Alternatively, the specimen could be stored at 4°C in the dark for about a week without increase of background fluorescence. Positive cell clones from connexin transfection experiments should show punctate fluorescent signals on contact plasma membranes to neighboring cells, as shown in **Fig. 1**.

4. Notes

1. For successful and stable transfection, the choice of the selection drug is important. We have found that HeLa cells show high spontaneous resistance to the selective antibiotics hygromycin and geniticine (G418). Thus, we prefer to select HeLa connexin transfectants using puromycin (1 µg/mL of medium). It is important to use puromycin of the highest quality available (e.g., Sigma, cat. no. P-8833). Furthermore, an overdose of puromycin should be avoided for selection, as it can lead to growth inhibition or cytotoxity toward HeLa connexin transfectants.
2. For efficient expression of the connexin coding region, the introduced sequence should contain only one start codon of translation. Thus, we usually remove all possible upstream start codons by appropriate genetic manipulation.
3. Before starting the calcium phosphate mediated transfection, the ability to form a fine precipitate should be checked with each aliquot of HeBS solution. For this purpose, 0.5 mL of 2× HeBS are mixed with 0.5 mL of 250 mM CaCl$_2$ and vortex-mixed. A fine precipitate should develop immediately and should be visible in the microscope.

4. Plasmid DNA for transfection of HeLa cells could be prepared with a commercial "Midi" kit based on silicagel. If only a low transfection yield is obtained, it may be due to impurity of the DNA. Contamination of the DNA with bacterial proteins or residual material from the columns used for the plasmid preparation could lower the yield of transfectants. To improve the yield of transfection, use CsCl-purified plasmid DNA or "endotoxin-free" commercial kits for plasmid preparation (e.g., QIA Endo free kit, Qiagen, Hilden, Germany).

5. Depending on the cell line, the optimal amount of plasmid DNA for transfection varies from 10 to 50 μg per tissue culture (10-cm diameter) when using the calcium phosphate protocol. For the lipofection protocol, only 2–15 μg/10 cm plate are necessary to obtain the same transfection efficiency as with the calcium phosphate protocol. Usually, transfection or lipofection works well by using nonlinearized supercoiled DNA, but if the yield of clones is low, try the transfection with linearized DNA.

6. Many factors can influence formation of the calcium phosphate precipitate and, therefore, the transfection efficiency. For example, the optimum pH range of the HeBS solution (between 7.05 and 7.12), the pH range of 7.2–7.4 of the culture medium during incubation with the precipitate on the cells, and finally, the size of the DNA calcium phosphate crystals. Only a fine crystallized precipitate on the cells yields good transfection efficiency. An alternative calcium phosphate transfection protocol with a 10-fold higher transfection efficiency was described by Ishiura et al. *(29)*, Chen and Okayama *(30)*, and Ausubel et al. *(26)*. The protocol was based on a different buffer system, called *N,N-bis*(2-hydroxyethy)-2-aminoethanesulfonic acid (BES) buffer. We have described the HeBS protocol instead of the high-efficiency BES protocol, as we noticed more multiple integrations of transfected DNA into the genome than by using the HeBS protocol. It has been reported that the BES protocol does not appear to work well for neuronal cell lines *(26)*. Thus, if the HeBS protocol gives poor yields of transfectants, the BES protocol could be a good alternative.

7. For lipofection we use the Tfx-20 reagents and protocol (Promega). The protocol was applied to HeLa cells, but, with slight modifications, it appears to be useful for several other established cell lines.

8. If steel or glass rings are not available for isolation of the transfected cell clones, the cells could also be removed from the bottom of the culture dish with a sterile pipet tip. The cells can be isolated under a microscope and a Gilson pipet (200 μL), and sterile Gilson tips are recommended for this procedure. This mechanical removal of the cells is more gentle than trypsinization and allows to pick cell clones on several days. After aspiration, the cloned cells should be transferred into a well of a 48-well culture dish and the cluster should be disrupted by pipetting the cells up and down several times.

9. We recommend to freeze aliquots of each transfectant cell clone during the first passages of the clone. Because it has been observed that transfected HeLa cell clones occasionally showed after many (>15) passages decreased expression of the exogenous connexin, it is advantageous to use the same clone at an early

passage number. If the cells do not survive the freezing protocol at high yield, it may help to modify the freezing conditions. For example, cool down the cells to –20°C for 1 d, then transfer the vial to –80°C for another day before storing the cells in a liquid nitrogen tank.

10. Screening of transfected and isolated cell clones by using the polymerase chain reaction (PCR) with genomic DNA as template, seems to be an easy and fast procedure to identify cell clones that have integrated the transfected connexin sequence. On the other hand, we experienced in some experiments that 60–90% of the isolated transfectants that showed resistance toward puromycin contained the exogenous connexin sequence (based on results of PCR or Southern blot hybridization), but expressed little or no connexin mRNA or protein. Perhaps, in these cases, integration may have occurred into silent regions of the genome. Thus, it can save time and frustration to analyze expression of connexin mRNA or protein as early as possible.

11. For further analysis of transfected cell clones, an immunoblot analysis may be useful, depending on the final goal of the experiment.

Acknowledgments

We thank Dr. Otto Traub for affinity-purified antibodies to Cx26, and Joachim Degen for his suggestions to improve the manuscript. Work in our laboratory has been supported by the Deutsche Forschungsgemeinschaft (SFB 284 and SFB 400), the Deutsche Krebshilfe, and the Fonds der Chemischen Industrie.

References

1. Bruzzone, R., White, T. W., and Goodenough, D. A. (1996) Connections with connexins: the molecular basis of direct intercellular signalling. *Eur. J. Biochem.* **238**, 1–27.

2. Condorelli, D. F., Parenti, R., Spinella, F., Salinaro, A. T., Belluardo, N., Cardile, V., and Cicirata, F. (1998) Cloning of a new gap junction gene (Cx36) highly expressed in mammalian brain neurons. *Eur. J. Neurosci.* **10**, 1202–1208.

3. Söhl, G., Degen, J., Teubner, B., and Willecke, K. (1998) The murine gap junction gene connexin36 is highly expressed in mouse retina and regulated during brain development. *FEBS Lett.* **428**, 27–31.

4. Manthey D., Bukauskas F., Lee L., Kozak, C., and Willecke K. (1999) Molecular cloning and functional expression of the mouse gap junction gene connexin57 in human HeLa cells. *J. Biol. Chem.,* **519**, 631–644.

5. Willecke, K. and Haubrich, S. (1996) Connexin expression systems: To what extent do they reflect the situation in the animal? *J. Bioenerg. Biomembr.* **28**, 319–326.

6. Fishman, G. I., Yang, G., Hertzberg, E. L., and Spray, D. C. (1995) Reversible intercellular coupling by regulated expression of a gap junction channel gene. *Cell Adhes. Commun.* **3**, 353–365.

7. Dahl, G., Miller, T., Paul, D., Voellmy, R., and Werner, R. (1987) Expression of functional cell–cell channels from cloned cat liver gap junction complementary DNA. *Science* **236**, 1290–1293.
8. George, C. H., Martin, P. E. M., and Evans, W. H. (1998) Rapid determination of gap junction formation using HeLa cells microinjected with cDNAs encoding wild-type and chimeric connexins. *Biochem. Biophys. Res. Commun.* **247**, 785–789.
9. Barrio, L. C., Suchuyna, T., Bargiello, T., Xu, L. X., Roginski, R. S., Bennett, M. L. V., and Nicholson, B. J. (1991). Gap junctions formed by connexin26 and -32 clone and in combination are differently affected by applied voltage. *Proc. Natl. Acad. Sci. USA* **88**, 8410–8414.
10. Holder, J. W., Elmore, E., and Barrett, J. C. (1993) Gap junction function and cancer. *Cancer Res.* **53**, 3475–3485.
11. Mesnil, M., Krutovsikh, V., Piccoli, C., Elfgang, C., Traub, O., Willecke, K., and Yamasaki, H. (1995) Negative growth control of HeLa cells by connexin genes: connexin species specificity. *Cancer Res.* **55**, 629–639.
12. Eghbali, B., Kessler, J. A., and Spray, D. C. (1990) Expression of gap junction channels in communication-incompetent cells after stable transfection with cDNA encoding connexin32. *Proc. Natl. Acad. Sci. USA* **87**, 1328–1331.
13. Zhu, D., Caveney, S., Kidder, G. M., and Naus, C. C. G. (1991) Transfection of C6 glioma cells with connexin43 cDNA: analysis of expression, intercellular coupling, and cell proliferation. *Proc. Natl. Acad. Sci. USA* **88**, 1883–1887.
14. Rup, D. M., Veenstra, R. D., Wang, H. Z., Brink, P. R., and Beyer, E. C. (1993) Chick connexin56, a novel lens gap junction protein. Molecular cloning and functional expression. *J. Biol. Chem.* **268**, 706–712.
15. Steinberg, T. H., Citivelli, R., Geist, S. T., Robertson, A. J., Hick, E., Veenstra, R. D., Wang, H. Z., Warlow, P, M., Westphale, E. M., and Liang, J. G. (1994) Connexin43 and connexin45 form jap junction with different molecular permeabilities in osteoblastic cells. *EMBO J.* **13**, 744–750.
16. Graham, F. L. and van der Eb, A. J. (1973) A new technique for the assay of infectivity of human adenovirus 5 DNA. *Virology* **52**, 456–459.
17. McCutchan, J. H. and Pegano, J. S. (1968) Enhancement of the infectivity of simian virus 40 desoxyribonucleic acid with diethylaminoethyl-dextran. *J. Natl. Cancer Inst.* **41**, 351–357.
18. Neumann, E., Schaefer-Ridder, M., Wang, Y., and Hofschneider, H. P. (1982) Gene transfer into mouse lyoma cells by electroporation in high electric fields. *EMBO J.* **1**, 841–845.
19. Fraley, R. (1980) Introduction of liposome-encapsulated SV40 DNA into cells. *J. Biol. Chem.* **255**, 10,431–10,435.
20. Felgner, P. L., Gadek, T. R., Holm, M., Roman, R., Chan, H. W., Wenz M., Northrop, J. P., Ringold, G. M., and Danielsen, M. (1987) Lipofection: a high efficient, lipid mediated DNA transfection procedure. *Proc. Natl. Acad. Sci. USA* **84**, 7413–7417.

21. Eckert, R., Dzarlieva-Barkowskarya, A., and Hülser, D. F. (1993) Biophysical characterization of gap junction channels in HeLa cells. *Pflügers Arch.* **205,** 404–407.

22. Elfgang, C., Eckert, R., Lichtenberg-Fraté, H., Butterweck, A., Traub, O., Klein, R. A., Hülser, D. F., and Willecke K. (1995) Specific permeability and selective formation of gap junction channels in connexin-transfected HeLa cells. *J. Cell Biol.* **129,** 805–817.

23. Cao, F., Eckert, R., Elfgang, C., Nitsche, J. M., Snyder, S. A., Hülser, D. F., Willecke, K., and Nicholson, B. J. (1998) A quantitative analysis of connexin-specific permeability differences of gap junctions expressed in HeLa transfectants and *Xenopus* oocytes. *J. Cell Sci.* **111,** 31–43.

24. Haubrich, S., Schwarz, H. J., Bukauskas, F., Lichtenberg-Frate, H., Traub, O., Weingart, R., and Willecke, K. (1996) Incompatibility of connexin 40 and 43 hemichannels in gap junctions between a mammalian cells is determined by intracellular domains. *Mol. Biol. Cell* **7,** 1995–2006.

25. Horst, M., Harth, N., and Hasilik. (1991) Biosynthesis of glycosylated human lysozyme mutants. *J. Biol. Chem.* **266,** 13,914–13,919.

26. Ausubel, F. M., Brent, R., Kingston, R. E., Moore, D. D., Seidman, J. G., Smith, J. A., and Struhl, K. (1989–1998) *Current Protocols in Molecular Biology,* Vol. 1–3. John Wiley & Sons, New York.

27. Sambrook, J., Fritsch, E. F., and Maniatis, T. (1989) *Molecular Cloning: A Laboratory Manual,* 2nd edit., Cold Spring Harbor Laboratory Press, Cold Spring Harbor, NY.

28. Chomczynski, P. and Sacchi, N. (1987) Single-step method of RNA isolation by acid guanidinium thiocyanate–phenol–chloroform extraction. *Anal. Biochem.* **162,** 156–159.

29. Ishiura, M., Hirose, S., Uchida, T., Hamada, Y., Suzuki, Y., and Okada, Y. (1982) Phage particle-mediated gene transfer to cultured mammalian cells. *Mol. Cell. Biol.* **2,** 607–616.

30. Chen, C. and Okayama, H. (1987) High efficiency transformation of mammalian cells by plasmid DNA. *Mol. Cell. Biol.* **7,** 2745–2752.

31. Traub, O., Look, J., Dermietzel, R., Brümmer, F., Hülser, D., and Willecke, K. (1989) Comparative characterization of the 21-kD and 26-kD gap junction proteins in murine liver and cultured hepatocytes. *J. Cell Biol.* **108,** 1039–1051.

II

ASSAYS FOR FUNCTION

12

Assaying the Molecular Permeability of Connexin Channels

Paolo Meda

1. Introduction

Channels made by connexin proteins at gap junctions share a number of similarities with other membrane channels. Thus, they are formed by the oligomerization of integral membrane proteins around a central hydrophilic space; they conduct current-carrying ions, although with much less selectivity than other ion-selective channels; and they show open-to-close transitions that can be modulated under a number of endogenous and exogenous conditions *(1,2)*. However, when compared to other membrane channels, the channels made by connexins also have two unique features. First, they are formed by two cells. Second, they are permeable to a variety of small molecules which, as judged from molecular weight, shape, and electrical charge, may include most of the endogenous metabolites and cytoplasmic factors that are needed for proper regulation of essential cell functions *(1,2)*.

Although the molecular basis and function(s) of this unique feature remain to be fully understood, permeability to exogenous and endogenous molecules has proven particularly useful to positively identify connexin channels in a wide variety of cell systems, and particularly within intact tissues that are hardly amenable to electrophysiological investigations of gap junctional conductance. Assaying the molecular permeability of gap junction channels has also been instrumental in mapping homo- and heterocellular communication territories within different systems *(1,2)*, and to show that, in spite of major similarities, channels made by distinct connexin isoforms may allow for a selective intercellular exchange of different molecules *(3–7)*. In this chapter the procedures that have gained wide acceptance in assaying the intercellular exchange of exogenous and endogenous molecules through connexin channels

From: *Methods in Molecular Biology, vol. 154: Connexin Methods and Protocols*
Edited by: R. Bruzzone and C. Giaume © Humana Press Inc., Totowa, NJ

are reviewed. The reader is referred to previous publications for insights into the general principles of interpretation of these approaches and for discussion of their potential artifacts and limitations *(8–11)*.

Most of the available methods for investigating intercellular communication mediated by connexin channels are dependent on the introduction into living cells of nontoxic tracers, which are then traced in their eventual intercellular movements. Molecules suitable for such experiments should be small enough to cross gap junction channels, whose cutoff size is approx 900 Da in vertebrate cells, and should not be able to leak across a normal nonjunctional membrane. As yet, a number of molecules with these characteristics have been found *(1–11)*. The choice of the most adequate tracer is dependent on several factors, including the scope of the experiment, the conditions under which the study is conducted, and the pattern of connexins in the system investigated. Most connexins are permeant to several tracers and, as yet, no molecule has been shown to permeate only one type of gap junction channel, that is, to be specific for any connexin species. However, it is also clear that different connexins form channels with distinct permeabilities and that these specific properties may allow for subtle discrimination between molecules that are only slightly different in size and/or electrical charge *(1–7)*. Hence, some connexin channels may allow for the passage of some, but not all, available tracers.

2. Materials

2.1. Specific Reagents

Tracer molecules that permeate connexin channels are now commercially available from several companies, for example, Molecular Probes (http://www.probes.com), Sigma Chemical (http://www.sial.com), and Eastman Organic Chemicals (http://www.eastman.com). The commercially available tracers that have gained widespread acceptance in tests of junctional communication are listed in the following subheadings.

2.1.1. Membrane-Impermeant Tracers

1. Lucifer Yellow CH (mol wt 443, two negative charges): this tracer has a high fluorescence efficiency, which ensures its detection in minute levels. It also binds to cell components after aldehyde fixation and photoactivation, a property exploited to identify the cell(s) containing the tracer after histological processing at both light and electron microscopy levels *(9,10)*. The tracer is usually injected as a 2–4% solution in either distilled water or 1 M LiCl, and can be stored for weeks in the dark at 4°C. The original dye does not dissolve in K^+-containing solutions. However, a potassium salt is now commercially available if there was such a need. Lucifer Yellow is observed using a barrier filter with an emission between 520 and 560 nm and an excitation filter with a range between 430 and 435 nm.

2. 6-Carboxyfluorescein (mol wt 376, two negative charges): this tracer has a lower fluorescence efficiency than Lucifer Yellow CH and does not resist histological processing. After recrystallization *(12)*, it can be stored in the dark at 4°C as a 20% solution in distilled water, pH 8–9. It is used as a 4% solution in either distilled water or 3 *M* KCl. Carboxyfluorescein is observed with the same barrier and excitation filters used for Lucifer Yellow.

3. 2',7'-Dichlorofluorescein (mol wt 401, one negative charge): has a permeability four to six times higher than 6-carboxyfluorescein *(3)*, presumably because of a smaller hydration radius that is due to its minor electrical charge. It has little affinity for binding to cytoplasmic proteins, and thus does not resist histological processing.

4. Neurobiotin (mol wt 287, one positive charge): has molecular dimensions that are less than those of Lucifer Yellow, and, hence, is expected to permeate connexin channels more easily. This expectation has been verified in some, although not all, coupled systems *(13,14)*. The tracer (Vector Laboratories, Burlingame, CA) is typically used at a concentration of 1–4% in 0.25 *M* lithium acetate, pH 7.3, and resists histological processing after fixation in 4% paraformalde-hyde. To be detected, neurobiotin should be visualized with either streptavidine coupled to a fluorochrome (cyanine dyes, fluorescein, or rhodamin) or to horse-radish peroxidase. When peroxidase-conjugated streptavidin is used, the activity of the enzyme is revealed after blockade of the endogenous peroxidase activity (by two successive 30 min incubations in a 0.1 *M* Tris-HCl buffer, pH 7.6, supple-mented with 0.5% H_2O_2) and membrane permeabilization (by a 10-min incuba-tion in a 0.1 *M* Tris-HCl buffer, pH 7,6, supplemented with 0.25% Triton-X). To this end, cells are incubated for 1–10 min in a 0.1 *M* phosphate buffer supple-mented with 1 mg/mL diaminobenzidine, 1.2 m*M* cobalt chloride, 0.8 m*M* nickel chloride, and 0.9 m*M* H_2O_2. The reaction is stopped by replacing the developing mixture with distilled water.

5. Other membrane-impermeant tracers that have been used less frequently and may be useful in particular cell systems or under specific conditions are 4', 6-diamidino-2-phenylindole dihydrochloride (DAPI) (mol wt 279, one positive charge) *(4)*, ethidium bromide (mol wt 314, one positive charge) *(4,5)*, propidium iodide (mol wt 414, two positive charges) *(4)*, biocytin (mol wt 373, no electrical charge) *(5)* and biotine-X cadaverin (mol wt 442, one positive charge) *(7)*. Most of these tracers can be dissolved in LiCl and used at a final concentration of 0.5–1%.

6. Dextrans have molecular weights and sizes that largely exceed the maximum cutoff limit of connexin channels. As such, these macromolecules are useful to ensure that the intercellular transfer of any of the earlier listed tracers is actually dependent on connexin channels, and is not accounted for by alternative path-ways, such as the persistence of cytoplasmic bridges at the end of mitosis *(15)*. To this end, we have used 10-kDa dextrans coupled to rhodamine *(16)* which, at the concentration of 100 mg/mL, may be introduced into cells as the other tracers.

7. It is obviously feasible to use the earlier listed tracers in combination. We have often combined Lucifer Yellow with either dextran-rhodamine or neurobiotin

(16,17). In the latter case, the two tracers were each used at a 2% concentration and injected into the same cell by electrical pulses of alternated polarity (negative pulses preferentially inject Lucifer Yellow, whereas positive pulses preferentially inject neurobiotin).

2.1.2. Membrane-Permeant Fluorogenic Esters

These nonfluorescent molecules freely diffuse across the cell membrane and are converted by cytoplasmic esterases into smaller, fluorescent compounds that are retained by living cells and pass through gap junction channels *(18–22)*. Three of these molecules are particularly useful:

1. Calcein-acetoxy-methyl ester (calcein-AM; mol wt 955, no charge): can be stored at 4°C as a 1 mg/mL solution in dimethyl sulfoxide (DMSO) and used in a concentration range of 2–10 µM *(18,19)*. It is converted by esterases into calcein (mol wt 623, four negative charges). This tracer does not resist histological processing. Its fluorescence is observed with the same barrier and excitation filters used for Lucifer Yellow.
2. Carboxyfluorescein diacetate (CFDA; mol wt 955, no charge): can be stored at 4°C as a 10 mg/mL solution in acetone and used in a concentration range of 4–40 µg/mL *(20)*. It is converted by esterases into 6-carboxyfluorescein (mol wt 376, two negative charges). This tracer does not resist histological processing. Its fluorescence is observed with the same barrier and excitation filters used for Lucifer Yellow.
3. *bis*-Carboxylethyl-carboxyfluorescein-acetoxy-methyl ester (BCECF-AM; mol wt 668, no charge): can be stored at 4°C as a 10 mg/mL solution in acetone and used in a concentration range of 0.9–1.5 µM *(21,22)*. It is converted by esterases into an impermeable acid form (BCECF; mol wt 520, four or five negative charges). This tracer does not resist histological processing. Its fluorescence is observed with the same barrier and excitation filters used for Lucifer Yellow.

2.1.3. Radiolabeled Compounds to Study the Permeability of Endogenous Molecules

Most radioactive precursor molecules are commercially available from major companies, for example, Amersham Pharmacia Biotech (http://www.apbiotech.com) or New England Nuclear (http://www.nnenlifsci.com). They should be preferably labeled by ^{3}H or ^{35}S and selected according to the scope of the experiment. The protocol that follows is given for [3H]uridine, which has been widely used *(47)*.

2.2. Apparatus

2.2.1. Apparatus for Microinjection

1. An antivibration table (e.g., VX95 Micro-Controle; http://www.newport.com).
2. An inverted microscope (e.g., Zeiss IM35 or Nikon Diaphot 300) for experiments on cell monoalyers or a regular upright microscope (e.g., Zeiss Universal)

for injection of thick specimens (*16,23–25*). All microscopes should be equipped for both fluorescence (we use either a HBO 50W or a XBO 75W Osram bulb and excitation and barrier filters for fluorescein and rhodamin detection) and phase-contrast views. Under both illuminations, ×25, ×40, and ×63 objectives are adequate for most purposes. A 6.3× objective and a dark-field condenser may also be required for injection of thick specimens.

3. A microscope stage that can be held at different temperatures (we use a home-made stage in which temperature is controlled by the circulation of water provided by a Minsitat pump, Huber, Germany).

4. One or two micromanipulators for holding and positioning the electrodes. For experiments on monolayer cultures, we use either MHV-3 oil-driven (Narishige, http://www.narishige.co.jp) or mechanical manipulators (Leitz GMBH, http://www.leitz.de). For experiments on tridimensional cell systems, we definitively prefer the latter ones.

5. A setup for iontophoretic or pressure injection (*see* **Notes 1** and **2**). For iontophoretic microinjection, we use a setup that includes a function generator that drives a pulse generator (e.g., PU8085 amplifier, School of Medicine, University of Geneva, Switzerland or Patch-clamp PC-501 A amplifier, Axon Instruments). Current pulses and recorded voltages are displayed on separate channels of a digital storage oscilloscope (2214, Tektronix) whose outputs are traced on a four-channel TA550 paper recorder (Gould; http://www.gould.co.uk). The PU8085 amplifier is based on a chop approach that allows for continuous monitoring of electrode resistance after each current pulse, and which is monitored on a separate 2205 Tektronix oscilloscope. For pressure injection (*see* **Note 2**), we use a Picospritzer II pulse generator (General Valve, Fairfield, NJ).

6. A micropipet puller (e.g., BB-CH puller from Mecanex, Geneva, Switzerland, PC-10 puller from Narishige, Tokyo, Japan).

7. A camera for the photographic (e.g., Nikon 501) or video recording (we use a high-sensitivity TM-560, Pulnix; http://www.pulnix.comin linked to a VO-5630 U-Matic recorder, Sony; http://www.sony.com) of the injections.

2.2.2. Apparatus for Scrape-Loading Experiments

1. An inverted microscope equipped as described in **Subheading 2.2.1.**
2. A photographic or video camera as described in **Subheading 2.2.1.**
3. A tool for scraping the culture (*see* **Note 3**).

2.2.3. Apparatus for Incorporation of Membrane-Permeant Fluorogenic Molecules

1. An inverted microscope equipped as described in **Subheading 2.2.1.**
2. A photographic or video camera as described in **Subheading 2.2.1.**

2.2.4. Apparatus for Assessing Permeability to Endogenous Molecules

A microscope equipped with phase-contrast optics as described in **Subheading 2.2.1.**

2.3. Solutions

2.3.1. Solutions for Microinjection

2.3.1.1. Cell Buffer

We perform most of the injections in a 4-(2-hydroxyethyl)-1-piperazine-ethanesulfonic acid (HEPES)-buffered modified Krebs–Ringer medium (*see* **Note 4**) which is prepared as follows:

1. Make a stock $NaHCO_3$ solution:

$NaHCO_3$	2.625 g
NaCl	5.347 g
H_2O	to 1000 mL

   Store at 4°C
2. Make a stock mixed salt solution:

NaCl	34.700 g
KCl	1.770 g
K_2HPO_4	0.808 g
$MgSO_4{\cdot}7H_2O$	1.463 g
$CaCl_2{\cdot}2H_2O$	1.870 g
H_2O	to 1000 mL

   Store at 4°C
3. Just before the experiment, make the final solution:

Stock $NaHCO_3$ solution	32 mL
Stock mixed salt solution	40 mL
HEPES	0.477 g
Glucose	0.100 g
H_2O	to 200 mL
4. Gas 10 min by bubbling air.
5. Add 1 *N* NaOH as needed to bring pH to 7.4.
6. Add 0.200 g of bovine serum albumin (fraction V; Sigma Chemical).
7. Use within the next 3 h.

2.3.1.2. TRACER SOLUTION

Most tracers are conveniently stored at the injection concentration (2–4%), in a HEPES (10 m*M*)-buffered solution of lithium chloride (1 *M*):

1. Make a solution of

$LiCl_2$	8.480 g
HEPES	0.477 g
KOH	As needed to bring pH to 7.2
H_2O	to 200 mL
2. Filter on a 0.22 µm pore filter (Sterile Acrodisk®13, cat. no. 4454, Gelman Sciences).
3. Store in the dark at 4°C.

2.3.1.3. INJECTION SOLUTION

The solution that fills the electrodes, and which will enter cells during micro-injection, should not compromise the most important conditions of cytoplasm. A simple HEPES (10 mM)-buffered solution of lithium chloride (150 mM) that meets these basic requirements is prepared as follows:

1. Make the following solution:
LiCl$_2$	1.272 g
HEPES	0.477 g
KOH	As needed to bring pH to 7.2
H$_2$O	to 200 mL
2. Filter on a 0.22 µm filter.
3. Store at room temperature.

2.3.1.4. FIXATION SOLUTION

Some gap junction tracers may be retained within microinjected cells and tissues by fixation at the end of the experiment. To achieve good preservation of structure, retention of tracers, and subsequent analysis by a variety of techniques, including immunofluorescence, without excessively increasing the background fluorescence of tissue, we use a 4% solution of paraformaldehyde:

1. Warm to 60°C 100 mL of 0.1 M phosphate buffer.
2. Add 4.0 g of paraformaldehyde (cat. no. 4005, Merck).
3. Allow to dissolve under continuous stirring in a ventilated environment.
4. Cool the solution to room temperature.
5. Add 1 N NaOH as needed to bring pH to 7.4.
6. Pass through a paper filter (cat. no. 595, Schleicher & Schuell).
7. Store in the dark at 4°C.
8. Use within the next 3 wk.

2.3.2. Solutions for Scrape Loading Experiments

1. Cell buffer: We perform most of the scrape-loadings in a HEPES-buffered modified Krebs–Ringer medium described in **Subheading 2.3.1.1.** (*see* **Note 4**).
2. Tracer solution: Essentially any of the tracers listed in **Subheading 2.1.1.** may be used for scrape-loading, and should be selected as per the same lines of thought discussed for microinjection (*see* also **Note 5**). We have typically used in combination Lucifer Yellow and dextran–rhodamine (*16*). The latter tracer cannot cross connexin channels gap junctions because of its large molecular mass (10 kDa), and thus labels only the cells that had incorporated it during the scraping. Dissolve the selected tracer as described in **Subheading 2.1.1.**
3. Fixation solution: Some gap junction tracers may be retained within scrape-loaded cultures by fixation at the end of the experiment. We routinely use a 4% solution of paraformaldehyde, as described in **Subheading 2.3.1.4.**

2.3.3. Solutions for Incorporation of Membrane-Permeant Fluorogenic Molecules

2.3.3.1. TRACER SOLUTIONS

Stock tracer solutions are made as follows:

1. Dissolve 1 mg of calcein-AM (or 10 mg of CFDA or 10 mg BCECF-AM) in 1 mL of DMSO (or acetone in the case of both CFDA and BCECF-AM).
2. Store at 4°C.

2.3.3.2. FIXATION SOLUTION

Some gap junction tracers may be retained within loaded cultures by fixation at the end of the experiment. We routinely use a 4% solution of paraformaldehyde, as described in **Subheading 2.3.1.4.**

2.3.4. Fixative Solution for Endogenous Molecules

3. Filter 5 mL 50% glutaraldehyde (cat. no. 49628, Fluka Chemie AG, Switzerland) through a paper (cat. no. 336251, Schleicher & Schuell, Germany).
2. Add the filtered glutaraldehyde to 95 mL of 0.1 M phosphate buffer, pH 7.4.
3. Correct the pH to 7.4 if needed.
4. Store (for weeks) at 4°C, in a tightly closed bottle protected from light.

3. Methods

3.1. Assessing the Permeability to Exogenous Molecules

3.1.1. Microinjection

Microinjection of membrane-impermeant, nontoxic tracers has been the first technique used to identify and map the communication network of a wide variety of cell systems. The approach is still widely used to study cell coupling in cultures (*see* **Fig. 1**) and is practically the only technique available to extend such studies to intact tissues *(17)*. Because it allows for a selective loading of tracers, microinjection permits correlation of morphological and functional data from individual cells. Furthermore, because the onset and duration of the tracer injection can be accurately controlled, the technique is also instrumental in kinetic studies aimed at evaluating the rate of transfer of a tracer from one cell to another. The major limitation of microinjection is that it requires some special equipment and skill to prevent damaging the cells, or at least to detect such damage during the injection. Another limitation is that only a few cells may be microinjected, usually not at the same time. Thus, the technique is not convenient when a large number of cells need to be simultaneously monitored for intercellular communication.

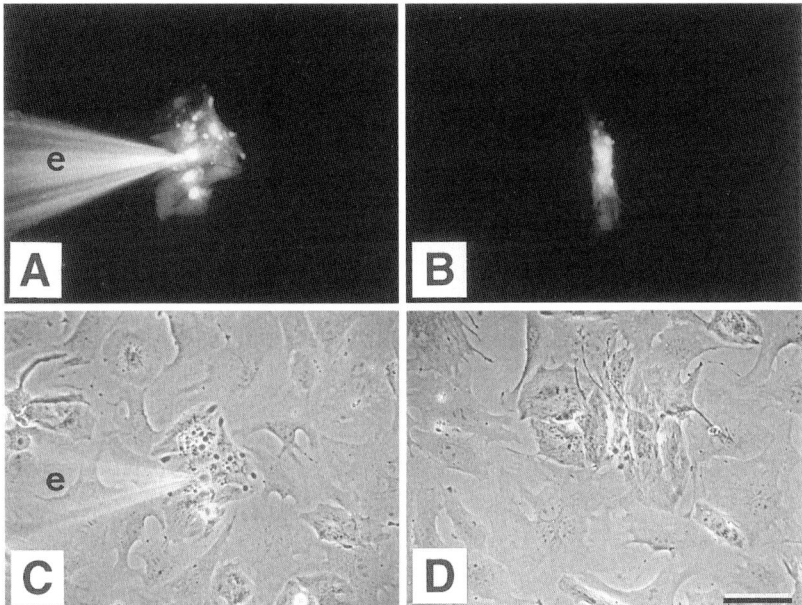

Fig. 1. Microinjection of a gap junction-permeant tracer. (**A**) Lucifer Yellow was introduced via a microelectrode (e) into a neonatal rat cardiomyocyte. Within seconds, the tracer diffused from the injected cell into several neighboring companion cells. Note, however, that adjacent fibroblasts did not receive the tracer. (**B**) The experiment was repeated, within the same culture dish, 15 min after addition to the culture medium of 1 mM heptanol, a blocker of connexin channels. As a result of this blockage, the intercellular exchange of Lucifer Yellow was drastically reduced. (**C,D**) Phase-contrast views of the fields shown in **A** and **B**, respectively. Bar, 20 µm.

3.1.1.1. Preparing the Cells to Be Injected

A preliminary condition for a successful microinjection is that the cell under study should not move during impalement. This is easily achieved with cells in monolayer culture that are grown on a dish suitable for the injection setup, and thus require no further preparation to be injected. The problem may be more complex for immobilizing on an adequate support tridimensional cell assemblies, such as intact tissues. To this end, we routinely use standard 35-mm culture dishes coated with Sylgard and polylysine (for alternatives, *see* **Note 6**).

1. Make Sylgard-coated dishes:
   Prepare 184 Sylgard silicone elastomer (Dow Corning) as directed by the manufacturer.
   Plate 0.5 mL of Sylgard per 35-mm tissue culture dish.
   Cure for 12 h at 60°C.
   Store at room temperature.

2. Make a poly-L-lysine solution:

> Dissolve 1–10 mg of poly-L-lysine hydrobromide (mol wt 150,000–300,000; Sigma Chemical) in 10 mL of phosphate-buffered saline (PBS), pH 7.2.
>
> Put in each dish 2 mL of polylysine solution.
>
> Incubate for 1 h at room temperature.
>
> Rinse 3× for 5 min each in PBS.

3. Just before the microinjection:

> Position a tissue fragment in the center of a Sylgard and poly-L-lysine-coated dish, within a 20-µL drop of albumin- and serum-free medium.
>
> After 5–30 min (a longer time period should be provided for attachment of small pieces) at room temperature and within a humidified chamber, remove the 20 µL of medium.
>
> Slowly add the medium to be used during the injection experiment.
>
> Cover the medium with a thin layer of light liquid paraffin (BDH Laboratory Supplies, Poole, UK) to avoid evaporation.
>
> Transfer the dish on the heated (37°C) stage of the injection microscope.

3.1.1.2. Preparing the Electrodes

Electrodes are pulled from filament-containing borosilicate glass capillary tubing (1.2 mm external diameter, World Precision Instruments ref. TW120F-4) using a micropipet puller that is adjusted to give the desired electrode resistance. Electrodes of 50–60 MΩ when filled with 3 M KCl are adequate for most uses (when filled with solutions of tracers, these electrodes show resistances 4–20× higher, depending on the solvent used). A few hours before the experiment:

1. Pull the electrodes using settings adapted to the type of electrode you need and puller equipment you use.
2. Bend slightly the electrode shaft on a small gas flame.
3. Fill only the tip of the electrode with a stock tracer solution using a 34-gauge nonmetallic syringe needle (MicroFil; WPI) fitted on a 0.22-µm pore filter (Sterile Acrodisk®13, cat. no. 4454, Gelman Sciences).
4. Store the filled electrodes in a clean, humidified container.
5. Use within the next 6–8 h.
6. Just before each injection, fill the shaft of the electrodes with the injection solution, using a 34-gauge nonmetallic syringe needle fitted on a 0.22-µm pore filter.

3.1.1.3. Preparing the Electrode on its Holder

The conducting silver wire of the electrode holder should be coated as follows.

1. Prepare the following coating solution:

KCl	7.456 g
HCl (1 N)	100 µL
H_2O	to 1000 mL

   Store at room temperature.

2. Put 10 mL of coating solution in a beaker.
3. Place in it the silver wire to be coated.
4. Connect the end of the silver wire that will link the electrode to the injection setup to the positive pole of 4.5 V battery (the negative pole of the battery is also linked to another silver wire that is immersed in the coating solution).
5. Check for the development of gas bubbles along the wire connected to the negative pole of the battery.
6. Disconnect the battery after 5 min, when the silver wire of electrode holder has darkened.
7. Mount the chlorinated wire on the electrode holder.
8. Place the tracer filled electrode in the holder.
9. Connect the electrode holder to the head stage of the micromanipulator.

3.1.1.4. PENETRATING A CELL

1. Under dark-field or phase-contrast illumination, position the electrode over the area (in tissues) or the perinuclear region of the cell (in monolayer cultures) to be injected, using a 6.3× objective.
2. Shift on the 25× or 40× objective lens.
3. Slowly lower the electrode against the cell membrane.
4. Compensate for series resistances using the balance control of the amplifier.
5. Compensate for electrode resistance using the neutral control of the amplifier.
6. Penetrate individual cells by briefly "vibrating" the electrode, using the negative capacitance control of an amplifier (this procedure also results in the ejection of some dye).
7. Check for correct penetration by a drop of resting membrane potential (for all tracers dissolved in a salt solution) and rapid filling of the injected cell (if the selected tracer is fluorescent).
8. As soon as the adequacy of the penetration has been ascertained, turn off the UV illumination to prevent photoabsorption which could damage the cell (if a video recording is made, the UV light should be decreased as much as possible with inert absorption filters and the signal amplified with an adequate high sensitivity camera) to minimize the binding of tracers to cytoplasmic and nuclear components.
9. Start injecting the tracer.

3.1.1.5. INJECTING A TRACER

1. Apply to the electrode hyperpolarizing (negative) or depolarizing (positive) current pulses, depending on whether the selected tracer has a net negative or a positive charge. The duration, amplitude, and frequency of these pulses can be varied depending on the type of electrode used (the amount of current that can be injected decreases per unit time with increasing electrode resistance), the amount of tracer to be injected, and the cell type. We routinely use square, 0.1-nA pulses of 900 ms duration and 0.5 Hz frequency for 3–5 min.
2. During the injection, check for maintenance of adequate penetration by the persistence of a stable resting membrane and, in the case of fluorescent tracers, by the persistence of a high cell fluorescence.

3. At the end of the experiment, interrupt the current pulses.
4. Gently pull the the electrode out of the cell.

3.1.1.6. RECORDING THE INJECTION

1. If you use a photographic camera, record each injection site before and/or after pulling out the injection electrode, using Kodak Ektachrome 400 daylight film for color slides and a constant exposure time of about 20 s.
2. If you use a video camera, record each injection online, using a high-sensitivity, intensifying equipment (e.g., TM-560, Pulnix; http://www.pulnix.com) and a VO-5630 U-Matic recorder (Sony; http://www.sony.com).

3.1.1.7. ANALYZING THE INJECTION

If a fluorescent tracer was chosen, the injected cell and its coupled neighbors may be immediately visualized under a microscope before (in the case of tracers that do not have affinity for cell components) or after fixation of the specimens (in the case of tracers that bind to cell components). If Lucifer Yellow or neurobiotin is chosen, it is further possible to identify the cells containing the tracer (*see* **Note 7**), after processing of the tissues for light and electron microscopy *(9,10,17,24,26)*.

3.1.1.7.1. For Light Microscopy

1. Fix the injected sample 30–90 min at room temperature in a 4% solution of paraformaldehyde.
2. Rinse 3× for 10 min in PBS.
3. Dehydrate by 10-min passages in a sequence of 30%, 50%, 70%, and 95% ethanol solutions.
4. Dehydrate by two 20-min passages in absolute ethanol.
5. Soak the specimens twice for 15 min in propylene oxide (absolute ethanol should be used if the preparation was grown in a plastic dish).
6. Infiltrate the specimen for 2 h in a 1:1 (v/v) mixture of Epon 812 and either propylene oxide or absolute ethanol.
7. Infiltrate the specimen for 12 h with pure Epon.
8. Embed the specimen in freshly prepared Epon 812.
9. Cure for 24 h at 60°C.
10. Cut serial sections of 1–5 μm thickness.
11. Screen the sections under a fluorescence microscope for injected and coupled cells which may be recognized by a persistent fluorescence labeling.

3.1.1.7.2. For Electron Microscopy

1. Fix the injected preparation 90 min at room temperature in the 4% paraformaldehyde solution, supplemented with 0.1–1% glutaraldehyde.
2. Proceed through **steps 2–9** as indicated for light microscopy in the previous section.
3. Cut section of about 600 nm thickness.
4. Screen sections under an electron microscope.

3.1.2. Scrape-Loading

Scrape-loading is a simple alternative to microinjection for introduction into cultured cells of tracers that cannot cross the cell membrane *(15)*. In this approach, monolayers of adherent cells are scraped in the presence of a gap junction permeant tracer, which becomes incorporated by cells along the scrape, presumably as a result of some mechanical perturbation of the membrane. As normal membrane permeability is reestablished, the tracer becomes trapped within the cytoplam and, with time, may move from the loaded cells into adjacent ones connected by functional connexin channels (*see* **Fig. 2**). The major advantages of the scrape-loading approach are that it does not necessitate most of the special equipment and skills that are needed for microinjection, and that it allows for the rapid and simultaneous assessment of junctional communication in large numbers of cells. It is therefore particularly useful when a large screening of multiple conditions is required or when different regions of a cell monolayer have to be compared within the very same culture dish *(16)*. However, the approach is hardly adequate to investigate small cells or cell assemblies (such as pairs), as well as low-density cultures. It is also complicated when the extent of junctional coupling is small or when selected cells have to be individually screened for coupling. Eventually, it is rarely applicable to tridimensional systems *(27)*.

3.1.2.1. PREPARING THE CULTURES TO BE SCRAPE-LOADED

Successful scrape-loading requires that cells are firmly adherent to the culture dish and form a homogeneous monolayer. To this end, plate the cells 2–3 d before the experiment at a density sufficient to obtain confluent cultures at the time of scrape-loading (plating density will vary depending on the growth rate of the cell type under study).

3.1.2.2. SCRAPE-LOADING A CULTURE

1. Check the culture dish under phase-contrast illumination to ensure that an adequate confluent monolayer has formed.
2. Remove the culture medium.
3. Rinse the cultures twice for 1 min with PBS.
4. Replace PBS with a volume of tracer solution sufficient to cover the culture (400 μL for a 35 mm Petri culture dish).
5. Immediately pass the scraping tool across the center of the monolayer (*see* **Note 3**) by moving it in a straight, rapid, and gentle way (there is no need to exert pressure against the bottom of the culture dish).
6. Place the dish in a humidified, dark environment (e.g., a plastic box wrapped in aluminum foil and containing a filter paper saturated with water) and at 37°C, for 2 min.
7. Remove all tracer solution (tracers may be saved and used again, except if the coupling test was performed in the presence of agents known or expected to alter junctional coupling).

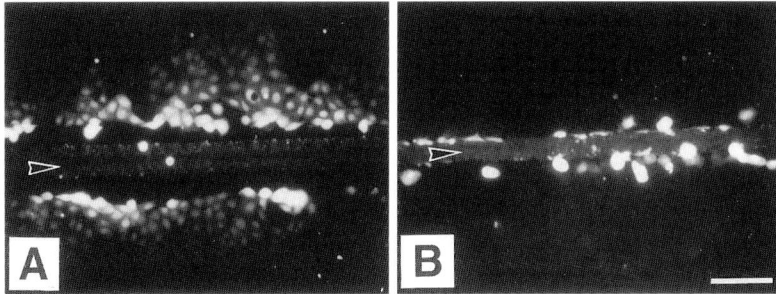

Fig. 2. Scrape-loading of a gap junction-permeant tracer. (**A**) Lucifer Yellow was intro-duced by making a scrape (*arrowhead*) across a monolayer of WB cells. After 2 min, the tracer had diffused from the first order of tracer-loaded cells, located along the scrape, to several orders of cells at distance from it. (**B**) After a few minutes of exposure to 1 mM heptanol, the intercellular exchange of Lucifer Yellow was blocked and the tracer was re-tained within the cells lining the scrape, which had been transiently permeabilized at the beginning of the experiment. Bar, 50 µm.

8. Rinse the cultures 3× for 1 min with PBS.
9. Replace PBS with 4% paraformaldehyde fixative for 30 min at room temperature.
10. Examine the result of the experiment or seal the dishes with parafilm and store them (for weeks if needed) at 4°C for subsequent analysis.

3.1.2.3. Recording (Un)Coupling in a Scraped Culture

1. Assess the tracer distribution under UV illumination, using filters for both fluo-rescein detection (to visualize Lucifer Yellow) and rhodamin (to visualize conju-gated dextran).
2. If you use a photographic camera, record several fields along the scrape line, using Kodak Ektachrome 400 daylight film for color slides and a constant expo-sure time of about 20–40 s. (As the fluorescence yield of a rhodamin signal is less than that of Lucifer Yellow, allow for a longer exposure time when recording the distribution of the rhodamin-labeled dextran.)
3. If needed, the extent of coupling may be further quantitated in scrape-loaded and paraformaldehyde-fixed dishes by scanning fluorescence intensity across the scrape *(16)*.

3.1.2.4. Analyzing the Results of a Scrape-Loading Experiment

The cells initially loaded during the scrape procedure are identified by their content of both the junctional permeant (e.g., Lucifer Yellow) and impermeant tracers (e.g., high molecular weight dextrans). Coupled partners are identified by their content of only the permeant tracer and by their position with respect to loaded cells, which they should always contact either directly (first-order coupled cells) or via intermediate coupled cells (*see* **Note 5**).

3.1.3. Incorporation of Membrane-Permeant Fluorogenic Molecules

The molecular permeability of connexin channels may be tested after loading cells with nonfluorescent, fluorogenic esters that are small and nonpolar molecules that freely diffuse across the cell membrane. Once within the cell, these molecules are converted by cytoplasmic esterases into fluorescent compounds that are highly hydrophilic, and therefore can no longer rapidly cross the cell membrane to exit living cells. However, the small size and hydrophilicity of these tracers permit them to easily pass through gap junction channels *(12,19–21)*. In this method, coupling is assessed by mixing the ester-loaded (donor) cells with unloaded (recipient) cells of the same or a different type, and by monitoring the transit of the now fluorescent species from donor to recipient cells. Loading with fluorogenic esters is easy and does not require the special equipment and expertise needed for microinjection. It may be preferred to microinjection and scrape-loading for investigating coupling in large numbers of cells, particularly if these are small and poorly (or not at all) adherent to standard culture supports. However, this method is not easily applicable to the vast majority of adherent cell types in culture, and cannot be used to investigate coupling within intact tissues. Furthermore, it may be complicated in the case of cells that do not incorporate the esters or retain their tracer derivatives homogeneously, for example, because of a variable expression of cytoplasmic esterases.

3.1.3.1. LOADING CELLS WITH THE ESTERS

1. Check for viability of the cultured cells under phase-contrast illumination.
2. Remove the medium.
3. Rinse the cultures twice for 1 min each with PBS.
4. Replace PBS with serum-free medium supplemented with the selected ester at the final concentration indicated in **Subheading 2.1.2.**
5. Incubate the cells for 15–30 min at 37°C.
6. Rinse 3× for 5 min each with culture medium containing 10% serum (to remove the nonfluorescent esters that may not have been incorporated, as well as the fluorescent derivatives that may have escaped from cells).
7. Plate the loaded cells in fresh, serum-containing medium and maintain them in the culture incubator.

3.1.3.2. ASSESSING (UN)COUPLING

1. Within 4 h following the loading, mix the loaded cells with nonlabeled partners in proportions appropriate for the experiment (typically 3:1).
2. Coculture the labeled and nonlabeled cells in complete culture medium, for the time required for the experiment. (Times ranging from 15 min to 24 h have been tested depending on the type of cell and experiment. However, it is convenient to keep the coculture period as short as possible, as the fluorescence of the ester derivatives steadily decreases with time.)

3. Examine the cocultures under UV illumination, using filters for fluorescein detection.
4. Record the (un)coupling events by photography or video recording.
5. If needed, cells may be separated by standard Ca^{2+} chelation and trypsin treatment procedures for quantitating coupling by fluorescence activated cell sorting *(18,22,28)*.

3.1.3.3. ANALYZING THE RESULTS OF AN ESTER-LOADING EXPERIMENT

In this type of experiment, coupling is demonstrated by the presence of fluorescence within the unloaded (recipient) cells, which therefore should be easily distinguishable from the (donor) cells that were initially loaded with the fluorescent tracer. In some systems, this has been achieved by labeling either the ester-loaded or the unloaded population with a marker such as a dialkylcarbocyanine (Di-I), that permanently labels cell membranes *(18)*. The further decision of whether cells that were not initially loaded with the tracer had actually accumulated it as a result of junctional transfer should be made only after ensuring that the putative recipient cells do contact donor cells, either directly or via other coupled cells. Nonloaded cells that had remained isolated should also remain nonfluorescent up to the end of the coculture period (*see* **Note 5**).

3.2. Assessing the Permeability to Endogenous Molecules

The technical approaches discussed previously allow for a direct evaluation of the permeability of connexin channels to exogenous tracers, which may not necessarily mimic the cell-to-cell exchange of endogenous molecules. Thus, many of the molecules that are expected to be exchanged in vivo by coupled cells have dimensions and charges (many of the physiologically relevant second messengers are negatively charged at physiological pH) different from those of several exogenous tracers, and hence may be restricted or facilitated in their passage through gap junction channels, depending on the type of connexin species *(3–7)*. Also, all the techniques listed previously evaluate intercellular exchanges after the experimental establishment of electrochemical gradients between "donor" and "recipient" cells, which may be orders of magnitude higher than those that could form within living cells *(29–31)*. Taking these limitations into account, various techniques have been devised to estimate the permeability of connexin channels to physiologically relevant molecules. Thus, a variety of endogenous, low molecular weight species have now been shown to cross junctional channels, including second messengers *(27,29–34)*; metabolism *(29–31,35,36)* and drug intermediates *(37)*; oligomaltosaccharides *(6)*; vitamins, amino acids, and nucleotides *(29–31,42)*. In view of the large size of the channel pore, it is surmised that morphogens could also pass connexin channels *(39)*.

In a few studies, the permeability of junctional channels to these endogenous signals was established after their direct introduction into cells by way of microinjection *(35)* or connexin-carrying liposomes *(6)*. In other experiments, attempts have been made to identify by HPLC screening entire set of native molecules that is exchanged between coupled cells *(43)*. However, the vast majority of the studies that have investigated the junctional-mediated exchange of endogenous molecules have taken advantage of the so-called metabolic cooperation approach.

In this approach, a population of "donor" cells is incubated in the presence of an excess of a radiolabeled precursor (typically uridine), and then cocultured with unlabeled "recipient" cells (*see* **Fig. 3**). Under such conditions, quantitative autoradiography allows for an evaluation of the transfer of the resulting metabolites from loaded to unloaded cells as a function of time *(29–32,39–42)*. Variations on this theme include "kiss of life" experiments, in which metabolite-rich cells rescue from death enzymatically deficient cells to which they are coupled, and "kiss of death" experiments in which recipient cells die after transfer of a noxious metabolite *(39–42)*. Recent studies on the so-called "bystander effect" *(44)* have provided evidence that the latter cell-to-cell exchange of toxic metabolites may actually be of therapeutic interest *(37,44–46)*.

3.2.1. Loading Cells with Radiolabeled Precursors

1. Check the viability of the cultured cells under phase-contrast illumination.
2. Disperse the cells using standard Ca^{2+} chelation and trypsin treatment protocols.
3. Wash the cells twice in complete culture medium.
4. Distribute aliquots of the dispersed cells within plastic culture dishes, preventing cell adherence (Falcon® cat. no. 1016, Becton Dickinson).
5. Culture one aliquot of (recipient) cells in fresh culture medium supplemented with 5 µCi/mL of [^{3}H]thymidine (15 Ci/mmol) for 22 h.
6. Culture another aliquot of (donor) cells for 19 h in normal medium and then for 3 h in medium supplemented with 25 µCi/mL of [^{3}H]uridine (25 Ci/mmol).
7. Collect donor and recipient cell separately and wash them 3× for 5 min each with culture medium containing 10% serum.

3.2.2. Assessing (Un)Coupling

1. Mix the thymidine- and the uridine-loaded cells in proportions appropriate for the experiment (typically 3:1) and plate the mixed cell population within 60-mm plastic culture dishes containing 5 mL of complete culture medium.
2. Coculture the cells for the time required for the experiment. (Times ranging from 3 to 24 h have been tested depending on the type of cell and experiment.)
3. Remove the medium.
4. Fix the cultures for 1 h at room temperature in the 2.5% glutaraldehyde solution.

3.2.3. Recording the Results of the Metabolic Cooperation

1. Wash the fixed cultures 3× 10 min each with PBS.
2. In a photographic darkroom, dilute 20 mL of Ilford L4 emulsion (Ilford, Basildon, England) in 20 mL of sterile, bidistilled water.
3. Remove the PBS from the culture dish.
4. Add 300 μL of emulsion per 35-mm tissue culture dish.
5. Cover homogeneously the cultures with a thin layer of emulsion by rolling the dish.
6. Discard excess emulsion.
7. Pack the emulsion-covered dishes in a dark plastic box containing a desiccant.
8. Expose for 10 d in the dark and at 4°C.
9. In a photographic darkroom, develop the autoradiographs for 5 min at 20°C with D19 (Eastman Kodak, Rochester, NY).
10. Rinse for 2 s in distilled water.
11. Fix for 5 min in a photographic fixative.
12. Seal a coverslip over the developed emulsion.
13. Examine the cultures using an inverted phase-contrast microscope.

3.2.4. Analyzing the Results of a Metabolic Cooperation Experiment

In this type of experiment, coupling is demonstrated by the autoradiographic labeling of the cytoplasm of recipient cells, due to the incorporation in their RNA of ^{3}H-nucleotides synthesized (from uridine) within donor cells and transferred across connexin channels. The protocol described previously permits positive identification of recipient cells, owing to the presence of the autoradiographic labeling of nucleus, which results from the incorporation of [^{3}H]thymidine into DNA. The pattern of labelling of recipient and donor cells may be assessed in cultures exposed only to either [^{3}H]thymidine and -uridine. Background staining is evaluated on the cytoplasm of recipient cells that had remained separated from all donors and positive recipients. All recipient cells that have acquired radiolabeled nucleotides as a result of coupling should contact donor cells, either directly or via other recipients (*see* **Note 9**).

4. Notes

1. As a function of the scope of the study, of the cell type investigated, and of the available funding, different setups may be found suitable for the intracellular microinjection and recording of gap junction tracers. Basically, a general-purpose setup should include a microscope, one or two micromanipulators, an electronic rack for performing and controlling the cellular microinjection, and a camera system to record the experiment. Most of the required equipment is commercially available from several companies, for example, Zeiss (http://www.zeiss.com), Nikon (http://www.nikon.com), Axon Instruments (http://www.axonet.com),

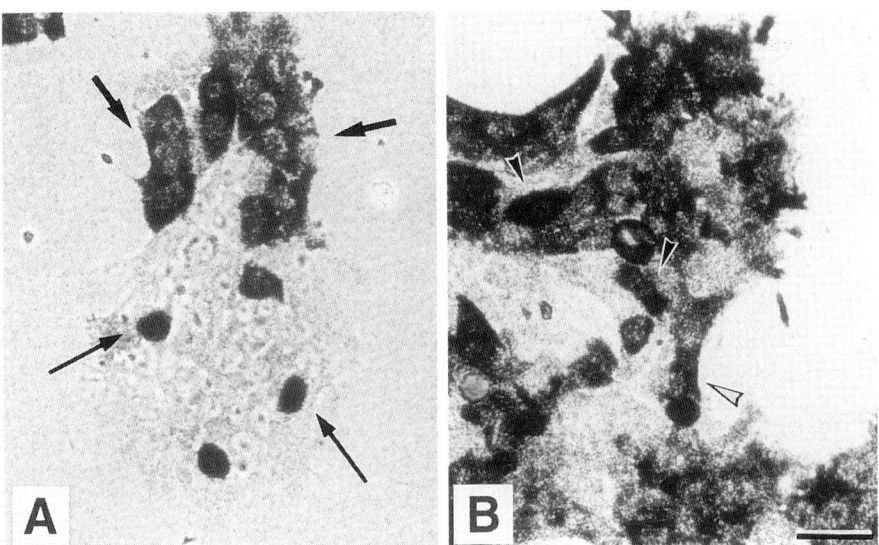

Fig. 3. Gap junction transfer of endogenous nucleotides. (**A**) Primary insulin-producing cells were allowed to incorporate either [^{3}H]uridine, a precursor of nucleotides, or [^{3}H]thymidine, which is incorporated into DNA. Immediately after mixing and plating, the two cell types were readily distinguished by their autoradiographic appearance. Thus, clusters were found to comprise both uridine-loaded cells, which are identified by an intense labeling throughout the cytoplasm (*thick arrows*) and thymidine-loaded cells, which are identified by an exclusively nuclear labeling (*thin arrows*). (**B**) After a few hours of culture, mixed clusters contained a third cell phenotype (*arrowheads*) that featured autoradiographic labeling over both cytoplasm and nucleus. This double labeling resulted from the connexin-dependent transfer of ^{3}H-nucleotides (mostly UDP and UTP), synthesized by the uridine-loaded (donor) cells, into the thymidine-loaded (recipient) cells. Bar, 10 μm.

World Precision Instruments (Aston, UK; http://www.wpiinc.com), and Tektronix (Wilsonville, OR; http://www. tek.com). Other necessary pieces of equipment (electrode puller, etc.) and materials may be found through the BioSupplyNet Source Book (http://www. biosupplynet.com) and the Axon Guide for Electrophysiology and Biophysics Laboratory Techniques (http://www. axonet.com).

2. Iontophoretic injection is usually amenable to a tighter control than pressure injection and is therefore preferred in studies of small cells. In particular, primary cells are usually more fragile than cells of permanent lines and may be easily damaged by pressure pulses and/or by an excessive injected volume. In contrast,

pressure injection may be preferred in experiments on large cells, whose proper filling requires large volumes of tracers, and becomes mandatory when large molecules, such as antibodies, have to be injected together with the gap junction tracers.

3. Several tools, ranging from a No. 10 surgical scalpel blade to a regular wooden tooth-pick may be suitable for making a scrape across a monolayer culture to be loaded with a gap junction tracer. In several cases, however, we have found that these tools result in the detachment of large portions of the cell monolayer from the tissue culture dish and/or in an excessive cell damage along the scraped line. To avoid these drawbacks, we have found most suitable to use a small diamond wheel glass cutter (Fletcher) that is gently passed across the cell monolayer *(16)*, in much the same way a wheel cutter is passed across a pizza pie to cut slices. Multiple scrape lines may be made across a single dish to increase the number of cells to be loaded. We usually perform two scrapings along two perpendicular axes passing through the center of the dish.

4. Microinjection and scrape-loading may be performed in any medium suitable for the cell and tissue under study. However, to standardize the experimental conditions (several components of media and serum may modulate connexin-dependent coupling) and to decrease the background autofluorescence to which several components of standard culture media (phenol red, etc.) contribute, we run all experiments in a HEPES-buffered modified Krebs–Ringer buffer without serum.

5. In interpreting the results of a scrape-loading experiment, several cautions must be taken. The tracers of functional connexin channels are selected for their very limited permeation across (nonjunctional) cell membrane, but may readily enter cells whenever this membrane loses its normal semipermability characteristics. This loss is temporarily achieved, on purpose, by the mechanical scraping of cells. However, it could also result from a permanent alteration due to cell aging or death, which is a virtually obligatory event in every culture. To differentiate such damaged cells from healthy cells containing the tracer as a result of either initial loading or subsequent junctional transfer, make sure that the gap junction permeant tracer is used in combination with a tracer known not to cross connexin channels (e.g., high molecular weight dextrans). Any cell showing labeling for both types of molecules may not have been coupled to neighbor cells. In addition, coupling evaluation is restricted to regions of the cell monolayer immediately lining the scrape line, which should preferentially pass across the center of the culture dish (even in healthy cultures, cells located at the periphery of the dishes may spontaneously take up gap junction permeant tracers, but not necessarily larger molecules, in the absence of any manipulation).

6. For successful microinjection, tridimensional cell assemblies, such as intact tissues, should be immobilized on an adequate support. When a Sylgard and poly-L-lysine coating is not adequate, tridimensional cell assemblies may be attached to standard culture dishes or coverslips using Tissucol Duo S (ImmunoAG, Vienna, Austria), a biological glue made of thrombin and a mixture of several human plasma proteins. To this end:

a. Mix in equal parts the two Tissucol components.
b. Layer 10 µL of the biological glue in the center of a tissue culture dish or a coverslip.
c. Rapidly stick the tissue to be injected on the biological glue.
d. Wait 5 min.
e. Cover with injection medium and process for microinjection.

7. If Lucifer Yellow was chosen, the injected and the adjacent coupled cells can be usually directly located on sections by the residual fluorescence of the tracer *(24)* or using specific antibodies against it *(9)*. If neurobiotin was chosen, the injected and the adjacent coupled cells can also be identified on sections after revealing the injected tracer with either peroxidase-labeled *(17)* or fluorochrome-tagged streptavidin.

8. In ester-loading experiments, nonloaded (recipient) cells that had remained isolated at the end of the experiment should not fluoresce, owing to their inability to establish functional connexin channels with primarily loaded donors. However, some fluorescence labeling of these cells may occur as a result of either the incorporation of residual fluorogenic esters from the medium or the leakage of not fully deesterified esters from donor cells. Both problems can be usually controlled by thoroughly washing donor cells before the coculture with nonloaded recipients. Alternatively, the labeling period of donor cells may be extended, providing sufficient time to cytoplasmic esterases for full transformation of the esters into membrane impermeant derivatives.

9. In metabolic cooperation experiments, coupling is demonstrated by the autoradiographic labeling of recipient cells in contact with donors, provided no such a labeling is seen in recipient cells that had remained isolated throughout the coculture period. It is possible, however, that these cells may incorporate some [^{3}H]uridine left in the medium or lost by donor cells. This event, which can be assessed by quantitating the cytoplasmic labeling of isolated recipients, is usually controlled by thoroughly washing donor cells before the coculture with non loaded recipients. The possible incorporation by recipient cells of labeled nucleotides and nucleic acids lost by damaged donor cells should also be controlled as follows:
a. Expose donor cells to distilled water (5 mL per 60-mm culture dish) for 20 min.
b. Lyophilize the resulting lysate.
c. Redissolve the lyophilized lysate in 5 mL of culture medium without radiochemical.
d. Use this medium to culture unlabeled (recipient) cells for 24 h.
e. Screen for the autoradiographic labeling of these cells (which should be at background levels) as described in **Subheading 3.2.3.**

Acknowledgments

The work of my group is supported by grants from the Swiss National Science Foundation (31-53720-98), the Juvenile Diabetes Foundation International (197124), the European Union (QGG-1-CT-1599-00516), and the Foundation de Reuter (265) and the Program Common de Recherche En Genie Biomedical, 1999–2000.

References

1. Kumar, N. M. and Gilula N. B. (1996) The gap junction communication channel. *Cell* **84,** 381–388.
2. Bruzzone, R., White, T. W., and Paul, D. L. (1996) Connections with connexins: the molecular basis of direct intercellular signaling. *Eur. J. Biochem.* **238,** 1–27.
3. Veenstra, R. D., Wang, H.-Z., Beyer, E. C., and Brink, P. R. (1994) Selective dye and ionic permeability of gap junction channels formed by connexin45. *Circ. Res.* **75,** 483–490.
4. Elfgang, C., Eckert, R., Lichtenberg-Fraté, H., Butterweck, A., Traub, O., Klein, R. A., Hülser, D. F., and Willecke, K. (1995) Specific permeability and selective formation of gap junction channels in connexin-transfected HeLa cells. *J. Cell Biol.* **129,** 805–817.
5. Little, T. L., Xia, J., and Duling, B. R. (1995) Dye tracers define differential endothelial and smooth muscle coupling patterns within the arteriolar wall. *Circ. Res.* **76,** 498–504.
6. Bevans, C. G., Kordel, M., Rhee, S. K., and Harris, A. L. (1998) Isoform composition of connexin channels determines selectivity among second messengers and uncharged molecules. *J. Biol. Chem.* **5,** 2808–2816.
7. Mills, S. L. and Massey, C. (1995) Differential properties of two gap junctional pathways made by AII amacrine cells. *Nature* **377,** 734–737.
8. Taylor, D. L. and Wang, Yu-Li (1980) Fluorescently labelled molecules as probes of the structure and function of living cells. *Nature* **284,** 405–410.
9. Stewart, W. W. (1981) Lucifer dyes—highly fluorescent dyes for biological tracing. *Nature* **292,** 17–21.
10. Stewart, W. W. (1978) Functional connections between cells as revealed by dye-coupling with a highly fluorescent naphthalimide tracer. *Cell* **14,** 741–759.
11. Brink, P. R. and Dewey, M. M. (1981) Diffusion and mobility of substances inside cells, in *Techniques in the Life Sciences (Part I. Techniques in Cellular Physiology)* (Baker, P. F., ed.), Elsevier/North-Holland, Limerick, Ireland, pp. 1–17.
12. Socolar, S. J. and Loewenstein, W. R. (1979) Methods for studying transmission through permeable cell-to-cell junctions. *Methods Membr. Biol.* **10,** 123–179.
13. Hatton, G. I. and Yang, Q. Z. (1994) Incidence of neuronal coupling in supraoptic nuclei of virgin and lactating rats: estimation by neurobiotin and Lucifer yellow. *Brain Res.* **650,** 63–69.
14. Peinado, A., Yuste, R., and Katz, L. C. (1993) Extensive dye coupling between rat neocortical neurons during the period of circuit formation. *Neuron* **10,** 103–114.
15. El-Fouly, M. H., Trosko, J. E., and Chang, C.-C. (1987) Scrape-loading and dye transfer. A rapid and simple technique to study gap junctional intercellular communication. *Exp. Cell Res.* **168,** 422–430.
16. Pepper, M. S., Spray, D. C., Chanson, M., Montesano, R., Orci, L., and Meda, P. (1989) Junctional communication is induced in migrating capillary endothelial cells. *J. Cell Biol.* **109,** 3027–3038.
17. Meda, P. (1999) Probing the function of connexin channels in primary tissues. *Methods,* **20,** 232–244.

18. Kiang, D. T., Kollander, R., Lin, H. H., LaVilla, S., and Atkinson, M. M. (1994) Measurement of gap junctional communication by fluorescence activated cell sorting. *In Vitro Cell. Dev. Biol.* **30A,** 796–802.

19. Frenzel, E. M. and Johnson, R. G. (1996) Gap junction formation between cultured embryonic lens cells is inhibited by antibody to N-cadherin. *Dev. Biol.* **179,** 1–16.

20. Goodall, H. and Johnson, M. H. (1982) Use of carboxyfluorescein diacetate to study formation of permeable channels between mouse blastomeres. *Nature* **295,** 524–526.

21. Guinan, E. C., Smith, B. R., Favies, P. F., and Pober, J. S. (1988) Cytoplasmic transfer between endothelium and lymphocytes: quantitation by flow cytometry. *Am. J. Pathol.* **132,** 406–414.

22. El-Sabban, M. E. and Pauli, B. U. (1991) Cytoplasmic dye transfer between metastatic tumor cells and vascular endothelium. *J. Cell Biol.* **115,** 1375–1382.

23. Meda, P., Bruzzone, R., Knodel, S., and Orci, L. (1986) Blockage of cell-to-cell communication within pancreatic acini is associated with increased basal release of amylase. *J. Cell Biol.* **103,** 475–483.

24. Salomon, D., Saurat, J.-H., and Meda, P. (1988) Cell-to-cell communication within intact human skin. *J. Clin. Invest.* **82,** 248–254.

25. Meda, P., Bosco, D., Chanson, M., Giordano, E., Vallar, L., Wollheim, C., and Orci, L. (1990) Rapid and reversible secretion changes during uncoupling of rat insulin-producing cells. *J. Clin. Invest.* **86,** 759–768.

26. Meda, P., Kohen, E., Kohen, C., Rabinovitch, A., and Orci, L. (1982) Direct communication of homologous and heterologous endocrine islet cells in culture. *J. Cell Biol.* **92,** 221–226.

27. Tsien, R. and Weingart, R. (1974) Cyclic AMP: cell-to-cell movement and inotropic effect in ventricular muscle, studied by a cut-end method. *J. Physiol. (London)* **242,** 95P–96P.

28. Lee, S. W., Tomasetto, C., Paul, D., Keyomarski, K., and Sager, R. (1992) Transcriptional downregulation of gap-junction proteins blocks junctional communication in human mammary tumor cell lines. *J. Cell Biol.* **118,** 1213–1221.

29. Gilula, N. B., Reeves, O. R., and Steinback, A. (1972) Metabolic coupling, ionic coupling and cell contacts. *Nature* **235,** 262–265.

30. Pitts, J. D. and Finbow, M. (1977) Junctional permeability and its consequences, in *Intercellular Communication* (DeMello, W. C., ed.), Plenum, New York, pp. 61–68.

31. Meda, P. and Spray, D. C. (2000) Gap junction function. *Adv. Mol. Cell Biol.*, **30,** 263–322.

32. Lawrence, T. S., Beers, W. H., and Gilula, N. B. (1978) Transmission of hormonal stimulation by cell-to-cell communication. *Nature* **272,** 501–506.

33. Saez, J. C., Connor, J. A., Spray, D. C., and Bennett, M. V. L. (1989) Hepatocyte gap junctions are permeable to the second messenger, inositol **1,4,5**–triphosphate, and to calcium ions. *Proc. Natl. Acad. Sci. USA* **86,** 2708–2712.

34. Berthoud, V. M., Ivanij, V., Garcia, A. M., and Saez, J. C. (1992) Connexins and glucagon receptors during development of the rat hepatic acinus. *Am. J. Physiol.* **263,** G650–G658.

35. Kohen, E., Kohen, C., Thorell, B., Mintz, D. H., and Rabinovitch, A. (1979) Intercellular communication in pancreatic islet monolayer cultures: a microfluorometric study. *Science* **204,** 862–865.

36. Giaume, C., Tabernero, A., and Medina, J. M. (1997) Metabolic trafficking through astrocytic gap junctions. *Glia* **21,** 114–123.

37. Li Bi, W., Parysck, L. M., Warnick, R., and Stambrook, P. J. (1993) *In vitro* evidence that metabolic cooperation is responsible for the bystander effect observed with HSV*tk* retroviral gene therapy. *Hum. Gene Ther.* **4,** 725–731.

38. Warner, A. (1992) Gap junctions in development—a perspective. *Semin. Cell Biol.* **3,** 81–91.

39. Hooper, M. L. and Subak-Sharpe, J. H. (1981) Metabolic co-operation between cells. *Int. Rev. Cytol.* **69,** 46–104.

40. Subak-Sharpe, H., Bürk, R. R., and Pitts, J. D. (1966) Metabolic co-operation by cell to cell transfer between genetically different mammalian cells in tissue culture. *Heredity* **21,** 342–343.

41. Subak-Sharpe, H., Bürk, R. R., and Pitts, J. D. (1969) Metabolic co-operation between biochemically marked mammalian cells in tissue culture. *J. Cell Sci.* **4,** 353–367.

42. Pitts, J. D. (1998) The discovery of metabolic co-operation. *Bioessays* **20,** 1047–1051.

43. Goldberg, G. S., Lampe, P. D., Sheedy, D., Stewart, C. C., Nicholson, B. J., and Naus, C. C. (1998) Direct isolation and analysis of endogenous transjunctional ADP from Cx43 transfected C6 glioma cells. *Exp. Cell Res.* **239,** 82–92.

44. Culver, K. W., Ram, Z., Wallbridge, S., Ishié, H., Oldsfield, E. H., and Blaese, R. M. (1992) *In vivo* gene transfer with retroviral vector-producer cells for treatment of experimental brain tumors. *Science* **256,** 1550–1552.

45. Elshami, A. A., Saavedra, A., Zhang, H., Kucharczuk, J. C., Spray, D. C., Fishman, G. I., Kaiser, L. R., and Albelda, S. M. (1996) Gap junctions play a role in the "bystander effect" of the herpes simplex virus thymidine kinase system *in vitro. Gene Ther.* **3,** 85–92.

46. Johnson, L. G., Olsen, J. C., Sarkadi, B., Moore, K. L., Swanstrom, R., and Boucher, R. C. (1992) Efficiency of gene transfer for restoration of normal airway epithelial function in cystic fibrosis. *Nat. Genet.* **2,** 21–25.

47. Meda, P., Amherdt, M., Perrelet, A., and Orci, L. (1981) Metabolic coupling between cultured pancreatic B-cells. *Exp. Cell Res.* **133,** 421–430.

13

Applying the *Xenopus* Oocyte Expression System to the Analysis of Gap Junction Proteins

I. Martha Skerrett, Mary Merritt, Lan Zhou, Hui Zhu, FengLi Cao, Joseph F. Smith, and Bruce J. Nicholson

1. Introduction

The utility of *Xenopus* oocytes as a system for the expression and electrophysiological analysis of many ion channels has been well documented. This system offers several advantages. The oocytes are robust in terms of their handling and care needs. This robustness, and their size allow for ready microinjection of cRNA into the cells without compromise to their health, or the use of sophisticated equipment. Thus, functional analyses can be accomplished within hours to days, as compared to the process of weeks to months that is involved in the isolation of mammalian cells transfected stably with the cDNA of interest. Transient transfections offer a somewhat more efficient means of expressing exogenous proteins in mammalian cells, but efficiencies are usually sufficiently low (0.1–10%) that identification of the cells expressing the introduced gene can be problematic. The size of oocytes is also an advantage for electrophysiological analysis, which can be performed by dual-electrode voltage clamp, rather than requiring the more demanding approach of patch clamp.

The application of the oocyte expression system to the analysis of gap junctional channels generally requires that the oocytes be stripped of their vitelline membrane (as is also done for patch-clamp studies of transmembrane channels expressed on oocytes). Two oocytes are then pushed together to form a pair from which intracellular currents can be recorded. This additional level of complexity is more than offset by the gain in rapidity of the assay. The efficiency problem of transient transfections noted in the foregoing is amplified in the case of gap junctional analysis that requires two cells, each expressing connexins, to interact (i.e., even if transient transfection efficiency is as high as

From: *Methods in Molecular Biology, vol. 154: Connexin Methods and Protocols*
Edited by: R. Bruzzone and C. Giaume © Humana Press Inc., Totowa, NJ

10%, then only 0.1% of contacts will occur between cells expressing connexins). This leaves stable transfections as the only viable expression alternative, and oocytes as the only reasonable way to do large, site-directed mutant screens. Although oocytes represent a rather different environment to a mammalian cell, in all cases in which the properties of connexins expressed in both oocytes and mammalian cells have been compared, they have proven to be identical.

Oocytes also provide two other unique advantages in the analysis of connexins and their gap junctional channels. Firstly, all vertebrate cell lines studied to date show some levels of coupling via endogenous gap junctions. In a few well-documented cases (e.g., HeLa and N2A cells), the endogenous connexins are expressed at low levels, although this tends to vary between subclones in different labs. Endogenous connexin RNA is also detected in *Xenopus* oocytes *(1)*, but only *Xenopus* connexin38 (Cx38) is expressed. This can be effectively and completely eliminated with the injection of antisense oligonucleotides (*see* **Subheading 3.3.2.**). This procedure is particularly effective in oocytes, as they possess a very active RNase H system *(2)*, and do not synthesize new RNA. A second advantage that is unique to the oocytes is the ease with which interactions between the various members of the connexin family can be studied. Heteromeric interactions within a cell can be investigated through the coinjection of cRNAs for different connexins, while studies of heterotypic interactions between connexins in apposed cells are even more straightforward and can be accomplished by simply manipulating together pairs of oocytes expressing different connexins.

Surprisingly, the oocyte expression system has also proven to be amenable to biochemical analyses of connexins. Specific regions of a connexin can be labeled in various ways and interactions with other cellular components—either endogenously or exogenously introduced—can be determined. Although individual cells have to be injected, the size and translational capacity of oocytes is such that 10–20 cells are usually sufficient for most analytical purposes, including coprecipitation. Sensitivity is increased further if radioactive amino acids are coinjected with the relevant RNAs. A particular advantage of doing these studies in oocytes is that biochemical analyses can be performed on the same cells on which functional assays are done.

Although the oocyte expression system is a remarkably powerful one for the study of gap junction structure and function, in some respects it is limited. The size of the cells, and hence the relatively low membrane resistance (R_m) prohibits the recording of single channels in paired oocytes, although for a few select connexins is it possible to study the activity of hemichannels in isolated membrane patches *(3)*. Immunodetection of subcellular distribution of connexins expressed in oocytes has also proven difficult. Although immunolocalization of connexin expression at the apposed surfaces of paired oocytes has been reported *(4)*, background due to the trapping of antibodies in the yolk or pigment granules often poses a problem and certainly prevents any possibility of identifying sites of intracellular accumulation. Dye transfer studies between oocytes have also proven limited. The system again offers the unique advantage of quantitative correlation of dye transfer with electrical conductance in the same cell pairs and ready measurements of directional dye passage through heterotypic gap junctions *(5)*. However, owing to the dimensions of the system, very high conductances are needed in order to detect dye spread between cells. Furthermore, the binding characteristics of the yolk have so far proven incompatible with the use of cationic dyes—significant transfer occurring only with anionic dyes such as Lucifer Yellow *(5)*. Despite the nonideal nature of the oocyte expression system for connexin localization and dye permeability, a large number of research articles and reviews *(6–8)* have been published over the last 10 years, attesting to its success as a functional assay for docking, gating, and regulation of connexins. We summarize in the following subheadings the techniques that are well adapted to the utilization of the oocyte expression system for a wide range of structure–function relationships.

2. Materials
2.1. Extraction and Preparation of Oocytes from Xenopus Laevis

1. Frogs: large, primed, and ovulating *Xenopus laevis* females (*Xenopus* 1, Ann Arbor, MI) should be maintained at a constant temperature of approx 20°C, on a 12-h light/dark cycle.
2. Anesthesia: 0.15% solution of MS222 (Sigma Chemical, St. Louis, MO) in dechlorinated water for submersion of frog.
3. Dissecting tools: Forceps, scissors, 3/8 circle surgical needle.
4. Collagenase, type 1A (Sigma Chemical, St. Louis, MO).
5. 17°C Incubator.
6. Dissecting microscope and light source (preferably fiberoptic).
7. Pasteur pipets, cut and flame polished for transferring oocytes.
8. 35-mm, 6-cm and 10-cm plastic culture dishes (no adhesive coating), 15-mL conical tubes (Miltex Instruments, Lake Success, NY).
9. 5–0 Silk, sterile surgical suture (Ethicon, Somerville, NJ).
10. OR2 media (*see* **Subheading 2.7.**).
11. L15 media (*see* **Subheading 2.7.**).

2.2. RNA Preparation

2.2.1. Preparation of DNA Template, In Vitro Transcription, Recovery, and Quantitation of RNA

1. Nuclease-free water (*see* **Note 1**).
2. Nuclease-free pipet tips (*see* **Note 1**).
3. At least 1 μg of linearized and cleaned template DNA at a concentration > 0.2 μg/μL.

4. In vitro RNA Transcription Kit (mMessage mMachine Kit from Ambion, Austin, TX).
5. Geneclean Kit (BIO101, Vista, CA) for purification of plasmid DNA.
6. RNAID Kit (BIO101, Vista, CA) for purification of transcribed RNA.
7. Gel apparatus for analysis of template DNA and transcribed RNA.
8. 1% Agarose (FMC Bioproducts, Rockland, MN) gel.
9. Ethidium bromide (International Biotechnologies, New Haven, CT). Make up a stock solution of 5 mg/mL in double-distilled water and store at 4°C. Add 5 µL of stock to 100 mL of agarose gel mix used for electrophoresis. **Caution: Ethidium bromide is a carcinogen and a mutagen. Avoid contact with skin**.
10. 1 kb DNA ladder (Gibco BRL, Grand Island, NY).
11. RNA ladder (0.24–1.95 kb ladder, Gibco BRL, Grand Island, NY).
12. Gel loading buffer for DNA analysis (*see* **Subheading 2.7.2.**).
13. TBE electrophoresis buffer for DNA analysis (*see* **Subheading 2.7.2.**).
14. Gel loading buffer for RNA analysis under denaturing conditions (*see* **Subheading 2.7.2.**).
15. TAE electrophoresis buffer for RNA analysis under denaturing conditions (*see* **Subheading 2.7.2.**).
16. Deionized 6 *M* glyoxal (*see* **Subheading 2.7.2.**).

2.2.2. In Vitro Translation

1. RNA template from **Subheading 3.2.3.**
2. Rabbit Reticulocyte Lysate Kit including RNasin and Amino Acid Mixture minus methionine (Promega, Madison, WI).
3. ^{35}[S]Methionine (NEN, Boston, MA).
4. X-ray film (X-ARS, Kodak, Rochester, NY).
5. 12% Sodium dodecyl sulfate (SDS) polyacrylamide gel.
6. Molecular weight protein markers (e.g., Rainbow Colored High Molecular Weight Protein Markers, Amersham Life Technologies, Arlington Heights, IL).
7. SDS polyacrylamide gel electrophoresis buffer (*see* **Subheading 2.7.2.**).
8. SDS polyacrylamide gel loading buffer (*see* **Subheading 2.7.2.**).

2.3. RNA Injection and Cell Pairing

1. Glass capillaries (Drummond, Broomall, PA; 7" 3-00-203-G/XL).
2. Pipet puller to make injection needles (Sutter, Novato, CA, e.g., model P-87).
3. Injection apparatus; needs to have the capability for 5–40 nl injection volume with high reproducibility (Drummond, Broomall, CA. e.g., Nanoject 1 Auto/Oocyte Injector).
4. Coarse micromanipulator (Brinkmann, Mississauga, ONT, e.g., model MM33; specify R or L hand).
5. RNA prepared to desired concentration (varies between 12.5 ng/µL for some wild-type constructs and 600 ng/µL for mutants that couple poorly).
6. A 27-base antisense oligonucleotide complementary to nucleotides 327–353 of the *Xenopus* Cx38 coding region *(9)*.

7. Injection dish (*see* **Note 2**).
8. Nuclease-free pipet tips (*see* **Note 1**).
9. Pasteur pipet, cut and flame polished for transferring the oocytes.
10. Agar-filled culture dishes with wells for pairing the oocytes along with special culture dish lids for making the wells in the agar (*see* **Note 3**).
11. Noncompressible liquid such as mineral oil.
12. 30G 2" Needle and syringe for back-filling injection pipets.

2.4. Electrical Recording

1. Two dual electrode voltage clamp amplifiers with the capacity to clamp large cells such as oocytes (Axon Instruments, Foster City, CA, e.g., GeneClamp Amplifier).
2. Silver chloride pellets (Axon Instruments, Foster City, CA, e.g., GeneClamp Amplifier), rather than silver-chloride coated wire, may help stabilize the voltage clamp of oocytes when transjunctional currents are large.
3. Four coarse micromanipulators (Brinkmann, Mississauga, ONT, e.g., model MM33, specify R or L hand).
4. Data acquisition software and computer. Any version of pClamp (Axon Instruments, Foster City, CA) is sufficient for the basic routines used to study transjunctional currents in oocytes. pClamp runs on IBM-compatible computers, the computer capacity required depends on the version. It is advisable to consult a representative in the Technical Assistance Department at Axon Instruments to ensure compatibility of hardware and software.
5. A to D converter with at least four input and two output channels (Axon Instruments, Foster City, CA, e.g., Digidata 1200).
6. A vibration-free table (e.g., TMC, Peabody, MA) and Faraday cage may be necessary depending on the stability of the floor and the adequacy of the grounding system.
7. Glass capillaries (WPI, Sarasota FL, e.g., TW150-4, fire polished).
8. 150 m*M* KCl solution and microFil™ syringe needles (e.g., WPI, Sarasota, FL) for electrode filling.

2.5. Cytoplasmic Perfusion of Paired Oocytes (in Addition to the Supplies and Equipment Required for Electrical Recording)

1. Perfusion chambers (*see* **Fig. 1** and **Note 4**).
2. Microdissection scissors (Fine Science Tools, Foster City CA e.g., 8 cm straight no. 15000-08).
3. Perfusion pump with an exchange rate of 0.5–10 mL per min (e.g., WIZ, ISCO, Lincoln, NE).
4. For cleaning the chambers: 2% solution of microwash detergent (e.g., Conrad 70, Fisher Scientific, Pittsburgh, PA), perchloric acid (Mallinckrodt, Paris, KY), and hydrogen peroxide (CVS, Woonsocket, RI).
5. For assembling the chambers: Vaseline and fine paintbrush.
6. Intracellular solution (*see* **Subheading 2.7.3.**).

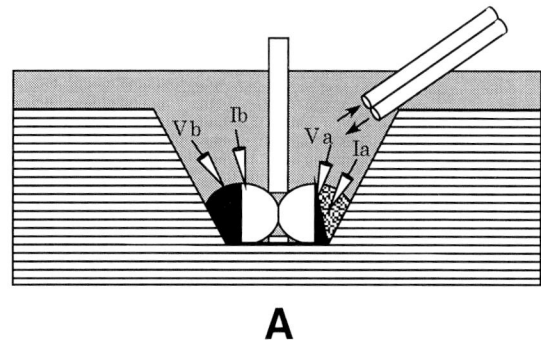

A

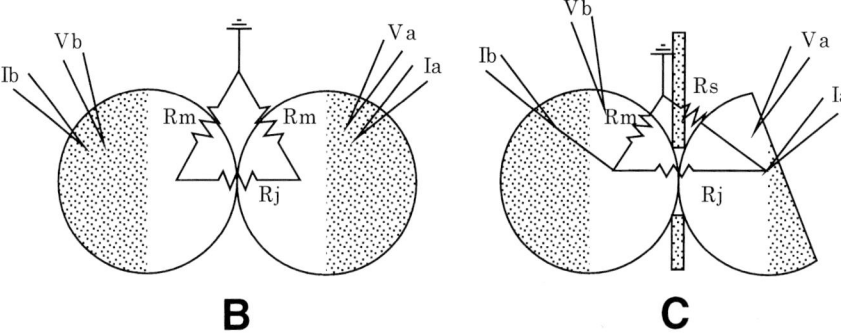

B   **C**

Fig. 1. **(A)** Schematic representation of the dual oocyte perfusion system. Two chambers are separated by a coverslip, designed to facilitate oocyte coupling and seal formation. The intact oocyte remains in standard extracellular solution (e.g. L15) while the other oocyte is perfused with intracellular solution (*see* **Subheading 2.5.**) and cut open with microdissecting scissors. Standard dual-electrode voltage clamp techniques are used to measure transjunctional currents. **(B,C)** Equivalent circuits for oocytes paired in the perfusion chamber before **(B)** and after **(C)** one of the oocytes is cut and perfused. The seal between the oocyte and the coverslip (R_s) replaces R_m (the membrane resistance) of the perfused oocyte in the equivalent circuit. Typical values of R_m, R_s, and R_j are 1 MΩ, 0.2–1 MΩ and 0.02–0.2 MΩ, respectively.

2.6. Immunoprecipitation

1. 1.5-mL Eppendorf tubes.
2. 1-mL Syringe and 25G 1.5-in. needles (Precision Glide, Becton Dickinson, Franklin Lakes, NJ).
3. Shaker at 4°C.
4. Protein A–sepharose (CL-4B) beads (e.g., Sigma Chemical, St. Louis, MO).
5. Monoclonal or polyclonal antibody against specific connexin sequence.
6. [^{35}S]Methionine (0.43 µCi/mL, NEN, Boston, MA).

7. Oocytes injected with an adequate quantity of RNA and radiolabeled methionine. About six oocytes are required for each lane of an analytical SDS polyacrylamide gel.
8. Immunoprecipitation buffers for denaturing or nondenaturing conditions (*see* **Subheading 2.7.4.** and **Note 5**).
9. Materials for SDS-polycrylamide gel electrophoresis (SDS-PAGE) (*see* **Subheadings 2.2.2.** and **2.7.2.**).

2.7. Buffers and Solutions

2.7.1. Oocyte Handling

1. OR-2 Solution: 82.5 mM NaCl, 2 mM KCl, 1 mM MgCl$_2$, 5 mM Tris base, pH 7.3 (Sigma, St. Louis, MO). Make up 1 L of OR-2 solution and add 20 mg each of streptomycin sulfate, penicillin G, and gentamicin sulfate (Sigma Chemical, St. Louis, MO) to make a final concentration of 20 µg/mL. Sterile filter and store at 4°C.
2. L15 medium (2 L of half-strength): One bottle of L15 (Sigma Chemical, St. Louis, MO), 1975 mL of H$_2$O, 25 mL of 1 M (HEPES pH 7.6 (Sigma, St. Louis, MO). This makes 2 L of half-strength L15. Add 40 mg each of streptomycin sulfate, penicillin G, and gentamicin sulfate (e.g., Sigma Chemical, St. Louis, MO) to make a final concentration of 20 µg/mL. Sterile filter and store at 4°C. The final pH should be 7.4, with no adjustment necessary.

2.7.2. RNA Preparation and Protein Analysis

Numbers in parenthesis represent final concentrations.

1. TBE electrophoresis buffer: prepare 1 L of 10× TBE and dilute it with double-distilled water to 1×, pH about 8.2, prior to use.
    108 g of Tris base (89 mM) (TRIZMA, Sigma, St. Louis, MO).
    55 g of boric acid (89 mM) (Sigma, St. Louis, MO).
    40 mL of (0.5 M) Na$_2$EDTA (20 mM), pH 8 (Sigma, St. Louis, MO).
    Double-distilled water to 1 L.
2. Gel loading buffer for DNA: prepare 10× solution and dilute to 1× with double-distilled water prior to use.
    20 mL of Ficoll 400 (20%) (Sigma, St. Louis, MO).
    20 mL of 0.5 M Na$_2$EDTA (20 mM), pH 8.0 (Sigma, St. Louis, MO).
    1 g of SDS (1%) (Sigma, St. Louis, MO).
    0.25 mL of bromophenol blue (0.25%) (Sigma, St. Louis, MO).
    0.25 mL of xylene cyanol (0.25%) (Sigma, St. Louis, MO), optional, helps monitor the gel.
    Double-distilled water to 100 mL.
3. TAE electrophoresis buffer: Prepare 1 L of 50× and dilute it to 1×, pH 8.5, with double-distilled water prior to use.
    242 g of Tris base (2 M) (TRIZMA, Sigma, St. Louis, MO).
    57.1 mL of glacial acetic acid (5.71%) (Sigma, St. Louis, MO).
    37.2 g of Na$_2$EDTA (0.1 M) (Sigma, St. Louis, MO).
    Diethylpyrocarbonate (DEPC)-treated water (*see* **Note 1**) to 1 L.

4. Denaturing buffer for RNA:

   1.7 μL of 100 *M* sodium phosphate, pH 7.0 (Sigma, St. Louis, MO).

   2.9 μL of 6 *M* deionized glyoxal (*see* **Subheading 2.2.1.**).

   8.6 μL of dimethyl sulfoxide (DMSO) (Sigma, St. Louis, MO).

   1–2 μL of RNA.

   Glyoxal is usually obtained as a 40% solution (Sigma, St. Louis, MO). Deionize it by passing it though a mixed-bed resin (e.g. Bio-Rad AG 501-X8 or X8D resin, Bio-Rad, Hercules, CA) until the pH is neutral. Store at –20°C in small aliquots in tightly capped tubes. Prepare the sodium phosphate by dissolving 0.1 mol in a minimal volume of water. Adjust the pH to 7.0 with concentrated phosphoric acid. Adjust the volume to 1 L and autoclave.

5. Loading buffer for denatured RNA *(4×)*: make up 10 mL of 4× buffer and store at 4°C. Add approx 4 μL of buffer to 12 μL of sample prior to analysis.

   5 mL of glycerol 50% (JT Baker, Phillipsburg, NJ).

   1 mL of 100 m*M* sodium phosphate, pH 7.0 (10 m*M*).

   40 μL of bromophenol blue (0.4%) (Sigma, St. Louis, MO).

   4 mL of double distilled water.

6. SDS PAGE buffer: prepare 1 L of 5× solution and dilute to 1× with double distilled water prior to use. Do not adjust the pH, as the required dilution results in a solution of pH 8.3.

   15.1 g of Tris base (203 m*M*) (TRIZMA, Sigma, St. Louis, MO).

   72 g of glycine (0.96 *M*) (Sigma, St. Louis, MO).

   5 g of 0.5% (w/v) SDS (0.96 *M*) (Sigma, St. Louis, MO).

   Double-distilled water to 1 L.

7. Loading (sample) buffer for protein: prepare approx 100 mL of 2× solution starting with 40 mL of water and adding the components in the order listed. Dilute loading buffer to 1× with the sample prior to use.

   40 mL of double-distilled water

   1.52 g of Tris base (203 m*M*) (TRIZMA, Sigma, St. Louis, MO).

   20 mL of glycerol (32%) (JT Baker, Phillipsburg, NJ).

   2.0 g of (w/v) SDS (32%) (Sigma, St. Louis, MO).

   2.0 mL of β-mercaptoethanol (5%) (Sigma, St. Louis, MO).

   1 mg bromophenol blue (approx 0.002%) (Sigma, St. Louis, MO).

   Adjust pH to 6.8 with 1 *N* HCl.

   Double-distilled water to 100 mL.

2.7.3. Cytoplasmic Perfusion of Oocytes

1. Intracellular solution (for perfusion experiments): 89 m*M* KCl, 2.4 m*M* NaHCO$_3$, 0.8 m*M* MgCl$_2$, 0.2 m*M* EGTA (Sigma, St. Louis, MO), 15 m*M* 4-(2-hydroxethyl)-1-piperazineethanesulfonic acid (HEPES) (Sigma, St. Louis, MO), adjust to pH 7.4 with KOH.

2.7.4. Immunoprecipitation

1. Immunoprecipitation (IP) buffer 1 (for cell lysis and IP under nondenaturing conditions): 50 m*M* Tris base, pH 7.6 (TRIZMA; Sigma, St. Louis, MO), 100 m*M*

NaCl, 1% Nonidet P-40 (NP-40) (or IPEGAL CA 6310; Sigma, St. Louis, MO), 0.5% sodium deoxycholate (Fisher Scientific, Fairlawn, NJ), 0.1% SDS (Sigma, St. Louis, MO), 2 mM EDTA (Sigma, St. Louis, MO), 1 mM NaVO$_4$ (Sigma, St. Louis, MO), 50 mM NaF (Sigma, St. Louis, MO), 40 mM β-glycerophosphate (Sigma, St. Louis, MO), 1 mM phenylmethylsulfonyl fluoride (PMSF) (Sigma, St. Louis, MO), 10 µg/mL of pepstatin A (Sigma, St. Louis, MO), 10 µg/mL of leupeptin (Sigma, St. Louis, MO), 10 µg/mL of aprotonin (Sigma, St. Louis, MO). The protocol requires 4 mL of buffer per six oocytes.

2. IP buffer 2 (for cell lysis under denaturing conditions): 50 mM Tris-HCl base, pH 7.6 (TRIZMA; Sigma, St. Louis, MO), 1% SDS, 100 mM NaCl, 1 mM NaVO$_4$, 6 mM EDTA, 50 mM NaF, 40 mM β-glycerophosphate, 10 µg/mL of leupeptin, 10 µg/mL of pepstatin A, 10 µg/mL of aprotonin, 1 mM PMSF. The protocol requires 800 µL per six oocytes.

3. IP buffer 3 (for IP under denaturing conditions): 50 mM Tris-HCl base, pH 7.6 (TRIZMA; Sigma, St. Louis, MO), 2% Triton X-100, 100 mM NaCl, 1 mM NaVO$_4$, 6 mM EDTA, 50 mM NaF, 40 mM β-glycerophosphate, 10 µg/mL of leupeptin, 10 µg/mL of pepstatin A, 10 µg/mL of aprotonin, 1 mM PMSF. The protocol requires 3.2 mL per six oocytes.

3. Methods

3.1. Extraction and Preparation of Oocytes from Xenopus laevis

Detailed instructions on the care of *Xenopus laevis* frogs are available elsewhere *(10)*. To maintain consistent oocyte quality and minimize seasonal variations, we have found it advisable to maintain the same animals over 2–3 yr in a constant temperature, 12-h day/light cycle. It is also advisable to keep new frogs in separate tanks to avoid synchronization of seasonal cycles. Oocytes should not be removed from each frog more often than once every 3 mo to ensure animal health and avoid depletion of the oocytes.

1. Submerge the frog in a chilled solution of 0.15% MS222. The anesthetic is absorbed through the skin. Place a wet paper towel on top of a bed of ice. When the frog is nonresponsive to touch, place it face up on the wet paper towel and ice.

2. Perform an abdominal laparotomy with a 2-cm horizontal incision to the left or right of the midline and remove three to five lobes of oocytes. Place the oocytes in a 35-mm Petri dish containing OR-2 solution (*see* **Note 6**).

3. Using suture, stitch the endodermal layer first. Use two or three stitches, connecting the first two stitches and doubling the suture on the second stitch. Stitch the skin layer. Return the frog to a cold, shallow aquarium tank to recover as the water slowly returns to room temperature.

4. Separate the oocytes into clumps (5–10 oocytes/clump is best). Use forceps and remove as much sac as possible. Place the clumped oocytes in a 15-mL conical tube containing 15 mg of collagenase dissolved in 5 mL of OR-2 solution. After the oocytes are added bring the volume to 10 mL with OR-2 solution and start timing while gently rocking the tube on a shaker. Collagenasing times will vary between

25 and 45 min depending on the collagenase and the oocytes. New collagenase should be tested on a few oocytes prior to use, as efficiency and toxicity vary between lots.

5. When the oocyte clumps look like they might be starting to separate, rinse them with OR-2 solution—first in the tube 4 to 5×, then in a 10-cm Petri dish another 4 to 5× using a shortened, polished pasteur pipet. Rinse once with L15 medium and store in sterile culture dishes containing L15 in a 17°C incubator.

6. Oocytes usually remain healthy for up to a week if the medium is changed daily. Removing the follicular membrane prolongs the health of the oocytes, as does removal of early- and late-stage oocytes from the preparation.

7. Using fine forceps remove the outer layer of follicle cells surrounding each oocyte. In most cases this layer is an opaque off-white color and is clearly visible. Often fine, branching capillaries can be seen in the follicular layer. The follicular layer is degraded during the collagenase treatment and may therefore be partially or totally absent from some oocytes. To preserve the health of the oocyte, it should be noted that while it is customary to remove the follicular layer manually after a short collagenase treatment, longer collagenase treatments can be used to completely defolliculate the oocytes. However, these prolonged treatments can also compromise the long-term health of the oocytes.

3.2. RNA Preparation

3.2.1. Preparation of the DNA Template

1. The DNA of interest should be cloned into a vector so that sense RNA can be transcribed from either SP6, T3, or T7 polymerase sites (SP6 is usually less efficient). Ideally, a vector with promoters at either end will allow for the synthesis of both sense and antisense (as a negative control) transcripts. The translation of some cRNAs in oocytes can be inefficient. This can be improved by cloning the gene of interest between the 5' and 3' untranslated regions of *Xenopus* β-globin to provide an endogenous ribosome binding site. Several vectors have been produced for this purpose from the basic SP64T (*11*) to other variants with multiple cloning sites inserted between 5' and 3' untranslated β-globin sequences. The orientation of the original DNA template should always be confirmed by selective restriction enzyme cuts (*see* **Note 7**).

2. Once the orientation is confirmed, linearize the template with a restriction enzyme that cuts at the 3' end of the insert (when making sense cRNA). Linearize up to 5 μg of template. (Each transcription reaction requires only 1 μg of DNA but starting with more DNA will save time later). Use 8–10 U of restriction enzyme per microgram of DNA to ensure complete digestion.

3. Incubate at the suggested temperature for 1–2 h. Following incubation briefly centrifuge the sample in a microfuge (5 s) and take an aliquot (approx 0.5 μg) for electrophoresis on a 1% agarose gel for quantitation and to ensure full linearization (*see* **Note 7**).

4. Purify the digested DNA using a Geneclean Kit (BIO101, Vista, CA) following the manufacturers' guidelines. Alternatively, the DNA can be prepared by stan-

dard phenol–chloroform and chloroform extractions followed by ethanol precipitation in 0.3 M sodium acetate. It is estimated that 80% of the linearized DNA template is recovered with the Geneclean Kit, and it is generally not necessary to quantitate the DNA again by running an aliquot (e.g., 1 μL) on a 1% agarose gel (*see* **Note 7**).

3.2.2. In Vitro Transcription

Before initiating the in vitro transcription reactions a number of precautions should be taken to avoid contamination of kit components with nucleases (*see* **Note 1**). Always wear gloves when working with RNA and store all kit components at –80°C. Any chemicals used in the transcription reaction or analysis of RNA should be maintained carefully (e.g., powdered chemicals should poured from their containers, rather than inserting spatulas).

1. Prepare the m*M*essage m*M*achine In Vitro Transcription Kit (Ambion, Austin, TX) by thawing, gently vortex-mixing, and briefly centrifuging (5 s in a microfuge) the Transcription Buffer, Ribonucleotide Mix, and nuclease-free water. Keep the Enzyme Mix on ice. The usual reaction volume is 20 μL and should include 1 μg of linearized template DNA (*see* **Subheading 3.2.1.**) in a volume of 6 μL or less. Add the reagents, in the order shown, to a 1.5-mL Eppendorf tube at room temperature. The instruction manual provided with the m*M*essage m*M*achine Kit (Ambion, Austin, TX) gives more detailed instructions and lists the constituents of the following buffers and mixes.
   4 μL of nuclease-free water for a final reaction volume of 20 μL.
   2 μL of 10× reaction buffer.
   10 μL of 2× ribonucleotide mix.
   1 μL of 1 μg linearized template DNA.
   2 μL of 10× enzyme mix (either SP6, T7, or T3 RNA polymerase).
   1 μL of RNasin (Promega, Madison, WI) can also be added if yield is low or degradation is suspected.
2. Mix contents by flicking the tube gently. Incubate at 37°C for 1–2 h. For SP6 polymerase reactions a 2-h incubation is preferred, whereas shorter incubation times (approx 1 h) are sufficient for T7 or T3 reactions.
3. Add 1 μL of RNase-free DNase (from the m*M*essage m*M*achine Kit, 2 U/μL) to the reaction. Mix thoroughly with a pipet. Incubate at 37°C for a further 15 min. If alternatives to the m*M*essage m*M*achine kit are used for RNA transcription add 1 μL of RQ1 DNase (an RNase free DNAase from Promega, Madison, WI).

3.2.3. Recovery and Quantitation of the RNA

The m*M*essage m*M*achine Kit (Ambion, Austin, TX) includes protocols for LiCl precipitation or phenol–chloroform extraction of RNA. However, for the purpose of expression in oocytes, we have more consistent success using the RNAID Kit (BIO101, Vista, CA).

1. Add 65 µL of RNA binding salt and 20 µL of RNAid mix. Incubate at room temperature for 20 min mixing occasionally.
2. Centrifuge for 1 min at maximum speed in a benchtop centrifuge (approx 17,000 g_{max}). The RNA should pellet with the beads but save the supernatant for analysis.
3. Wash the beads with 700 µL of RNA wash solution and resuspend with a pipet tip. Centrifuged for 1 min as previously.
4. Repeat the wash step twice.
5. Dry the pellet by removing as much solution as possible with a fine-tipped pipet.
6. Elute the RNA by resuspending the pellet in 20 µL of nuclease-free water and incubating at 45–55°C for 3 min. After gentle mixing, centrifuge for two min at 17,000 g_{max} to pellet the beads and collect the supernatant, which contains the RNA.
7. Perform a second elution on the pellet using only 5 µL of nuclease-free water. Incubate and harvest the supernatant as before. Store stock solutions of RNA at –80°C.
8. Quantitate the RNA under denaturing conditions by gel electrophoresis (but see **Note 7**). Prepare a 1% agarose gel in 1X TAE electrophoresis buffer (*see* **Subheading 2.7.2.**). Since glyoxal reacts with ethidium bromide, the gels are poured and run in the absence of ethidium bromide. Prepare the RNA for gel electrophoresis by adding 1–2 µL of RNA to the approx 13 µL of denaturing buffer for RNA (*see* **Subheading 2.7.2.**). Incubate at 50–55°C for 1 h. Also prepare a known quantity of RNA ladder using the same techniques. Add 4 µL of RNA loading buffer (4×) (*see* **Subheading 2.7.2.**) to the 15 µL of denatured RNA and analyze the sample on the prepared gel.
9. After the gel electrophoresis is complete, stain the gel for 10 min in 50 m*M* NaOH by adding 5 µg/mL of ethidium bromide.
10. Destain the gel for 15 min in 100 m*M* sodium phosphate, pH 7.0 (*see* **Note 8**).

3.2.4. In Vitro Translation Reaction

It may be useful, particularly for mutagenesis studies where the coding region has been changed, or when no functional expression of the cRNA is detected in oocytes, to test the RNA template in an in vitro translation system. The results of the translation will provide information about the quality of the RNA and the size and quantity of the protein product that can be produced.

1. Assemble the following components in a 1.5-mL eppendorf tube:
   25 µL of Rabbit Reticulocyte Lysate (Promega, Madison, WI).
   1 µL of Amino Acid Mixture minus methionine (Promega, Madison, WI).
   4 µL of [^{35}S]methionine (10 mCi/mL).
   1 µL of 40U/µL RNasin ribonuclease inhibitor.
   1 µg of RNA template.
   Nuclease-free water to final volume of 50 µL.
2. Mix gently by pipeting the reaction up and down.
3. Incubate the reaction at 30°C for 2 h.
4. Add 3 µL of the product to 8 µL of 2× SDS polyacrylamide gel loading buffer (*see* **Subheading 2.7.2.**) and 5 µL of double-distilled water. Run on a 12% poly-

acrylamide gel using the SDS polyacrylamide gel electrophoresis buffer (*see* **Subheading 2.7.2.**) and view the results using autoradiography.

3.3. RNA Injection

3.3.1 Preparation of the Injection Setup

1. Set the injection apparatus (*see* **Note 9**) to deliver 40 nL of solution at each injection.
2. Prepare the injection needles by pulling glass capillaries with a pipette puller. Typically needles are fused at the tip after pulling, and the tip is then broken with fine forceps under a dissecting microscope. The tip should be gradually tapered and broken to a diameter that minimizes damage to the oocyte, but allows free flow of oil and solution. Diameters of 20–30 μm are ideal. Completely fill the injection capillary with mineral or paraffin oil using a 30G, 2-in. needle and syringe. Secure it to the injection apparatus, ensuring that the needle is free of small air bubbles.
3. Immediately before injection, place a drop of RNA on a nuclease-free surface such as the protected side of parafilm or the inside of a sterile cell culture dish. The concentration of the connexin RNA can range from 12 ng/μL to 10 μg/μL, possibly higher depending on the construct. Antisense oligonucleotide to *Xenopus* Cx38 (1 μg/μL) can be coinjected with connexin RNA to minimize endogenous coupling, in which case the RNA and *Xenopus* Cx38 antisense DNA are mixed prior to injection, For connexins that couple with Cx38 (e.g., Cx43) it may be necessary to preinject antisense oligonucleotide (*see* **Subheading 3.3.2.**). The volume of RNA loaded into the injection capillary will depend on how many oocytes are to be injected (40 nL/oocyte plus a little extra). Although the injector will have the capacity to take up larger volumes (5–10 μL) of solution, complications can arise during the injection procedure, in particular, bubbles may be sucked into the injection capillary while the RNA is being taken up. To avoid risking the loss of large quantities of RNA it is best to limit each fill of the needle to a volume of approx 2 μL.
4. Carefully position the tip of the injection needle within the drop of RNA and depress the "fill" button on the injector. Hold the button down until the drop has been taken up, ensuring that air bubbles do not enter the capillary.
5. Retract the needle and begin injecting the oocytes as soon as possible.

3.3.2. Preparation of the Oocytes for Injection

1. Transfer about 25 oocytes (*see* **Subheading 3.1**), to the injection dish using the cutoff pasteur pipet. The injection dish (described in **Note 2**) should be rinsed prior to use with 70% ethanol, distilled water, and then half-filled with L15 media. The preferred site of injection is the vegetal pole, although satisfactory cell coupling is also achieved through injection at the equator. The oocytes should be positioned with the preferred injection site facing upward.
2. It may be necessary to eliminate, or limit the contribution of endogenous *Xenopus* connexins to cell coupling. Typically, *Xenopus* Cx38 is the only endogenous

connexin to interfere with studies of mammalian connexins expressed in oocytes. In cases in which *Xenopus* Cx38 is capable of functionally coupling with the connexin of interest (e.g., Cx43, Cx30.3, Cx37), it is advisable to preinject the oocytes with antisense oligonucleotide against *Xenopus* Cx38. The antisense oligonucleotide (1 µg/µL or 4 ng/oocyte) should be injected 2–4 d prior to injection of connexin RNA. The half-life of Cx43, determined through [^{35}S]methionine chase studies in the presence of cyclohexamide, is about 20 h, indicating that a 2–4 d preinjection allows ample time for the turnover of *Xenopus* Cx38.

3. Prior to the first injection, ensure that the apparatus is working by focussing the microscope on the tip of the needle and pressing the injection button. A small drop of RNA should be expelled from the tip. Slowly lower the injection needle, until the needle touches the membrane of the oocyte. Push firmly to penetrate the membrane, taking care not to crush or overly distort the oocyte. Press the inject button, wait a second or two, and then retract the needle. Check periodically to ensure that the needle is not clogged, either by watching the meniscus between the RNA and the oil move, observing a slight swelling of the oocytes, or lifting the needle out of solution and focussing on the tip while pressing the inject button. After the oocytes are injected transfer them back to a culture dish containing L15.

3.3.3. Pairing the Oocytes

1. Allow the oocytes at least 3–4 h to recover after injection before manually stripping the vitelline membrane. Stripping the vitelline membrane is a difficult and arduous task and the success rate varies from about 30% to nearly 100%, depending on the batch of oocytes. Use two pairs of fine forceps. Pinch a small portion of the outer membrane together with one pair of forceps and attempt to pull the vitelline membrane away from this section without grabbing the plasma membrane. The vitelline membrane is clear and colorless and sometimes adheres strongly to the plasma membrane (*see* **Note 10**). After the membrane is removed, the oocytes become fragile, sticky, and will burst upon contact with air. They also tend to stick to the base of the culture dish, and it is advisable to transfer them to a dish lined with 1% agar in L15 solution either prior to, or immediately after, stripping. Removal of the vitelline membrane is not essential for the formation of gap junctions between oocytes as the oocytes send out processes that penetrate the vitelline envelope. However, it is necessary for the reliable development of significant junctional conductances within hours or days of oocyte pairing.

2. The vegetal poles of two oocytes, with vitelline membranes removed, are paired by gently rolling two oocytes together in an agar well. This same well is used for subsequent electrical recording of junctional coupling (*see* **Note 11**). Minimize disturbance to the dish once the oocytes are paired as they are easily washed out of the wells or uncoupled if the dish is tilted, particularly within the first few hours after pairing.

3.4. Electrical Recording

Measurable junctional conductance (G_j) may develop as early as six h after cell pairing. On the other hand it may take several days for coupling to occur depending on the RNA concentration, the construct, and the batch of oocytes (*see* **Notes 12–15**). Typically cells are paired in the afternoon, left at room temperature overnight, and recordings are made the next day. Varying the concentration of injected RNA is probably the simplest way to control the development of junctional conductance, although reducing the incubation time (to 6–8 h) or the temperature (to 19°C) can also significantly reduce coupling levels (*see* **Notes 12–14**).

1. Prepare the recording electrodes using the specified glass capillaries and the electrode puller. The electrodes should taper gradually at the tip and have a resistance in L15 media of 0.5–2.0 MΩ when filled with 150 mM KCl. Ensure that the appropriate voltage pulse protocols are set up and running correctly (*see* **Table 1**).
2. Ground the L15 media, preferably using a silver chloride pellet connected to a virtual ground headstage. Ensure that the amplifier is in "setup" mode at all times except during the voltage pulse protocols. Half-fill the electrodes with 150 mM KCl and test the tip resistance in L15 media. Zero the offset potentials for all four electrodes and ensure that the offset is not drifting.
3. Impale one oocyte with both a current injecting (I) and voltage sensing (V) electrode and check that the membrane potential reading is similar for both electrodes. The resting membrane potential of the oocytes will depend on the oocyte batch and can range from –20 mV to –60 mV for healthy oocytes. When positioning two electrodes within the same oocyte keep the tips as far apart as possible. Follow the same guidelines for impaling the second oocyte.
4. Voltage clamp the oocytes by switching from "setup" to "voltage clamp" mode. Clamp the oocytes either to their initial resting membrane potential or a more negative potential ensuring that both are clamped at the same potential.
5. Check G_j using a simple, 1 s voltage pulse protocol (*see* **Table 1**).
6. Measure and calculate G_j from the recorded current and voltage traces by measuring the magnitude of the current step in the clamped cell (e.g., I_b) and the magnitude of the voltage step in the pulsed cell (V_a). Ensure that both V_a and V_b remain clamped (at the pulsed and holding potential, respectively,) during the pulse by observing the recorded voltage trace (*see* **Fig. 2**).
7. If voltage sensitivity is to be measured, use a longer pulse protocol (*see also* **Note 12**). The pulse length will depend on the connexin being expressed. For instance, Cx40 inactivates rapidly ($\tau = 0.5$ s at $V_j = 50$ mV) and a pulse length of 5–10 s is sufficient for analysis of voltage sensitivity. On the other hand Cx26 gates slowly ($\tau = 5.6$ s at $V_j = 104$ mV) and pulse lengths of 30–40 s are required for analysis of voltage gating *(8)*. It is also possible to assess voltage-dependent inactivation by extrapolation of I_j, as long as the pulse length is long enough to obtain an accurate exponential fit of the current decay. In any case, given the limited time resolution for voltage-dependent changes in current with large cells such as oocytes, it is advantageous to extrapolate exponential fits to zero and infinity to obtain accurate measurements of initial and steady state conductances, respectively. One standard method for obtaining values for a number of parameters associated with voltage gating involves the generation of a G_j vs, V_j plot with con-

Table 1
Standard Pulse Protocols for Dual-Cell
Two-Electrode Voltage Clamp Analysis of Gap Juntions[a]

Duration	V_a	V_b
Quick test of junctional conductance[b]		
250 ms	V_{hold}	V_{hold}
1 s	$V_{hold} - 100$ mV	V_{hold}
250 ms	V_{hold}	V_{hold}
1 s	$V_{hold} + 100$ mV	V_{hold}
Voltage pulse protocol to assess voltage sensitivity[c]		
1 s	V_{hold}	V_{hold}
0.5 s	$V_{hold} + 20$ mV	V_{hold}
1 s	V_{hold}	V_{hold}
10 s[d]	$V_{hold} + V_{step}$	V_{hold}
15 s[d]	V_{hold}	V_{hold}
0.5 s	$V_{hold} - 20$ mV	V_{hold}
1 s	V_{hold}	V_{hold}
10 s[d]	$V_{hold} - V_{step}$	V_{hold}
5 s	V_{hold}	V_{hold}

[a]For this protocol, "a" is designated as the pulsed cell and "b" is the clamped one. Thus, I_b corresponds to I_j.
[b]Interepisode duration = 1 s; Sampling rate = 1 KHz.
[c]V_{hold} = −20 to −40 mV (as close as possible to the resting membrane potential of the oocytes). V_{step} = 10 mV, increasing in 10 mV steps to 100 mV. An episode is acquired for each V_{step} (e.g., ten episodes acquired for ten 10 mV steps). Interepisode duration = 30 s. Sampling rate = 100 Hz.
[d]For channels that inactivate slowly (e.g., Cx26 and Cx32) the pulse duration can be extended to 40 s and the interpulse holding duration can be extended to 45 s.

ductance normalized to the initial conductance interpolated at 0 mV. The drop in G_j with increasing V_j is fit with a modified Boltzmann equation of the form:

$$G_j = G_{j\,(min)} + [1 - Gj_{\,(min)}] / [(1 + e^{A_1\,(-V-V_{01})}) (1 + e^{A_2\,(V-V_{02})})] \tag{1}$$

This formulation (**Eq. 1**), modified from the original Spray, Harris and Bennett *(12)* equation by Ye Chen, allows a seamless fitting of the gating associated with each hemichannel (subscripts 1 and 2), whether in homotypic or heterotypic configurations. V_{01} and V_{02} represent the transjunctional voltages at which half maximal conductance occurs for each polarity of V_j and reflect the transition energy between open and closed states. For homotypic channels that do not display sen-

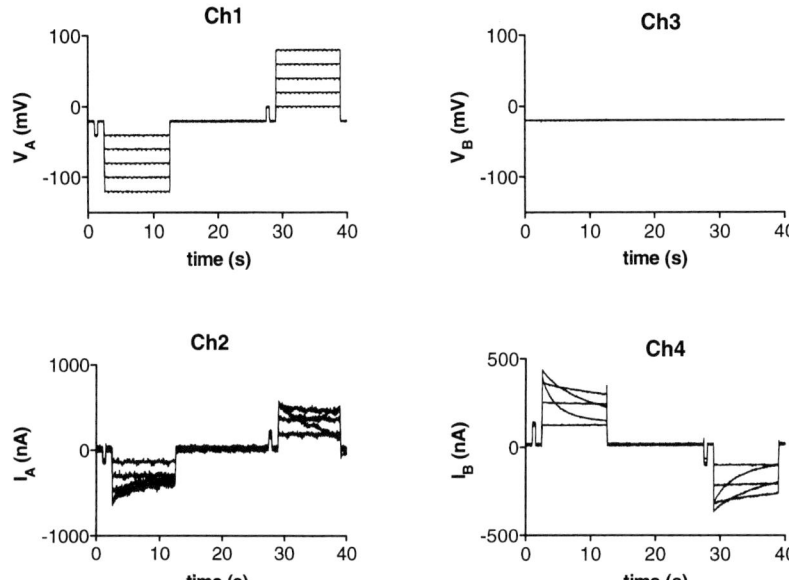

Fig. 2. Recordings from the four channels of a dual electrode voltage clamp during a series of voltage pulses applied to paired oocytes. Channels 1 and 2 represent V_m and I_m, respectively for the oocyte held at –40 mV and pulsed in 20 mV steps to –140 mV and +60 mV. Channels 3 and 4 represent V_m and I_m respectively for the oocyte continuously voltage clamped at –40 mV. In this case G_j, calculated from the "instantaneous" current, measured about 10 ms after the change in V_m, was about 4.5 μS. The oocytes were injected with 4 ng of Cx32 RNA plus 4 ng of *Xenopus* Cx38 antisense oligonucleotide and were paired for 28 h.

sitivity to transmembrane voltage, the absolute values of V_{01} and V_{02} are equal. A_1 and A_2 are directly proportional to the gating charge (*n*) as follows in (**Eq. 2**):

$$n = Akt/q \qquad (2)$$

where *q*, *T* and *k* represent the elementary charge, temperature, and Boltzmann constant respectively. **Fig. 3** shows the normalized G_j vs. V_j relationship for Cx32 expressed in *Xenopus* oocytes with a Boltzmann curve fit to the data.

3.5. Cytoplasmic Perfusion of Paired Oocytes

3.5.1. Preparation of the Perfusion Chambers

For the perfusion experiments, oocytes are paired in perfusion chambers (*see* **Note 4**) rather than in agar wells. Always clean and dry the chambers immediately prior to use. A 2% solution of microwash detergent is usually sufficient. If

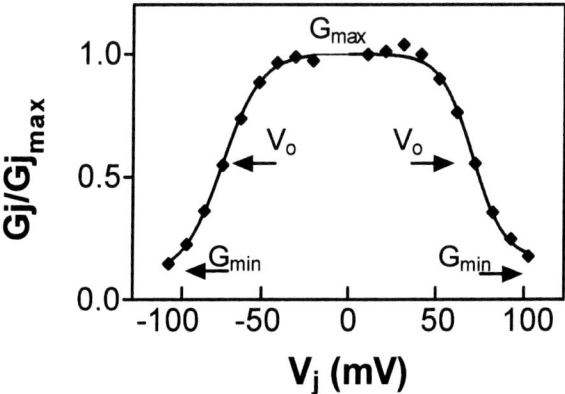

Fig. 3. Junctional conductance (G_j) plotted as a function of transjunctional voltage (V_j) for paired oocytes injected with Cx32 cRNA. Gj was normalized to the initial conductance (G_{max}). The steady state Gj, calculated at the end of the voltage pulse was then plotted against Vj and the data were fit to a modified Boltzmann equation (*see* text) with a V_0 of approx 70 mV.

material has previously dried in the chambers or if they have been contaminated, do not wash them with alcohol as this will cause cracking of the Perspex components through dehydration. An alternative to microwash detergent is a solution of 3% hydrogen peroxide, which must be rinsed thoroughly before introducing the oocytes. Although the perfusion chambers are not sterile, if they are treated carefully and the L15 is supplemented with antibiotics contamination of the wells will not occur within the 24 h required for the development of junctional conductances. The main concern is that R_{seal} decays with multiple uses of the chambers. However, treatment of the coverslips that divide the chambers with perchloric acid as described in the following, will enhance adhesion when required.

1. Make a fresh saturated solution of KCl.
2. Mix 1 vol of the KCl with 3 vol of concentrated perchloric acid. Immediately immerse the coverslips. Stir gently for 10 min, make sure the coverslips do not stack on each other and all surfaces are treated evenly.
3. Discard the spent mixture with large volumes of running water and rinse the coverslips for at least 10 min.
4. Wash the Perspex chamber with 2% microwash solution, rinse well with double-distilled water, and air-dry.
5. Reassemble the chambers with the treated coverslips. Apply small amounts of vaseline to the interface between the two chamber halves and the coverslip to prevent leakage, although avoid vaseline squeezing into the chamber itself. Rinse and air-dry the assembled chambers.

6. Add 600 μL of L15 kept at room temperature to each well. Wipe dry the surfaces around the chamber. If required, expel air bubbles from the wells with a syringe and needle, particularly those that may contact the oocytes. Cover the chambers with sterile culture dish lids.

3.5.2. Preparation of the Oocytes

Prepare the oocytes as described previously for electrical recording. Carefully transfer the stripped oocytes to the wells, attempt to roll them into the lower part of the chamber so the vegetal poles will face each other. Recordings are typically made 12–24 h after pairing (*see* **Notes 12–15**).

3.5.3. Preparation of the Electrical Recording Apparatus

Preparation of the electrical recording apparatus, micorelectrodes and pulse protocols is the same as for intact oocytes. The ground electrode should be placed in the chamber with the pulsed intact oocyte.

1. Test G_j as described previously. Junctional conductances in the 5 μS–10 μS range are ideal. Intact oocytes can be clamped at their resting Vm or at -40 mV. A series of long voltage pulses (*see* **Table 1**) should be applied to ensure that the transjunctional currents show the characteristic voltage sensitivity of the expressed connexin. This recording can be used as a reference to ensure that the gap junction plaques and R_{seal} are not damaged when the oocyte is cut and perfused.
2. Once G_j and voltage sensitivity have been noted, remove the recording electrodes and perfuse one oocyte with intracellular solution using a peristaltic pump at a flow rate of 0.5–1 mL/min. It is best to perfuse the oocyte that will be clamped rather than the oocyte that will be pulsed. Care should be taken to maintain an equal volume of solution on the two sides of the chamber. When the L15 medium has been completely replaced with intracellular solution, use fine microdissecting scissors and cut open the perfused oocyte under a dissecting microscope. Cut away as much of the oocyte as possible (usually one third to one half of the oocyte), being careful not to disrupt the seal between the oocyte and the coverslip. To minimize the problem of having the scissors stick to the oocyte, keep them extremely clean—they can be washed in a sonicator with some detergent, rinsed with distilled water and ethanol, and air-dried.
3. After dissection, test G_j again to determine whether the seal has been damaged. On the side of the cut and perfused oocyte, place the electrodes in contact with the remaining plasma membrane. If possible use the electrodes to pry and hold the oocyte open, increasing the volume of yolk that is in contact with the perfusate. Ground is maintained in the bath of the intact oocyte. Pulses are usually also applied to the intact oocyte. The resting membrane potential of the intact oocyte will have dropped to near zero if the cells are well coupled. After one oocyte has been cut and perfused, it is best to clamp the intact oocyte and the perfused chamber at –20 mV during the voltage pulse protocol. If the cut oocyte cannot be clamped

at –20 mV it is probably because the seal between the oocyte and the coverslip has been broken. A series of long voltage pulses (**Table 1**) should be acquired and compared to those recorded prior to cutting. Do not proceed if G_j has significantly increased or if a large proportion of the voltage sensitivity is lost (indicating that much of the current between the chambers is a leak through R_{seal}).

4. A perfusion pump can be used to add reagents to the cytoplasmic face of the oocyte once it has been successfully cut and perfused. However, influx and efflux rates must be carefully monitored to avoid even minor changes in fluid level within the chamber that could compromise the oocyte seal. It is also possible to add reagents to the chamber from a stock solution, mixing the solution by pipetting 10 μL or 20 μL up and down a few times. Again it is critical to ensure the levels in the perfused and intact chambers remain equal.

3.6. Immunoprecipitation

1. Prepare the oocytes for injection as described previously. Approximately six oocytes are required for each lane that will be run on the gel but it is best to inject 8 or 10 oocytes/lane depending on the health of the oocytes. The quantity of RNA injected will vary between connexins and constructs. In general, the amount of RNA required for immunoprecipitation (IP) is the same as that required for measurements of G_j in oocyte pairs. Ideally, coincident measurements of G_j can be obtained from paired oocytes of the same batch. We use approx 0.5 ng/oocyte and 4 ng/oocyte of wild-type rat Cx43 and rat Cx32 RNA respectively for immunodetection. For detection purposes, [^{35}S]methionine should be injected along with the RNA at this stage. Use the same procedure as for cytoplasmic injections of RNA, injecting 40 nL/oocyte of a 0.43 μCi/mL solution. It is best to do this 20 min after injection of the RNA and to work in a fume hood, as the [^{35}S]methionine solution is volatile. Ensure that correct authorization and procedures for the use of radioactive materials are followed.

2. Let the injected oocytes sit for 6–24 h in L15 (half-strength) at room temperature.

3. Transfer the oocytes to 1.5 mL Eppendorf tubes (six oocytes per tube) using a fire-polished Pasteur pipet

4. It is best to follow the "denaturing protocol" to maximize connexin yield, unless protein interactions are being studied.
   a. Denaturing protocol:
      Add 200 μL of IP buffer 2.
      Pass the oocytes 18× through a 25G needle.
      Boil for 5 min. Because, some connexins (Cx26, Cx32) tend to aggregate in SDS when heated, in these cases, solubilize at room temperature for 20 min.
      Add 800 μL of IP Buffer 3 to each tube and mix by inversion.
   b. Nondenaturing protocol:
      Add 1 mL of IP buffer 1.
      Pass the oocytes 18× through a 25G needle.
      Mix my inversion.

5. Centrifuge at 17,000g_{max} in a benchtop microcentrifuge for 5 min at 4°C.

6. Carefully transfer the supernatant to a 1.5-mL Eppendorf tube. Discard the pellet. Add 6 µL of crude serum containing the specific antibody to the supernatant (1 µL of antibody/oocyte). For this protocol the concentrations of nonspecific and specific antibody in the crude serum are approx 4 mg/mL and 1.5 mg/mL, respectively.

7. Incubate at 4°C overnight, shaking at a slow speed.

8. While incubating with the primary antibody, preswell the protein A–Sepharose beads overnight. This protocol requires 100 µL of preswollen Sepharose beads per six oocytes. Make up a 10% solution of Protein A–Sepharose in PBS buffer, mix well, and leave on a shaker at 4°C overnight (*see* **Note 16**).

9. The next morning, add 100 µL of preswollen Protein A–Sepharose to each tube, making sure to vortex-mix the Protein A–Sepharose before removing each 60-µL aliquot. Mix by inversion and incubate at 4°C for 1 h.

10. Centrifuge 15 s at $17,000g_{max}$ in a benchtop microcentrifuge.

11. Discard the supernatant and then resuspend the pellet in 1 mL of buffer. For *nondenaturing conditions* use buffer 1. For *denaturing conditions* make up a wash solution consisting of 1 vol of IP buffer 2 diluted in 4 vol of IP buffer 3.

12. Repeat this wash step two more times through successive centrifugations and resuspensions.

13. Allow 3 min for the last centrifugation. Remove the supernatant of the last wash as completely as possible, without disturbing the beads.

14. Add 8 µL of SDS-polyacrylamide gel loading buffer and 8 µL of double-distilled water to the pellet and vortex-mix.

15. Elution is achieved by boiling for 10 min (e.g., Cx43) or incubating at room temperature for 30 min in the cases of connexins that tend to aggregate on heating (e.g., Cx26, Cx32).

16. Centrifuge the sample for 4 min at $17,000g_{max}$ in a benchtop microcentrifuge.

17. Load the supernatant on a 12 1/2% SDS-PAGE gel and proceed with normal electrophoresis. The scale of this procedure is for normal "minigels" of approx 7 × 8 cm.

18. Detect the ^{35}S-labeled connexin using autoradiography (*see* **Note 17**).

4. Notes

1. Only nuclease-free water, pipet tips, and microcentrifuge tubes should be used during RNA preparation. Prepare DEPC treated water by adding 0.5 mL DEPC to 1 L of double-distilled water (0.05%). Stir well, incubate overnight, and autoclave at least 45 min. To treat pipet tips and microcentrifuge tubes, immerse them in 0.05% DEPC in double-distilled water and stir overnight. Autoclave at least 45 min. DEPC-treated water is also available commercially (Quality Biological, Gaithersburg, MD) as are nuclease-free pipet tips (VWR, Westchester, PA) and Eppendorf tubes (Marsh Biomedical Products, Rochester, NY). **Caution: DEPC can be explosive prior to autoclaving, so treatment should be performed in open containers in a fume hood and autoclaving should be performed immediately**.

2. A 35-mm tissue culture dish can be modified to secure the oocytes during injection. Line the bottom with 20–40 layers of parafilm and make 30 or so divots in the parafilm with the tip of a ballpoint pen.

3. Prepare the agar-filled pairing dishes by dissolving 1% agar in L15. Pour the solution into sterile 6-cm culture dishes to a depth of about 5 mm. Place the special well-making lids on the dishes until the agar has set. The special lids can be prepared by cutting the tips off 0.5-mL Eppendorf tubes and fusing three or four of them to the inside of the culture dish lid with super glue, ensuring that they do not quite touch the base of the culture dish when assembled.

4. Perfusion chambers are currently available from B.J. Nicholson, Dept. of Biological Science, SUNY at Buffalo, Buffalo, NY, USA, 14260. A perfusion chamber is constructed from two Perspex blocks with small wells (volume approx. 1 mL) in the center. When screwed together, the wells are separated by a coverslip that has a small hole (diameter = 0.79 mm), positioned strategically to allow pairing of the oocytes on opposite sides of the coverslip partition. One of the oocytes forms a seal with the coverslip (R_{seal} = 0.2–1 MΩ), the value of which is critical for two reasons. Firs, only one of the wells is electrically grounded and second, R_{seal} replaces R_m (membrane resistance of the oocyte) when one of the oocytes is cut and perfused (*see* **Fig. 3**).

5. If stock solutions are to be used, prepare all in double-distilled water except for PMSF (200 mM in ETOH) and pepstatin A (5 mg/mL in ETOH). Adjust the EDTA stock solutions to a pH of 8.0. The IP buffers can be prepared several days in advance and stored at 4°C but the protease inhibitors and PMSF should be added directly before use.

6. If using an OR-2 solution containing calcium, be sure to remove calcium and rinse the oocytes thoroughly before collagenasing, as proteases are activated in the presence of calcium, increasing damage to the oocytes during treatment.

7. A standard procedure for DNA analysis by electrophoresis is as follows: Prepare a 1% agarose gel in 1× TBE buffer containing 0.05 µg/mL of ethidium bromide for visualization of DNA. Fill the gel apparatus with 1× TBE to approx 1 mm above the level of the gel. Ethidium bromide is highly carcinogenic and mutagenic, so avoid contact with skin. Add 1 µL of 10× DNA loading buffer to 9 µL of double-distilled water containing about 0.5 µg of the DNA to be analyzed. Run next to a 1 kb DNA ladder (Gibco-BRL, Grand Island, NY). Visualize the DNA with an ultraviolet light.

8. It is possible to obtain a rough estimate of RNA quantity under nondenaturing conditions (e.g., as for DNA analysis by electrophoresis). However, under nondenaturing conditions RNA may appear as multiple bands or as smeared bands. The detection limit for RNA under denaturing and nondenaturing conditions is in the 500 ng range with ethidium bromide.

9. The injection protocol is described for an automated injection system, namely the Nanoject 1 Auto/Oocyte Injector (Drummond, Brooomall, PA). Fixed and variable volume models are available, both are remote controlled. The fixed volume model delivers 46 nL/oocyte. The variable volume model can be set to deliver volumes between 4.6 nL and 73.6 nL. The injection volume is commonly preset to 40 nL but this can be decreased for small oocytes. An updated model, the Nanoject 2, is now available from Drummond (Broomall, PA) and features dual

injection speed and the capacity to hold flared glass which forms a superior seal. A less expensive option for injecting RNA into the cytoplasm of ooctyes is the manually controlled Oocyte Microinjection Pipet (Drummond, Broomall, PA). This manual injector is postioned on a coarse micromanipulator in a similar way to the automated injector but injection is controlled by a knob on the back of the dispenser. Ten microliters of solution can be taken up and dispensed manually in volumes > 30 nL. If accurate volumes are required the automated version is recommended, especially for inexperienced hands.

10. Inexperienced investigators may find it useful to expose the oocytes to a slight hypertonic solution (e.g., full-strength L15) for 5 min as this causes shrinkage of oocyte and separation of the vitelline membrane from the plasma membrane of the oocyte, facilitating initial stripping. However, this process can further compromise the health of the oocyte and should be minimized and eventually eliminated once experience is gained.

11. Pairing oocytes with their vegetal poles in apposition results in higher junctional conductance than when animal poles are paired. Distribution of connexin proteins is determined by the net positive charge on the protein, the membrane potential, and the electrical field generated along the animal–vegetal axis of the oocyte *(13)*.

12. It is best to check G_j sooner rather than later, as the characteristic voltage sensitivity of the connexins is lost at high G_js. Voltage sensitivity appears to decrease as G_j increases because the series resistance (R_s) contributed by the electrode and cytoplasm becomes proportionally larger with respect to the junctional resistance (R_j). A large proportion of the voltage drop then occurs across R_s and the actual transjunctional voltage (V_j) is smaller than the difference that is recorded between the two voltage clamp command potentials ($V_a - V_b$) *(14)*. For this reason, studies of voltage-gating should be carried out on oocyte pairs with low G_js (e.g, < 5 μS) with the understanding that the G_j- V_j curves may differ slightly from those obtained for single channels or small cells.

13. Many mutants that show minimal or no coupling in homotypic pairings, couple more efficiently when paired heterotypically with wild-type. This is particularly apparent for mutants with altered voltage-gating characteristics that appear to activate rather than inactivate in response to V_j *(15)*.

14. Addition of micromolar concentrations of lectins to the media surrounding the oocytes stimulates cell–cell adhesion and facilitates electrical coupling, but is not required in most cases. Treatment with lectins is usually performed as follows. Approximately 18 h after injection of RNA oocytes are stripped of the vitelline membrane and incubated with the lectin glycine max (soybean agglutinin, Sigma, St. Louis, MO) for 20 min, washed extensively, and then paired. After 2 h, G_j is approximately 10-fold higher in lectin-treated oocyte pairs than in untreated control pairs *(16)*.

15. It may take slightly longer for G_j to develop when the oocytes are paired in the perfusion chambers. If this is the case, inject a higher RNA concentration rather than waiting longer, as the paired oocytes usually survive only for 30 h or so in the chambers.

16. If using a mouse monoclonal antibody, Protein G should be used. It is also possible to use a secondary antibody (e.g., anti-rabbit or anti-mouse as appropriate)

conjugated to Sepharose beads. Pan sorbin (processed *Staphylococcus aureus* bacteria) is a less expensive alternative, but does not pellet as well.

17. Immunoprecipitated connexin bands can also be detected by other methods if preferred. If the protein is radiolabeled, phosphoimage analysis allows for more rapid detection and ready quantitation of the bands. If the protein is not radiolabeled, there is typically insufficient protein for direct staining, particularly given the high background from the heavy and light IgG bands from the immunoprecipation. Western blot analysis is possible but an antibody from a different species must be used to avoid the same background problem from IgG. Surface biotinylation *(17)* can also be used in conjunction with immunoprecipitation as a method of detecting connexin expression at the plasma membrane, as the biotin is readily detected by an avidin conjugate.

References

1. Gimlich, R. L., Kumar, N. M., and Gilula, N. B. (1990) Differential regulation of the level of three gap junction RNA's in *Xenopus* embryos. *J. Cell Biol.* **11,** 597–605.
2. Dash, P., Lotan, I., Knapp, M., Kandel, E., and Goelet, P. (1987) Selective elimination of mRNA's in vivo: complementary oligonucleotides promote RNA degradation by RNaseH-like activity. *Biochemistry* **84,** 7896–7900.
3. Trexler, E. B., Bennett, M. V. L., Bargiello, T. A., and Verselis, V. (1996) Voltage gating and permeation in a gap junctional hemichannel. *Proc. Natl. Acad. Sci. USA* **93,** 5836–5841.
4. Swenson, K. I., Jordan, J. R., Beyer, E. C., and Paul, D. L. (1989) Formation of gap junctions by expression of connexins in *Xenopus* oocyte pairs. *Cell* **57,** 145–155.
5. Cao, F., Eckert, R., Elfgang, C., Nitsche, J. M., Snyder, S. A., Husler, D. F., Willecke, K., and Nicholson, B. J. (1998) A quantitative analysis of connexin-specific permeability differences of gap junctions expressed in HeLa transfectants and *Xenopus* oocytes. *J. Cell Sci.* **111,** 31–43.
6. Dahl, G. (1992) The *Xenopus* oocyte cell–cell channel assay for functional analysis of gap junction proteins, in *Cell–Cell Interactions: A Practical Approach* (Stevenson, B. R., Gallin, W. J., and Paul, D. L., eds.), Oxford University Press, New York, pp. 143–165.
7. Ebihara, L. (1992) Expression of gap junctional proteins in *Xenopus* oocyte pairs. *Methods Enzymol.* **207,** 376–380.
8. Nicholson, B. J., Suchyna, T., Xu, L. X., Hammernick, P., Cao, F. L., Fourtner, C., Barrio, L., and Bennett, M. V. L. (1993) Divergent properties of connexins expressed in *Xenopus* oocytes. *Prog. Cell Res.* **3,** 3–13.
9. Barrio, L. C., Suchyna, T., Bargiello, T., Xu, L. X., Roginski, R. S., Bennett, M. V. L., and Nicholson, B. J. (1991) Gap junctions formed by connexin 26 and 32 alone and in combination are differently affected by applied voltage. *Proc. Natl. Acad. Sci. USA* **88,** 8410–8414.
10. Goldin, A. L. (1992) Maintenance of *Xenopus laevis* and oocyte injection. *Methods Enzymol.* **207,** 266–278.

11. Krieg, P. A. and Melton, D. A. (1984) Functional messenger RNAs are produced by SP6 in vitro transcription of cloned cDNA's. *Nucleic Acids Res.* **12,** 7057–7070.

12. Spray, D. C., Harris, A. L., and Bennett, M. V. L. (1981) Equilibrium properties of a voltage-dependent junctonal conductance. *J. Gen. Physiol.* **77,** 75–94.

13. Levine, E., Werner, R., Neuhaus, I., and Dahl, G. (1993) Asymmetry of gap junction formation along the animal–vegetal axis of *Xenopus* oocytes. *Dev. Biol.* **156,** 490–499.

14. Van Rijen, H. V. M., Wilders, R., Van Ginneken, A. C. G., and Jongsma, H. J. (1998) Quantitative analysis of dual whole-cell voltage clamp determination of gap junctional conductance. *Pflügers Arch.* **436,** 141–151.

15. Suchyna, T. M., Xu, L. X., Gao, F., Fourtner, C. R., and Nicholson, B. J. (1993) Identification of a proline residue as a transduction element involved in voltage gating of gap junctions. *Nature* **365,** 847–849.

16. Levine, E., Werner, R., and Dahl, G. (1991) Cell-cell channel formation and lectins. *Am. J. Physiol.* **261,** C1025–C1032.

17. Musil, L. S. and Goodenough, D. A. (1991) Biochemical analysis of connexin43 intracellular transport, phosphorylation and assembly into gap junctional plaques. *J. Cell Biol.* **115,** 1357–1374.

14

Mutagenesis to Study Channel Structure

Gerhard Dahl and Arnold Pfahnl

1. Introduction

Recently the crystal structure of a potassium channel has been determined
(1). The emerging structural features turned out to be a beautiful illustration of
both the power and limitation of the approach of structure–function analysis by
mutagenesis. The crystallographic data essentially confirmed models developed on the basis of mutagenesis data; in particular, the lining of a critical part
of the channel pore, the selectivity filter, by the P segment was verified. It also
exemplifies how much more detail the crystal structure can give as compared
to the crude information even the most sophisticated mutagenesis approach
can yield.

Structure–function analysis by mutagenesis is based on the conjecture that
mutations without functional consequences occur in positions that serve no
specific function but may merely act as spacers. On the other hand, alteration
of function including total loss of function as a result of mutation is typically
seen to be indicative of a specific functional role of the mutated amino acid(s).
Interpretation of mutagenesis data, however, is far from being that simple. For
example, even if a string of amino acids serves a spacer function, amino acids
in that region may not be replaced indiscriminately if the spacer has to approximate domains exactly to each other. It is also feasible that mutation of a single
amino acid may exert global structural changes by long-range effects and thus
may erroneously indicate the position of a functional domain. Thus, structure–
function analysis needs to be an iterative process in which proper interpretation requires complementary data from different approaches.

The crystal structure of gap junction channels will be solved eventually.
Strides toward this goal have already been made *(2–4)*. However, even when
all nooks and crannies of the channel structure are revealed, the assignment of

From: *Methods in Molecular Biology, vol. 154: Connexin Methods and Protocols*
Edited by: R. Bruzzone and C. Giaume © Humana Press Inc., Totowa, NJ

function to the structural details will require alternative techniques. Mutagenesis can be expected to play a leading role in this endeavor, only that mutations then can be designed on a more rational basis than is possible now.

The general strategies of structure–function analysis by mutagenesis and functional expression are described. In addition, detailed techniques for cysteine-scanning mutagenesis are given.

1.1. Mutation Strategies

1.1.1. Evolutionary Guidance/Sequence Diversity

The fastest and cheapest way to analyze functional domains of proteins is to search databases for proteins with the same function in different species. This approach will readily identify functional domains, provided the protein in question is present in many branches of the phylogenetic tree and has no species-specific functional peculiarities. With a sufficiently large database, the sequence alignment will identify variable and invariant regions within the amino acid sequence. Functional domains and folding domains will likely be found in the invariant regions, while the variable regions are likely to represent spacer segments.

If the protein of interest is known to cause an inherited disease, the causative mutations are to be found in structurally/functionally important domains. For example, the mutations causing Charcot–Marie–Tooth disease are clustered in the first transmembrane segment and the extracellular loops of connexin32 (Cx32). The first transmembrane segment contains pore-lining amino acids *(5)*, and most of the amino acids in the extracellular loops are essential in the docking process between gap junction hemichannels *(6–8)*.

1.1.2. Site-Directed Mutagenesis

Without guidelines from other approaches, the change of single amino acids by site-directed mutagenesis and functional comparison between exogenously expressed mutants and wild-type proteins is an inefficient process. To seriously test a protein by site-directed mutagenesis one would need to change the amino acid in every position into all other amino acids, express these mutants exogenously, and then determine the functional consequences of the various mutations. Such an exhaustive mutational study actually has been done with a model system, a phage λ repressor *(9)*, where the possibility of a functional selection made this approach feasible. This example shows that the designations "variant" and "invariant" are not as clear-cut as one might expect. Although certain positions were found to be truly variant, allowing substitution by almost any amino acid without functional loss, others allowed only small groups of amino acids to replace those of the wild-type sequence. Such groups may be quite heterogeneous, for example, a glutamate could be replaced by the other negatively charged amino acid, aspartate, but also by serine, threonine, and alanine. On the other hand, other amino acids could be replaced only by amino acids belonging to the same class (e.g., hydrophobic).

The success rate for site-directed mutagenesis to yield relevant data can be boosted by consulting mutability tables *(10)*, which indicate the odds at which an amino acid mutated during evolution. Cysteine is one of the amino acids least likely to mutate without loss or alteration of function. Based on the conservation of the extracellular cysteines in connexins and the observation that gap junction formation was affected by thiol reagents, one of the first applications of site-directed mutagenesis to connexins targeted the six conserved cysteines in the extracellular loops *(6,11)*. Substitution of any of the extracellular cysteines by serines resulted in loss of the mutant connexins to form patent gap junction channels. Subsequently, it has been shown that they are involved in intramolecular disulfide bonds *(12,13)*.

Other means to increase the success rate for site-directed mutagenesis to affect protein function is to identify target areas in the form of consensus sequences for secondary modification. For example, potential phosphorylation sites can be eliminated by site-directed mutagenesis and the effect on function assayed. This approach has identified a functionally important tyrosine phosphorylation site in Cx43 *(14)* and excluded two potential phosphorylation sites in Cx32 as elements indispensable for basic channel function *(15)*.

Typically mutational analysis of proteins is done with the assumption that side chain properties are the main determinants of an amino acid's role in the protein. However, this assumption may not always be valid: in potassium channels, for example, it is the backbone of amino acids that is channel lining *(1)*.

1.1.3. Deletion Mutagenesis

A way to screen for functional domains is to delete groups of amino acids. Because the amino terminus of membrane proteins is a key determinant of membrane insertion and folding it customarily is spared from the deletion process. Carboxy (C)-terminal deletions, on the other hand, by and large, are inconsequential. Connexin 32 can be truncated by 58 amino acids at the C-terminus without effect on basic channel function *(15)*. Similarly Cx43 can be truncated to a similar extent without loss of basic channel function *(16)*. The interpretation of this finding is straightforward, it shows that for all the functions tested the C-terminus is dispensable. Of course, it does not mean that the C-terminus exerts no function in a physiological setting. For example, a mutant Cx32 (d219) forms apparently regular channels in oocytes *(17)* as well as in cultured cells *(18)*, yet the same mutation in the human Cx32 gene results in Charcot–Marie–Tooth disease. Also, a deletion mutant of Cx43 was found to have altered pH sensitivity, indicating that in Cx43 the C-terminus contains a sequence responsible for pH sensitivity *(19)*. This activity subsequently was shown to be contained within a short peptide, and it was suggested that pH gating of Cx43 has aspects of a ball-and-chain mechanism *(20)*.

1.1.4. Mutational Protein Scans

For mutational scans, successive amino acids within a target area of a protein are substituted one at a time by the same amino acid. The substituting amino acid is chosen for certain properties. For example, a popular choice is cysteine for its reactivity with thiol reagents. This approach is labor intensive; even with a preselection of a target area, tens of mutants have to be analyzed.

1.1.4.1. CYSTEINE SCAN

Cysteine replacement mutagenesis has been successfully applied to a number of membrane channel proteins for identification of pore-lining amino acids. Pioneering work was done by Akabas and Karlin on the nicotinic acetylcholine receptor *(21)*. Channels to which this technique has been applied to evaluate pore-lining segments include γ-aminobutyric acid (GABA) receptor channels *(22)*, potassium channels *(23,24)*, the cystic fibrosis tranmembrane conductance regulator (CFTR) *(25)*, and gap junction channels *(5)*. This approach is also called the substituted-cysteine-accessibility method (SCAM) *(25)*.

In most cases the cysteine scans confirmed the location of pore-lining segments as deduced from prior data on toxin binding sites, as determined by site directed mutagenesis, or by data on domain swapping. Application of cysteine scans to connexins was not as forthcoming owing to the inaccessibility of the pore to chemical reagents in complete gap junction channels. However, the discovery of open hemichannels formed by certain connexins when expressed in *Xenopus* oocytes made the approach feasible. Contrary to other channels, no guidance to the pore by independent data was available for gap junction channels. Speculation that the putative amphipathic helix in the third transmembrane segment could form the channel pore had dominated the thinking in the field even though no known channel has an amphipathic helix as pore lining. However, the SCAM approach identified the first transmembrane segment as an important part of the channel lining, with an additional contribution possibly by the third transmembrane segment *(5)*.

An elegant variant of a cysteine scan was applied by Nicholson and co-workers to test the folding properties of the extracellular loops of connexins *(26)*. Shifting the position of the extracellular cysteines, with and without preservation of spacing, resulted in a pattern of active connexin mutants consistent with a β conformation of the extracellular loops.

1.1.4.2. Sugar Scan

N-linked glycosylation of proteins occurs exclusively in the exoplasmic compartment of the Golgi apparatus and the rough endoplasmic reticulum. This feature can be used to determine whether a particular amino acid sequence is located extracellularly. N-linked glycosylation takes place at the consensus sequence NXS/T, where X stands for any amino acid. Replacing the endstanding amino acids of a triplet by N and S or T will insert a potential glycosylation site into the protein of interest. Provided that this site is accessible to glycosylation enzymes and not buried, it will be used for addition of sugar moieties. Glycosylation results in retardation of the protein on sodium dodecyl sulfate-polyacrylamide gel electrophoresis (SDS-PAGE). Insertion of artificial glycosylation sites in different parts of a protein by site-directed mutagenesis, followed by in vitro translation or exogenous expression and analysis of the protein by SDS-PAGE, thus can be used to determine whether glycosylation occurs at the introduced sites. This "sugar scan" was developed independently in three laboratories for mapping of extracellular sequences of Cx32 *(7)* of CFTR *(27)*, and of a glutamate receptor *(28)*. Of course only positive results are interpretable; if the artificial site is used, the segment is located extracellularly. Lack of glycosylation of such an artificial glycosylation site on the other hand does not exclude an extracellular localization because this amino acid stretch could be extracellular but buried, in addition to being located in the membrane or intracellularly.

In theory, any secondary modification of proteins involving consensus sequences should be useful for similar scanning procedures. For example, phosphorylation sites could be used for mapping of intracellular segments.

1.1.5. Accessibility Test for Localizing Channel Gates

A reactive amino acid such as cysteine, native or engineered, can serve as a reference point for localization of channel gates. This approach is based on the premise that access of a test reagent to the reactive amino acid is determined by the activation of the gate to be tested. Furthermore, the test substance must be able to access the reference site in the open channel when applied either extracellularly or intracellularly. Intracellular application of reagent is possible with excised inside-out membrane patches and extracellular application can be done with outside-out membrane patches or intact cells. If, after activation of the channel's gate, the test substance is not accessible to the reactive site when applied from either side, the gate must collapse all of the channel or the part containing the reactive site. If the test substance cannot access the site extracellularly but can react when applied intracellularly the gate is located extracellular to the reactive site. An intracellular location of the gate to the reference point would allow the test substance to react when applied extracellularly but not intracellularly. Recently, a voltage gate of Cx46 hemichannels has been mapped with this approach extracellularly to position 35 *(29)*.

1.1.6. Domain Swapping

Domains of a membrane protein may be defined by topology: extracellular, transmembranous, or intracellular. Alternatively, domains may be defined by invariant or conserved regions in aligned sequences, or they may be indicated by other (mutational) procedures such as cysteine scanning, for example. Exchange of putative functional domains between related proteins can serve as a vigorous test for a functional assignment to a domain. The rationale of this approach is to transfer a specific functional property from the donor to the recipient protein. For example, cysteine scanning has identified equivalent amino acids in the first transmembrane segment of two connexins as likely contributors to the pore lining of gap junction channels *(5)*. The two connexins, Cx46 and $Cx32E_143$, form channels that can easily be discriminated from each other by their characteristic channel kinetics and by their single-channel conductance. Replacement of the first transmembrane domain in $Cx32E_143$ by the corresponding Cx46 sequence resulted in a channel with properties exhibited by Cx46 channels *(30)*. Thus, the recipient channel acquired the channel properties from the donor, strongly suggesting that the first transmembrane segment in connexins indeed provides the critical portion of the pore lining.

Earlier applications of domain swapping to connexins were done to test voltage gating *(31)*, docking specificity of connexins *(32)*, and the docking gate *(33)*.

2. Materials

2.1. Media for Bacteria

1. LB Medium: 10 g of Bacto-tryptone (from DIFCO), 5 g of yeast extract (from DIFCO), and 10 g of NaCl; add H_2O to 1 L and autoclave.
2. LB Plates: 10 g of Bacto-tryptone, 10 of g yeast extract, 15 of g Bacto-agar (from DIFCO), autoclave. Add 50 mg/L of ampicillin (Sigma) when LB has cooled down.
3. SOC Medium: 20.0 g of Bacto-tryptone, 0.5 g of NaCl, 5.0 g of yeast extract, 10 mL of 250 mM KCl, add double-distilled H_2O (ddH_2O) to 1 L, and autoclave. When cool add: 10 mL of sterile 1 M $MgCl_2$ (autoclaved) and 20 mL of filtered-sterilized 1 M glucose.

2.2. Reagents

Chemicals and reagents are available from the following sources:

1. Expand™ high-fidelity polymerase chain reaction (PCR) system from Boehringer Mannheim (Indianapolis, IN).
2. Ambion mMESSAGE mMACHINE™ from Ambion (Austin, TX).

3. PCR primers from Life Technologies (Gaithersburg, MD).
4. Maleimidobutyryl biocytin (MBB) from Calbiochem (San Diego, CA).
5. Collagenase type 1 from Worthington (Lakewood, NJ)
6. Restriction enzymes from New England Biolabs (Beverly, MA).
7. [2-(Trimethylammonium)ethyl]methanethiosulfonate bromide (MTSET) from Toronto Research Chemicals (North York, Ontario, Canada).
8. All other chemicals can be purchased from Sigma (St. Louis, MO).

2.3. General-Purpose Solutions and Buffers

1. TE buffer (1×): 10 mM Tris-HCl, pH 8.0, 1 mM Na$_2$-EDTA, pH 8.0.
2. P1: RNase A (100 µg/mL) in 50 mM Tris-HCl, 10 mM EDTA (free acid), pH 8.0.
3. P2: 200 mM NaOH, 1% SDS.
4. P3: 3.0 M KAc, pH 5.5.
5. QBT: 750 mM NaCl, 50 mM 15% Tris-HCl, ethanol, pH 7.0.
6. QF: 1.25 M NaCl, 50 mM Tris-HCl, 15% ethanol, pH 8.5.

2.4. Solutions Used for Electrophysiology Experiments

1. Oocyte Ringers (OR2)

Reagent	mM	MW	[Stock], mM	g/L (stock)	Amount to use per liter (per 4 L)
NaCl	82.5	58.44			4.82 g (19.27 g)
KCl	2.5	74.55	250	18.7	10 mL (40 mL)
CaCl$_2$	1.0	147.0	100	11.1	10 mL (40 mL)
MgCl$_2$	1.0	203.31	100	20.3	10 mL (40 mL)
Na$_2$HPO$_4$	1.0	120.0	100	26.8	10 mL (40 mL)
HEPES	5.0	238.3	500	119.2	10 mL (40 mL)

Add NaOH until the solution is clear, then adjust pH to 7.5.

Polyvinylpyrollidone					0.5 g (2.0 g)
Antibiotics (penicillin, 100 U/mL; streptomycin, 10 mg/mL)					5 mL (20 mL)

2. Patch-clamp solution (KG)

Reagent	mM	MW	Amount to use/L
Potassium gluconate	140	234.2	31.7880 g
KCl	10	74.55	0.7455 g
TES	5	229.2	1.1460 g

pH 7.5 (6.0 for testing pH gate)

2.5. Equipment

1. Glass capillary tubing with filament (cat. no. GC150F-15, Warner Instrument).
2. Micropipet puller P-97 (Sutter Instrument).

3. Microforge MF-830 (Narashige).
4. Oscilloscope (Tektronics).
5. Chart recorder (Soltec).
6. Axopatch-1B amplifier (Axon Instruments).
7. VR-10B digital data recorder (Instrutech Corporation).
8. Power Macintosh computer (Apple).
9. ITC-18 Computer Interface (Instrutech)
10. Acquire and TAC (both from Bruxton).
11. Gene Amp 2400 (Perkin Elmer).
12. Microwave (Sears).
13. Centrifuge (Eppendorf).
14. Centrifuge (Sorvall).
15. Spectrophotometer Genesys 5 (Spectronics Instruments).

3. Methods
3.1. Generation of Mutant Connexins

The following protocol describes the specific steps involved in the production of one specific cysteine mutant (Cx46L35C) for the cysteine-scanning mutagenesis approach. The same basic procedures are applicable to any other mutagenesis approach; in particular, PCR cassette techniques can be used to generate mutants containing single amino acid changes as well as to make complex chimeric connexins.

3.1.1. Cassette Mutagenesis

1. Wild-type Cx46, obtained from Dr. D. Paul *(34)*, is in the expression vector pSP64T. For mutagenesis in the first transmembrane segment, a cassette flanked by the restriction sites *Nhe*I (in vector) and *Pml*I is chosen.
2. Use the vector containing wild-type Cx46 as a template to generate two partially overlapping fragments (**a** and **b**) by PCR using the primers I and II, and III and IV (*see* **Fig. 1**).
3. Primers I and IV are located upstream and downstream of the *Nhe*I and *Pml*I sites, respectively. Primers II and III are complementary and contain the base changes for the cysteine replacement and, in addition, a silent base change for introduction of a new restriction site (*Avr*II) to be used for diagnostic purposes.

3.1.2. Protocol for Generation of Cassette by PCR

1. Label two PCR tubes.
2. Add to the first PCR tube 2 ng of primer I and 2 ng of primer II.
3. Add to the second PCR tube 2 ng of primer III and 2 ng of primer IV.
4. Add to each PCR tube:
   a. Template DNA (wild-type plasmid, 5 ng).
   b. dNTPs (200 ng of each dNTP).
   c. ddH$_2$0 (to yield total volume of 50 μL).
   d. 10× PCR enzyme buffer with 15 m*M* MgCl$_2$.
   e. DNA polymerase mix (e.g., Expand™ high-fidelity PCR system).

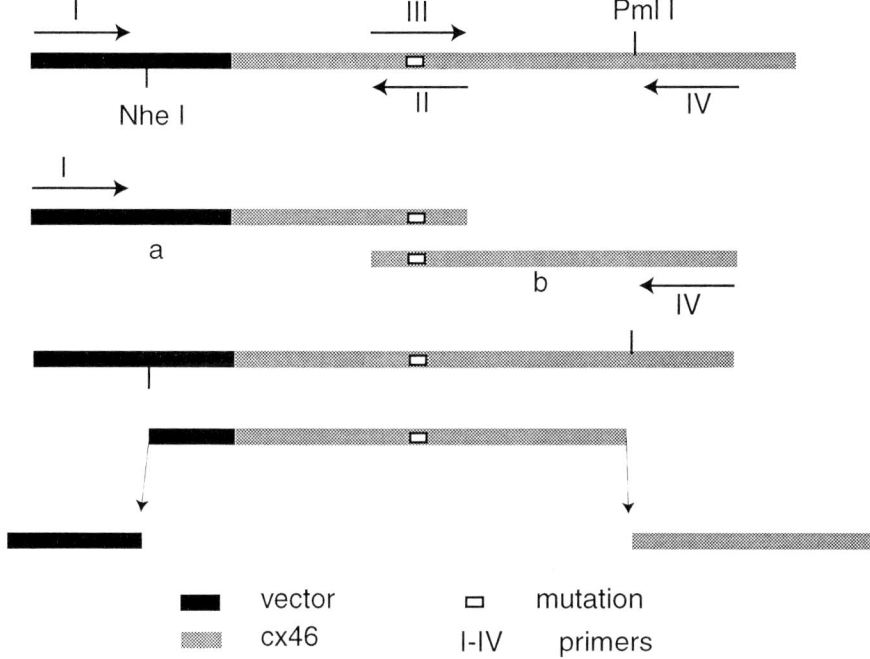

Fig. 1. Construction of mutation-containing cassette by PCR. Primers are indicated by *arrows* and have the following sequence: **(I)** GGCCACGATGCGTCCGG CGTAGAGG **(II)** CGCCCCTAGGACACAAATGCGGAAGATG **(III)** CATCT TCCGCATTTGTGTCCTAGGGGCG **(IV)** CCTCCCGCTCTTTCTTCTTCTCCTCC.

5. Mix well and start the reaction immediately (*see* **Note 1**).
6. Overlay with mineral oil (not necessary if using a thermocycler with a heated lid).
7. The thermocycler protocol includes one cycle at 94°C for 1 min, 15 cycles of 94°C for 30 s (melting temperature), 45–65°C for 30 s (annealing temperature), 68°C for 2 min (extension temperature). This is followed by one cycle at 68°C for 7 min. Finally, hold at 4°C until tubes are retrieved from the thermocycler.
8. Both fragments are gel purified in low melting agarose and subsequently separated from the agarose with QIAquick PCR Purification kit (Quiagen).
9. Analyze 5 µL of the PCR product by agarose–ethidium bromide (EtBr) gel electrophoresis. Ensure the DNA has not been lost in the purification procedure and is the proper size.

3.1.3. Fusion of PCR Fragments

1. Fragments a and b are then fused by PCR. The reaction mixture contains: 2 ng of primers I and IV, 5 ng of purified fragments a and b, 200 ng of dNTPs, water, PCR enzyme buffer, and DNA polymerase mix.
2. The mixture is then placed into a thermocycler using the same program as above (*see* **Subheading 3.1.2.**, **step 7**).

3. Test-digest 5 μL of the fusion product for the new restriction site and analyze by gel electrophoresis for proper sizes.
4. The fused fragments are then digested with *Nhe*I and *Pml*I for generation of the insertion cassette.
5. The vector containing wild-type Cx46 is cut with the same restriction enzymes, and both vector and insertion cassette are gel purified. If the restriction enzymes used create cohesive ends the vector should be treated with alkaline phosphatase to prevent ligation of the ends.
6. To ligate the cassette into the vector mix the two fragments at a 3:1 molar ratio (insert/vector) and add ligase buffer, T4 DNA ligase, and water.
7. Incubate the ligation mixture at 15°C overnight. After ligation, use the reaction mixture to transform bacteria, which are then plated on ampicillin plates (the vector codes for ampicillin resistance).

3.1.4. PCR Screen for Mutant Clones

Usually 30 colonies should be analyzed.

1. Pick one colony at a time with a toothpick.
2. Rub the toothpick on the bottom of PCR tube. (Take the same toothpick and place it into 3–5 mL of Luria broth [LB] broth with ampicillin and grow culture overnight for preparation of a Maxiprep in case the PCR test should be positive for the desired clone.)
3. Place the PCR tube with sample of bacterial colony into the microwave and set on high for 30–45 s.
4. Add the PCR reaction mixture to each microwaved PCR tube. The PCR reaction mixture is at a total volume of 25 μL/PCR tube and contains:
   a. Primer I (1 ng of ~ 30 mer).
   b. Primer IV (1 ng of ~ 30 mer).
   c. 10× PCR enzyme buffer.
   d. dNTPs (200 ng of each dNTP).
   e. ddH$_2$0 (up to 25 μL).
   f. *Taq* DNA polymerase.
5. Mix ingredients well and overlay with mineral oil (not necessary if using a thermocycler with a heated lid). The PCR thermocycler protocol is performed as described in **Subheading 3.1.2., step 7**.
6. Analyze 5 μL of the PCR product by agarose–EtBr gel electrophoresis.
7. Perform digest with screening restriction enzyme *Avr*II.
8. Analyze 5 μL of the digest product by agarose–EtBr gel electrophoresis. Load both digested and undigested DNA. If PCR product is cut then the clone is potentially positive. The next day take the corresponding miniprep and pore into 100 mL of LB broth for large-scale DNA preparation.
9. Verify mutation by sequencing through the cassette (*see* **Note 2**).

3.1.5. Large-Scale Preparation of Plasmid DNA (Maxiprep) and Purification of DNA

1. Prepare overnight culture in 150–500 mL of LB medium with 100 µg/mL of ampicillin.
2. Harvest the bacterial cells by centrifugation at 4°C for 15 min at approx 3000g in Beckman JA-10 rotors.
3. Remove all traces of supernatant by inverting the open centrifuge tube until all medium has been drained. Freeze the pellet at –20°C for storage or proceed directly to the protocol.
4. Resuspend the bacterial pellet in 10 mL of buffer P1. Ensure that the RNase A has been added to buffer P1. The bacteria should be resuspended completely, leaving no cell clumps.
5. Add 10 mL of buffer P2, mix gently (invert 4–5×), and incubate at room temperature for 5 min.
6. Add 10 mL of chilled buffer P3, mix immediately but gently (by inverting the tube 5–6×), and incubate on ice for 20 min.
7. Centrifuge at 4°C for 30 min at approx 18,000g. Remove supernatant promptly.
8. Centrifuge at 4°C for 15 min at 18,000g (optional). Remove supernatant promptly. The supernatant should be clear.
9. Equilibrate a QIAGEN-tip 500 by applying 10 mL of buffer QBT, and allow the column to empty by gravity flow.
10. Apply the supernatant from **step 8** onto the QIAGEN-tip and allow it to enter the resin by gravity flow.
11. Wash the QIAGEN-tip with 2 × 30 mL of buffer QC.
12. Elute DNA with 15 mL of buffer QF and collect the samples in a 30-mL tube.
13. Precipitate DNA with 10.5 mL of isopropanol, previously equilibrated to room temperature. Invert 4–5×. Centrifuge immediately at 8000g at 4°C for 30 min, and carefully remove the supernatant.
14. Wash DNA with 15 mL of cold 70% ethanol, air-dry for 5 min, and redissolve in a suitable volume (200–500 L) of buffer (TE). Overdrying the pellet will make the DNA very difficult to redissolve.

3.1.6. Transcription

1. The plasmid has to be linearized with restriction enzyme (*Eco*RI) to result in plasmid cleavage distal to the coding region.
2. About 10 µg of linearized plasmid is then used as template for the large-scale synthesis of in vitro transcribed capped RNA (Ambion mMESSAGE mMACHINE™ In Vitro Transcription Kit) for injection and expression in *Xenopus* oocytes (*see* Chapter 13). The expression vector SP64T incorporates an SP6 promoter site for use with bacteriophage SP6 RNA polymerase.
3. The RNA is recovered from the transcription reaction by lithium chloride precipitation.

4. The RNA pellet is washed in 70% EtOH and is resuspended in nuclease-free distilled water at 1 mg/mL.
5. Typically the yield of a large scale preparation following the manufacturer's instructions is 200–250 μg of RNA.
6. The RNA is aliquoted and stored at –70°C.

3.2. Oocyte Assay

Xenopus oocytes are the preferred exogenous expression system because of its speed, simplicity, reliability, and efficiency. The oocyte expression system was pioneered by Gurdon and co-workers *(35)* and has been adopted for functional expression of a large number of proteins including gap junction proteins *(36)*. For expression of complete gap junction channels oocytes expressing the same or different connexin(s) are paired *(37)* (*see* also Chapter 13 by Skerrett et al.). For expression of gap junction hemichannels single oocytes can be used *(33,34)*.

Oocytes are isolated from the ovary by collagenase treatment (type 1, Worthington, 2 mg/mL). Stage V and VI oocytes are collected and injected with ~20 nl of synthetic mRNA. If open hemichannel forming connexins are expressed, the oocytes are incubated in OR-2 solution with the calcium concentration increased to 5 mM to keep the channels closed *(33,38)*.

3.2.1. Two-Electrode Voltage Clamp of Oocytes

1. For measuring effects of test substances on the whole ensemble of gap junction hemichannels in the membrane, an oocyte is impaled with two electrodes. One electrode records the membrane potential and the other is for current injection. The electrodes are connected to a voltage clamp circuit, which allows to clamp the potential arbitrarily at various potentials and also provides a measure of the clamp current (*see* **Note 3**). Application of small test pulses allows determination of the membrane conductance at the various holding potentials.
2. The reversible effect of the thiol reagent MTSET on the cysteine mutant Cx46L35C is shown in **Fig. 2**. On application of MTSET the test-pulse-induced currents diminish in size, indicating that the membrane conductance is reduced by the thiol reagent. This could be due to fewer channel openings or to a reduction in channel conductance. On washout of the thiol reagent the membrane currents reverse to their original value, consistent with the reversibility of the disulfide bond formed between the thiosulfonate and the thiol group of cysteine. As a control wild-type Cx46 channels are subjected to the same procedure. **Figure 2A** shows that application of MTSET to wild-type channels remains without effect, indicating that the effect seen with Cx46L35C is attributable to the engineered cysteine in position 35 of Cx46 (*see* **Note 4**).

3.2.2. Patch-Clamp of Oocytes

1. For recording of effects of test substances on single connexin hemichannels the patch-clamp technique *(39)* is applied. The Axopatch-1B amplifier (Axon Instruments) is used for this purpose.

A

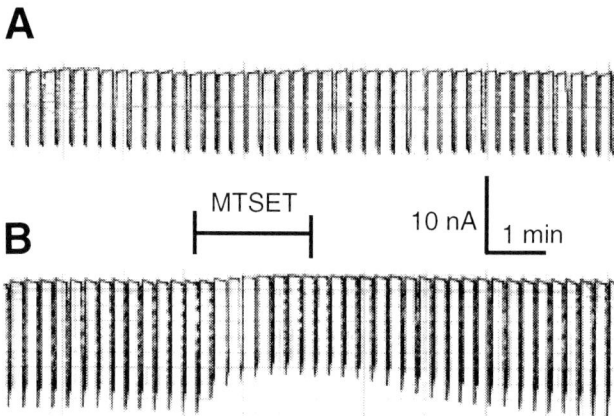

MTSET

10 nA | 1 min

B

Fig. 2. The thiol reagent MTSET (1 m*M*) applied at the time indicated reversibly reduces the membrane currents induced by 5 mV hyperpolarizing voltage steps in Cx46L35C (**B**) but not in wild-type Cx46 (**A**) channels.

2. For establishment of a tight seal between the patch pipet and the plasma membrane the vitelline layer has to be removed from the oocyte. This can be done with forceps (*see* Chapter 13).
3. Recordings are filtered at 5 kHz and digitized using an VR-10B digital data recorder and stored on videotape.
4. The recordings are subsequently transferred to a Power Macintosh (Apple) computer using an ITC-18 Computer Interface (Instrutech) and analyzed. The acquisition and analysis softwares are Acquire and TAC (both from Bruxton).
5. All recordings are made at room temperature (21–23°C).
6. Patch-pipets are made from glass capillary tubing with filament (#GC150F-15, Warner Instrument). Patch-pipets are pulled using a P-97 Micropipette Puller (Sutter Instrument) and the tips are polished with a microforge (Narishige Scientific Instruments) to a resistance of 10–20 MΩ in standard solution.
7. After a patch is excised and a potential hemichannel identified, the patch is moved into a microperfusion chamber.
8. The microperfusion chamber has a diameter of 1.5 mm and can be made by blowing out the sealed end of a glass pipet (*see* **Fig. 3**). The chamber is continuously perfused with solution. The standard pipet and bath solution consists of 140 m*M* potassium gluconate (*see* **Note 5**), 10 m*M* KCl, 5 m*M* TES, pH 7.5 (KG).
9. Voltage ramps can be applied by connecting to the external command of the patch-clamp amplifier a custom-designed voltage ramp generator.
10. Once a patch is established and channel activity is observed, gap junctional hemichannels (vs. other channels) are identified by the following criteria:
    a. Conductance and gating characteristics. Cx46 channels have a conductance of ~300 pS at negative holding potentials. The conductance is larger at negative than at positive holding potentials, the channels rectify.

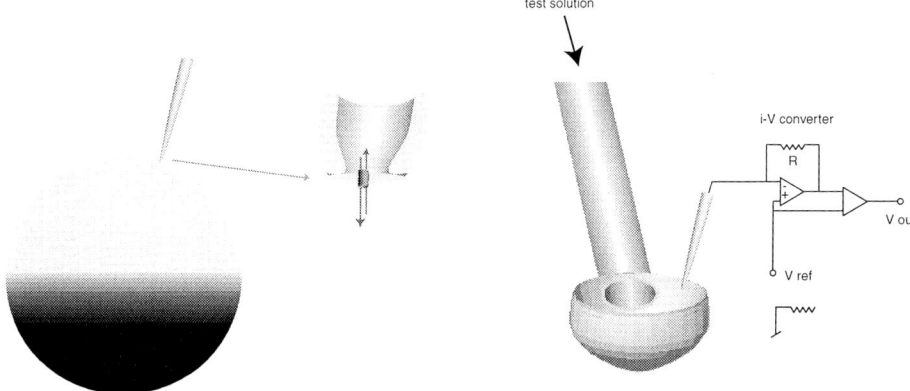

Fig. 3. Excision of inside-out membrane patch and perfusion in microchamber. The pipet is approached to the oocyte surface until a jump in resistance is noticed. Suction is applied to the pipet until resistance reaches a gigaohm range. The pipet is pulled back and transferred to the microperfusion chamber. The chamber is produced by first sealing the end of a glass pipet in the flame of a Bunsen burner. While the sealed end is reheated in the flame, pressure is applied by blowing into the pipet from the open end until a hole is created.

   b. Sensitivity to pH and calcium. A pH of 6.4 closes Cx46 channels. A 5.0 mM concentration of calcium added to the potassium gluconate solution closes Cx46 channels.
   c. Partial block of Cx46L35C channel with MBB (applicable only to few, pore-lining, cysteine substituted amino acids) (*see* **Note 6**).
11. The effect of the thiol reagent MBB on an excised membrane patch containing a Cx46L35C channel is shown in **Fig. 4**. The channel is held at –30 mV and channel activity is seen as a series of opening and closing events. After application of MBD the channel currents are five times smaller, yet other channel properties such as voltage dependence and sensitivity to acidification remain essentially unaltered *(29)*. This is interpreted in terms of modification of channel conductance by MBB, consistent with a pore lining location of the cysteine in position 35 (*see* **Notes 7** and **8**).

4. Notes

1. The DNA polymerase mix Expand™ high-fidelity PCR system contains two polymerases, one of which is Pwo polymerase. If Pwo (which has a proofreading function) is mixed with primers or template without dNTPs, then partial degradation of primer and template could occur because Pwo also has 3' to 5' exonuclease activity.
2. The two outside primers, I and IV (**Fig. 1**) are used for sequencing. They are located outside the insertion cassette; thus the complete cassette is sequenced on

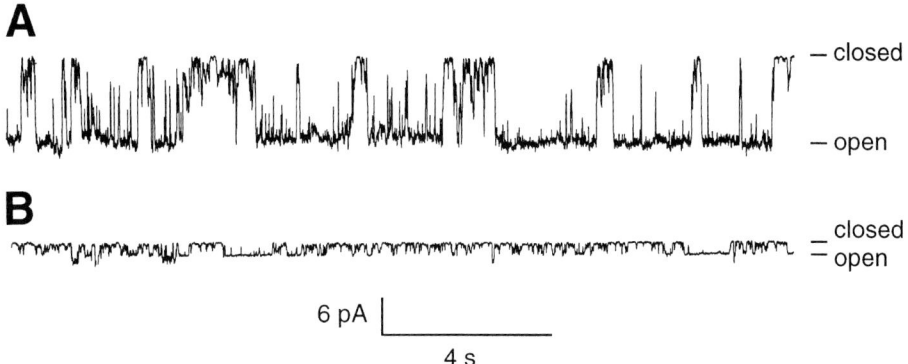

Fig. 4. Single-channel activity in a membrane patch excised from an oocyte expressing Cx46L35C before (**A**) and after (**B**) application of MBB.

both strands. However, in the specific example given here we used the SP6 primer instead of primer I to obtain more sequence of the coding region (primer I is located in the vector). Sequencing is done with an automated system at a central facility.

3. We use a custom-made oocyte voltage-clamp apparatus (*37*). However, a number of commercial products are now available, for example, the "GeneClamp" (Axon Instruments). *See* Chapter 13.

4. For cysteine scanning of connexins we prefer to use maleimides rather than the methanethiosulfonates because they are available in a larger range of sizes that are more appropriate for the large gap junction channels. Furthermore, maleimides are more stable than thiosulfonates, which have a half-life of only a few minutes. A disadvantage of maleimides is the irreversibility of the maleimide-thiol reaction. **Caveat:** small reducing agents, including methanethiosulfonates, can enter cells through gap junction hemichannels and affect other channels from the cytoplasm and thereby change membrane conductance.

5. The oocytes' autochthonous channels are well characterized and include a stretch-activated channel, the most prevalent one, and a chloride channel. By using gluconate instead of chloride the chloride channel remains "invisible." The stretch-activated channel is easily recognized by its response to mechanical stress and its characteristic features.

6. The majority of these criteria yield statistical arguments; channels with properties not seen in control oocytes are observed upon mRNA injection. The most compelling argument for channel identity, however, comes from specific changes of channel properties as a consequence of mutation. Cx46L35C is a prime example; this mutant exhibits channel properties indistinguishable from wild-type Cx46 except that it has acquired sensitivity to thiol reagents. For identification of channels formed by other connexins the same criteria can be used, only that different values apply, for example, the single channel conductance of

Cx32E$_1$43 channels is five times smaller than that of Cx46 channels and the channel kinetics are considerably faster *(30)*. Mutations equivalent to Cx46L35C or other mutations may be used for identification of other connexin channels.

7. There is a narrow window of opportunity to harvest patches containing single connexin channels. As after injection of mRNA into oocytes the concentration of the channels increases with time, the channels tend to cluster. As a consequence "empty" and "leaky" patches are obtained when collected from oocytes containing large amounts of connexin. Typically the best chance to find patches with single channels is between 4 h and 16 h after injection of ~20 nL of mRNA (1 mg/mL).

8. Noise usually is not a problem when recording from connexin channels owing to their large conductance. However, certain connexins and certain experimental conditions may require reduction of electrical noise. This can be done with coating the pipet with an insulating agent such as Sylgard. Alternatively, quartz glass may be used for fabrication of the pipets.

References

1. Doyle, D. A., Cabral, J. M., Pfuetzner, R. A., Kuo, A., Gulbis, J. M., Cohen, S. L., Chait, B. T., and MacKinnon, R. (1998) The structure of the potassium channel: molecular basis of K$^+$ conduction and selectivity. *Science* **280,** 69–77.

2. Makowski, L., Caspar, D. L., Phillips, W. C., and Goodenough, D. A. (1977) Gap junction structures. II. Analysis of the x-ray diffraction data. *J. Cell. Biol.* **74,** 629–645.

3. Perkins, G. A., Goodenough, D. A., and Sosinsky, G. E. (1998) Formation of the gap junction intercellular channel requires a 30 degree rotation for interdigitating two apposing connexons. *J. Mol. Biol.* **277,** 171–177.

4. Unger, V. M., Kumar, N. M., Gilula, N. B., and Yeager, M. (1999) Three-dimensional structure of a recombinant gap junction membrane channel. *Science* **283,** 1176–1180.

5. Zhou, X. W., Pfahnl, A., Werner, R., Hudder, A., Llanes, A., Luebke, A., and Dahl, G. (1997) Identification of a pore lining segment in gap junction hemichannels. *Biophys. J.* **72,** 1946–1953.

6. Dahl, G., Werner, R., Levine, E., and Rabadan-Diehl, C. (1992) Mutational analysis of gap junction formation. *Biophys. J.* **62,** 172–180.

7. Dahl, G., Nonner, W., and Werner, R. (1994) Attempts to define functional domains of gap junction proteins with synthetic peptides. *Biophys. J.* **67,** 1816–1822.

8. Warner, A., Clements, D. K., Parikh, S., Evans, W. H., and DeHaan, R. L. (1995) Specific motifs in the external loops of connexin proteins can determine gap junction formation between chick heart myocytes. *J. Physiol.* **488,** 721–728.

9. Bowie, J. U., Reidhaar-Olson, J. F., Lim, W. A., and Sauer, R. T. (1990) Deciphering the message in protein sequences: tolerance to amino acid substitutions. *Science* **247,** 1306–1310.

10. Dayhoff, M. O., Barker, W. C., and McLaughlin, P. J. (1974) Inferences from protein and nucleic acid sequences: early molecular evolution, divergence of kingdoms and rates of change. *Orig. Life* **5,** 311–330.

11. Dahl, G., Levine, E., Rabadan-Diehl, C., and Werner, R. (1991) Cell/cell channel formation involves disulfide exchange. *Eur. J. Biochem.* **197,** 141–144.

12. John, S. A. and Revel, J. P. (1991) Connexon integrity is maintained by noncovalent bonds: intramolecular disulfide bonds link the extracellular domains in rat connexin43. *Biochem. Biophys. Res. Commun.* **178,** 1312–1318.

13. Rahman, S. and Evans, W. H. (1991) Topography of connexin32 in rat liver gap junctions. Evidence for an intramolecular disulphide linkage connecting the two extracellular peptide loops. *J. Cell Sci.* **100,** 567–578.

14. Swenson, K. I., Piwnica-Worms, H., McNamee, H., and Paul, D. L. (1990) Tyrosine phosphorylation of the gap junction protein connexin43 is required for the pp60v-src-induced inhibition of communication. *Cell. Regul.* **1,** 989–1002.

15. Werner, R., Levine, E., Rabadan-Diehl, C., and Dahl, G. (1991) Gating properties of connexin32 cell–cell channels and their mutants expressed in Xenopus oocytes. *Proc. R. Soc. Lond. B. Biol. Sci.* **243,** 5–11.

16. Fishman, G. I., Moreno, A. P., Spray, D. C., and Leinwand, L. A. (1991) Functional analysis of human cardiac gap junction channel mutants. *Proc. Natl. Acad. Sci. USA* **88,** 3525–3529.

17. Rabadan-Diehl, C., Dahl, G., and Werner, R. (1994) A connexin-32 mutation associated with Charcot–Marie–Tooth disease does not affect channel formation in oocytes. *FEBS Lett.* **351,** 90–94.

18. Omori, Y., Mesnil, M., and Yamasaki, H. (1996) Connexin 32 mutations from X-linked Charcot–Marie–Tooth disease patients: functional defects and dominant negative effects. *Mol. Biol. Cell.* **7,** 907–916.

19. Liu, S., Taffet, S., Stoner, L., Delmar, M., Vallano, M. L., and Jalife, J. (1993) A structural basis for the unequal sensitivity of the major cardiac and liver gap junctions to intracellular acidification: the carboxyl tail length. *Biophys. J.* **64,** 1422–1433.

20. Ek-Vitorin, J. F., Calero, G., Morley, G. E., Coombs, W., Taffet, S. M., and Delmar, M. (1996) pH regulation of connexin43—molecular analysis of the gating particle. *Biophys. J.* **71,** 1273–1284.

21. Akabas, M. H., Stauffer, D. A., Xu, M., and Karlin, A. (1992) Acetylcholine receptor channel structure probed in cysteine-substitution mutants. *Science* **258,** 307–310.

22. Xu, M. and Akabas, M. H. (1993) Amino acids lining the channel of the gamma-aminobutyric acid type A receptor identified by cysteine substitution. *J. Biol. Chem.* **268,** 21,505–21,508.

23. Kurz, L. L., Zuhlke, R. D., Zhang, H. J., and Joho, R. H. (1995) Side-chain accessibilities in the pore of a K^+ channel probed by sulfhydryl-specific reagents after cysteine-scanning mutagenesis. *Biophys. J.* **68,** 900–905.

24. Lu, Q. and Miller, C. (1995) Silver as a probe of pore-forming residues in a potassium channel. *Science* **268,** 304–307.

25. Akabas, M. H., Kaufmann, C., Cook, T. A., and Archdeacon, P. (1994) Amino acid residues lining the chloride channel of the cystic fibrosis transmembrane conductance regulator. *J. Biol. Chem.* **269,** 14,865–14,868.

26. Foote, C. I., Zhou, L., Zhu, X., and Nicholson, B. J. (1998) The pattern of disulfide linkages in the extracellular loop regions of connexin 32 suggests a model for the docking interface of gap junctions. *J. Cell. Biol.* **140**, 1187–1197.

27. Chang, X. B., Hou, Y. X., Jensen, T. J., and Riordan, J. R. (1994) Mapping of cystic fibrosis transmembrane conductance regulator membrane topology by glycosylation site insertion. *J. Biol. Chem.* **269**, 18,572–18,575.

28. Hollmann, M., Maron, C., and Heinemann, S. (1994) N-glycosylation site tagging suggests a three transmembrane domain topology for the glutamate receptor GluR1. *Neuron* **13**, 1331–1343.

29. Pfahnl, A. and Dahl, G. (1998) Localization of a voltage gate in connexin46 gap junction hemichannels. *Biophys. J.* **75**, 2323–2331.

30. Hu, X. and Dahl, G. (1999) Exchange of conductance and gating properties between gap junction hemichannels. *FEBS Lett.* **451**, 113–117.

31. Rubin, J. B., Verselis, V. K., Bennett, M. V. L., and Bargiello, T. A. (1992) A domain substitution procedure and its use to analyze voltage dependence of homotypic gap junctions formed by connexins 26 and 32. *Proc. Natl. Acad. Sci. USA* **89**, 3820–3824.

32. White, T. W., Paul, D. L., Goodenough, D. A., and Bruzzone, R. (1995) Functional analysis of selective interactions among rodent connexins. *Mol. Biol. Cell* **6**, 459–470.

33. Pfahnl, A., Zhou, X. W., Werner, R., and Dahl, G. (1997) A chimeric connexin forming gap junction hemichannels. *Pflügers Arch.* **433**, 773–779.

34. Paul, D. L., Ebihara, L., Takemoto, L. J., Swenson, K. I., and Goodenough, D. A. (1991) Connexin46, a novel lens gap junction protein, induces voltage-gated currents in nonjunctional plasma membrane of Xenopus oocytes. *J. Cell. Biol.* **115**, 1077–1089.

35. Gurdon, J. B., Lane, C. D., Woodland, H. R., and Marbaix, G. (1971) Use of frog eggs and oocytes for the study of messenger RNA and its translation in living cells. *Nature* **233**, 177–182.

36. Werner, R., Miller, T., Azarnia, R., and Dahl, G. (1985) Translation and functional expression of cell–cell channel mRNA in Xenopus oocytes. *J. Membr. Biol.* **87**, 253–268.

37. Dahl, G. (1992) The *Xenopus* oocyte cell-cell channel assay for functional analysis of gap junction proteins, in *Cell–cell Interactions. A Practical Approach* (Stevenson, B., Gallin, W., and Paul, D., eds.), IRL, Oxford, pp. 143–165.

38. Ebihara, L. and Steiner, E. (1993) Properties of a nonjunctional current expressed from a rat connexin46 cDNA in *Xenopus* oocytes. *J. Gen. Physiol.* **102**, 59–74.

39. Hamill, O. P., Marty, A., Neher, E., Sakmann, B., and Sigworth, F. J. (1981) Improved patch-clamp techniques for high-resolution current recording from cells and cell-free membrane patches. *Pflügers Arch.* **391**, 85–100.

15

Dual Patch Clamp

Harold V. M. Van Rijen, Ronald Wilders, Martin B. Rook, and Habo J. Jongsma

1. Introduction

The main feature of gap junctions is the passage of ions and small molecules between adjacent cells. In the heart, for example, junctional currents between adjacent cells are elicited by the membrane potential difference when one of the cells generates an action potential, while the other cell is still at resting membrane potential, thereby allowing propagation of the action potential. Experimentally, such gap junctional currents can be directly evoked and measured by independently controlling the membrane voltage of both cells of a cell pair. This is accomplished by "dual voltage clamping" a cell-pair, a method introduced by Spray et al. (1). Using separate electrodes for voltage control and current injection, this technique allowed accurate determination of kinetics and conductance of gap junctions (1–3). At present, the dual voltage-clamp technique is usually performed with two patch pipets (4), each on one cell of a cell pair, which control voltage and measure current simultaneously (5–8).

The aim of this chapter is to describe how gap junction characteristics are determined using the dual patch-clamp technique (9,10). The experimental environment is discussed. We also discuss how gap junctional conductances on both the macroscopic and microscopic levels are measured and calculated.

2. Materials
2.1. Experimental Setup

1. Two identical patch-clamp amplifiers, with headstages suitable for both single-channel and whole-cell measurements. These amplifiers can be purchased from, for example, Axon Instruments (www.axon.com), World Precision Instruments

From: *Methods in Molecular Biology, vol. 154: Connexin Methods and Protocols*
Edited by: R. Bruzzone and C. Giaume © Humana Press Inc., Totowa, NJ

(www.wpiinc.com), HeKa Electronics (www.heka.com), Bio-Logic (www.bio-logic.fr), or custom built.

2. Two three-axis micromanipulators, preferably hydraulic (e.g., Narishige [www.narishige.co.jp] or Newport [www.newport.com]).

3. An inverted microscope equipped with phase-contrast or similar optics with a 40× objective is used for visually positioning the electrodes on the cells with the manipulators. Dedicated microscopes are available from several vendors, for example, Nikon (www.nikon.com), Olympus (www.olympus.com), Leica (www.leica.com) and Zeiss (www.zeiss.com).

4. A programmable voltage stimulator, either analog (e.g., Grass [astro-med.com/grass]) or digital using voltage clamp software installed on a computer. Such software is available from, for example, Axon Instruments (pClamp, www.axon.com) or HeKa Electronics (Pulse/PulseFit, www.heka.com). Analog/digital (A/D) boards are manufactured by, for example, National Instruments (www.natinst.com) or Advanech (www.advantech.com). Note that each software package will work only with selected A/D-boards.

5. A storage oscilloscope to monitor voltages and currents.

6. Data storage device. Data may be stored on a (digital) tape recorder (e.g., Bio-Logic [www.bio-logic.fr]), but direct storage on a computer, equipped with an A/D board and proper software (commercially available, such as pClamp by Axon Instruments [www.axon.com], or custom written) has become the standard. Today fast computers allow high sampling rates and massive digital storage and allow combination of command voltage protocols, sampling, storage, and analysis at low cost.

7. Borosilicate glass capillaries preferably with inner filament (World Precision Instruments, www.wpiinc.com).

8. Pipet puller (e.g., Narishige [www.narishige.co.jp], Sutter [www.sutter.com]). For details on the fabrication of electrodes *see* **Note 1**.

2.2. Solutions and Chemicals

1. Internal pipet solution (IPS). Usually, this solution contains high K^+, low Na^+, Ca^{2+}–EGTA, Mg^{2+}, KOH–HEPES, and some form of ATP as an energy source for the cell. Reagents may be obtained from, for example, Sigma (www.sigma-aldrich.com). **Table 1** summarizes some recipes for IPS found in the literature (*see* **Note 2**).

2. Bathing solution. Usually protein-free salt solutions that mimic the normal cellular environment. A typical composition of a bathing solution is Tyrode's solution *(11)* (in mM): 140 NaCl, 5.4 KCl, 1 $CaCl_2$, 1.8 $MgCl_2$, 5 HEPES, 5 glucose, pH 7.4 (NaOH). As with IPS, many modifications to the composition of the bathing solutions have been made (for some details *see* **Note 3**).

3. Heptanol or halothane, for reduction of junctional coupling. Heptanol and halothane are both used in concentrations of 2–3 mM (for preparation of stock solutions *see* **Note 4**).

Table 1
Typical IPS for Whole-Cell Recording as Used in the Literature

Type of IPS	Composition (in mM)	Features	Reference
Normal IPS for whole-cell recording	130 K-gluconate 10 KCl 10 MgCl$_2$ 0.6 CaCl$_2$ 10 EGTA 5 Na$_2$ATP 10 HEPES pH 7.2 (KOH)	IPS with large counterion (gluconate); the low [Cl$^-$] prevents osmotic swelling	*(23)*
CsCl IPS for whole-cell recording	135 CsCl 1–2 MgCl$_2$ 0.5 CaCl$_2$ 5.5–10 EGTA 5 Na$_2$ATP 5 HEPES pH 7.2 (CsOH)	IPS which strongly reduces K-channel activity. May induce osmotic problems due to high [Cl$^-$]	*(40,41)*
pH-clamp IPS for whole-cell recording	100 K-gluconate 8 KOH 25 (NH$_4$)$_2$SO$_4$ 1 MgCl$_2$ 0.5 BAPTA 5 Na$_2$ATP 5 PIPES pH 7.0	Allows control of cytoplasmic pH in combination with the appropriate bathing solution: 100 HEPES, 1 CaCl$_2$, 1 MgCl$_2$, 52.5–74.25 K-gluconate, 7.5–0.5 (NH$_4$)$_2$SO$_4$	*(42)*

2.3. Cell Preparation

1. Pairs of regenerative or immortalized cells are formed by seeding cells in culture dishes at low density. After one cell cycle, single cells become cell pairs.
2. For nonregenerative primary cells, such as adult cardiomyocytes, incomplete dissociation of tissue during the isolation procedure will result in cell pairs.
3. Another technique to generate mammalian cell pairs was devised by Loewenstein et al. *(12)* and used by Rook et al. *(8)*. Two freshly dissociated single neonatal rat cardiomyocytes were sealed to a patch pipet. Subsequently, the cells were gently pushed together, and after some time new gap junction channels were formed successively. This method was also successfully applied for other cell types *(13–16)*.

3. Methods

3.1. The Dual Patch Clamp Principle

1. Select a cell pair under the microscope.
2. Fill both pipets with IPS and manipulate the electrodes into the bath and close to the cells (*see* **Note 5**).
3. Compensate for fast capacitive currents and electrode offset, using the compensation circuitries of the patch-clamp amplifiers (*see* **Note 6**).
4. Gently lower the electrodes onto the cells (*see* **Note 7**).
5. Apply gentle suction to the pipet via a side port in the pipet holder. The tip strongly seals to the cell membrane, thereby forming an electrical resistance of many gigaohms (gigaohm seal) between the electrode interior and the bathing solution *(17)* (*see* **Note 8**).
6. By applying more suction the membrane under the pipet tip is ruptured and the whole-cell configuration is achieved. When both pipets are in the whole-cell configuration, an equivalent resistive circuit as depicted in **Fig. 1A** is present. Electrical access to the cell interior is achieved via the combined resistance of the pipet and the broken membrane patch. This resistance is commonly referred to as the *series resistance* (R_s), because it is in series with both the membrane and junctional resistance. Each cell has its membrane resistance (R_m), and both cells are connected by the junctional resistance (R_j).
7. To determine junctional conductance, junctional currents are elicited by applying a potential difference across the gap junction. Usually, both cells are kept at a common holding potential close to the resting membrane potential (V_{rest}) to minimize holding, that is, membrane currents, followed by a stepwise change of the holding potential of one of the cells, for example, cell A (*see* **Fig. 1C**). This results in amplifier currents I_a and I_b (Fig. 1A), which are the sum of the membrane current (I_{m1}) and the junctional current (I_j) in the stepped cell A ($I_a = I_{m1} + I_j$), and the membrane current minus the junctional current in the nonstepped cell B ($I_b = I_{m2} - I_j$).

3.2. Determination of Macroscopic Characteristics of Gap Junctional Currents

3.2.1. Determination of Macroscopic Gap Junctional Conductance

The most commonly used way of calculating the gap junctional conductance (g_j) from dual voltage-clamp data is by ignoring the voltage drop across the series resistances and the membrane currents. If cell A is the nonstepped cell, g_j is calculated from:

$$g_j = I_a/(V_a - V_b) \tag{1}$$

Similarly, if cell B is the nonstepped cell,

$$g_j = -I_b/(V_a - V_b) \tag{2}$$

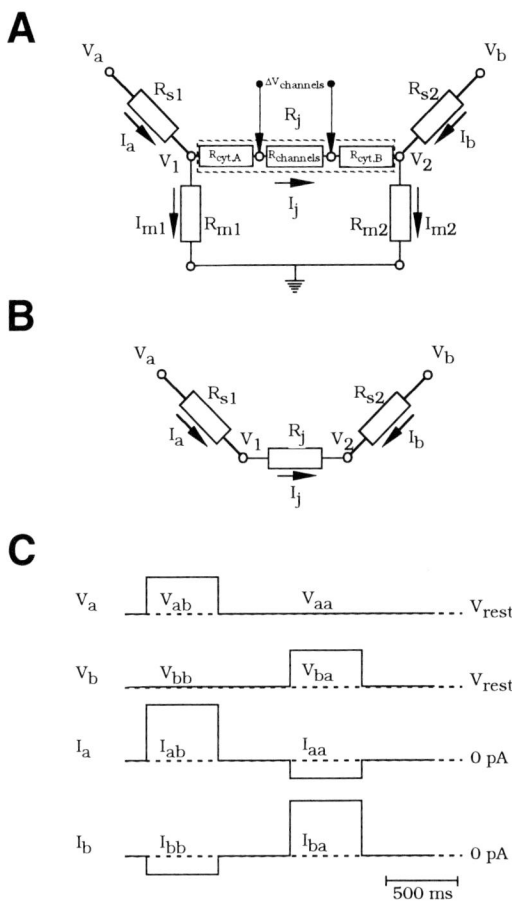

Fig. 1. **(A)** Equivalent resistive circuit of a cell pair under dual whole-cell voltage clamp conditions. The junctional current (I_j) flows across the junctional resistance $(R_j$, which is the sum of the two cytoplasmic access resistances $R_{cyt,A}$ and $R_{cyt,B}$, and the resistance of the gap junction channels, $R_{channels}$) as a result of the difference between the membrane potential of cell A (V_1) and cell B (V_2). The difference between the membrane and pipet potentials $(V_a$ and $V_b)$ of the cells is determined by the series resistances and pipet currents of cell A and B $(R_{s1}$ and I_a, and R_{s2} and I_b, respectively). **(B)** Equivalent resistive circuit of a cell pair under dual voltage-clamp conditions with infinitely large membrane resistances. **(C)** Diagram of the command potential protocol used for experiments on cell pairs in a dual voltage-clamp setup and the resulting pipet currents. First the pipet potential of cell A was changed for 500 ms. Next, after 500 ms the pipet potential of cell B was changed for 500 ms. A step in the pipet potential of cell A, from V_{aa} to V_{ab}, evokes pipet current steps in both cells: from 0 to I_{ab} in cell A, and from 0 to I_{bb} in cell B. Similarly, a step in cell B from V_{bb} to V_{ba} evokes pipet currents I_{aa} and I_{ba} in cell A and B, respectively.

When junctional resistance is low, as compared to the series resistances, the voltage drop across the series resistances is no longer negligible (*see* **Subheading 3.4.1.**). Hence, numerical correction for series resistance is accomplished by subtracting the voltage drop across these resistances. **Equations 1** and **2** are then replaced with:

$$g_j = I_a/[(V_a - I_a \cdot R_{s1}) - (V_b - I_b \cdot R_{s2})] \tag{3}$$

and

$$g_j = -I_b/[(V_a - I_a \cdot R_{s1}) - (V_b - I_b \cdot R_{s2})] \tag{4}$$

respectively. Of course, estimates of R_{s1} and R_{s2} are required (cf. **Fig. 7**).

However, under conditions of low membrane resistance (discussed in detail in **Subheading 3.4.3.**), it is necessary to use a correction method that takes both membrane and series resistance into account. According to a recently published method *(18)*, gap junctional conductance is calculated by:

$$g_j = I_a - [(V_a + I_a \cdot R_{s1})/R_{m1}]/[(V_a - I_a \cdot R_{s1}) - (V_b - I_b \cdot R_{s2})] \tag{5}$$

if cell A is the nonstepped cell. Similarly if cell B is the nonstepped cell:

$$g_j = -I_b + [(V_b - I_b \cdot R_{s2})/R_{m2}]/[(V_a - I_a \cdot R_{s1}) - (V_b - I_b \cdot R_{s2})] \tag{6}$$

Evidently, for this correction method, estimates of both series and membrane resistances are required . This issue is discussed in **Subheading 3.4.4.**

3.2.2. Determination of Macroscopic Voltage Sensitivity

When small transjunctional voltage steps are applied to gap junctions, their conductance is rather constant in time. However, upon larger voltage steps, most gap junctions exhibit voltage sensitivity, and tend to decrease their conductance, which is seen as a relaxation of the junctional current as shown in **Fig. 2A**. At the start of the 100 mV command voltage step on a computer-simulated cell pair, containing six connexin43 (Cx43) gap junction channels, an instantaneous current $I_{j,0}$ is seen in the nonstepped cell, which strongly diminishes during the 2-s step, reaching a quasi-steady-state value ($I_{j,\infty}$) within 1–2 s. A common way to present and quantify voltage-dependent parameters of gap junctions is by plotting all the G_j values, normalized for the junctional conductance measured at a small voltage step (e.g., 10 mV), for voltage steps from, for example, –100 to 100 mV as presented in **Fig. 2B**. The data are fitted to a Boltzmann equation *(2,3)*:

$$G_j = (G_{max} - G_{min}) / [1 + \exp A \cdot (V_j - V_0)] + G_{min} \tag{7}$$

This fitting procedure gives the value for G_{max} and G_{min} (the normalized maximal and residual junctional conductance, respectively), A (= $n \cdot q/k \cdot T$, where n is the equivalent number of electron charges q that move through the

A

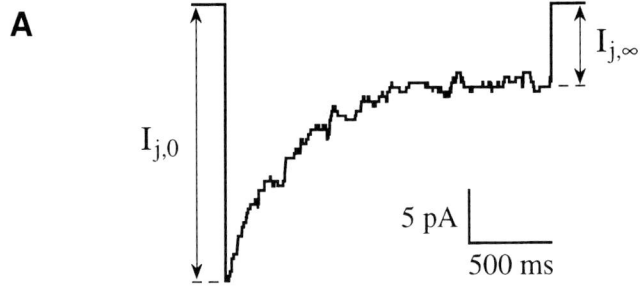

B

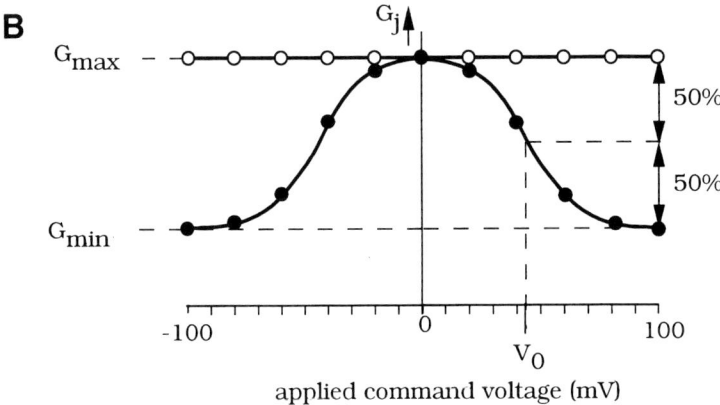

applied command voltage (mV)

Fig. 2. **(A)** Current trace of the nonstepped cell of a computer-simulated cell pair, containing six Cx43 gap junction channels, as a result of a 100-mV command voltage step. $I_{j,0}$ and $I_{j,\infty}$ are the instantaneous and the quasi-steady-state junctional current, respectively. **(B)** Diagram of instantaneous *(open circles)* and quasi-steady-state *(closed circles)* values of normalized conductance (G_j) vs the applied command voltages and a Boltzmann fit (**Eq. 7,** *solid line*).

entire transjunctional field between open and closed states, k is Boltzmann's constant, and T is absolute temperature), and V_0 (the value of half-maximal inactivation).

3.3. Determination of Microscopic Characteristics of Gap Junctional Currents

3.3.1. Determination of Single-Channel Conductance

In the literature, three ways of measuring single-channel conductances can be found. The first is by performing experiments on very poorly coupled cells that are connected by only a few gap junction channels *(19–22)*. The second is to use well-coupled cell pairs, exposed to uncoupling agents such as halothane or heptanol (**Fig. 3A**) *(23,24)*. A third way is by use of "induced cell pairs," in which dissociated cells are pushed together, thereby allowing *de novo* formation of gap junction channels *(8,13–16)*.

All three methods result in current traces as shown in **Fig. 3B**. At a driving force of 50 mV gating of junctional channels is seen as stepwise current changes in traces of both the stepped and nonstepped cell (lower and upper trace, respectively). The distribution of single gap junction channel events can be visualized by converting the current transitions to conductance steps and displaying those values in a frequency histogram (**Fig. 3C**). Mean amplitudes of the conductance steps can be calculated by means of gaussian curve fitting. A second method is by constructing an all-points histogram. With this method, all digitized points of the current in the nonstepped cell are used, and the current values of the points are binned, plotted, and fitted to a gaussian function as shown in **Fig. 3D** (taken from the same experiment as panels **B** and **C**). The first peak, at 0.21 pA, represents the baseline level, that is, when the channels are all closed. The second peak is found at 6.23 pA, which is the current level when one Cx40 channel is in the open state. The difference between the two peaks divided by the transjunctional voltage (50 mV) gives the conductance of the channel, 120.4 pS. For limitations of these measurements and analyses *see* **Note 9**.

3.3.2. Determination of Single-Channel Kinetics

The conductance of a gap junction between a cell pair can be expressed as

$$g_j = n \cdot P_o \cdot \gamma_j \tag{8}$$

with n, P_o, and γ_j being the total number of channels in a gap junction, the open probability of a single gap junction channel, and the conductance of a single gap junction channel, respectively. The open–close kinetics of single gap junction channels can be determined only in cell pairs that are coupled with few channels by nature. Halothane and heptanol, widely used to measure single-channel conductances, exert their uncoupling effect by reducing the open probability of the channels *(24)*. Evidently such a method interferes with determination of channel kinetics.

The open probability of gap junction channels in cell pairs containing few channels can be determined by a statistical approach, using a binomial distribution. One approach is by determination of the fraction of time spent by steady-state I_j at different current levels *(25,26)*. A second approach is to construct all-points current amplitude histograms and fit these histograms to a probability density function *(27,28)*. *See* **Note 10** for limitations of either approach.

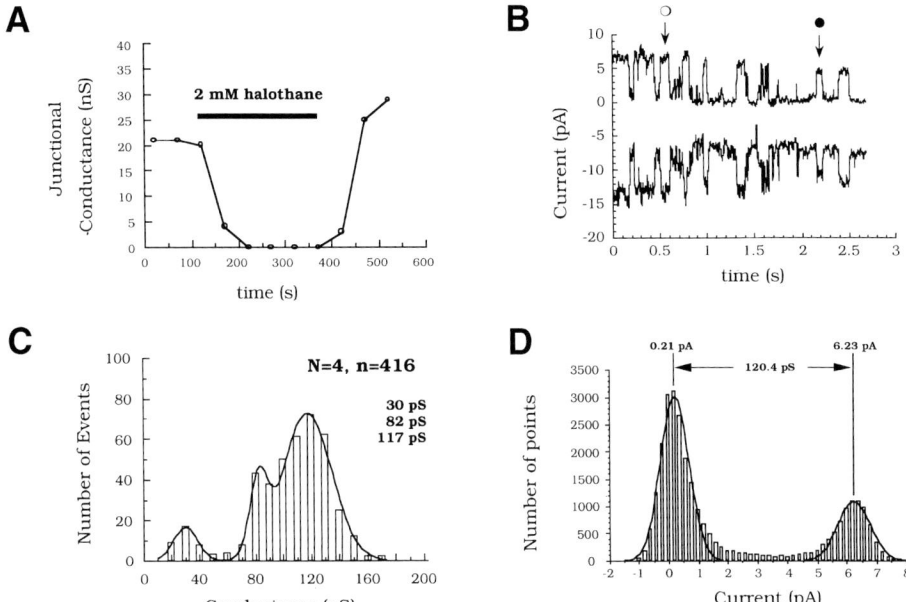

Fig. 3. **(A)** Dual voltage-clamp experiment on a SKHep/hCx40 cell pair in which 2 m*M* halothane is applied at *t* = 100 s. Junctional conductance rapidly decreases to near-zero values. After removal of the halothane, coupling is restored. **(B)** Single-channel current measured under halothane conditions at 50 mV driving force in SKHep1/hCx40 cell pairs. The upper and lower traces represent the current in the nonstepped and stepped cell, respectively. Two unitary current steps of Cx40 are detected, reflecting single-channel conductances of approx 80 *(closed circle)* and 120 pS *(open circle)*. **(C)** Frequency histogram of 416 single channel conductance steps from four experiments. Gaussian curve fitting *(solid line)* revealed three channel sizes: 30, 82, and 117 pS. The smallest size of 30 pS represents the endogenous Cx45 channel of SKHep1 cells. **(D)** All-points histogram of the same SKHep1/hCx40 cell pair as shown in **B**. Gaussian curve fitting *(solid line)* revealed two states, 0.21 pA (baseline level) and 6.23 pA (one Cx40 channel open). The conductance of the channel was 120.4 pS (50 mV driving force).

3.4. Corrections for Nonjunctional Resistances

3.4.1. Errors Introduced by Series Resistance

The introduction of two additional series resistances (R_s), caused by the resistance of the patch pipet and residual resistance of the disrupted membrane under the patch electrode, can induce considerable errors in measuring the junctional conductance. Because the patch pipet is used for recording the membrane potential and injecting current simultaneously, a voltage drop will occur across each series resistance, by result of which the actual membrane potentials V_1 and V_2 will always differ from the clamp potentials V_a and V_b, respectively.

To investigate the effect of series resistance on the determination of steady-state gap junctional conductance, we used a computer model approach to quantify the effect of series resistance on the measured fraction of true junctional conductance:

$$F_j = g_{j,m}/g_{j,t} \tag{9}$$

where $g_{j,m}$ is the junctional conductance calculated from **Eq. 1** or **2**, and $g_{j,t}$ is the true junctional conductance.

If the membrane resistances of both cells are infinitely high, the equivalent resistive circuit reduces to that of **Fig. 1B**. The measured current is then determined by:

$$I_a = I_j = -I_b = (V_a - V_b)/(R_{s1} + R_j + R_{s2}) \tag{10}$$

Figure 4 shows F_j at different series resistances (R_s, where $R_s = R_{s1} + R_{s2}$, and $R_{s1} = R_{s2}$) as a function of $g_{j,t}$, with $g_{j,t}$ ranging between 0.1 and 100 nS. In the ideal case of zero series resistance ($R_s = 0$), $G_{j,m}$ does not deviate from $g_{j,t}$, and $F_j = 1$. At low intercellular conductance values of 0.1 nS, the series resistance induced error is very close to 0, and up to a few percent at 1 nS ($F_j = 0.996$ and $F_j = 0.962$ at 40 MΩ series resistance, respectively). With increasing intercellular conductance, however, significant errors occur. Even at very favorable experimental conditions, with total series resistance as low as 10 MΩ, an intercellular conductance of 20 nS will be measured as 16.7 nS (17% error); at a total series resistances of 20 MΩ, only 14.2 nS will be measured (29% error). Conversely, a measured conductance of 25 nS with 20 MΩ total series resistance results from a true intercellular conductance of 50 nS.

An example of a series resistance induced error is shown in **Fig. 5**. The conductance between a SKHep cell pair transfected with human Cx40 is measured, and the effect of the membrane permeable protein kinase A activator 8-Br-cAMP on the conductance is determined. Without correction for R_s (4 MΩ on both sides) g_j increases from 62 to 88 nS, an increase of 42%. However, after correction for R_s, the increase amounts to 66% (84–139 nS). The presence of R_s resulted in an underestimation of the effect by 36%!

The determination of dynamic properties of gap junctions, such as voltage dependence, is also affected by series resistance *(29,30)*. As a result of series resistance, the actual voltage drop across the gap junction is smaller than the applied voltage difference. Virtually all patch-clamp amplifiers are equipped with series resistance compensation circuitry. This electronic circuit increases the command voltage to a value equal to $I \cdot R_s'$, where I is the amplifier current

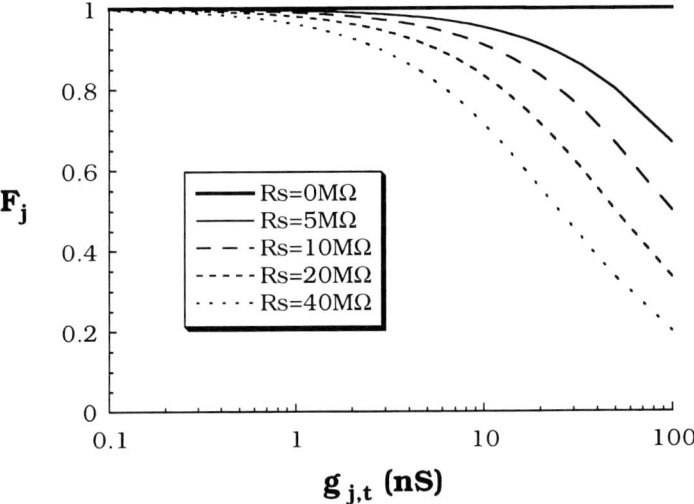

Fig. 4. Relationship between the measured fraction (F_j) of the true junctional conductance ($g_{j,t}$), and $g_{j,t}$, as calculated using Eq. 3 or 4, at total series resistance ($R_s = R_{s1} + R_{s2}$) of 0, 5, 10, 20, or 40 MΩ, in case of negligible membrane currents ($R_{m1} = R_{m2} = \infty$).

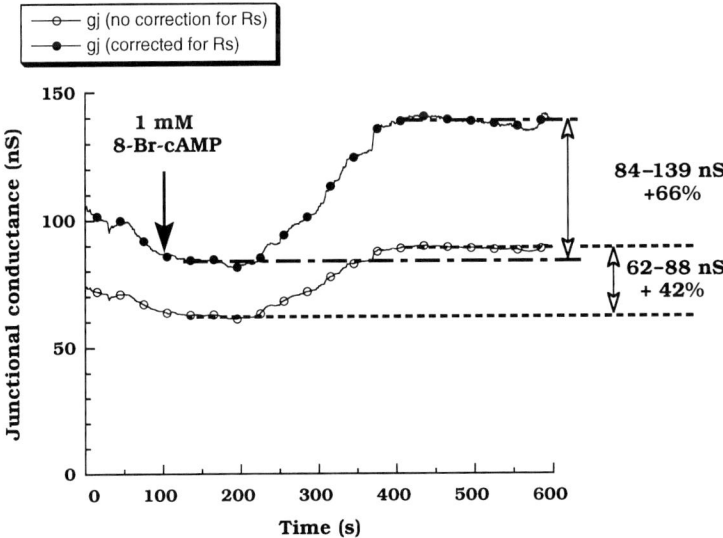

Fig. 5. Modulation of coupling in a SKHep1/hCx40 cell pair by 1 mM 8-Br-cAMP (*arrow* indicates time of application), with and without correction for series resistance ($R_{s,a} = R_{s,b} = 4$ MΩ). Without any correction, junctional conductance increases from 62 to 88 nS (+42%); *open symbols*. After correction for series resistance, however, junctional coupling increases from 84 to 139 nS (+66%; *closed symbols*).

and R_s' is the resistance in the feedback circuit. However, this positive feedback circuit tends to oscillate at high correction values, that is, at R_s' values close to R_s, and practically, correction for R_s during the experiment is limited to 70% (R_s' cannot be higher than 70% of R_s). Therefore voltage dependence is always underestimated in the presence of series resistance.

3.4.2. Errors Introduced by Cytoplasmic Access Resistance

A general observation when accessing gap junctional voltage dependence is that gap junctions with high conductances (composed of many channels) exhibit less voltage sensitivity than gap junctions with low conductance values (composed of fewer channels). This effect was first attributed to pipet series resistance *(29)*, because large gap junctions conduct more current at the same voltage difference, which results in a larger voltage drop across the series resistances than in the case of small gap junctions. However, Wilders and Jongsma *(30)* calculated that the effect of series resistance alone is not large enough to explain the observed effect. Rook et al. *(8,31)* suggested that the tight packing of the gap junction channels resulted in a cytoplasmic access resistance, thereby causing a cytoplasmic potential drop that masks the voltage sensitivity. Wilders and Jongsma *(30)* constructed a biophysical model in which the actual voltage across each channel was calculated and in which the open and close kinetics of the individual channels were incorporated. They showed that even in the case of complete compensation for series resistance, the number of channels in the gap junction may be largely underestimated, and that the overestimation of normalized steady-state junctional conductance easily masks transjunctional voltage dependence of the individual channels.

At any rate, cytoplasmic access resistance is, unlike pipet series resistance, not artificial. Under normal physiological conditions, voltage differences between cells, for example, in the conducting heart, will also be attenuated by the cytoplasmic access resistance to the gap junction. This will result in less intercellular current than expected from the number of gap junction channels, the open probability, and their individual conductance, that is, **Eq. 8** no longer holds.

3.4.3. Errors Introduced by Membrane Resistance

Intuitively, it is obvious that gap junctional conductance is determined very reliably in cell pairs with infinitely low series resistance and infinitely high membrane resistances. Unfortunately, these conditions are never met. The presence of finite values of series and membrane resistances therefore always results in errors in measured junctional conductance. To investigate these errors, we performed computer simulations. In these simulations, amplifier currents I_a and I_b were calculated with V_a and V_b arbitrarily set to 100 and 0 mV, respectively, while one of the resistances was incremented. From these cur-

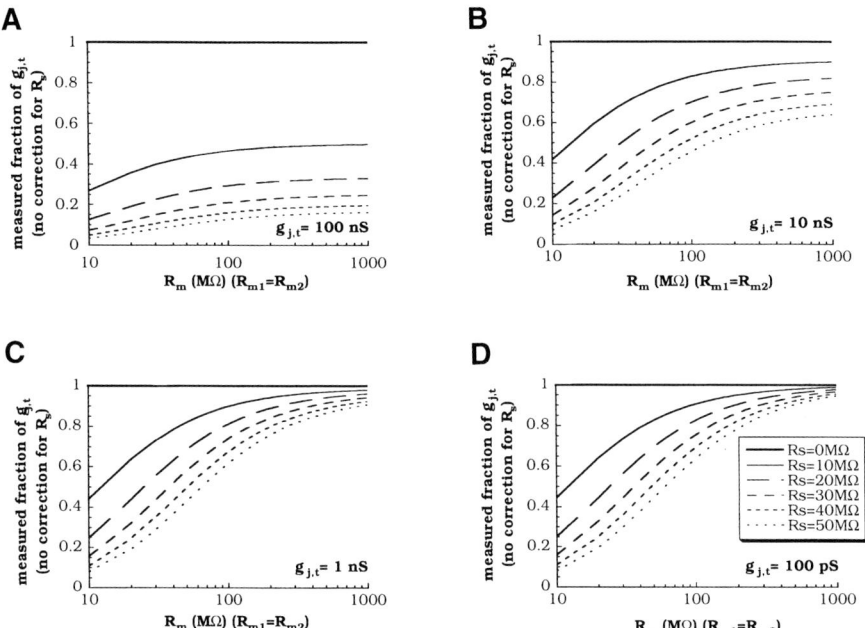

Fig. 6. Relationship between the measured fraction of true junctional conductance ($g_{j,t}$) and the membrane resistance of both cells, as determined by computer simulations. The membrane resistance of the stepped cell (R_{m1}) was set equal to that of the nonstepped cell (R_{m2}). Total series resistance (R_s) was set to 0–50 MΩ as indicated. No correction for R_s was applied. (**A**) $g_{j,t}$ = 100 nS. (**B**) $g_{j,t}$ = 10 nS. (**C**) $g_{j,t}$ = 1 nS. (**D**) $g_{j,t}$ = 100 pS.

rents, junctional resistance was calculated according to **Eq. 1** or **2**. Furthermore, it was assumed that $R_{s1} = R_{s2}$. Thus, a total series resistance of, for example, 30 MΩ, corresponds to a series resistance of 15 MΩ for each of the two pipets.

Figure 6 shows F_j as a function of R_m if $R_{m1} = R_{m2}$ at $G_{j,t}$ values of 100 nS, 10 nS, 1 nS, and 100 pS (**A–D**, respectively). When R_s equals 0, the pipet potential is identical to the membrane potential ($V_a = V_1$ and $V_b = V_2$). Thus the junctional current (I_j) is elicited by the command voltage. Also, $V_2 = 0$, resulting in zero current flow across R_{m2}, so that I_b equals $-I_j$. As a consequence, no errors occur (*cf.* **Eq. 1**): $g_{j,m} = g_{j,t}$ and $F_j = 1$ (*bold solid lines*). When R_m is 1 GΩ, F_j has values very similar to those presented in Fig. 4, where $R_{m1} = R_{m2} = \infty$. When R_m is lowered, a significant decrease in F_j is observed. With decreasing values of $g_{j,t}$, the errors induced by R_s become smaller, as can be seen from the values for F_j at R_m = 1 GΩ. Similarly the errors induced by R_m become smaller with decreasing $g_{j,t}$. However, even at values for $g_{j,t}$ as low as 100 pS, considerable errors are still observed.

3.4.4. Correction for Series and/or Membrane Resistance

To investigate in which respects F_j benefits from correction for R_s and R_m under experimental conditions, we performed experiments on equivalent resistive circuits in a dual voltage clamp setup, using resistors having resistance values in the range normally present in actual dual whole-cell voltage-clamp experiments. The command voltage protocol of alternately stepping cell A and cell B (*see* **Fig. 1C**) was applied to obtain estimates of R_{m1} and R_{m2} through (**18**):

$$R_{m1} = [R_{s2} \cdot (V_{ba} \cdot I_{ab} + V_{ab} \cdot I_{ba} + R_{s1} \cdot \{I_{aa} \cdot I_{bb} - I_{ba} \cdot I_{ab}\}) - V_{ab} \cdot V_{ba}] /$$
$$[R_{s2} \cdot (I_{ba} \cdot I_{ab} - I_{aa} \cdot I_{bb}) - V_{ba} \cdot (I_{ab} + I_{bb})] \tag{11}$$

and

$$R_{m2} = [R_{s1} \cdot (R_{s2} \cdot \{I_{ab} \cdot I_{ba} - I_{bb} \cdot I_{aa}\} - I_{ab} \cdot V_{ba}) + V_{ab} \cdot (V_{ba} - I_{ba} \cdot R_{s2})] /$$
$$[R_{s1} \cdot (I_{bb} \cdot I_{aa} - I_{ab} \cdot I_{ba}) + V_{ab} \cdot (I_{aa} + I_{ba})] \tag{12}$$

Next, **Eqs. 1, 2, 5,** and **6** were used to calculate g_j. Series resistance was determined by applying the protocol as presented in **Fig. 7** on both cells simultaneously.

Figure 8 shows that the experimental results (*solid, dashed,* and *dotted lines with markers*) are highly similar to the simulation results (*solid, dashed,* and *dotted lines without markers*). From **Fig. 8A–D**, it is evident that, with decreasing membrane resistance, correction for R_s alone is not sufficient, and that correction for both R_s and R_m is necessary: When R_m in both cells is as low as 100 or 20 MΩ, correction for both R_s and R_m can bring values for F_j close to 1 (Fig. **8C** and **D**, *diamonds*), while F_j amounts to ~0.9 and ~0.7, respectively, with correction for R_s alone (**Fig. 8C** and **D**, *squares*).

The beneficial effect of correction for R_s and/or R_m is also apparent from **Fig. 8.** If , for example, a pair of adult ventricular cells with membrane resistances of 20 MΩ is coupled by a junctional conductance of 100 nS, with R_{s1} and R_{s2} both being 10 MΩ, a junctional conductance of 20 nS (80% error) will be measured (**Fig. 8D**). After correction for R_s, $g_{j,m}$ will amount to 67 nS (33% error). After correction for both R_s and R_m, $g_{j,m}$ is 97 nS (3% error). In this last case, R_{m1} and R_{m2} have calculated values of 21.4 and 22.2 MΩ, respectively, which are very close to the actual values.

In **Fig. 8E** and **F**, very asymmetrical membrane resistances were used. If R_m of the stepped cell is low ($R_{m1} = 20$ MΩ), while R_m of the nonstepped cell is high ($R_{m2} = 1$ GΩ), F_j responds to $g_{j,t}$ in a way similar to the configuration in which R_m of the nonstepped cell is low ($R_{m2} = 20$ MΩ) and R_m of the stepped cell is high ($R_{m1} = 1$ GΩ), if no correction for R_s or R_m is applied (**Fig. 8E** and **F**). If corrected for R_s only, however, F_j becomes 1 if R_m of the stepped cell was

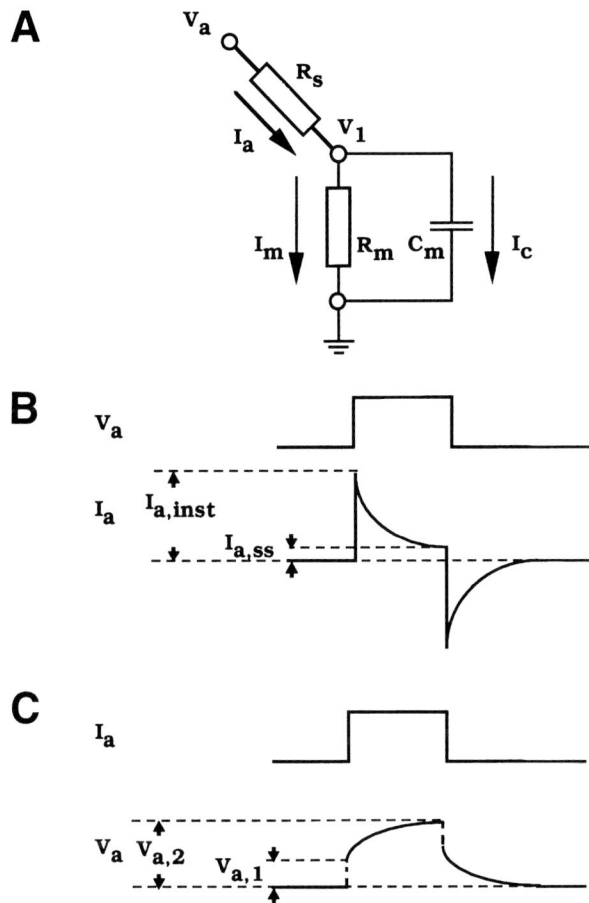

Fig. 7. (**A**) Equivalent circuit of a single cell under whole-cell patch-clamp conditions: an access resistance to the cell (R_s) in series with an RC-circuit, representing the membrane resistance (R_m) and the membrane capacitance (C_m). (**B**) Determination of R_s under voltage-clamp conditions. A stepwise change in V_a will result in a large instantaneous current ($I_{a,inst}$), which rapidly decreases to a steady-state value ($I_{a,ss}$). At the end of the voltage step a similar pattern of opposite sign is seen. The large instantaneous current at the beginning of the voltage step is solely determined by R_s, because the membrane capacitance effectively short-circuits the membrane resistance: $R_s = V_a/I_{a,inst}$. (**C**) Determination of R_s under current clamp conditions. A stepwise change in I_a will evoke a voltage deflection that consists of a fast rise of V_a to $V_{a,1}$, followed by a slower rise to a steady-state value, $V_{a,2}$. The opposite is seen at the end of the current step. The fast rise to $V_{a,1}$ at the beginning of the current step is again caused by an effective short-circuiting of the membrane resistance by C_m. The voltage deflection $V_{a,1}$ thus represents the voltage drop across R_s due to I_a. Therefore, R_s is determined by: $R_s = V_{a,1}/I_a$.

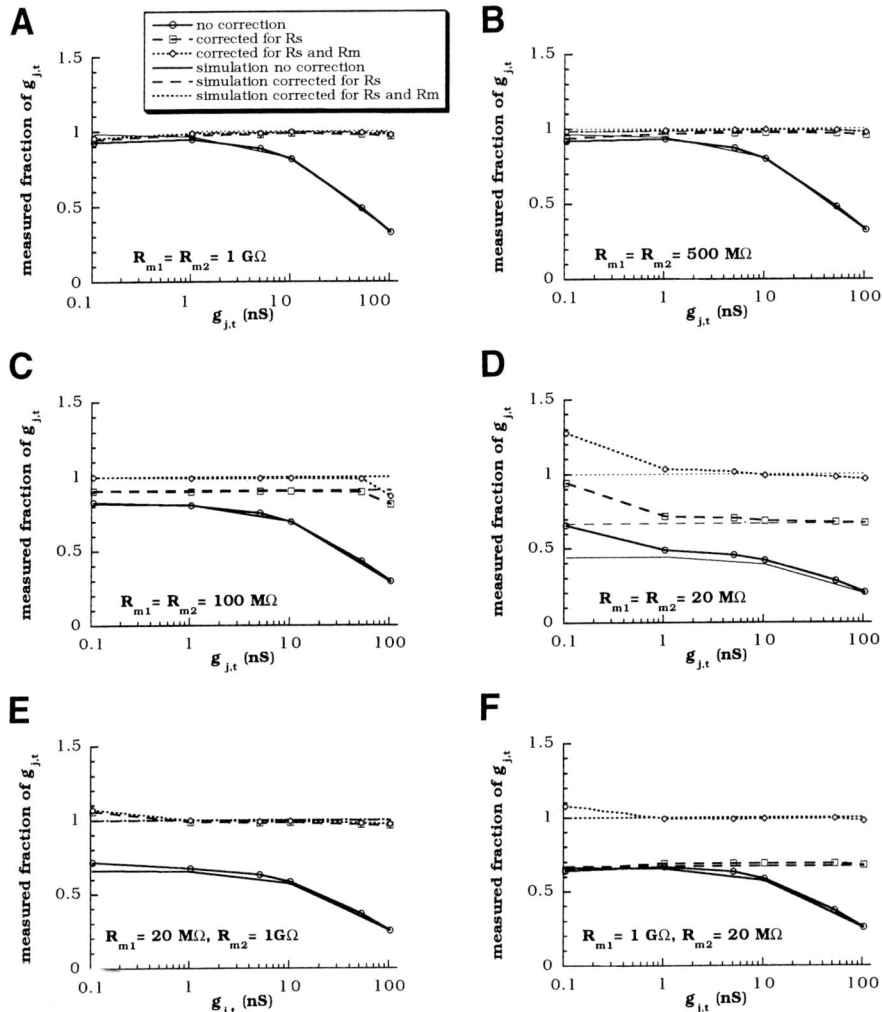

Fig. 8. (*Opposite page*) Relationship between the measured fraction (F_j) of true junctional conductance ($g_{j,t}$) and $g_{j,t}$, as determined in dual voltage-clamp experiments on electrical resistive circuits. In all experiments, series resistance was 9.6 MΩ on each side. First the membrane resistance of the stepped cell (R_{m1}) and that of the nonstepped cell (R_{m2}) were set to 1 GΩ, 500 MΩ, 100 MΩ, or 20 MΩ, as indicated. Next, $g_{j,t}$ was set to values near 0.1, 1, 5, 10, 50, or 100 nS. Junctional conductance was calculated, using the currents elicited in the nonstepped cell, in three ways: (1) Without any correction (*solid lines with circles*); (2) with offline correction for series resistance (*dashed lines with squares*); and (3) with offline correction for both series and membrane resistance (*dotted lines with diamonds*). The solid, dashed, and dotted lines without markers represent the numerical data taken from computer simulations

low (**Fig. 8E**, *dashed lines*). Conversely, if R_m of the nonstepped cell was low (**Fig. 8F**, *dashed lines*), correction for R_s only makes F_j independent of $g_{j,t}$, but still results in an error of about 30%. In both configurations correction for both R_s and R_m results in values for F_j close to 1 (**Fig. 8E** and **F**).

An alternative recording method, which also might reduce errors by nonjunctional resistances, is discussed in **Note 11**.

3.5. Double Perforated Patch Recording of Gap Junctional Currents

3.5.1. The Perforated Patch Technique

Horn and Marty *(32)* introduced a method that allows for electrical access to the cell interior, without perfusing the cytoplasm. For this method, the antimycotic drug nystatin or amphotericin B is added to the IPS. Then the pipet is sealed to the membrane, but the membrane patch is not disrupted. The drug in the IPS forms small pores, which are permeant for monovalent cations and anions, and impermeant for multivalent ions and molecules >0.8 nm in diameter *(33,34)*. Nystatin and amphotericin B have comparable properties, but amphotericin B perforates better, and results in lower series resistances.

3.5.1.1. NYSTATIN-PERFORATED PATCH

1. Add 70 µg/mL of nystatin (Sigma) and 70 µg/mL of pluronic F-127 (Molecular Probes) from a stock solution (*see* **Note 12**) to IPS. Store this solution in a dark place, as nystatin is light sensitive.
2. Fill the patch pipets with nystatin/pluronic–IPS.
3. Make gigaohm seals on both cells according to normal procedures (*see* **Subheading 3.1.**, **steps 3–5**), but do not break the membrane under the patch electrodes.
4. Wait for the membrane patches to be perforated. Usually this takes from several seconds up to 1 or 2 min (*see* **Note 13**).
5. Perform a dual voltage-clamp procedure analogous to the whole-cell method.

3.5.1.2. AMPHOTERICIN B-PERFORATED PATCH

1. Add 0.2 mg/mL of amphotericin B (Sigma) to IPS from a stock solution (*see* **Note 14**). Store this solution in the dark for a maximum of 1 h.
2. Front-fill the patch pipets with normal IPS, by placing the tip of the pipet in normal IPS (*see* **Note 15**).

---

under the same conditions, and represent the calculated junctional conductance (1) without any correction, (2) with correction for series resistance, and (3) with correction for both series and membrane resistance, respectively. (**A**) $R_{m1} = R_{m2} = 1$ GΩ. (**B**) $R_{m1} = R_{m2} = 500$ MΩ. (**C**) $R_{m1} = R_{m2} = 100$ MΩ. (**D**) $R_{m1} = R_{m2} = 20$ MΩ. (**E**) $R_{m1} = 20$ MΩ, $R_{m2} = 1$ GΩ. (**F**) $R_{m1} = 1$ GΩ, $R_{m2} = 20$ MΩ.

3. Back-fill the patch pipets with amphotericin B–IPS.
4. Make gigaohm seals on both cells according to normal procedures (*see* **Subheading 3.1., steps 3–5**), but do not break the membrane under the patch electrodes.
5. Wait for the membrane patches to be perforated. Usually this takes from several seconds up to 1 or 2 min (*see* **Note 13**).
6. Perform a dual voltage-clamp recording analogous to the whole-cell method.

3.5.2. Whole-Cell vs Perforated Patch

As mentioned previously, the whole-cell patch-clamp recording mode brings the cytoplasm in continuity with the IPS. This is advantageous if one wants to change the composition of the cell interior (*see* **Table 1**). However, if one wants to study the effect of for example, extracellular signals on the behavior of gap junction channels via intracellular processes, the disruption of the cell machinery might largely interfere with the experiment. The advantage of the perforated patch method is that it leaves the cell interior relatively intact. However, its application is more laborious than conventional whole-cell recording, and more importantly, the success rate, that is, actual, perforation of the membrane patch to series resistance values low enough to perform efficacious voltage clamp, is significantly lower.

3.6. Concluding Remarks

Dual patch-clamp recording is a quantitative technique to determine static and dynamic electrical characteristics of gap junction channels. However, as with most experimental techniques, dual patch-clamp recording has several drawbacks, which vary from quantitative errors in measured conductance or underestimation of voltage dependence to unwanted perfusion of the cytoplasm. With proper experimental design and use of correction methods, as discussed in this chapter, many of these limitations can be surmounted, allowing the dual patch-clamp technique to be a powerful method for quantitative determination of junctional conductance.

4. Notes

1. Pipet pullers usually have a two-step pulling mechanism in which the first step pulls the capillary over a short distance to thin the glass. The second step (with lower heat) separates the capillary. Increasing or decreasing the heat of the first step changes the shape of the electrode to either more sharp or blunt, while an increase or decrease of the second pulling step results in electrodes with smaller or wider tip diameters, respectively. The fabrication of patch pipets is based on trial and error, and usually is a balance between obtaining good seals and having low access resistances. Subsequently, the tip of the pipet is heat polished by bringing it in the vicinity of a red-hot filament under a microscope (for detailed discussion *see* **ref. 35**) . The typical electrode resistance when filled with pipet solution,

as measured in the bathing solution, is between 2 and 15 MΩ (*see* **Table 1** in **ref. *30*** for an overview)

2. To avoid clogging problems, the internal pipet solution (IPS) is usually filtered using a 0.22-μm syringe filter. Afterwards the IPS can be aliquotted and stored at –20°C.

3. Especially when cells with low input resistances, such as ventricular myocytes, are used, noise from nonjunctional membrane channels may interfere with measurement of small junctional currents. Addition of $NiCl_2$, CsCl, or $BaCl_2$ results in the reduction of Ca- and K-channel activity, and improves the resolution of small gap junctional events.

4. Heptanol can be directly added to the bathing solution. Halothane, however, is first thoroughly mixed 1:1 with bathing solution (stock solution, which can be stored at room temperature). The upper fraction of this mix (halothane-saturated bathing solution) is subsequently added 1:4 with normal bathing solution just before addition to the cells. This procedure results in a halothane concentration of approx 2 m*M*.

5. In preparations of primary cells, the surface of the bathing solution may contain some debris. To avoid contamination of the pipet tip with debris, slight overpressure to the pipet interior can be applied when entering the solution. Be sure never to apply suction to the pipet in the bathing solution before it is on the cell, as this will hinder seal formation.

6. The electrode offset is mostly determined by the liquid junction potential. This potential is caused by the unequal motility of large and small ions in the pipet solution. If the IPS is mainly composed of K-gluconate, the small potassium ions diffuse more easily from the pipet to the bathing or intracellular solution than the negatively charged large gluconate molecules. The precise value of the liquid junction potential can be calculated using JPCalc (*36*); also refer to www.axonet. com/junction.htm. Please note that this potential is also determined by the composition of the bathing solution, so that the value of the liquid junction potential in the bathing solution is different from that in the whole-cell configuration. The recorded membrane potential is therefore not correct. This problem can be avoided by using an IPS with ions of equal motility, such as KCl.

7. This process can be monitored very easily by applying small voltage steps (e.g., 2 mV for 50 ms) to the electrode. When the electrode touches the cell, the electrical resistance of the electrode will increase which is seen as a decrease of the electrode current during the voltage step.

8. If the gigaohm seal does not form very easily, one might try to depolarize the electrode potential somewhat, to values between –20 and –50 mV. Otherwise, try to change the shape of the electrode tip, for example, by making it somewhat larger in combination with more heat polishing.

9. Information about the conductance distribution of single gap junction channels is usually collected by measuring current transitions, acquired at low junctional conductance values, as induced by halothane or heptanol. After calculation of the corresponding conductance steps, frequency histograms are constructed (*see* **Fig. 3C**). This method has several limitations. First, different conductance states of gap

junction channels might not be equally sensitive to the junctional uncouplers, so that the distribution of γ_j is distorted. Second, by measuring conductance transitions only, all information about time is lost. Although this seems to be a minor point, as information about channels spending time in the open or closed state is not reliable because uncouplers that influence the open probability are used, it might be of importance for the distribution of γ_j in frequency histograms. If, for example, phosphorylation of a gap junction channel increases the gating frequency of a channel conductance, in a way that the channel just opens and closes more often, but overall spends equal time in the open state, no change in macroscopic conductance will be measured between cells. In a frequency histogram, however, an increased frequency of these events will be seen, suggesting that the preferential open state is changed.

All points histograms (**Fig. 3D**) have the advantage that factor time is still included, thus allowing more detailed description of single-channel kinetics. However, they are rather sensitive to drift in baseline, which makes it harder to find different channel conductances. Comparing **Fig. 3C** and **D** shows that the 30 pS and 82 pS peaks in the frequency histogram (**C**) cannot be detected in the all-points histogram (**D**).

10. Binomial statistical approaches on poorly coupled cell pairs do give information on (changes) in open probability of junctional channels. The basic assumptions, that is, identical channels that gate independently between one closed and one open state, are, however, almost never met *(27)*. Gap junction channels usually have more than one conductance, sometimes with residual conductances *(13,16,20,22,37–39)*. These properties hamper proper modeling and thus fitting of the all-points histograms, and interfere with reliable determination of gap junction channel kinetics.

11. In normal (continuous) voltage clamp with one patch pipet per cell, voltage monitoring and current injection are performed simultaneously. This results in a voltage drop across the pipet series resistance, and the actual membrane potential differs from the command voltage. Some amplifiers are able to overcome the interaction between the voltage recording and current injection, using a time-sharing technique. Here voltage monitoring and current injection are separated in time. A sample of the membrane potential is taken and compared with the desired command voltage. Then the amplifier switches to current injection mode and injects a current proportional to the difference between membrane and command potential. The rate of rise of the membrane potential in the current injection phase is limited by the parasitic capacitance of the electrode. At the end of the current injection phase, the amplifier switches to voltage recording. At the end of the voltage recording period a new sample of the membrane potential is taken, and the amplifier switches to the current injection phase. The cycling rate must be set to a value of 10 or more cycles per membrane time constant, so that the membrane capacitance "smoothens" the membrane voltage response to the current pulses.

Although this recording method seems a very good way to overcome series resistance induced clamp errors, it has several pitfalls. The effect of series resistance is not ruled out. High series resistance values increase the electrode time constant and will slow the electrode response and result in an error in the membrane potential, because the current does not completely decay to baseline before the next voltage measurement. Changes in parasitic capacitance of the recording electrode during the experiment, due to changing bath solution levels, also influence the time constant of the electrode and thus the value of the measured membrane potential.

The errors in measured gap junctional conductance using a dual discontinuous voltage clamp setup are mainly determined by a combination of electrode time constant (and effective compensation thereof), sampling rate, and the membrane time constant. If all are properly set and compensated (and parameters do not change during the experiment), no errors due to series resistance will occur.

12. Nystatine perforated patch stock solution: Add 50 mg of pluronic F-127 to 1 mL of dimethyl sulfoxide (DMSO), at 37°C if pluronic dissolves poorly. Subsequently, add 50 mg of nystatin to this solution, and dissolve well by vortex-mixing and sonication of the solution. Aliquots of this stock (to avoid repeated freezing and thawing) can be stored at –20°C for several months. Add 1.4 µL of stock solution to 1 mL of IPS, to obtain the final concentration of 70 µg of nystatin and pluronic per milliliter of IPS.

13. To improve the perforation of the patch, observe the perforation process in voltage clamp, and slightly hyperpolarize the patch to approx –20 mV. Wait until the series resistance reaches acceptable values (*see* **Fig. 7B**).

14. An amphotericin B stock solution is prepared by dissolving 3 mg of amphotericin B in 50 µL of DMSO. This stock solution can be stored for several days at –20°C.

15. Patch pipets filled with amphotericin B–IPS seal very poorly to the membrane. Therefore the pipet is frontfilled with normal IPS. However, the amount of frontfilling (i.e., the time that the pipet tip is held in IPS) has to be determined experimentally. With too much normal IPS in the tip, the patch will not perforate; too little IPS will result in poor sealing of the electrode to the membrane and perforation of the membrane during the sealing process, so that the seal resistance cannot be determined. Start with 10 s of frontfilling.

References

1. Spray, D. C., Harris, A. L., and Bennett, M. V. L. (1979) Voltage dependence of junctional conductance in early amphibian embryos. *Science* **204,** 432–434.
2. Spray, D. C., Harris, A. L., and Bennett, M. V. L. (1981) Equilibrium properties of a voltage-dependent junctional conductance. *J. Gen. Physiol.* **77,** 77–93.
3. Harris, A. L., Spray, D. C., and Bennett, M. V. L. (1981) Kinetic properties of a voltage-dependent junctional conductance. *J. Gen. Physiol.* **77,** 95–117.
4. Hamill, O. P., Marty, A., Neher, E., Sakmann, B., and Sigworth, F. J. (1981) Improved patch-clamp techniques for high resolution current recording from cells and cell-free membrane patches. *Pflügers Arch.* **391,** 85–100.
5. Neyton, J. and Trautmann, A. (1985) Single-channel currents of an intercellular junction. *Nature* **317,** 331–335.

6. White, R. L., Spray, D. C., Campos de Carvalho, A. C., Wittenberg, B. A., and Bennett, M. V. L. (1985) Some electrical and pharmacological properties of gap junctions between adult ventricular myocytes. *Am. J. Physiol.* **249**, C447–C455.
7. Weingart, R. (1986) Electrical properties of the nexal membrane studied in rat ventricular cell pairs. *J. Physiol.* **370**, 267–284.
8. Rook, M. B., Jongsma, H. J., and Van Ginneken, A. C. G. (1988) Properties of single gap junctional channels between isolated neonatal rat heart cells. *Am. J. Physiol.* **255**, H770–H782.
9. Giaume, C. (1991) Application of the patch-clamp technique to the study of junctional conductance, in *Biophysics of Gap Junction Channels* (Peracchia, C., ed.), CRC Press, Boca Raton, FL, pp. 175–190.
10. Veenstra, R. D. and Brink, P. R. (1992) Patch-clamp analysis of gap junctional currents, in *Cell-Cell Interactions: A Practical Approach* (Stevenson, B., Galling, W. J., and Paul, D. L., eds.), IRL Press, Oxford UK, pp. 167–201.
11. Tyrode, M. V. (1910) The mode of action of some purgative salts. *Arch. Int. Pharmacodyn.* **20**, 205–223.
12. Loewenstein, W. R., Kanno, Y., and Socolar, S. J. (1978) Quantum jumps of conductance during formation of membrane channels at cell-cell junction. *Nature* **274**, 133–136.
13. Bukauskas, F. F. and Weingart, R. (1993) Multiple conductance states of newly formed single gap junction channels between insect cells. *Pflügers Arch.* **423**, 152–154.
14. Weingart, R. and Bukauskas, F. F. (1993) Gap junction channels of insect cells exhibit a residual conductance. *Pflügers Arch.* **424**, 192–194.
15. Bukauskas, F. F. and Weingart, R. (1994) Voltage-dependent gating of single gap junction channels in an insect cell line. *Biophys. J.* **67**, 613–625.
16. Bukauskas, F. F., Elfgang, C., Willecke, K., and Weingart, R. (1995) Biophysical properties of gap junction channels formed by mouse Connexin40 in induced pairs of transfected human HeLa cells. *Biophys. J.* **68**, 2289–2298.
17. Opsahl, L. R. and Webb, W. W. (1994) Lipid-glass adhesion in giga-sealed patch-clamped membranes. *Biophys. J.* **66**, 75–79.
18. Van Rijen, H. V. M., Wilders, R., Van Ginncken, A. C. G., and Jongsma, H. J. (1998) Quantitative analysis of dual whole-cell voltage-clamp determination of gap junctional conductance. *Pflügers Arch.* **436**, 141–151.
19. Reed, K. E., Westphale, E. M., Larson, D. M., Wang, H. Z., Veenstra, R. D., and Beyer, E. C. (1993) Molecular cloning and functional expression of human connexin37, an endothelial cell gap junction protein. *J. Clin. Invest.* **91**, 997–1004.
20. Veenstra, R. D., Wang, H. Z., Beyer, E. C., Ramanan, S. V., and Brink, P. R. (1994) Connexin37 forms high conductance gap junction channels with subconductance state activity and selective dye and ionic permeabilities. *Biophys. J.* **66**, 1915–1928.
21. Veenstra, R. D., Wang, H. Z., Beyer, E. C., and Brink, P. R. (1994) Selective dye and ionic permeability of gap junction channels formed by Connexin45. *Circ. Res.* **75**, 483–490.

22. Beblo, D. A., Wang, H. Z., Beyer, E. C., Westphale, E. M., and Veenstra, R. D. (1995) Unique conductance, gating, and selective permeabilities properties of gap junction channels formed by Connexin40. *Circ. Res.* **77,** 813–822.

23. Burt, J. M. and Spray, D. C. (1988) Intropic agents modulate gap junctional conductance between cardiac myocytes. *Am. J. Physiol.* **254,** H1206–H1210.

24. Bastiaanse, E. M. L., Jongsma, H. J., Van der Laarse, A., and Takens-Kwak, B. R. (1993) Heptanol-induced decrease in cardiac gap junctional conductance is mediated by a decrease in the fluidity of membranous cholesterol-rich domains. *J. Membr. Biol.* **136,** 135–145.

25. Rook, M. B., Van Ginneken, A. C. G., De Jonge, B., El Aoumari, A., Gros, D., and Jongsma, H. J. (1992) Differences in gap junction channels between cardiac myocytes, fibroblasts, and heterologous pairs. *Am. J. Physiol.* **263,** C959–C977.

26. Chanson, M., Chandross, K. J., Rook, M. B., Kessler, J. A., and Spray, D. C. (1993) Gating characteristics of a steeply voltage-dependent gap junction channel in rat schwann cells. *J. Gen. Physiol.* **102,** 925–946.

27. Manivannan, K., Ramanan, S. V., Mathias, R. T., and Brink, P. R. (1992) Multichannel recordings from membranes which contain gap junctions. *Biophys. J.* **61,** 216–227.

28. Ramanan, S. V. and Brink, P. R. (1993) Multichannel recordings from membranes which contain gap junctions. II. Substates and conductance shifts. *Biophys. J.* **65,** 1387–1395.

29. Moreno, A. P., Eghbali, B., and Spray, D. C. (1991) Connexin32 gap junction channels in stably transfected cells. Equilibrium and kinetic properties. *Biophys. J.* **60,** 1267–1277.

30. Wilders, R. and Jongsma, H. J. (1992) Limitations of the dual voltage clamp method in assaying conductance and kinetics of gap junction channels. *Biophys. J.* **63,** 942–953.

31. Rook, M. B., De Jonge, B., Jongsma, H. J., and Masson-Pévet, M. A. (1990) Gap junction formation and functional interaction between neonatal rat cardiocytes in culture: a correlative physiological and ultrastructural study. *J. Membr. Biol.* **118,** 179–192.

32. Horn, R. and Marty, A. (1988) Muscarinic activation of ionic currents measured by a new whole-cell recording method. *J. Gen. Physiol.* **92,** 145–159.

33. Marty, A. and Finkelstein, A. (1975) Pores formed in lipid bilayer membranes by nystatin. Differences in its one-sided and two-sided action. *J. Gen. Physiol.* **65,** 515–526.

34. Holz, R. and Finkelstein, A. (1970) The water and nonelectrolyte permeability induced in thin lipid membranes by the polyene antibiotics nystatin and amphotericin B. *J. Gen. Physiol.* **56,** 125–145.

35. Penner, R. (1995) A practical guide to patch clamping, in *Single-Channel Recording* (Sakmann, B. and Neher, E., eds.), Plenum Press, New York and London, pp. 3–30.

36. Barry, P. H. (1994) JPCalc, a software package for calculating liquid junction potential corrections in patch-clamp, intracellular, epithelial and bilayer measure-

ments and for correcting junction potential measurments. *J. Neurosci. Methods* **51,** 107–116.

37. Kwak, B. R., Van Veen, T. A. B., Analbers, L. J. S., and Jongsma, H. J. (1995) TPA increases conductance but decreases permeability in neonatal rat cardiomyocyte gap junction channels. *Exp. Cell Res.* **220,** 456–463.

38. Moreno, A. P., Fishman, G. I., and Spray, D. C. (1992) Phosphorylation shifts unitary conductance and modifies voltage dependent kinetics of human connexin43 gap junction channels. *Biophys. J.* **62,** 51–53.

39. Moreno, A. P., Sáez, J. C., Fishman, G. I., and Spray, D. C. (1994) Human connexin43 gap junction channels. *Circ. Res.* **74,** 1050–1057.

40. Spray, D. C., Chanson, M., Moreno, A. P., Dermietzel, R., and Meda, P. (1991) Distinctive gap junction channel types connnect WB cells, a clonal cell line derived from rat liver. *Am. J. Physiol.* **260,** C513–C527.

41. Kwak, B. R., Hermans, M. M. P., De Jonge, H. R., Lohmann, S. M., Jongsma, H. J., and Chanson, M. (1995) Differential regulation of distinct types of gap junction channels by similar phosphorylating conditions. *Mol. Biol. Cell* **6,** 1707–1719.

42. Grinstein, S., Romanek, R., and Rotstein, O. D. (1994) Method for manipulation of cytosolic pH in cells clamped in the whole cell or perforated patch configurations. *Am. J. Physiol.* **267,** C1152–C1159.

16

Determining Ionic Permeabilities of Gap Junction Channels

Richard D. Veenstra

1. Introduction

The dual whole-cell patch-clamp technique has enabled investigators to measure single gap junction channel currents from a variety of primary cell cultures and communication-deficient connexin-transfected cell lines *(1)*. Over the past 10 years, dual whole-cell recordings of junctional currents have contributed significantly to our understanding of the conductance and gating of gap junctions formed by one or more connexins. Only recently have direct approaches to the quantitative measurement of ionic or molecular permeabilities of these same connexin gap junction channels been achieved *(2,3)*. One difficulty of obtaining reliable and reproducible measurements of junctional ionic permeabilities results from the requirement of establishing an internal bionic environment while maintaining the integrity of both coupled cells. Since a true bionic potential cannot be established between two living cells in comparison to an isolated channel in a membrane patch or artificial lipid bilayer, the actual ionic reversal potential results predominately from the asymmetric ionic gradients for the two primary ions X and Y being investigated and the symmetrical ionic gradients of all other permeant ions in the internal solution that serve to keep the ionic potential near 0 mV. Another difficulty in measuring the ionic reversal potential from which the relative permeability ratios are calculated is that one does not have an absolute reference to the bath ground when recording in the dual whole-cell configuration. This adds another degree of complexity to the accurate measurement of junctional currents using the dual whole-cell patch-clamp technique.

To begin to determine ionic permeability ratios requires the ability to accurately assess the junctional currents and voltage in a cell pair under symmetrical

From: *Methods in Molecular Biology, vol. 154: Connexin Methods and Protocols*
Edited by: R. Bruzzone and C. Giaume © Humana Press Inc., Totowa, NJ

ionic conditions as the oppositely directed X and Y ionic potentials provide an additional junctional ionic current independent of the applied command potentials of the two voltage clamp amplifiers to both cells. It is the accurate assessment of the net ionic current flow resulting from ions X and Y in the presence of any other permeant ions W, Z, etc. that one must achieve to successfully calculate the relative ionic permeabilites of any two ions. This necessitates the determination of the ionic permeability ratios of all permeant ions present in the internal solution relative to one ion chosen to be the common ion present in all asymmetric ionic potential experiments. The best check to determine if the experimental ionic permeability ratios calculated for each ionic pair satisfy the condition of accounting for all other permeant ions is to solve for all of the relative ionic permeability ratios simultaneously. In this chapter, the methods we have used to determine the monovalent cation permeability ratios of rat connexin43 and connexin40 gap junction channels and the potential difficulties of determining ionic reversal potentials in gap junction channels permeant to both cations and anions simultaneously are described.

2. Materials

The three basic components needed to perform the ionic permeability measurements of gap junction channels are a source of in vitro cell pairs, a dual whole-cell recording system, and a series of internal pipet solutions necessary to create the asymmetric ionic conditions for all permeant ions to be investigated and any permeant counterion that must be present. The possible sources of in vitro cell pairs are too numerous to mention in specific detail, but there are three principal sources of in vitro cell cultures. One source is the primary culturing of native tissues and is desirable if one wishes to investigate the physiological permeability of native gap junctions. There are two other sources of communication-competent cells, often with more defined levels of connexin expression: communication-competent established cell lines or connexin-transfected communication-deficient established cell lines. One important consideration is that the in vitro cell pair must be amenable to whole-cell voltage clamp by patch-clamp amplifiers, that is, have a high input resistance in the whole-cell recording configuration (parallel combination of gigaohm [$G\Omega$] seal resistance and cellular membrane resistance) and a moderate to high junctional resistance relative to the electrode series resistance to reduce transjunctional voltage errors (*1*). The latter can be corrected for by calculating the accuracy of the voltage clamp of both cellular membrane voltages which requires knowledge of the series resistance of each electrode and cellular input resistances. Large errors in membrane voltage control, due to a high series resistance relative to the cell input resistance caused by either a high series resistance or low input resistance in one cell, are not readily compensated for in permeability measurements, as the partner cell will partially clamp the membrane voltage of the poorly clamped cellular potential. This condition produces a DC current across the junction at all membrane potentials other than the resting potential for both cells (i.e., 0 mV for a "leaky cell"). Internal dialysis is vital to establishing the proper asymmetric ionic gradients; therefore geometries that are too large or complex in three dimensions should be avoided. Small spherical cells are best suited to these investigations on gap junction channels. Because the highest possible resolution of any ionic reversal potential (E_{rev}) is obtained from single-channel currents, it is desirable to use cell pairs with a high junctional resistance (R_j). This is not readily controlled in primary cell cultures or stable transfection systems to date because the use of inducible promotors for connexin expression remains unpublished. Average junctional resistances of 150–500 MΩ were obtained using the eukaryotic expression vector pSFFV-neo for stable transfection of rat neuro2A (N2A) neuroblastoma cells employed routinely in this laboratory *(2–4)*. Single gap junction channel recordings were obtained in 10–20% of connexin-transfected N2A cell pairs without pharmacological intervention using this approach. Pharmacological uncouplers should be avoided initially because the effect on the connexin channels is undetermined and may affect permeability (by altering conductance states in addition to gating properties, i.e., open probability). A list of necessary equipment and ionic salts to perform the necessary experiments is provided in the following subheading. The choice as to the exact type or model of amplifier, micromanipulator, patch clamp glass, etc. is generally determined by the preference of the electrophysiologist performing the experiments. The only major requirement is that they adequately perform their specified task effectively (*see* **Note 1**).

2.1. Equipment for Dual Whole-Cell Patch-Clamp of Transjunctional Voltage

1. Two patch clamp amplifiers capable of either single-channel or whole-cell voltage clamp.
2. Two three-dimensional micromanipulators with fine and coarse movements. For spherical cells, a 45° angle of approach with the patch electrodes is optimal. Low drift, especially in the vertical plane, and vibration isolation during translocation of the patch electrode are important features to ensure stable recordings over tens of minutes and enhance the success of junctional current recordings. Either DC motor-driven stages or dampened hydraulic (preferably water) are acceptable.
3. One inverted phase-contrast microscope for viewing the preparation and both patch electrodes. Ultralong working distance 40X objective lenses are best for the final establishment of gigaohm (GΩ) seals onto each cell membrane. Establishment of the whole-cell configuration can be achieved by either transient capacitance overcompensation ("buzz" circuit feature optional on some patch-clamp amplifiers) or mechanical disruption by negative pressure (suction) applied to the patch electrode.

Perforated patch recording techniques may not be suitable for ionic permeability ratio experiments since the permeability of a test ion through the nystatin or amphotericin B channels may be limiting, particularly when larger diameter ions are to be used (*see* **Note 2**).

4. One voltage stimulator for running the voltage-clamp protocols to be used. Two synchronized voltage command channels are preferred, as this enables the investigator to control the holding potential of both cells independently without having to switch on and off the external command input to one of the cells to achieve a transjunctional voltage clamp.

5. A digital recording device capable of sampling at least a 5-kHz analog bandwidth signal. A minimum of two channels are required and four recording channels is optimal so that current and voltage can be stored for both cells. Whether one chooses a computer-driven or standalone off-line A/D recorder is up to the discretion of the investigator (*see* **Note 3**).

6. One digital storage oscilloscope for monitoring the real-time current recordings. Display of both whole-cell currents is required for dual whole-cell recording of junctional currents, so at least two channels are required.

7. One vibration isolation table on which the microscope and micromanipulators for patch electrode positioning is required. This is standard equipment for any patch-clamp system, as is shielding of external electrical noise by proper grounding and containment of the headstage amplifiers and recording chamber with bath reference and patch recording electrodes within a Faraday cage.

2.2. Solutions for Ionic Permeability Measurements

2.2.1. External Solutions

A serum and protein-free external bath solution is required for any patch clamp recording from live cells in vitro. A physiological saline is typically prepared for use as this will maintain the viability of all cells in the recording chamber for a few hours. It is preferable to keep plasma membrane resistance high to maximize the junctional current signal-to-noise ratio in the dual whole-cell recording configuration, so high K^+ solutions in general should be avoided. The precise composition of the saline will depend on what membrane currents are present in the cells being used for the experiments. For mammalian cells 140 mM NaCl, 1.3 mM KCl, 1.8 mM CaCl$_2$, 0.8 mM MgSO$_4$, 0.9 NaH$_2$PO$_4$, 5.5 mM dextrose, 10 mM 4-2(hydroxyethyl)-1-piperazineethanesulfonic acid (HEPES), and pH 7.4 (titrated with 1 N NaOH) is a typical saline composition. The osmolarity of the standard saline is ≈290 milliosmoles (mOsm) when low [K^+] is used in place of physiological plasma [K^+] of 4.3 mM. This reduces background K^+ currents in the whole-cell membrane. Trace amounts of ionic blockers of background membrane currents such as 10 mM CsCl and/or tetraethylammoniumCl (TEACl) for K^+ currents and 100 μm ZnCl$_2$ for Cl$^-$ currents can be added if necessary.

2.2.2. Internal Pipet Solutions

1. A standard internal pipet solution (IPS) consists of a 120 mM K$^+$ monovalent salt, ≈ 200 nM free Ca^{2+}, ≈ 1 mM free Mg^{2+}, 3 mM ATP, and pH 7.2–7.4. Physiological Cl$^-$ and Na$^+$ are 12–14 mM inside mammalian cells, but it is optimal to use as few monovalent cations and anions in the standard IPS as is necessary when performing ionic reversal potential experiments. For elimination of Cl$^-$ and Na$^+$ from your IPS, *see* **Note 4**. If the experimental objective is to compare monovalent cation permeabilities, a single primary anion should be chosen. As 6–12 mM Cl$^-$ is present from the 1 mM free MgCl$_2$ and the 3–5 mM CaCl$_2$-EGTA or BAPTA buffer, Cl$^-$ is the best choice for the standard monovalent anion. Another reason for using Cl$^-$ salts is that an Ag/AgCl$_2$ wire junction is typically used to make electrical contact with the IPS in the patch electrode and high Cl$^-$ will improve the stability and minimize the electrode voltage offset. Because the external saline is predominately Cl$^-$ (≈ 150 mM), the electrode voltage offset with the bath is also minimized by using a Cl$^-$ IPS. This ensures that only a single permeant monovalent anion is present, and therefore only one cation/anion permeability ratio needs to be determined.

2. To determine the Eisenman selectivity sequence for the earth alkali metals, ≥ 98% pure RbCl, CsCl, KCl, NaCl, and LiCl are needed. The ammonium series of organic cations beginning with ammoniumCl (NH$_4$Cl), tetramethylammoniumCl (TMACl), and tetraethylammoniumCl (TEACl) are tolerated at intracellular concentrations of 120–140 mM and are suitable candidates for an IPS, but tetrabutylammoniumCl (TBACl) is too hydrophobic at these concentrations to maintain plasma membrane integrity for more than a few minutes. High-quality crystalline salts of these chemicals are readily available from any commercial supplier (e.g., Sigma Chemical, Fisher Scientific, Fluka). TetrapropylammoniumCl (TPACl) is also readily available but I do not know if it can be used as an IPS, as TPA was not previously tested experimentally. All of the tetraalkylammonium ions are irritants and become increasing hygroscopic with increasing molecular weight, so gloves and goggles should be used when handling these compounds and they must be stored in a desiccator at room temperature.

3. To achieve submicromolar [Ca^{2+}] in the IPS requires the use of a calcium chelator such as EGTA or BAPTA. The advantages for using BAPTA are that BAPTA binds Ca^{2+} faster, is more selective for Ca^{2+} over Mg^{2+}, and is less sensitive to pH changes. The major disadvantage of using BAPTA instead of EGTA is that it costs at least 10 times more than EGTA. The ionic strength and the monovalent and divalent ions in the IPS plus any other chelators must be considered when using either of these Ca^{2+} buffers. Hence, it is advisable to obtain a computer program to calculate the amount of EGTA or BAPTA to use and how much CaCl$_2$ should be added to achieve the desired free [Ca^{2+}] in the IPS *(5)*. In general, for a 290–300 mOsm solution at room temperature and pH 7.4, adding 3 mM CaCl$_2$ and 5 mM BAPTA with 1 mM free Mg^{2+} will yield 200–300 nM free [Ca^{2+}]. Both EGTA and BAPTA can be purchased as a free acid or as a Na$_4$ or K$_4$ salt, respectively (*see* **Note 5**).

4. To achieve 1 mM free Mg^{2+}, 1 mM $MgCl_2$ should be added to the IPS.

5. Because ATP chelates Mg^{2+}, it is advantageous to use the MgATP salt instead of the Na_2ATP salt for any internal IPS. This should be added fresh to each IPS on the day of the experiment. MgATP is either added fresh as a powder (stored at –20°C in a desiccator) or prepared as a 200X stock solution dissolved in the appropriate IPS and stored frozen at –20°C until use.

6. HEPES is commonly used as a pH buffer and 25 mM HEPES is used in the IPS to ensure that this remains the predominant pH buffer once the IPS internally dialyzes the cell. A 10 mM concentration of HEPES is more commonly used in IPSs for whole-cell recording. Again, the pH should be titrated with 1 N base or acid corresponding to the primary monovalent ionic salt (e.g., 1 N KOH or HCl for a KCl-containing IPS).

7. Final preparation of each stock IPS begins with 3 mM $CaCl_2$, 5 mM BAPTA (free acid or Na_4 or K_4 salts, *see* **Note 5**), 1 mM $MgCl_2$, and 10 or 25 mM HEPES plus the desired concentration of the monovalent ionic salt. For each IPS, 140 mM monovalent salt is the standard. Use only ultrapure deionized water (18 MΩ-cm) for dissolving the salts to achieve the final volume of 100 mL (or multiples thereof). Carefully titrate the pH to 7.4 with the appropriate base ($\approx$ 2 mL of 1 N XOH per 100 mL of IPS for alkali earth metals). The osmolarity of this IPS will approximate 290 milliosmoles and should be adjusted to match the osmolarity of the external saline. If ionic blockers are to be used, they should be added in equal amounts to the saline and the IPS to maintain osmotic balance. The effect of the final addition of 3 mM MgATP to the IPS on the day of the experiment is neglected although it will raise the osmolarity and lower the pH slightly ($\leq$ 1%).

8. To create the asymmetric ionic gradients for two monovalent cations, a known amount of each ion X and Y should be added to the opposite side to establish a known Nernst potential for each ion. It is important that this low ionic concentration remain constant during the experiment; otherwise the Nernst potential will change owing to electrodiffusion. At room temperature a 10:1 concentration gradient for a monovalent ion will produce an equilibrium potential of $\approx$ ±58 mV at 20°C. The larger the opposing ionic gradients, the more accurate the measurement of the relative permeability ratio will be, so smaller ionic gradients will reduce the sensitivity and larger ionic gradients may be too large to measure if the relative permeability of one ion is low relative to the other. For example, a 10:1 relative permeability ratio will result in a bionic reversal potential equal to 70% of the Nernst potential for the more permeant ion, or ±40 mV for a 10:1 ionic gradient. Since all other ions in the IPS are constant, only the two ions of interest will be diluted by reciprocal 10:1 dilutions and the final concentrations of X and Y can be obtained by multiplying the initial concentration by the volume fraction in the final experimental IPS. This procedure can be used for any series of ions of equal valence, that is, all cations or all anions.

9. If a channel is simultaneously permeable to both cations and anions, then the cation/anion permeability ratio must be determined for the ionic salt used as the standard reference for the relative ionic permeability ratio sequence to be deter-

mined for the solutions prepared in **step 8**. To perform this measurement, the concentration of the primary ionic salt must be varied by a known factor such as 2, 4, or 10. Osmotic equilibrium must then be maintained by adding an inert solute to one IPS. A 50% reduction in the primary 140 mM monovalent salt will require $\geq$ 140 mM sugar to sustain the original osmolarity of the IPS. The Nernst potential for a 2:1 monovalent salt gradient is $\pm$17.5 mV and $\pm$34.9 mV for a 4:1 salt gradient at 20°C. Raffinose (a trisaccharide, Gal-Glu-Fru) and stachyose (a tetrasaccharide, Gal-Gal-Glu-Fru) were found to be impermeant in all gap junction experiments performed to date whereas mannose or sucrose may be permeant in some connexin channels. At least a trisaccharide is recommended for low-salt IPS preparations to ensure that a permeant sugar does not displace ions from the channel pore and have possible undesirable effects on ionic conductance and permeability in addition to creating an unstable asymmetric ionic gradient. All of these sugars are commercially available (e.g., Sigma Chemical).

3. Methods

3.1. Ionic Reversal Potential Measurements from Homotypic Gap Junctions

Conventional dual whole-cell recording methods are all that is required to determine an ionic reversal potential under asymmetric ionic conditions. Proper voltage clamp control of the junction requires that the series resistance (R_s) to input resistance (R_{in}) ratio (R_s/R_{in}) be equal for the two cells in order for $V_{mA} = V_{mB}$ when $V_A = V_B$. This condition is essential; otherwise a voltage offset will exist across the junction independent of the established ionic gradient. This voltage offset will be additive to the ionic reversal potential of interest in the dual whole-cell configuration and will result in errors in the reversal potential measurement if not corrected for experimentally using the expression $V_j = V_A - V_B - [(I_A \cdot R_{sA}) - (I_B \cdot R_{sB})]$. Hence, it is critical to maintain a low series resistance throughout the experiment to accurately measure the ionic reversal potential. In general, an R_s of $\leq$5% of R_{in} or R_j is preferable and <1% is optimal (i.e., a 10 MΩ whole-cell patch electrode requires an R_{in} and $R_j \geq 1$ GΩ) (*see* **Note 6**). The whole-cell capacitive transient should be monitored occasionally by simultaneously stepping V_A and V_B by 5–10 mV from a common negative holding potential within the linear range of R_{in} (e.g., –40 or –80 mV). The time constant (τ_{cap}) of a single whole-cell capacitive transient equals $R_s \cdot C_{in}$ and the input capacitance can be determined by integrating the area under the capacitive transient for a passive membrane (*6*). The R_{in} must also be assessed to determine the ratio of R_s/R_{in} necessary in calculating junctional current as the difference (ΔI) in I_B when $V_A = V_B$ and when $V_A - V_B = \Delta V_A$ according to the expression $I_j = -\Delta I_B \cdot (1 + R_{sB}/R_{inB})$ (*see* **Note 7**). It is advantageous to determine R_{in} over the entire voltage range encountered during the experiment, so a voltage ramp from -100 mV to $+60$ mV should be applied simultaneously to both cells and R_{in} can be determined from the slope conductance of the $I–V$ relationship. Another advantage this procedure provides is the ability to chose a common holding potential $\pm\Delta V$ that corresponds to the linear part of the $I–V$ relationship for both cells. When these conditions are met, a dual whole-cell recording of unitary junctional currents will result in a linear $I–V$ plot where the slope conductance equals the single-channel conductance (γ_j) of the gap junction channel under the specified ionic conditions. An example is illustrated in **Fig. 1**. There should be no voltage offset in this graph under symmetrical ionic conditions if one is to be successful in accurately determining ionic reversal potentials (E_{rev}) under asymmetric ionic conditions in future experiments.

Once these conditions are established, the only other experimental variable that needs to be introduced is the use of an asymmetric ionic gradient. The disadvantage of the dual whole-cell recording configuration is that without the ability to exchange the IPS inside of one patch electrode (test electrode) (*see* **Note 8**), one cannot immediately determine the value of $I_j = E_{rev}/R_j$ when $V_A = V_B$ and all other dual whole-cell recording conditions are satisfied. Under true bionic conditions, $E_{rev} = +(RT/F) \cdot \log_{10}[(\{P_X/P_Y\} \cdot X_A + Y_A)/(\{P_X/P_Y\} \cdot X_B + Y_B)]$ when X and Y are the concentration of both cations (and $-RT/F$ when both X and Y are anions) or $E_{rev} = +(RT/F) \cdot \log_{10}[(\{P_X/P_Y\} \cdot X_A + Y_B)/(\{P_X/P_Y\} \cdot X_B + Y_A)]$ when X is a cation and Y an anion. Unfortunately, the concentration of all other permeant monovalent cations and anions in both IPSs must also appear in these expressions according to their valence. The value of E_{rev} is determined experimentally from the single-channel $I_j–V_j$ plot. An example is illustrated in **Fig. 2**. If γ_j or P_{ion} of either ion varies by more than one order of magnitude from the other, this plot will be noticeably curvilinear and the data

---

Fig. 1. (*Opposite page*) Whole cell currents from an N2A cell pair transfected with rat connexin43 (rCx43). Gap junction channel currents recorded using an internal pipet solution (IPS) containing 135 mM monovalent cations and 136 mM monovalent anions. For this example $K^+ = 116$ mM, $Cs^+ = 5$ mM; $Na^+ = 12$ mM; $TEA^+ = 2$ mM; and $Cl^- = 136$ mM. (**A**) A ΔV of -50 mV was applied to cell A from a common holding potential of $V_A = V_B = 0$ mV for 2 min. A 20-s interval of the resulting I_A and I_B currents, where $I_j = -\Delta I_B$ as indicated by the difference between the current levels 1 and 2 in this recording. (**B**) All points amplitude histogram of all 120 s of the net junctional current, I_j ($= -\Delta I_B$), from the -50 mV transjunctional voltage (V_j) pulse in **A**. Single-channel current, i_j, was -5.0 pA. Open-channel noise variance was 0.35 pA and open probability was 0.085 ($= 10.2$ s cumulative open time). (**C**) A ΔV of $+50$ mV was applied to the same cell pair as in **A** for 2 min. This 20-s interval illustrates two open channels with amplitudes indicated by the difference currents between levels 1, 2, and 3. (**D**) All

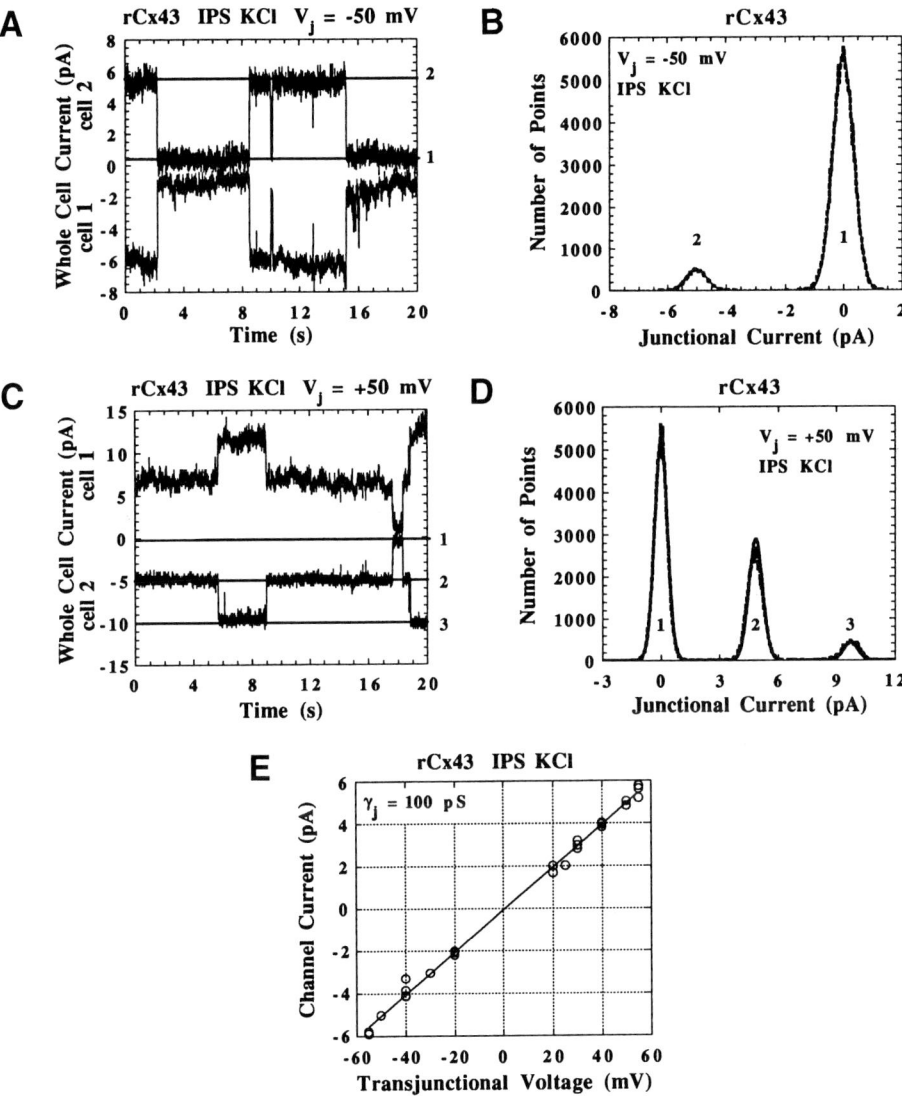

points amplitude histogram of all 120 s of net I_j from the +50 mV V_j pulse in **C**. Single channel current amplitudes were +4.8 and +4.95 pA. Open-channel noise variance was 0.25 pA and open probabilities were 0.20 and 0.25 (cumulative open times of 24 and 30 s). (**E**) Single gap junctional channel current–voltage relationship for the rCx43 cell pair shown in this figure. Every i_j value obtained from the all-points histogram for each 2-min V_j pulse is represented by an *open circle*. The line is a linear regression fit of the experimental data ($R = 0.99$) with a slope conductance (γ_j) of 100 pS. (Reproduced with permission from **ref. 2**.)

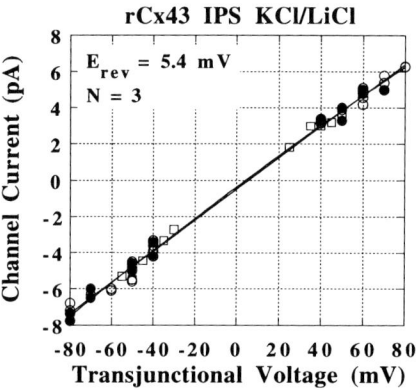

Fig. 2. Single gap junctional channel current–voltage relationship from three different cell pairs. Cell A contained LiCl IPS with a 100:1 dilution of KCl IPS and cell B contained the opposite KCl IPS with a 100:1 dilution of LiCl IPS (*see* **Table 1**). Each datum point represents the i_j value obtained from a 2-min V_j pulse as illustrated in **Fig. 1**. The *solid line* is a linear regression fit of the experimental data ($R = 0.9975$) with an *x*-intercept (E_{rev}) of +5.4 mV. This corresponds to a relative K^+/Li^+ permeability ratio of 1.35 (*see* **Table 1**). A second-order polynomial fit, assuming a curvilinear *I–V* relationship, is also shown as a slightly downward concave curve ($R = 0.9976$) and has an E_{rev} of +4.0 mV. The corresponding relative K^+/Li^+ permeability ratio is 1.27 assuming a modified relative Cl^-/Li^+ permeability ratio of 0.30 (*see* **Fig. 3**). (Reproduced with permission from **ref. 2**.)

should be fit with a second-order polynomial rather than a straight line. The correlation coefficient of both linear regression and second-order polynomial can be used to determine if one fit is statistically better than the other. The proper fit is important because the intercept point on the voltage axis is the experimental determination of E_{rev}. This procedure must be performed for every test and permeant ion present in the standard IPS. In the example shown in **Fig. 2**, there is no statistical difference in the correlation coefficients for the linear and second-order polynomial fits of the single-channel current–voltage plot, yet extrapolation of the two fits to the zero current value reveals a 1.4 mV difference in the E_{rev} value, which translates into a slight change in the published $P_{K/Li}$ value from 1.35 to 1.27.

The relative cation/anion permeability ratio must be determined using the low-salt IPS described in **step 9** in **Subheading 2.2.** of this chapter. An example of one unilateral low-salt experiment is illustrated in **Fig. 3**. Again the single-channel I_j–V_j plot is generated using the same procedures as before, only now there exists a unidirectional salt gradient for the primary cation and anion which results in equivalent Nernst potentials of opposite sign owing to the opposite

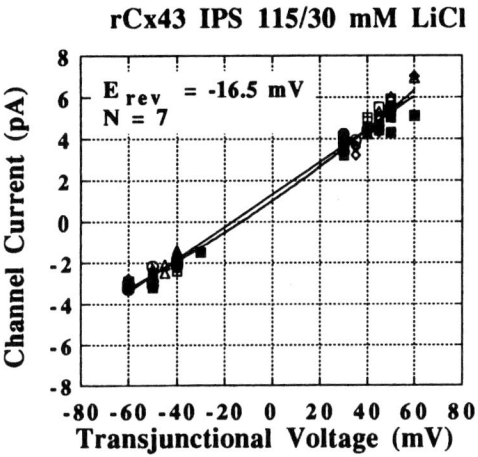

Fig. 3. Single gap junctional channel current–voltage relationship from seven different cell pairs. Cell A contained LiCl IPS and cell B contained 30 mM LiCl IPS with 169 mM raffinose added to maintain osmotic balance (*see* **Table 1**). The linear regression fit of the data had an E_{rev} of -16.5 mV ($R = 0.9939$). This corresponds to a relative Cl$^-$/Li$^+$ permeability ratio of 0.18 (*see* **Table 1**). A second-order polynomial fit , assuming a curvilinear I–V relationship, is also shown as a slightly upward concave curve ($R = 0.9944$) and has an E_{rev} of -12.5 mV. The corresponding relative Cl$^-$/Li$^+$ permeability ratio is 0.30 assuming a modified relative K$^+$/Li$^+$ permeability ratio of 1.27 (*see* **Fig. 2**). (Reproduced with permission from **ref. 2**.)

valence of ions X and Y. Again the data should be fitted with a straight-line or second-order polynomial to provide the best determination of E_{rev} from the voltage axis intercept. Once these experimental E_{rev} values are obtained for all of the ions of interest, all that remains is to calculate the relative ionic permeability ratios for each ionic combination. In the example shown in **Fig. 3**, there is no statistical difference in the correlation coefficients for the linear and second order polynomial fits of the single-channel current–voltage plot, yet extrapolation of the two fits to the zero current value reveals a 4.0 mV difference in the E_{rev} value which translates into a change in the published $P_{Cl/Li}$ value from 0.18 to 0.30. The difference between the polynomial and linear regression fits is more obvious in this case, as one side contains only 25% of the monovalent ions yet a curvilinear I–V relationship is not perceptible by eye. The only way to address this issue is to obtain a continuous channel I–V by running a voltage ramp during a stable channel opening to obtain channel currents at values too small to measure from discrete channel openings according to the procedures illustrated in **Fig. 1**.

3.2. Calculation of Relative Ionic Permeability Ratios

To solve for the exact set of relative ionic permeabilities from the experimental E_{rev} values obtained for the Cl⁻ salts of Rb⁺, Cs⁺, K⁺, Na⁺, Li⁺, TMA⁺, and TEA⁺ a matrix such as that shown in **Table 1** should be constructed. This is a rather conventional set of cations to use to determine the cation selectivity sequence for a gap junction channel, as the largest ion, TEA⁺, has a radius of approx 8 Å *(7)*. Ammonium, NH_4^+, probably should be included in this series as it has an aqueous mobility identical to that of K⁺ and is the precursor molecule for the other tetraalkylammonium cations. In the example provided in **Table 1**, Na_2ATP, 5 m*M* CsCl, and 2 m*M* TEACl were used in every IPS. LiCl was used as the standard IPS because its known aqueous mobility was intermediate in value. In accordance with the procedures outlined in **Subheading 2.2.**, 5 m*M* CsCl and 2 m*M* TEACl were also added to the external bath saline and the final osmolarity was 310 milliosmoles for all solutions. MgATP is recommended and, when used in place of Na_2ATP, reduces the amount of added $MgCl_2$ necessary to yield 1 m*M* free Mg^{2+} from 4 to 1 m*M*, thus reducing the final IPS [Cl⁻] by 6 m*M*. Divalent cations and anions were not considered in the calculations because the free [ion] of any divalent cation (Ca^{2+}, Mg^{2+}) or anion (HPO_4^{2-}, SO_4^{2-}) was 1.0 m*M* or less. Osmotic balance was achieved by the addition of 169 m*M* raffinose to the 30 m*M* LiCl IPS and was confirmed by the use of stachyose in one experiment. A 100:1 IPS_A/IPS_B dilution was used in this example.

Initial estimates of P_{ion} are required for the initial calculations of E_{rev}[a] and it is convenient to use the relative aqueous mobilities or diffusion coefficients ($D_{ion} = (RT/F) \cdot \mu_{ion}$) of the standard and test ions as a starting point. These relative values are: Rb⁺, 2.01; Cs⁺, 2.00; K⁺, 1.90; Na⁺, 1.29; NH_4^+, 1.90; TMA⁺, 1.16; TEA⁺, 0.84; and Cl⁻, 1.97 *(8)*. It is evident from the close agreement between the measured E_{rev} and calculated E_{rev} that a single value of P_{ion} satisfies all of the ionic conditions encountered by the rat Cx43 channel. This approach yields a more rigorous determination of P_{ion} for a gap junction channel than an estimate obtained for one pair of ions. The use of the Goldman–Hodgkin–Katz (GHK) voltage equation implies that electrodiffusion occurs independently of any other permeant ion. This paradigm is often untrue and other methods must be invoked to calculate an E_{rev} (and P_{ion}) value should the GHK equation fail to reproduce the observed set of E_{rev} values (*see* **Note 9** and **Table 2**).

4. Notes

1. The preference for the choice of equipment is also influenced by the cost and availability of the item. Hence, the choice of antivibration table, patch-clamp amplifiers, patch electrode glass, vibration isolation table, inverted microscope, etc. varies dependent on location and vendor reliability and price contracts, etc. as well as on the electrophysiologist's training and experience.

Table 1
Matrix for Calculating P_{ion} in the Cx43 Gap Junction Channel

Ion	Measured E_{rev} (mV)	Calculated E_{rev}^a (mV)	z	[ion]$_A$ (mM)	[ion]$_B$ (mM)	P_{ion}	Numerator[b]	Denominator[c]
Rb$^+$	9.5 ± 1.3	+9.2	+1	0 or 1.2	0 or 115	1.63	see below	see below
Cs$^+$	8.6 ± 0.7	+7.9	+1	5 or 6.2	5 or 120	1.53	see below	see below
K$^+$	5.5 ± 0.1	+5.4	+1	1 or 2.2	1 or 116	1.35	see below	see below
Na$^+$	1.2 ± 0.1	+0.8	+1	12 or 13.2	12 or 127	1.05	see below	see below
Li$^+$	(0)	(0)	+1	115	1.2 or 30	(1)	see below	see below
NH$_4^+$	—	—	+1	—	—	—	—	—
TMA$^+$	−4.9 ± 0.9	−5.0	+1	0 or 1.2	0 or 115	0.74	see below	see below
TEA$^+$	−12.7 ± 0.7	−12.7	+1	2 or 3.2	2 or 117	0.43	see below	see below
Cl$^-$	−16.3 ± 1.0	−16.5	−1	136	136 or 51	0.18	see below	see below

[a]Calculated E_{rev} = +25.26 ln (numerator/denominator) at 20°C.
[b]Numerator = $P_K[K^+]_A + P_{Na}[Na^+]_A + P_{Cs}[Cs^+]_A + P_{TEA}[TEA^+]_A + P_{Li}[Li^+]_A + P_X[X^+]_A + P_{Cl}[Cl^-]_B$.
[c]Denominator = $P_K[K^+]_B + P_{Na}[Na^+]_B + P_{Cs}[Cs^+]_B + P_{TEA}[TEA^+]_B + P_{Li}[Li^+]_B + P_X[X^+]_B + P_{Cl}[Cl^-]_A$.
Reproduced with permission from **ref. 2**.

Table 2
Matrix for Calculating Anionic P_{ion} in the Cx43 Gap Junction Channel

Ion Cx43	Measured E_{rev} (mV)	Calculated $E_{rev}{}^a$ (mV)	z	$[ion]_A$ (mM)	$[ion]_B$ (mM)	P_{ion}	Numerator[b]	Denominator[c]
K+	—	—	+1	116	116	1.35	see below	see below
Cs+	—	—	+1	5	5	1.53	see below	see below
Na+	—	—	+1	12	12	1.01	see below	see below
TEA+	—	—	+1	2	2	0.43	see below	see below
Br–	−0.9 ± 1.2	−1.5	−1	0 or 1.2	0 or 115	(1.08)	see below	see below
Cl–	(0)	(0)	−1	136	21	0.18 (1)	see below	see below
Acetate	+3.8 ± 0.5	+3.6	−1	0 or 1.2	0 or 115	(0.85)	see below	see below
Glutamate	+13.2 ± 0.5	+13.1	−1	0 or 1.2	0 or 115	(0.52)	see below	see below

[a]Calculated $E_{rev} = +25.26 \ln$ (numerator/denominator) at 20°C.

[b]Numerator $= (1/135) \cdot (P_K[K^+]_A + P_{Na}[Na^+]_A + P_{Cs}[Cs^+]_A + P_{TEA}[TEA^+]_A) + P_X[X^-]_A + P_{Cl}[Cl^-]_B$.

[c]Denominator $= (1/135) \cdot (P_K[K^+]_B + P_{Na}[Na^+]_B + P_{Cs}[Cs^+]_B + P_{TEA}[TEA^+]_B) + P_X[X^-]_A + P_{Cl}[Cl^-]_A$ where $135 = [K^+] + [Cs^+] + [Na^+] + [TEA^+]$.

Reproduced with permission from **Ref. 12**.

2. The perforated patch technique is most advantageous for experiments involving physiological regulation, as intracellular second messengers and protein kinases are not dialyzed out of the cell using this approach. Successful dialysis with organic ions using this approach may be problematic as a typical 10 MΩ whole-cell patch pipet has a radius of ≈ 1.0 μm (assuming 150 m*M* KCl) and the pore diameter of amphotericin B is 8 Å *(9)*.

3. On-line real-time recordings are possible using a variety of hardware/software packages available (e.g., pClamp 8.0 from Axon Instruments, Foster City, CA) and offer the advantage of storing data in a binary format that is readily analyzed by computer. The disadvantage is that should a file be lost or erased, you have no backup and you must also select an appropriate low-pass filter frequency for digital sampling during data acquisition. Thus, there remains an advantage for maintaining an off-line data storage device (digitized data recorder using storage disk or VCR-based format) that can record 2 h or more of continuous data at a wider analog bandwidth (e.g., 5 or 10 kHz depending on sample rate/channel).

4. Elimination of all Cl^- is essentially impossible to achieve given the amount of monovalent and divalent cations that must be present for cellular homeostasis. Na^+ can be eliminated from the standard IPS if 3 m*M* MgATP salt is used instead of 3 m*M* Na_2ATP salt without consequence and is therefore desirable. The complete elimination of Cl^- by using other available monovalent salts (e.g., nitrate, acetate) or polyvalent salts (e.g., citrate, sulfate) have drawbacks. For polyvalent salts, the osmolarity changes must be calculated accordingly. For all Cl^--free solutions, the basis for the electrode junction potential must be considered. One potential source of error in measuring an ionic E_{rev} begins with the dual whole-cell recording configuration. The initial conditions that $V_{mA} = V_{mB}$ when $V_A = V_B$ must be true if an accurate measurement of the ionic E_{rev} is to be obtained using the procedure described in this chapter. Therefore, it is imperative that $R_{sA}/R_{inA} = R_{sB}/R_{inB}$ and that $R_s << R_{in}$ to eliminate series resistance errors during dual cell voltage clamp and monitoring of junctional currents from whole-cell current recordings. What is not readily apparent from the equivalent resistive circuit of the dual whole-cell patch clamp configuration is that there are virtual grounds for both voltage clamp amplifiers and a bath ground that must be common for $V_A = V_B$ when a common command potential is applied to both amplifiers simultaneously. This is usually achieved by using a single Ag/AgCl junction and a reference solution that is in contact with the bath solution that serves as the common ground for the head stages of both amplifiers. This contact is usually achieved by using a salt bridge between a ground well and the recording chamber because the silver ions are toxic to the cells. The second part of this circuit is the Ag/AgCl junction with the IPS in each recording electrode (patch pipet) and the liquid junction potential of the patch electrode with the bath solution. Any voltage offsets due to different ionic compositions of the two IPSs and the bath saline and the liquid–wire Cl^- junctions must be compensated or recorded prior to whole cell recording from either cell or else a voltage offset will exist that is independent of the command potentials of the voltage clamp amplifiers ($V_{electrode} - V_{bath}$)

– $V_{command}$, i.e., virtual ground. This will affect the measured biological permeability differences of the two ions under asymmetric ionic conditions of the IPSs which is the basis for measuring an E_{rev} in the first place. Beginning with a 2–6 MΩ patch electrode when filled with IPS KCl, the chlorided silver wires for each electrode and headstage can be checked and maintained by determining the pipet offset relative to the bath reference. This value should be minimal (0–2 mV) when using IPS KCl provided that the silver wire is properly chlorided. If not, the silver wire should be replated in 1 *N* HCl before continuing the experiment. The bath reference is achieved by using a bath saline salt bridge (3% agar) between the recording chamber and the grounding well. The grounding well can be filled with IPS KCl, thereby making all wire–solution and electrode-bath junctions identical except when it is necessary to create an experimental ionic gradient by using a second IPS in electrode B. Alternatively, a 3M KCl reference electrode can be used. The differences between the two approaches are provided in an example to follow. By using Cl⁻ salts, the two Ag/AgCl wire-liquid junctions should be negligible provided that the IPSs were prepared and diluted properly since the [Cl⁻] should be constant. If this condition is true, the difference in the offset potentials of the 140 m*M* KCl and XCl electrodes should reflect the difference in aqueous diffusion potentials between ions X^+ and K^+ relative to the bath ground. This value should be subtracted from the final experimental results.

For example, the liquid junction potential of a diffusion limited 140 m*M* KCl pipet with a 140 m*M* NaCl bath saline will be 4.3 mV. If $X^+ = Na^+$, there will be no liquid junction potential difference between the test pipet and the bath. If 140 m*M* KCl is used as the ground reference electrode, it will also have a 4.3 mV liquid junction potential and the ground and KCl electrode and pipet liquid junction potentials will cancel out. In this case, the value of the amplifier junction potential null reflects the Ag/AgCl wire to liquid KCl potential which should be ≈ 0 mV. Since the liquid junction potentials of the NaCl pipet and the ground reference electrode differ by 4.3 mV, the value of the amplifier junction potential null will equal 4.3 mV plus the Ag/AgCl wire to liquid NaCl junction potential. If a 3M KCl bridge is used as the ground reference electrode, the difference in liquid junction potentials between the KCl pipet and the reference electrode will be (4.3 – 1.9) = 2.4 mV. The liquid junction potential difference between the NaCl electrode and the reference electrode will be 1.9 mV. The resulting liquid junction potential difference between the NaCl and KCl electrodes is 0.5 mV plus any Ag/AgCl to KCl or NaCl wire/liquid junction potential difference. This presumably stable offset potential is in series with $V_A - V_B$ at all times and will produce small errors in the junctional voltage clamp (i.e., the applied $V_j \neq 0$ mV when $V_A = V_B$) unless compensated for by either nulling this potential difference prior to GΩ seal formation or substraction it from $V_A - V_B = V_j$ in the resulting junctional I–V curve.

The latter condition of maintaining the a constant [Cl⁻] when comparing test cations reveals another potential source of error when anionic reversal potentials are to be measured. Comparing test anions to Cl⁻ necessitates replacing Cl⁻ with another test anion (e.g., Br⁻ or other halide ions, NO_3^-, acetate, or other organic anions) and this produces a wire–liquid junction potential equivalent to the Cl⁻ diffusion potential of the IPS relative to the standard Cl⁻-containing IPS. Furthermore, and this is especially true for the halide ions Br⁻, F⁻, and I⁻, the current passing through the recording electrode will replate the silver wire at a rate determined by the magnitude of the current and the time duration of the current flow. Hence, larger junction potentials will develop during the experiment especially when using halide anions. The effect with Br⁻ is small because its diffusion coefficient is similar to that of Cl⁻, but F⁻ and I⁻ present a more difficult situation. F⁻ and I⁻ ions are also not as well tolerated at high concentrations inside a cell and are best avoided for dual whole-cell experiments. Nitrate and organic anions such as acetate do not replate the silver wire significantly and stable measurements can be obtained with these anions. We always used the same wire for a given set of Cl⁻ substitution experiments, thus producing a stable offset potential which we recorded for each experiment. The relative diffusion coefficients for some other anions are: Br⁻, 1.02; Cl⁻, 1.00; NO_3^-, 0.94, and acetate, 0.54 *(8)*.

5. The free acid is desirable for ionic permeability ratio experiments unless K⁺ or Na⁺ is to be used as the primary monovalent cation in that specific IPS. The free acids of BAPTA or EGTA will not dissolve until the pH is adjusted to 7.0 or higher with the base (1 *N* XOH) specific to the monovalent cation to be used in each specific IPS.

6. Even with two 10 MΩ whole-cell patch electrodes, a series resistance relative to the junction of 20 MΩ will be present. This automatically results in a 10% error for a 5 nS junctional conductance cell pair even if the calculated error in the respective membrane voltages is 2% (R_{in} = 1 GΩ). Therefore, low junctional conductances (high junctional resistances) are preferred.

7. In the dual whole cell voltage clamp configuration, $I_j = -\Delta I_2 \cdot [1 + (R_{e12}/R_{in2})]$ as defined by Veenstra and Brink *(1)* where $\Delta I_2 = I_2(V_A + \Delta V_A, V_B) - I_2(V_A = V_B)$. Recently, I_j was defined as $-I_2 \cdot [1 + (R_{e12}/R_{in2})] - (V_B/R_{in2})$ where the latter term defines the nonjunctional membrane current of the postjunctional cell *(10)*. Since V_B is constant when $V_A = V_B$ and when ΔV_A is applied, this term will cancel out if $I_2(V_A = V_B)$ is subtracted from $I_2(V_A + \Delta V_A, V_B)$ provided that R_{in2} remains constant. Hence, $-\Delta I_2 \cdot [1 + (R_{e12}/R_{in2})] \approx -\Delta I_2$ provides an accurate on-line estimate of I_j provided that the holding current for cell 2 does not change during the ΔV_A step.

8. There are published methods for internally perfusing a patch pipet in the whole-cell configuration *(11)*. This technique requires internal dialysis of a patch electrode near the tip using negative pressure as the driving force for fluid flow. Technical difficulties associated with this approach include plugging of the internal dialysis tube by small air bubbles (micrometer diameter tubing near the electrode tip), increased noise due to the fluid column extending to the IPS fluid reservoir near the amplifier headstage, and possible disruption of the GΩ seal if turbulence develops during internal perfusion. Secondarily, switching IPS solutions near the tip of the patch electrode will cause shifts in the offset potential for that electrode as described in **Note 4**. The value of this diffusion potential must be determined or

compensated for during the experiment to achieve an accurate measurement of the ionic E_{rev}. The advantage of attempting this technique is the added ability to test several ions in one experiment and the ability to determine E_{rev} from macroscopic recordings by switching from symmetric to asymmetric IPS solutions and recording the voltage shift of the junctional I–V relationship. To accomplish the latter requires being able to return to the control condition to determine any changes in the junctional recording (junctional conductance changes should affect only the slope, but input resistance changes will also shift the I–V relationship) and being able to account for the fluid diffusion potential shift that develops during internal electrode perfusion as already mentioned.

9. The last consideration does not involve potential sources of error in measuring an ionic E_{rev}, but rather to calculating the corresponding relative P_{ion} using conventional GHK theory. It is highly probable that one may find it impossible to determine a set of ionic reversal potentials using the standard GHK voltage equation because it depends on the principles of ionic independence and bidirectional fluxes. If either of these conditions is not true, the GHK voltage equation will not be valid and one is faced with the task of deriving an equation based on the biophysical principles of electrodiffusion. One example of a previously derived permutation of the GHK voltage equation was used to determine the anionic P_{ion} values (relative to Cl^-) for the rat Cx43 channel *(2)*. **Table 2** contains the matrix necessary to calculate the anionic P_{ion} values for the rat Cx43 channel using the same experimental procedures *(2,12)*. The basis for dividing the cationic P_{ion} terms for the Cx43 channel in **Table 2** is derived from the theory that anion permeability is obligatorily dependent on the presence of a single bound cation at a site within the channel pore. This expression is analogous to what was derived previously to explain the cation permeability of the anionic amphotericin B channel *(9)*. This formulation was necessitated by the observation that with a P_{Cl}/P_K ratio of 0.13 from **Table 1**, anionic reversal potentials could not exceed +2.7 mV even if acetate or glutamate were not permeant ($P_X = 0$) according to the conventional GHK voltage equation.

This methodology has allowed the calculation of any cation or anion P_{ion} value from the dual whole-cell asymmetric ionic E_{rev} encountered to date for connexin gap junction channels. The procedures outlined in this chapter should permit investigators to determine their own ionic permeabilities that would be directly comparable to previous published values obtained from the rat Cx43 and Cx40 channels *(2,3)*.

References

1. Veenstra, R. D. and Brink, P. R. (1992) Patch clamp analysis of gap junctional currents, in *Cell–Cell Interactions: A Practical Approach* (Stevenson, B. R., Gallin, W., and Paul, D. L., eds.), IRL, Oxford, pp. 167–201.
2. Wang, H.-Z. and Veenstra, R. D. (1997) Monovalent ion selectivity sequences of the rat connexin43 gap junction channel. *J. Gen. Physiol.* **109,** 491–507.
3. Beblo, D. A. and Veenstra, R. D. (1997) Monovalent cation permeation through the connexin40 gap junction channel: Cs, Rb, K, Na, Li, TEA, TMA, TBA, and

effects of anions Br, Cl, F, acetate, aspartate, glutamate, and NO_3. *J. Gen. Physiol.* **109,** 509–522.

4. Veenstra, R. D., Wang, H.-Z., Westphale, E. M., and Beyer, E. C. (1992) Multiple connexins confer distinct regulatory and conductance properties of gap junctions in developing heart. *Circ. Res.* **71,** 1277–1283.

5. Bers, D. M., Patton, C. W., and Nuccitelli, R. (1994) A practical guide to the preparation of Ca^{2+} buffers, in *A Practical Guide to the Study of Ca^{2+} in Living Cells, Methods in Cell Biology,* Vol. 40 (Nuccitelli, R., ed.), Academic, New York, pp. 3–29.

6. Sakmann, B. and Neher, E. (1995) *Single-Channel Recording,* 2nd ed., Plenum, New York and London, p. 33.

7. Robinson, R. A. and Stokes, R. H. (1965) *Electrolyte Solutions,* revised 2nd ed., Butterworths, London, UK.

8. Hille, B. (1992) *Ionic Channels of Excitable Membranes,* 2nd ed. Sinauer, Sunderland, MA.

9. Borisova, M. P., Brutyan, R. A., and Ermishkin, L. N. (1986) Mechanism of anion-cation selectivity of amphotericin B channels. *J. Membr. Biol.* **90,** 13–20.

10. Van Rijen, H. V. M., Wilders, R., Van Ginneken, A. C. G., and Jongsma, H. J. (1998) Quantitative analysis of dual whole–cell voltage-clamp determination of gap junctional conductance. *Pflügers Arch.* **436,** 141–151.

11. Soejima, M. and Noma, A. (1984) Mode of regulation of the ACh-sensitive K-channel by the muscarinic receptor in rabbit atrial cells. *Pflügers Arch.* **400,** 424–431.

12. Veenstra, R. D. and Wang, H.-Z. (1998) Biophysics of gap junction channels, in *Heart Cell Communication in Health and Disease* (DeMello, W. C. and Janse, M. J., eds.), Kluwer Academic, Boston/Dordrecht/London, pp. 73–103.

17

Fluorescence Recovery After Photobleaching

**Jean Délèze, Bruno Delage, Olfa Hentati-Ksibi,
Franck Verrecchia, and Jean-Claude Hervé**

1. Introduction

As was first demonstrated by the cell-to-cell spread of a fluorescent diffusion tracer introduced into one cell by microinjection *(1)*, the permeability of the gap junctions is not restricted to the small intracellular electrolytes that carry the junctional currents. Quantitative data on the permeability of gap junctions for larger solutes, which include second messengers and other signaling molecules, contribute to the understanding of their function. The microinjection method (*see* Chapter 12) provides information on the presence of functional gap junctions and on the size of the permeating molecules, but permeability coefficients have usually not been obtained with this technique. Indeed, for reasons related to the complex geometry of the diffusion system from one progressively microinjected cell toward a variable number of adjacent and remote cells, long and difficult calculations are involved even in the most simple case of one-dimensional diffusion *(2)*.

With the fluorescence recovery after photobleaching (FRAP) method, the initial and boundary conditions of the diffusion system can usually be reduced, with a precision sufficient for practical purposes, to those of a simple system of two compartments separated by a permeable membrane. The first compartment comprises the cell or the set of cells connected by gap junctions to the photobleached cell, which constitutes the second compartment. Furthermore, with appropriate caution (*see* **Note 1**), the risk of cell injury, which in microinjection experiments requires constant attention and may be difficult to recognize, is usually negligible with the FRAP method.

From: *Methods in Molecular Biology, vol. 154: Connexin Methods and Protocols*
Edited by: R. Bruzzone and C. Giaume © Humana Press Inc., Totowa, NJ

1.1. The FRAP Method

The irreversible photochemical bleaching reaction of fluorescent molecules exposed to a light beam of sufficient intensity and of wave length within the absorption spectrum of the fluorochrome forms the basis of the fluorescence recovery after photobleaching techniques. The FRAP method has been devised for measuring the two-dimensional diffusion of fluorescently labeled macromolecules in membranes *(3)*. After monitoring the basal fluorescent emission excited by means of an attenuated laser beam focused on a small membrane area, the light emission of the laser is briefly raised about 10,000 times, then the subsequent fluorescence intensity, stimulated again by the attenuated laser beam, is recorded. Fluorescence recovery occurs in the bleached spot if the labeled molecules are mobile in the plane of the membrane. The transport processes involved (random diffusion or directed flow) are identified, and the corresponding diffusion constants or flow velocities are determined by analysis of the spatial and temporal kinetics of fluorescence recovery *(4,5)*.

1.2. Measurement of the Cell-to-Cell Transfer of Cytosolic Molecules by FRAP

The FRAP method has been adapted to measure the cell-to-cell exchange of a fluorescent diffusion tracer and to obtain quantitative information on the permeability of gap junctions for small molecules (gap-FRAP *[6]*). Ideally, the permeability coefficient (P) of the gap junctional membrane should be obtained. As in other biological membranes, P, which measures the ease of passage of a solute (s) driven by its kinetic energy, is equal to the diffusion constant (D_s) of the substance inside the membrane, divided by the thickness (x) of the membrane; in a gap junction, x is the length of the unit channel. The permeability coefficient is the flux of substance that crosses the unit area of membrane in unit time when the concentration gradient at the edges of the membrane is equal to unity. To measure a net flux of diffusible substance across a membrane separating two compartments, a concentration gradient must exist between both sides, either in natural conditions or after establishment by the experimenter. In the gap-FRAP technique, a fluorescent diffusion tracer is first introduced at uniform concentration into all the cells in a tissue culture dish, and the concentration gradient is created by photobleaching the fluorescence in one cell. If the bleached cell communicates with neighbor cells by means of functional gap junction channels, relaxation of this concentration gradient is detected by monitoring the fluorescence signal, using low-intensity excitation.

1.3. Determination of the Transfer Constant of a Solute Across Gap Junctions

Diffusion of a small molecule such as fluorescein is much more rapid in the cytosol than across the gap junction membrane and the concentrations inside small source and sink cells can be considered spatially uniform at all times (the correctness of this condition can be controlled by the FRAP method). With this simplification, the concentration gradient of the substance across the length of the gap junction channels is simply the difference of concentrations ($C_1 - C_2$) between the unbleached and bleached cells, and the amount of solute (ds) crossing the junction in time (dt) is given by the well-known solution of Fick's general equation for a system of two compartments separated by a membrane:

$$ds/dt = -PA\,(C_1 - C_2) \qquad (1)$$

where A is the area of the gap junction membrane. Because the flux of substance (ds/dt) is equal to the change in concentration with time in the bleached cell multiplied by the cell volume (V), **Eq. 1** becomes:

$$dC_2/dt = -PA(C_1 - C_2)/V \qquad (2)$$

For a diffusion driven transport, the flux of substance across a concentration boundary is proportional to the concentration gradient at each instant. Therefore

$$dC_2/dt = e^{-kt} \qquad (3)$$

where $k = PA/V$ (in units of time^{-1}) is the rate constant of the change in concentration. Determination of (k), which hereafter will be referred to as the transfer constant or relative permeability constant, is sufficient when studying regulations of the gap junction permeability in the same type of cell. For quantitative comparisons of the permeabilities of different cell types, the cell volume and the gap junction surface or the number of gap junction channels should also be evaluated.

2. Materials

2.1. Dye Loading Procedure for Gap-FRAP Experiments

To measure the gap junction permeability by the FRAP technique, a fluorochrome must first be introduced into cells in a tissue culture dish. To do this without the need of multiple microinjections, an ester form of the fluorescent molecule chosen as a diffusion tracer, usually 5(6)-carboxyfluorescein (CF), is added for a time to the bath solution. The nonpolar and hydrophobic esters are membrane permeant and diffuse into the cells, where they are hydrolyzed by cytoplasmic esterases, releasing their fluorescent unesterified polar moieties (7). The latter compounds are hydrophilic and accumulate inside the cells because of their very low membrane permeabilities.

2.2. Preparation of Solutions

The composition of the solutions that have proved convenient for performing FRAP experiments in cultured mammalian Sertoli or heart cells is given below. Unless otherwise stated, all concentrations are given in millimoles per liter.

1. Replace the culture medium, which contains fluorescent organic compounds, by a physiological saline adapted to the cell type under study. The modified Tyrode solution employed in our laboratory for this purpose contains: 144 NaCl, 5 KCl, 2.5 $CaCl_2$, 2.5 $MgCl_2$, 0.3 NaH_2PO_4, and 5.6 N-(2-hydroxyethyl) piperazine-N'-2-ethanesulfonic acid (HEPES). This solution is buffered to pH 7.2–7.4 with NaOH.
2. Prepare a stock solution of 5(6)-carboxyfluorescein diacetate (CFDA) in dry dimethyl sulfoxide (DMSO), containing 5 mg of CFDA/mL of DMSO (which in molar concentration gives 10 mM CFDA DMSO). Keep this solution for no more than 1 wk in a refrigerator and protect it from light.
3. Dilute the CFDA stock solution to 1:1000 (v/v) with Tyrode to obtain a final CFDA concentration of about 10 μM (e.g., by mixing 2 μL of stock solution per 2 mL of Tyrode). Sonicate for 30 s and use it immediately.

The final concentration of the solvent DMSO, 1/1000 the volume of the physiological solution, has no effect on the intercellular coupling of mammalian cells.

The mentioned ingredients are readily available from most small and from all major suppliers of chemicals for biological research.

2.3. Technical Requirement for Gap-FRAP Experiments

The most efficient procedure to direct the stimulating or photobleaching light beam onto the object is by means of the lenses that also collect the fluorescent emission (epiillumination geometry, commonly known as epifluorescence). To perform gap-FRAP experiments, an epifluorescence microscope must be equipped with a sufficiently powerful light source. Devices permitting rapid large-scale variations of the duration and intensity of the emitted light output and switching from photobleaching to monitoring levels are compulsory.

For photobleaching, the intense light beam focused on the object plane by the microscope objective must be precisely directed onto selected target cells, a task that is conveniently performed by means of a computer-driven microscope stage or galvanometric mirror. Data acquisition is controlled by observing the fluorescent emission on a video screen and quantitative analysis of the stored images is usually performed off line.

Facilities for quantitative measurements of the rate of cell-to-cell transfer of fluorescent diffusion tracers by gap-FRAP have been incorporated in the multipurpose laser workstation for cytophotometry Anchored Cell Analysis and Sorting (ACAS, Meridian Instruments, Okemos, MI). This system includes the technical features of an equipment suitable to efficiently perform gap-FRAP analysis.

2.4. Apparatus

2.4.1. Fluorescence Microscope

Select a commercially available inverted epifluorescence microscope. To excite the fluorescent emission of the dye-loaded specimen, use a 40× objective lens with numerical aperture 0.55 or higher. With these optics, a coherent light beam is focused to a theoretical diffraction diameter of 1 μm or smaller and the depth of field will not exceed 1.5 μm. Low-power objective lenses (10 or 20×) are useful for observing the preparation and choosing an object field, but they give fluorescence images of insufficient brightness for useful work.

2.4.2. Light Source for Stimulation and Photobleaching

The argon ion laser tuned at the 488 nm wavelength is the most convenient light source available at present for stimulating the fluorescent emission and for photobleaching fluorescein based dyes. Fast photobleaching of whole cells is an essential requisite for gap-FRAP measurements and a sufficient light power must therefore be available. At the present time, the laser power of most commercially available confocal microscopes may be adequate for spot photobleaching of membranes *(8)*, but unless customized it is not sufficient for rapid photobleaching of whole cells as required by the gap-FRAP technique. According to our practice, an argon ion laser with a total light output of 5 W and delivering 1500 mW at λ 488 nm is satisfactory.

2.4.3. Filters and Detectors for the Emitted Light

The fluorescence of CF is stimulated at 488 nm by the attenuated laser light. The emission spectrum, which at pH 7 presents a sharp peak at 516 nm, is collected by the objective and directed to a photomultiplier tube, after separation from the incident laser beam by a 510 nm long pass chromatic beam splitter (a "dichroic mirror") and a 530 ± 30 nm bandpass filter.

2.4.4. Devices for Rapid Switching from Monitoring to Bleaching Light Levels

These operations could be performed manually by exchanging density filters, but with an important loss of the time resolution of data acquisition. A substantial improvement could be reached by the use of an electromechanical filter wheel driven by programmable pulse generators. In the ACAS Meridian, continuous-amplitude variation of the light beam and very rapid switching from monitoring to bleaching levels is computer controlled by specialized software acting on an acousticooptic modulator.

2.4.5. Image Acquisition

A scanning microscope stage moving the biological specimen in front of a laser beam focused to a diffraction spot about 1 μm in diameter considerably reduces the total irradiation of the cells and makes it possible to record data from areas much larger than the field of view of the microscope objective lens. The stage is driven by a computer-controlled high-speed step motor. The light intensities measured at each step by the photomultiplier are stored in the computer memory together with their (x, y) coordinates. The fluorescent image of the cells is reconstructed on a video screen by the software, as a map of the color-coded intensities of emitted light.

Directing the incident light onto the cells by means of a scanning mirror, instead of moving the object stage, also reduces illumination, but viewing of the specimen is then restricted to the objective field.

In principle, it should be possible to perform gap-FRAP experiments using whole field illumination instead of a scanning device to stimulate the fluorescent emission, but a very sensitive solid-state imager should then be employed to permit limiting the intensity and duration of illumination during post-bleach data acquisition.

3. Methods

3.1. Determination of CFDA Concentration and Loading Time

The loading time varies with the type of cell, depending on the surface/volume ratio and on the amount of intracellular esterases. To avoid excessive loading, which may result in incomplete hydrolysis, compartmentalization in organelles, and phototoxicity from byproducts of photobleaching (*see* **Note 1**), the CFDA concentration and loading time should be adapted to the investigated cells and kept to the minimum compatible with practical data acquisition. Starting with a CFDA solution containing 10 μ*M* as described in **Subheading 2.2.**, the increase of fluorescence while the cells are taking up and hydrolyzing CFDA should be recorded in preliminary trials to find the appropriate ester concentration and loading time (usual range 10–30 min at room temperature with a solution containing 10 μ*M* CFDA). As a practical indication, loading is sufficient when a fluorescent image of the cells with an intensity scale graded from 1 to approx 5000 can be recorded with a sensitive light detector (e.g., a photomultiplier tube) without excessive background noise.

Hydrolysis of CFDA by cytoplasmic esterases releases CF, a gap junction permeant and highly fluorescent fluorochrome. An excessively slow dye loading may be caused by an insufficient activity of the esterases at room temperature. In this case, replacement of the cells into an incubator at 37°C for short times (2–3 min) is usually adequate to accelerate hydrolysis of the ester.

3.2. Washing Off the Fluorochrome Ester

When the loading process is completed, the cell culture dishes are washed several times with Tyrode solution to remove the fluorochrome-ester and prevent further dye loading during subsequent measurements.

3.3. Reloading the Same Cells in Long-Term Experiments

The culture dishes loaded with CF generally lose there fluorescence within 2–3 h. The same cells can be reloaded with CFDA and it is possible to study long-term effects on gap junctions for several days.

3.4. Image Analysis

3.4.1. Determination of the Rate Constant (k) of Fluorescence Recovery

An increase of fluorescence intensities in a bleached cell in contact with other cells indicates the presence of open gap junction channels (*see* **Note 2**). The fluorescence redistribution from the unbleached (source) cells to the bleached (sink) cell as a function of time is displayed from the recorded data. In thin layers (up to 100 μm) and at low dye concentrations (range $10^{-8} – 10^{-3} M$), the intensity (F) of fluorescent emission varies linearly with the dye concentration *(9)*. Therefore, the integrated fluorescence intensity in a bleached cell is proportional to the total amount of fluorophore, and its rise during recovery is a measure of the changing concentration of the fluorescent molecules. When the concentration of the diffusible substance in the source cell is constant, the previously defined transfer constant or relative permeability constant (k) can be directly obtained from **Eq. 2**, which becomes *(10)*:

$$(F_i – F_t)/(F_i – F_0) = e^{-kt} \tag{4}$$

F_i, F_0, and F_t are the integrated fluorescence intensities in a bleached cell before, immediately after and at time (t) after photobleaching. The rate constant (k), the inverse value of the time constant, is determined from the slope of the change in fluorescence intensity in the bleached cell vs time, represented in semilogarithmic coordinates (*see* **Note 3**).

3.4.2. Normalizing (k) According to the Number of Connected Cells

Equation 4 strictly applies to the case of a constant concentration in the source cells. If this concentration decreases substantially during diffusional transport to the sink cell, the transfer constant (k) obtained by fitting **Eq. 4** to the data obtained for the bleached cell will appear faster and should be normalized according to the number of connected cells. Consider the three sets of cells represented in **Fig. 1**. In **Fig. 1A**, the bleached cell is interconnected to several neighbors in a confluent layer. In **Fig. 1B** and **C** the bleached cell is paired to one unbleached neighbor, and in **Fig. 1D** the bleached cell, although in a confluent culture, is connected to only two other cells (cell triplet). A photobleached cell in a confluent layer (**Fig. 1A**) communicates directly with several contacting (first-order) cells, which constitute a first compartment. These cells are themselves connected to more remote cells (second- and third-order cells), which build up a number of successive compartments separated by diffusion barriers. In this multicompartment system, the recovery curve is a sum of exponential functions with decreasing time constants. The change in (F_i) is small, because the volume of the first-order compartment is large relative to that of the bleached cell, and because (F_i) is also maintained by diffusion from more remote compartments. In such cases, the initial part of the fluorescence recovery curve has been found to present a monoexponential course of sufficient duration *(11)* to allow obtaining the (k) value of interest, that of the bleached cell, without the necessity of a multiexponential fitting of the data.

At equilibrium, the tracer concentrations in the sink (bleached) cell and in the source (unbleached) cell are equal. In a pair of cells of nearly equal volumes, the final concentration reached after diffusion equilibrium will be close to half the concentration gradient created by photobleaching, as shown in **Fig. 1B, C**. For the same number of junctional channels, the time constant of fluorescence recovery in a bleached cell belonging to a pair will therefore appear twice as fast as that measured if the same cell were part of an interconnected system with constant concentration in the source cells.

Similarly, in a system of three cells of nearly equal volumes (**Fig. 1D**), the final equilibrium concentration will amount to about two thirds of the

Fig. 1. (*Opposite page*) Cell-to-cell diffusion of CF and fluorescence levels reached at equilibrium in different sets of cells. The three successive color-coded pictures of fluorescent intensities in each row correspond to a prebleach scan, to a scan performed immediately after photobleaching (at t_0), and to a scan recorded when diffusion equilibrium is approached. The fluorescence recovery curves of the selected cells are plotted, in percent of the prebleach emission *vs* the time after photobleaching, in the graphs on the *right*. In **A**, the bleached cell (1) belongs to an extended layer of interconnected cells (a Cx26 transfected clone of HeLa cells *[21]*). The fluorescence emission of the bleached cell increased close to prebleach values while the change of fluorescence intensity integrated in surrounding cells (2) was very small. In **B** and **C**, the bleached cells belong to a pair of ventricular myocytes from an adult rat (**B**) and from a newborn rat (**C**), cultured for 2–3 d. The fluorescence intensities integrated in the source and sink cells vary in a symmetrical manner and the fluorescence levels reached at equilib-

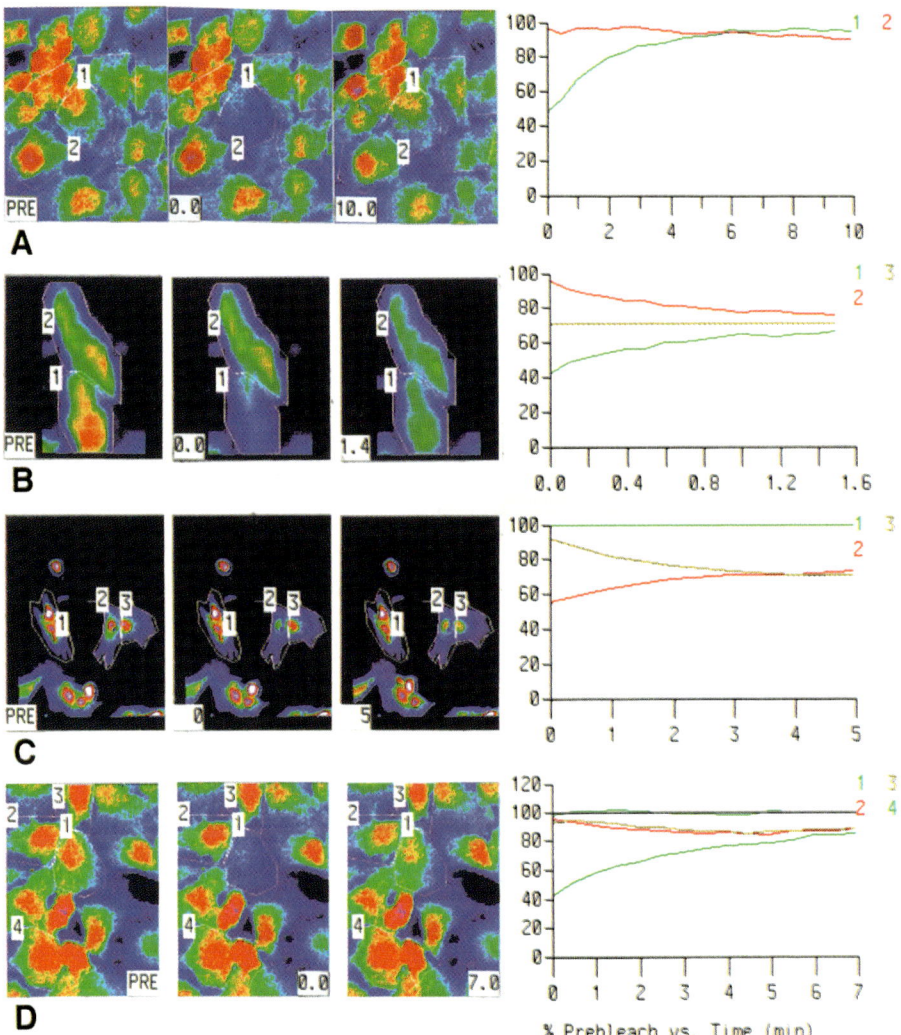

rium in both cells are halfway between the prebleach and postbleach (t_0) values, as expected. Curve *3* in **B** represents the added intensities of the source and sink cells. In **D**, the bleached cell in a confluent culture (same HeLa-Cx26 transfectant as in **A**) receives fluorescent molecules from two neighbors only (cell triplet). In those conditions, the fluorescence intensity at equilibrium levels off in both cells to about two thirds of the concentration gradient created by photobleaching, as expected for cells of nearly equal volumes. The fluorescent emission of an unbleached control cell is shown in rows **C** (cell *1*) and **D** (cell *4*). Those values can be used to correct the recovery curves for a decrease of fluorescent emission during successive scans, as done in **B** and **C**. The horizontal sides of the scanned fields in **A, B, C, D** measure 100, 75, 160, and 100 μm, respectively.

Table 1
Relative Permeability Constants (k) of 5(6)-Carboxyfluorescein in the Gap Junctions of Different Cell Types

Cultured cell types	k (min^{-1}) $\pm$ SD	Reference
Human fibroblasts	0.133 ± 0.066	*(6)*
Rat Sertoli cells	0.088 ± 0.013	*(11)*
Human trophoblasts	0.108 ± 0.011	*(12)*
Human myometrial cell line (PHM1-41)	0.169 ± 0.006	*(13)*
Neonatal rat ventricular myocytes	0.24 ± 0.12	*(14)*

initial concentration gradient at t_0 and the apparent time constant will be shortened accordingly by a factor 3/2. It is therefore advisable to use similar sets of cells for measurements (e.g., cell pairs or cells in confluence), or else to normalize the (k) values according to the tracer concentrations reached at equilibrium.

Table 1 presents the transfer constants (k) obtained for CF in different cultured cells. It can be seen that, even without normalization for cell volume and gap junction surface, or number of junctional channels, these values allow to compare directly the rapidity of the cell-to-cell exchange of molecules of size and weight similar to that of CF.

3.5. Controls

3.5.1. Correction for a Spontaneous Loss of Fluorescence

A spontaneous decay of fluorescence, usually <10% of the initial intensity level, occurs in isolated unbleached cells during a 10-min data acquisition, owing to a slow outflux of CF to the extracellular medium and to some photobleaching. The recovery curves of the bleached cells can be corrected for this fluorescence decay, using the data recorded in unbleached cells (e.g., **Fig. 1B–D**).

3.5.2. Testing the Innocuousness of Photobleaching

The importance of avoiding cell injury by an excessive intensity or duration of the photobleaching light pulses has already been stressed (*see* **Note 1**). It is therefore a good idea to control the innocuous character of the photobleaching beam by performing several successive gap-FRAP measurements on the same cells, after diffusion equilibrium of the tracer has been reached. If the rate constant (k) of fluorescence recovery does not change significantly during a number of consecutive bleaches of the same cells, as illustrated in **Fig. 2**, the light intensity can safely be deemed adequate. With this precaution, several measurements of (k) in control and in experimental conditions can be obtained on

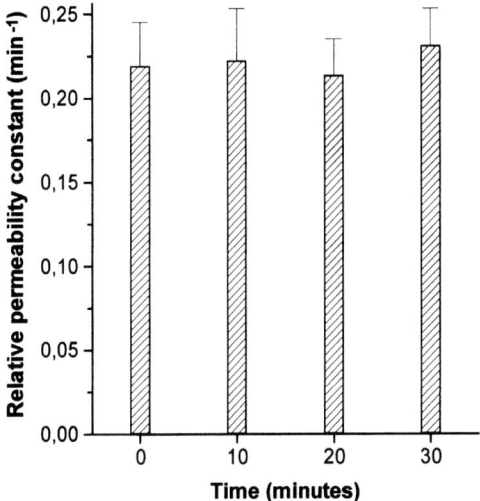

Fig. 2. Comparison of the results of four consecutive determinations of the relative permeability constants (k) of dye transfer between cultured ventricular myocytes of neonatal rat. The relative permeability constant (k) was determined at 10-min intervals in 23 pairs of cells. Bars indicate mean (k) values (min^{-1}) $\pm$ SEM.

the same cells, and the degree of reversibility of agents acting on the gap junctional communication can be examined. Also, pairing of data obtained for the same cells allows one to establish a statistical significance with a reduced number of measurements.

3.6. Conclusions

3.6.1. Comparison with Other Methods

3.6.1.1. FLUORESCENCE RECOVERY VS DYE-INJECTION

Obtaining information on the cell-to-cell transport by diffusion through gap junction is much more rapid and simple by the gap-FRAP than by the microinjection method. Performing intracellular microinjections is time consuming, the fluorochrome concentration necessary to detect dye diffusion into several cells away from one microinjected cell is much higher than that used for recording fluorescent recovery into one cell, and the amount of microinjected fluorochrome is difficult to standardize. Even for well-trained skillful experimenters, the risk of cell injury during the injection process is always present. In comparison, loading cells with the membrane permeant ester of a fluorochrome is a straightforward repetitive process. With appropriate equipment, directing the photobleaching light beam on target cells necessitates no special skill. With proper attention to the fluorochrome concentration and to the intensity of the bleaching light beam, the risk of cell injury is minimized. The quantitative analysis of the gap junction permeability can usually be modeled by a simple two-compartment diffusional system comprising the photobleached cell, separated from one or several first-order unbleached cells by the gap junction membrane. A further advantage of the gap-FRAP technique is the possibility to obtain paired data in control and test conditions and to perform long-term observations on the same cells.

3.6.1.2. Permeability or Conductance Measurements?

Permeability of the gap junctions is a measure of the cell-to-cell exchange of molecules driven by their kinetic energy, and should not be confused with the electrical conductance, which describes the ionic currents driven by a gradient of electrical potential. The information acquired by these two kinds of data is therefore not equivalent, in the sense that knowledge of the conductance of a unit junctional channel for a particular ion cannot be used to predict the permeability of a given molecule.

A number of recently published data indicate that, in many functional or pharmacological studies, data acquisition by gap-FRAP is more rapid and convenient than by probing electrical coupling or measuring the junctional conductance. For examples, the gap-FRAP technique has been used to quantify the decrease in gap junctional communication during mitosis *(15)*, or to probe the effect of site-directed anti-connexin antibodies *(16)*. An increase in the frequency of cells coupled by gap junctions, which precedes cell fusion and is enhanced by human chorionic gonadotrophin, has been observed during the in vitro differentiation of human trophoblastic cells *(12)*. Similarly, the percentage of coupled GT1-7 mouse neurons *(17)* or of human trophoblasts *(18)* was shown to increase after stimulation of the cAMP pathway. Fluorescence recovery recorded in a small area at the surface of the rat parietal cortex was slowed, but not suppressed, during long-lasting ischemia *(19)*.

3.6.2. Limitations of the Method

Using a scanning microscope stage extends the investigated area and reduces the total irradiation of the specimen at the cost of a limited time resolution of data acquisition. With the ACAS Meridian, the shortest possible interval between successive scans of a small field is approx 8 s. This is not sufficient to obtain accurate time constants when the cell-to-cell exchange suddenly becomes much faster, as is for instance the case in trophoblastic cells connected by gap junctions and in the same cells after fusion *(12)*. Much faster acquisition rates, but limited to the field of the microscope objective lens, can be obtained by means of solid-state imagers. But whatever the gain in speed of the apparatus, the gap-FRAP method is not appropriate for measuring fast changes of the gap junction properties during application of a chemical or physical agent, as can be done by recording the electrical conductance *(22)*, because of the time necessary for determination of the rate constant (k).

4. Notes

1. To avoid photochemical and thermal cell injury, a highly sensitive light detector should be used. This will allow to minimize dye loading and illumination, and thereby reduce the amount of energy absorbed by the specimen. Excessive dye loading may also cause cell damage or impairment of gap junctional communication by increasing the release of acids during intracellular hydrolysis of the ester, and of free radicals during photobleaching.

2. An artifactual increase of the fluorescence level may also happen in an isolated bleached cell. The experiment should then be rejected and appropriate measures taken to prevent this circumstance. Possible causes are the presence of the ester form of the dye in the bath (insufficient washing of the ester) or in the cytosol (insufficient time after loading the ester or too low temperature), and excessive loading (too high a cytosolic concentration can result in loading of organelles, autoabsorption and concentration-dependent quenching of the fluorescent emission, which may increase after photobleaching).

3. In a modification of **Eq. 4** *(15,20)*, (F_i) is replaced by the fluorescence intensity reached at diffusion equilibrium (F_e), which simply amounts to adding a constant number to the experimental data. The time constants of fluorescence recovery and the corresponding (k) values, obtained by fitting the same set of data to **Eq. 4** or to its modified form ($F_e - F_t)/(F_e - F_0)$ *(15)*, are therefore identical.

References

1. Kanno, Y. and Loewenstein, W. R. (1964) Intercellular diffusion. *Science* **143**, 959–960.
2. Brink, P. R. and Ramanan, S. V. (1985) A model for the diffusion of fluorescent probes in the septate giant axon of earthworm. Axoplasmic diffusion and junctional membrane permeability. *Biophys. J.* **48**, 299–309.
3. Peters, R., Peters, J., Tews, K. H., and Bahr, W. (1974) A microfluorometric study of translational diffusion in erythrocyte membranes. *Biochim. Biophys. Acta* **367**, 282–294.
4. Axelrod, D., Koppel, D. E., Schlessinger, J., Elson, E., and Webb, W. W. (1976) Mobility measurement by analysis of fluorescence photobleaching recovery kinetics. *Biophys. J.* **16**, 1055–1067.
5. Koppel, D. E. (1979) Fluorescence redistribution after photobleaching. A new multipoint analysis of membrane translational dynamics. *Biophys. J.* **28**, 281–292.
6. Wade, M. H., Trosko, E. J., and Schindler, M. (1986). A fluorescence photobleaching assay of gap junction-mediated communication between human cells. *Science* **232**, 525–528.

7. Rotman, B. and Papermaster, B. W. (1966) Membrane properties of living mammalian cells as studied by enzymatic hydrolysis of fluorogenic esters. *Proc. Natl. Acad. Sci. USA* **55,** 134–141.

8. Gribbon, P. and Hardingham, T. E. (1998) Macromolecular diffusion of biological polymers measured by confocal fluorescence recovery after photobleaching. *Biophys. J.* **75,** 1032–1039.

9. Barrows, G. H., Sisken, J. E., Allegra, J. C., and Grasch, S. D. (1984) Measurement of fluorescence using digital integration of video images. *J. Histochem. Cytochem.* **32,** 741–746.

10. Schindler, M., Trosko, J. E., and Wade, M. H. (1987) Fluorescence photobleaching assay of tumor promoter 12-*O*-tetradecanoylphorbol 13-acetate inhibition of cell–cell communication. *Methods Enzymol.* **141,** 447–459.

11. Pluciennik, F., Joffre, M., and Délèze, J. (1994) Follicle-stimulating hormone increases gap junction communication in Sertoli cells from immature rat testis in primary culture. *J. Membr. Biol.* **139,** 81–96.

12. Cronier, L., Bastide, B., Hervé, J. C., Délèze, J., and Malassiné, A. (1994) Gap junctional communication during human trophoblast differentiation: influence of human chorionic gonadotrophin. *Endocrinology* **135,** 402–408.

13. Burghardt, R. C., Barhoumi, R., Stickney, M., Monga, M., Ku, C. Y., and Sanborn, B. M. (1996) Correlation between connexin43 expression, cell–cell communication, and oxytocin-induced Ca^{2+} responses in an immortalized human myometrial cell line. *Biol. Reprod.* **55,** 433–438.

14. Verrecchia, F. and Hervé, J. C. (1997) Reversible blockade of gap junctional communication by 2,3-butanedione monoxime in rat cardiac myocytes. *Am. J. Physiol.* **272,** C875–C885.

15. Stein, L. S., Boonstra, J., and Burghardt, R. C. (1992) Reduced cell-cell communication between mitotic and nonmitotic coupled cells. *Exp. Cell Res.* **198,** 1–7.

16. Bastide, B., JarryGuichard, T., Briand, J. P., Délèze, J., and Gros, D. (1996) Effect of antipeptide antibodies directed against three domains of connexin43 on the gap junctional permeability of cultured heart cells. *J. Membr. Biol.* **150,** 243–253.

17. Matesic, D. F., Hayashi, T., Trosko, J. E., and Germak, J. A. (1996) Upregulation of gap junctional intercellular communication in immortalized gonadotropin-releasing hormone neurons by stimulation of the cyclic AMP pathway. *Neuroendocrinology* **64,** 286–297.

18. Cronier, L., Hervé, J. C., Délèze, J., and Malassiné, A. (1997) Regulation of gap junctional communication during human trophoblast differentiation. *Microsc. Res. Techn.* **38,** 21–28.

19. Lin, J.H-C., Weigel, H., Cotrina, M. L., Liu, S.,Bueno, E., Hansen, A. J., Hansen, T. W., Goldma, S., and Nedergaard, M. (1998) Gap-junction-mediated propagation and amplification of cell injury. *Nat. Neurosci.* **1,** 494–500.

20. Stein, L. S., Stein, D. W. J., Echols, J., and Burghardt, R. C. (1993) Concomitant alterations of desmosomes, adhesiveness, and diffusion through gap junction channels in a rat ovarian transformation model system. *Exp. Cell Res.* **207,** 19–32.

21. Elfgang, C., Eckert, R., Lichtenbergfrate, H., Butterweck, A., Traub, O., Klein, R. A., Hülser, D. F., and Willecke, K. (1995) Specific permeability and selective formation of gap junction channels in connexin-transfected HeLa cells. *J. Cell Biol.* **129,** 805–817.
22. Bastide, B., Hervé, J. C., Cronier, L., and Délèze, J. (1995) Rapid onset and calcium independence of the gap junction uncoupling induced by heptanol in cultured heart cells. *Pflügers Archiv.* **429,** 386–393.

18

Capture of Transjunctional Metabolites

Gary S. Goldberg and Paul D. Lampe

1. Introduction

By providing for the direct, intercellular exchange of small (<1000 daltons) molecules, gap junctions mediate intercellular communication and a variety of physiological functions. The intercellular exchange of molecules through these channels is crucial for processes ranging from fertilization and development to cell homeostasis and differentiation (*1–3*). However, identification of the specific transjunctional molecules that control such processes has remained somewhat elusive.

Even though it is a commonly used method, measurement of dye transfer may not necessarily represent the transfer of biologically meaningful compounds. To address this problem, we have developed techniques to "capture," identify, and quantitate the transfer of endogenous transjunctional metabolites (*4,5*). For this procedure, "donor" cells are metabolically labeled with a radioisotope, fluorescently marked, and plated with nonlabeled "receiver" cells. After communicating for a specified period of time, the donors and receivers are separated from each other by fluorescence-activated cell sorting (FACS). Radioactive transjunctional molecules that traveled from the donors to the receivers are then resolved from the lysed cells by filtration and chromatography.

An important consideration in selection of the tracer and the labeling time is the rate of its metabolism. In this example, donors are labeled overnight with [^{14}C]glucose. Because glucose exchange is relatively rapid, the cells reach a "steady state" of incorporated isotope by the time of the co-culture procedure where junctional transfer occurs. Of course this does not eliminate the possibility that a molecule that passes through a gap junction could be metabolized by the receiving cell. To reduce this possibility, we suggest that cells be allowed to transfer for a relatively short period and that the FACS and lysis steps occur

From: *Methods in Molecular Biology, vol. 154: Connexin Methods and Protocols*
Edited by: R. Bruzzone and C. Giaume © Humana Press Inc., Totowa, NJ

as rapidly as possible. A preloading support protocol is given here that can help ensure that the cells to be used communicate well under the co-culture conditions employed *(6)*.

A diagrammatic presentation of the labeling/capture procedure is presented in **Fig. 1**. The method consists of six steps:

1. Metabolic and fluorescent labeling of donor cells.
2. Co-culture of donor cells with receiver cells to allow for transfer of radioactive transjunctional metabolites.
3. Separation of the donors and receivers by FACS.
4. Filtration of cell lysates to isolate hydrophilic transjunctional molecules.
5. High-performance liquid chromatography (HPLC) resolution of metabolites.
6. Thin-layer chromatography (TLC) resolution of metabolites.

2. Materials

1. Common chemicals are from Sigma unless otherwise stated. All solutions are stored at 4°C unless otherwise noted.
2. Sterile 10-cm tissue culture dishes, 50 mL and 6 mL (12 × 74 mm) tubes with caps, cell strainers (Falcon cat. no. 2235), 12 × 75 sterile culture tubes, cell culture medium (e.g., Dulbecco's modified Eagle medium [DMEM] [1 g/L glucose] with 10% fetal bovine serum [FBS]), 0.25% trypsin/1 mM EDTA, phosphate-buffered saline (PBS), and/or any other items needed for the culture of the target cells. Gap junction competent cells: connexin43 (Cx43) transfected C6 glioma cells are used here.
3. Calcein-AM stock solution (5 mM): Dissolve 5 mg/mL of calcein-acetoxymethyl ester (AM) (C3100, Molecular Probes) in dimethyl sulfoxide (DMSO). Store at –20°C.
4. DiI stock solution: 10 mg/mL of DiI (D282, Molecular Probes) in DMSO stored at –20°C.
5. Preloading labeling solution: Add 1 µL of Calcein stock solution to 1 mL medium, mix, then add 1 µL of DiI stock solution and mix again. Vary volumes accordingly. Make 1.5–2 mL for each 6-cm plate of cells to be labeled. Use fresh.
6. Fluorescence microscope equipped with rhodamine and fluorescein optics.
7. Isotope: This example uses D-[U-^{14}C]glucose from Amersham stored at –20°C.
8. Sterile, low protein binding nonpyrogenic 13-mm syringe filters with a 0.2 µm pore size (e.g., Millex from Millipore) and 5- or 10-mL syringes with 25-gauge needles.
9. A nontoxic, specific gap junction blocker (*see* **Note 2**): α-carbenoxolone (ACO, from Biorex Labs, Enfield UK) is used here—stock solution 100 mM in H$_2$O. PBS supplemented with a gap junction blocker is also needed. PBS with 100 µM ACO is used here. This is made by adding 50 µL of ACO stock solution to 50 mL of PBS.
10. FACS Solution D: PBS with 10 mM glucose, 30 mM n-(2-hydroxyethyl) piperazine-N'-2-ethanesulfonic acid (HEPES), 0.1% bovine serum albumin, 0.75% sodium citrate, 1 mM EDTA.

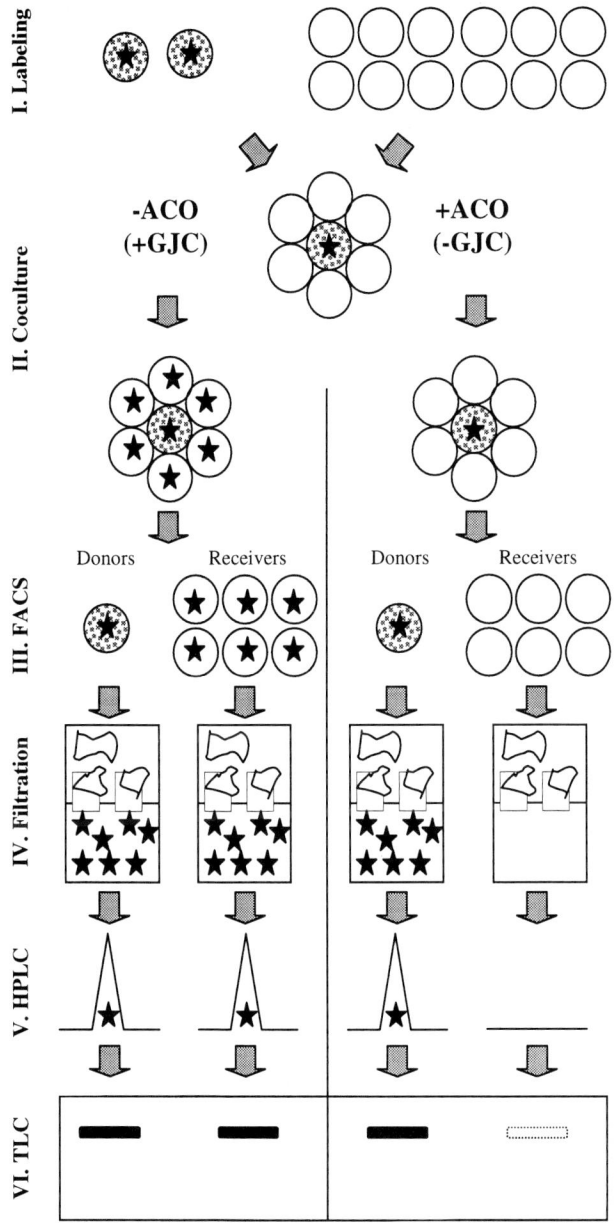

Fig. 1. Outline of the capture protocol. Donor cells are labeled with a fluorescent marker to allow separation *(noted by stippling)* and with radioisotope to metabolically label transjunctional molecules *(filled stars)*. The cells are co-cultured and allowed to communicate with receiver cells. The two cell types are separated by FACS and their contents analyzed by HPLC and TLC.

11. A fluorescence-activated cell sorter, tubes, filters, and an operator. A modified FACStar plus (Becton-Dickinson, San Jose, CA) was used here.

12. Lysis Solution: 10 mM Tris-HCl, pH 8.0, 10 mM EDTA, pH 8.0, 1% Nonidet P-40 (NP-40), 1 mM phenylmethylsulfonyl fluoride (PMSF); PMSF is added immediately prior to use from a 100 mM stock in ethanol kept at –20°C.

13. Dilution solution: 10 mM Tris-HCl, pH 8.0, 10 mM EDTA, pH 8.0.

14. Filtration devices, such as Centricons (Amicon) with 50-kDa and 3-kDa nominal molecular mass cutoffs (NMWCOs).

15. Whatman Linear-K preabsorbent TLC plates (4866–821) or plates best suited to your application.

16. TLC ascension buffer: Freshly made 0.6% NH$_4$OH, 70% isopropanol is used here.

17. TLC detection reagents. Fresh P-anisadine–phthalate reagent: 1.23 g of p-anisadine and 1.66 g of phthalic acid in 100 mL of methanol. Fresh ninhydrin reagent: 0.2 g of ninhydrin in 100 mL of ethanol. Migration standards such as nucleotides and amino acids at 2 mg/mL in water.

18. High performance liquid chromatograph, appropriate HPLC columns, and solvents. A Hewlett Packard 1050 LC with a diode array detector was used here with Bio-Rad Microsorb C18 and Aminex HPX-87H ion exchange columns.

19. Equipment able to detect captured radioactive transjunctional metabolites within required sensitivity ranges. Storage phosphor screens and a Molecular Dynamics PhosphorImager equipped with ImageQuant software (Molecular Dynamics, Sunnyvale, CA) was used here.

3. Methods

3.1. Preloading Support Protocol

This technique is used to verify conditions that allow the cells to communicate, as well as to determine the efficacy of negative controls such as the use of gap junction blockers, under the culture conditions to be employed. Doing these preliminary experiments will allow you to determine the amount of time necessary for your particular cells to form gap junctions and communicate during co-culture of the donors and receivers. Basically this protocol utilizes calcein instead of radioactive glucose metabolites to measure dye transfer.

1. For standard cells, split one 10-cm plate into one 6-cm plate and a number of 10 cm plates determined by the number of conditions you wish to test and grow to confluence (e.g., 3 d). The 6-cm plate will be used to label the donor cells with calcein and DiI.

2. Prepare a 12 × 75 mm sterile culture tube with cap for each condition to be tested. As a control, add an appropriate volume of GJC blocking agent (such as ACO) to one or more of these tubes to achieve the desired final concentration at a final volume of 3.05 mL. To label donors, completely remove medium from a confluent 6-cm plate of cells by aspiration. Add 1.5–2.0 mL of labeling solution. Incubate at 37°C, 100% humidity, and 5% CO$_2$ for 15–30 min.

3. While cells are incubating, remove medium from each confluent 10-cm plate of cells to be used as receivers, wash once with PBS, add trypsin, then remove the trypsin by aspiration after the appropriate time. After the cells round up and separate, suspend them in 7 mL of medium/plate. This volume can be varied depending on the density of the cells and their plating efficiency. However, it is very important that the cells be separated from each other and suspended well. Add 3 mL of this suspension to each tube from **step 1**.

4. Wash the labeled plate of donors twice with PBS and once with trypsin. Suspend the cells in 5 mL of medium after they separate from the plate and each other. Add 50 µL of this suspension to each of the tubes containing unlabeled cells from **step 3**. You can also add some to a tube without unlabeled cells as a control

5. Mix the cells well by inverting the tubes and, if necessary, vortex-mix. Pour the contents of each tube onto a 6-cm culture dish. The last drops can be transferred by tapping the tube onto a dry area of the plate. Swirl the dish and incubate. After the cells settle on the plate (about 1.5–3 h) examine them by fluorescence microscopy. The DiI and calcein are visualized using filter sets suitable for rhodamine and fluorescein, respectively. The resulting ratio of donors to receivers enables quantitation of GJC as the number of receivers getting calcein, but not DiI, from an individual donor cell. This enables determination of the rate and extent of recovery from ACO washout.

3.2. Labeling of Donor Cells for FACS

Start with six 10-cm plates of healthy confluent cells (*see* **Note 1**). For standard cells, split one 10-cm plate into six 10-cm plates with 10 mL of DMEM per plate and grow for 3 d. One of these plates will be used to label the donor cells, while the other five will be used as receivers. Donors and receivers are plated at a 1:6.25 ratio as described below. The experimental design is outlined schematically in **Fig. 2**. This enables analysis of donors and receivers plated in the absence, presence, and after washout of ACO, before and after sorting, as well as analysis of medium from these cells (*see* **Note 2**). Donor cells must be labeled both metabolically with the radionuclide of choice and fluorescently to allow cell sorting. Unlike the support protocol described in **Subheading 3.1.**, however, donors here are not labeled with calcein.

3.2.1. Radioactive Labeling of Donors

1. Gently remove the cell culture DMEM medium from the plate of donor cells and replace with 5 mL of fresh medium with a 10-mL pipet. Discard the remaining medium. In this way, cells are metabolically labeled in the same medium they have been growing in, rather than fresh medium (*see* **Note 3**).

2. Tip the plate and add the radionuclide (50 µCi of glucose, i.e., 250 µL of D-[U-^{14}C]glucose from Amersham) while gently swirling. This results in a final activity of 10 µCi/mL.

3. Incubate overnight. (Beware of volatile radionuclides that may be produced from this labeling and take suitable precautions.)

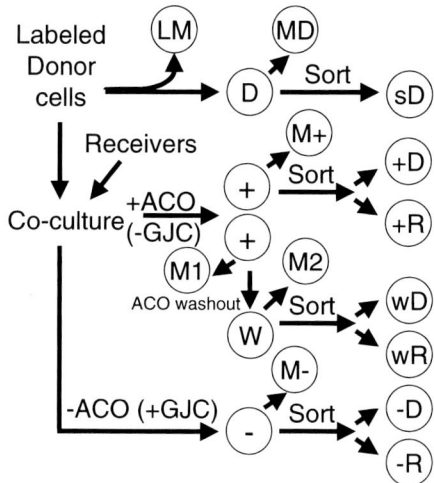

Fig. 2. Experimental design. After the radioactive labeling medium (LM) is removed, donor cells (labeled with the tracer molecule of interest and DiI) are co-cultured with nonlabeled receivers in the absence or presence of ACO. After communication, the cells are separated by FACS and analyzed for transjunctional molecules. –, +, W are cells co-cultured in the absence, presence, or after washout of gap junction blocker (ACO) before sorting. D are presorted donors plated without any receivers. –D, –R, M– are sorted donors, receivers, and medium from cells co-cultured without ACO. +D, +R, M+ are sorted donors, receivers, and medium from cells co-cultured with ACO. wD, wR, M2 are sorted donors, receivers, and medium after ACO washout from cells co-cultured with ACO, while M1 is medium from the same cells before ACO washout. sD and MD are sorted donor cells and medium from the donor cells plated without receivers.

4. Remove the radioactive medium from the labeled donor cells with a 5- or 10-mL syringe and 25-gauge needle. Pass this medium through a 13-mm, 0.2-µm low protein binding nonpyrogenic syringe filter into a sterile 6-mL polystyrene tube with cap. This filtration aids in cell labeling and dye removal (*see* **Note 4**).

3.2.2. Fluorescent Labeling of Donors

1. Add 25 µL of DiI stock solution to the filtered medium in this tube (50 µM final) and mix well by inversion.
2. Aspirate any residual medium from the plate, add the DiI labeled radioactive medium to the cells, and return to the incubator for 25 min.

3.3. Co-Culture

Co-culture allows the molecule of interest to pass from the donors to receivers via gap junctions. As a control, this procedure should also be performed in parallel in the presence of gap junction channel blockers. Our conditions include

Table 1
Preparation of Cells for Co-Culture

Plate label	Donors	Receivers	Medium	ACO stock
Donors alone	2 mL	0 mL	12.5 mL	0 μL
Mix without ACO	4 mL	25 mL	0 mL	0 μL
Mix with ACO	4 mL	25 mL	0 mL	30 μL

donors alone, as well as receivers and donors with and without the channel blocker ACO. Media is also collected here and in the next section to examine labeling efficiency and the possibility of radioactive material leaking and diffusing between cells.

1. While the donor cells are being labeled with DiI, suspend the receivers in 50 mL of medium. Wash each of the five plates once with 2 mL of PBS (2 mL per plate) and then with 2 mL of trypsin/EDTA. Tilt the plate for 30 s and aspirate completely to remove as much PBS and trypsin as possible (*see* **Note 5**). Monitor the cells under a microscope. When the cells separate well from the plate and each other, tap the plates sharply against a hard surface to release the cells. Suspend the cells from each plate in 10 mL of medium by vigorously passing up and down a 10-mL pipet several times before transferring into a sterile 50-mL polypropylene tube with cap.
2. Collect the labeling medium (LM in **Fig. 2**) from the donor cells. Wash the plate 4 times with 4 mL of PBS, 2 mL of trypsin, suspend in 10 mL of medium, and transfer to a 50-mL tube as done for the receivers (*see* **Notes 4** and **5**).
3. Prepare cells according to **Table 1**, mix well, and add 12.5 mL from each tube to a 10-cm plate. Swirl the plates and begin incubating.
4. After the cells have formed monolayers (*see* **Note 5**), collect medium from the washout plate (M1 in **Fig. 2**). Wash once with 5 mL of ACO-free medium. Replace with 12.5 mL of fresh ACO-free medium and incubate another 20 min with the other cells. This allows for only 20 min or less of communication between the cells of this treatment group (i.e., wD and wR in **Fig. 2**).

3.4. Separation

Separation includes releasing the cells from the dish and sorting the cells by FACS. Media are also collected here to examine the possibility of radioactive material leaking and diffusing between cells.

1. Collect media from the plates (MD, M–, M+, and M2 in **Fig. 2**) and save for analysis. Wash the plates once with PBS supplemented with the gap junction blocker and once with trypsin/EDTA (2 mL per plate).
2. When the cells separate well from the plates and each other, tap them sharply against a hard surface and suspend them vigorously in FACS solution D (2 mL/ plate) and transfer to 6-mL tubes with caps. This should commonly work out to

around 3 million cells/mL. This yields four groups (tubes) of cells: D, –, +, and W, as shown in **Fig. 2**.

3. Sort cells by Fluorescence Activated Cell Sorting based on several criteria including FSC (forward scatter), SSC (side scatter), and at least one parameter of fluorescence (FL1 and FL2) (*see* **Note 6**). This yields seven groups of cells: sD, –D, –R, +D, +R, wD, and wR, as shown in **Fig. 2**.

4. Transfer 150 μL of each sorted sample to a new tube for reanalysis to verify purity (approx 5% of donors and 2% of receivers) by FACS. Also examine cells by microscopy and plating to verify their purity and viability. Donors, but not receivers, should exhibit DiI fluorescence, while all cells should adhere and grow when replated on tissue culture dishes.

5. Pellet sorted cells and 0.5 mL of nonsorted cells at 3000*g* for 10 min. This can be done in the same tubes used to collect the cells. Carefully aspirate the liquid completely from the cell pellets. Freeze immediately in dry ice and store at –70°C.

3.5. Filtration

Filtration involves lysing the cells and filtering through progressively smaller molecular weight cutoff filters to separate out transjunctional candidates.

1. Lyse each tube of cells (i.e., sD, –D, –R, +D, +R, wD, and wR) completely in 75 μL of lysis solution for 20 min at 4°C. This is enough solution to efficiently suspend and lyse well over 2 million cells.

2. Add 423 μL of dilution solution to each tube and mix well.

3. Centrifuge through 50-kDa Centricons. Collect retentates and assay radioactivity (*see* **Note 7**).

4. Centrifuge filtrate through 3-kDa Centricons. Collect retentates and filtrates and assay for radioactivity (*see* **Note 7**). Freeze samples at –70°C.

5. Because all of these samples are analyzed by scintillation counting, it is convenient to count an aliquot of all medium samples also at this point (*see* **Note 7**).

3.6. HPLC

The HPLC consists of a pump with mixing valves (or multiple pumps), an injection port, a column, and a detector (usually UV absorbance but refractive index is also useful). Most systems utilize a computer to coordinate the various pieces and track the absorbance changes. Given the diversity of small biological molecules that may pass through gap junctions, defining a particular HPLC technique to separate all of these molecules is impossible. Reversed-phase HPLC utilizes a gradient of increasing hydrophobicity (usually acetonitrile or methanol) to elute molecules from the column. Reversed phase columns have several distinct advantages: C18 columns have incredible resolving power (some chiral forms of amino acids can be separated from each other). One effectively desalts the eluted material. The solvents usually are 100% volatile. The latter two points can allow much easier subsequent purification/molecule identification. For example, electrospray mass spectrometry can utilize HPLC separation to prepare the material for mass spectrometry without derivatization. Reverse phase is particularly well suited to use for amino acids/peptides. The main disadvantage is that several potential transjunctional molecules are highly hydrophilic and do not have any affinity for the C18 matrix and thus pass through without separation. For this reason, we utilized the Aminex HPX-87H column from Bio-Rad which separates small molecules using multiple mechanisms including ion exclusion, ion exchange, and reversed phase. Many different columns exist to separate a variety of types of molecules, and if you are planning to test the junctional transfer of a specific molecule or class of molecules, speak to a number of manufacturers (e.g., Bio-Rad, Vydac, Hewlett-Packard, Waters). We will describe our procedure below but it will have to be adapted for different columns and molecules of interest.

1. Program the HPLC to perform the necessary gradient. For the C18 column we use a water–acetonitrile–trifluoroacetic acid gradient from 99.9:0:0.1 to 79.9:20:0.1 over 20 min at 0.6 mL/min and 37°C.
2. After blank runs are performed, inject known standards such as the specific tracer molecule and any other metabolites. This can allow you to modify the gradient to allow for better separation of these molecules and note their absorbance/refractive index properties to aid with subsequent identification.
3. Inject the filtered samples onto the C18 column and run the gradient. These samples should run well because detergents present in the lysis solution are diluted by the dilution solution. The entire sample remaining after scintillation counting can normally be injected, but this may depend on your needed detection levels and HPLC apparatus such as detection loops, etc. (*see* **Note 7**).
4. Fractions are taken every minute and 5% of each fraction is scintillation counted. The fractions are then frozen and can be lyophilized.
5. For our studies, several of the glucose metabolites were not well separated by C18-HPLC, so we also utilized an Aminex HPX-87H column (Bio-Rad) with isocratic elution using 5 mM H_2SO_4 at 0.35 mL/min and 40°C. For this column, the lyophilized fractions from the C18 column are suspended in 5 mM H_2SO_4 before being injected (*see* **Note 7**).
6. Fractions are taken every minute and 50% is scintillation counted.
7. For both columns elution times and absorbance/refractive index properties are compared with known standards and any tentative identifications are made. Further separation (e.g., TLC) may or may not need to be performed.

3.7. TLC

Many transjunctional molecules chromatograph similarly on the HPLC. Therefore, further separation of each fraction can help purify and identify specific compounds. Owing to the small size and relative hydrophilicity of transjunctional molecules, TLC can be a valuable separation technique.

1. Add standards to an aliquot of each sample to be analyzed. It is best to include standards in the samples to be analyzed to directly compare migration of radioactive bands with known compounds. Every quality of the bands, including shape, size, and migration should be identical with a standard to identify a molecule. Nucleotides (2.5 µg per lane) can be easily be detected by UV quenching. Sugars (3 µg per lane) and amino acids (2 µg per lane) can also be used and detected as described below in **step 6**.

2. Fifty microliters of the HPLC fractions can be spotted onto the preabsorbent zone of a TLC plate and allowed to dry completely. These samples should also contain internal standards (e.g., glucose, ADP, ATP) to allow chemical detection as described below **step 6**. More sample can be applied again to the same lane to increase the level of sample for more rapid detection. However, some samples may contain salt that can retard migration and resolution when concentrated by drying.

3. Place precut Whatman 3MM paper into a standard glass TLC chromatography tank. Add enough ascension buffer to fill the tank to about 1 cm depth from the bottom of the tank (TLC with harmful solvents should be run in a fume hood). Equilibrate for 20 min. Place the plate into the chamber and run until the solvent front travels 1–2 cm from the top of the plate (about 5 h).

4. Remove the plate(s), air-dry, and heat 110°C for 5–10 min. Visualize nucleic acid standards with a hand-held UV lamp and outline bands with a soft pencil.

5. Expose to a PhosphorImager storage screen to detect labeled metabolites. If you have enough radioactivity, you may be able to detect these by standard autoradiography. However, we have routinely exposed some samples obtained from receiver cells for 3 wk to a Molecular Dynamics PhosphorImager storage screen to obtain acceptable detection of signal.

6. To detect sugars or amino acid standards, spray the plates with p-anisadine–phthalate reagent or ninhydrin reagent, respectively. Air-dry and heat at 110°C for 5–10 min until a color forms. Sugars and amino acids show up as yellow-green or purple, respectively, by these methods *(7)*.

4. Notes

1. Grow and maintain cells in the appropriate medium that will allow efficient GJC. Take great care to keep cells healthy during all procedures by working quickly, efficiently, and gently. It is essential to have all tubes and materials ready, labeled, and in place before starting.

2. It is important to use negative controls to identify transjunctional molecules. Communication deficient cells as closely related as possible to the communicating cells that are examined are useful here. Nontoxic gap junction blockers, such as ACO and AGA, can also be employed. The reversible nature of these compounds enables the capture of molecules that travel within a specified length of time. A final concentration of 100 µ*M* ACO is sufficient to inhibit Cx43-mediated GJC between Cx43-transfected C6 cells *(4–6,8)*. Other gap junctions and cells may require more or less of a particular blocker *(6)*.

3. For metabolic labeling, pick an isotope that will give the best chance of labeling an unknown molecule, or, possibly, targeting a suspected molecule. D-[U-^{14}C]glucose from Amersham has been used successfully with this procedure *(4,5)*. Every carbon is labeled in this form of glucose. Therefore, this labeling is relatively general with respect to metabolites derived from this food source. Safety and cleaning procedures must be rigorously adhered to, particularly to deal with aerosol produced during cell sorting procedures, incubations, and metabolic labeling. Also, care must be taken to contain media and solutions that are aspirated. This protocol uses the same medium that the cells have been growing in to best simulate the ongoing cellular environment.

4. Fluorescent labeling depends on the parameters required and available on the cell sorter. A series of nonisotopic runs should be performed to determine the best parameters to separate the donor and receiver cells from each other. It is VERY important to remove all dye by aspirating completely. Otherwise receivers will also become labeled. In addition, the filtration procedure removes cellular debris and other material that can incorporate DiI and potentially allow some labeling of the receivers.

5. For co-culture, it is **very** important to aspirate all trypsin completely. Otherwise, residual trypsin will inhibit prompt formation of communicating monolayers. Examine the cells under a microscope to determine when to remove medium for the washout plate. However, do not go into the incubator too early or often.

6. FACS may be the most challenging and critical step in this procedure. To begin with, the cells must be completely separated from each other in suspension prior to sorting. Cell strainers (*see* **Subheading 2.**) can be used for this purpose. In line filters can be used during sorting, but should be changed between each sorting sample. It is also advisable to vortex-mix often. Because the cells are sorted based on fluorescence, a donor and receiver pair of cells that were not completely separated by trypsin/EDTA treatment will be sorted into the donor pool. Therefore, the receivers can reach a very high level of purity more easily than the donors. Obviously, quality and accurate quantitation are imperative for this procedure. So, windows must be cut very conservatively. The vast majority of cells do not meet this sort criteria and are not collected. Of course, it is also essential to carefully reanalyze collected cells to verify their purity and numbers. It is important to maintain the cells on ice and sort as fast as possible to prevent metabolism of transjunctional molecules. In addition, reverse the sorting order in different experiments to verify the absence of effects of any time differences due to sorting. Culture cells as close as possible to the sorter to make the process as efficient as possible. It should be possible to routinely sort more than 3000 cells per second to collect about 200,000 donors and 1 million receivers in approx 15 min with > 99% and 85% purity for receivers and donors, respectively. As mentioned above, more than two thirds of these cells may not meet sorting criteria and go uncollected.

7. It is important to keep track of volumes during all manipulations, including filtration and chromatography, as all radioactive measurements should be converted to values per cell.

Acknowledgments

We would like to thank Dave Sheedy and Carleton Stewart (Roswell Park Cancer Center) for cell sorting. We are also very grateful to Christian Naus (University of Western Ontario) and Bruce Nicholson (SUNY Buffalo) for their help and support. This work was funded in part by grants from the NIH (GM55632) to PDL, International Agency for Research on Cancer (WHO), Wendy Will Case Cancer Research Fund, and PachTech to GSG.

References

1. Bruzzone, R., White, T. W., and Paul, D. L. (1996) Connections with connexins: the molecular basis of direct intercellular signaling. *Eur. J. Biochem.* **238,** 1–27.
2. Goodenough, D. A., Goliger, J. A., and Paul, D. L. (1996) Connexins, connexons, and intercellular communication. *Annu. Rev. Biochem.* **65,** 475–502.
3. Kumar, N. M. and Gilula, N. B. (1996) The gap junction communication channel. *Cell* **84,** 381–388.
4. Goldberg, G. S., Lampe, P. D., Sheedy, D., Stewart, C. C., Nicholson, B. J., and Naus, C. C. G. (1998) Direct identification and analysis of transjunctional ADP from Cx43 transfected C6 glioma cells. *Exp. Cell Res.* **239,** 82–92.
5. Goldberg, G. S., Lampe, P. D., and Nicholson, B. J. (1999) Selective transfer of endogenous metabolites through gap junctions composed of different connexins. *Nat. Cell Biol.* **1,** 457–459.
6. Goldberg, G. S., Bechberger, J. F., and Naus, C. C. G. (1995) A pre-loading method of evaluating gap junctional communication by fluorescent dye transfer. *Biotechniques* **18,** 490–497.
7. Krebs, K. G., Heusser, D., and Wimmer, H. (1969) Spray reagents, in *Thin-Layer Chromatography* (Stahl, P., ed.),   Springer-Verlag, New York, pp. 854–908.
8. Goldberg, G. S., Moreno, A. P., Bechberger, J. F., Hearn, S. S., Shivers, R. R., MacPhee, DJ, Zhang, Y. C., and Naus, C. C. G. (1996) Evidence that disruption of connexon particle arrangements in gap junction plaques is associated with inhibition of gap junctional communication by a glycyrrhetinic acid derivative. *Exp. Cell Res.* **222,** 48–53.

III

PHYSIOLOGY AND BIOLOGY OF CONNEXINS

19

The Study of Connexin Hemichannels (Connexons) in *Xenopus* Oocytes

E. Brady Trexler and Vytas K. Verselis

1. Introduction

Studies of ion channels often utilize patch clamping in excised patch configurations together with rapid solution exchange to examine the kinetics of drug and/or agonist action at the single-channel level. However, studies of these sorts have not been possible for gap junction (GJ) channels, which are formed by the coupling of two hemichannels, one from each cell. Accessing a GJ channel directly with a patch electrode would require disruption of one cell, and excision of that patch would require pulling it off as an intact double-membrane structure. Direct patching onto a junctional membrane has been accomplished in the earthworm ventral nerve chord by teasing out the median giant septate axon and cutting it close to a septum. The septum bulges out and appears to maintain a high density of GJ channels, making it a preparation amenable to patching *(1)*. However, invertebrates express a family of GJ proteins unrelated in primary sequence to the vertebrate connexins *(2)*, and a direct patch approach has not proved applicable to mammalian cells expressing native or exogenous connexins owing to technical difficulties such as membrane resealing and/or vesicle formation on cell disruption. Nature, however, has provided an alternative that circumvents these technical difficulties. Some connexins, most notably Cx46, are capable of forming functional hemichannels, or connexons, in addition to forming GJ channels between cells *(3–5)*. Instead of patching onto a junctional membrane, one can patch onto the surface membrane of a cell expressing one of these connexins and record from single hemichannels. Although hemichannels are not full GJ channels, they display many of the conductance and gating properties expected of unapposed connexin hemichannels *(6)* and

From: *Methods in Molecular Biology, vol. 154: Connexin Methods and Protocols*
Edited by: R. Bruzzone and C. Giaume © Humana Press Inc., Totowa, NJ

thus, are likely to contribute significantly to our understanding structure–function relationships of full GJ channels.

For the most part, functional hemichannels have been observed using the *Xenopus* oocyte expression system, but there are reports of functional hemichannels in transfected mammalian cells as well *(7)*. In addition, connexins in isolated horizontal cells from teleost retina form functional hemichannels *(8)*. In two-electrode voltage clamp experiments, depolarization of *Xenopus* oocytes expressing Cx46, Cx38, or Cx56 to inside positive voltages elicits slowly activating outward currents *(4,6)*. Although it is uncertain why these connexins readily open in the surface membrane when unapposed, whereas other connexins do not, the slowly activating outward currents seem to be a hallmark of functional connexin hemichannel expression. Indeed, we and others have seen these same slowly activating outward currents in *Xenopus* oocytes expressing hemichannels formed of the chimerical connexin, Cx32*43E1, in which the first extracellular loop of Cx32 is replaced with that of Cx43 *(9)*.

Electrophysiological studies of connexin hemichannels are much like studies of other ion channels, and some of the methods described here are not new. However, there are some peculiarities associated with recording from connexin hemichannels and we provide the methodologies needed to successfully obtain and excise membrane patches from *Xenopus* oocytes expressing Cx46 hemichannels. We also describe the methods we have employed to rapidly exchange solutions to excised patches while acquiring data in a gap-free mode. The ability to rapidly and uniformly apply connexin modulators to either face of a hemichannel and the greater time resolution and lower noise afforded by recording from single hemichannels in small membrane patches are new to the connexin field. The use of hemichannels is likely to be an approach that will routinely accompany structure–function and regulation studies of GJ channels.

2. Materials

2.1. Reagents

All reagents are available from Sigma (St. Louis, MO), except where noted.

1. XOS and XOS+Ca^{2+} (*Xenopus* oocyte isolation and storage solutions). Prepare 10× stocks with the following concentrations: for XOS: 880 mM NaCl, 10 mM KCl, 10 mM MgCl$_2$, 100 mM n-(2-hydroxy ethyl) (HEPES), 10 g/L of glucose; adjust pH to 7.6 with NaOH. For XOS+Ca^{2+}: add 18 mM CaCl$_2$ in addition to XOS components. If filter sterilized, these solutions may be stored for up to 3 mo at 4°C. On dilution, 2.5 mM pyruvate is added to XOS+Ca^{2+} and pH is readjusted to 7.6 with NaOH. We usually make 250–500 mL of both 1X solutions per frog.
2. Type IV Collagenase (Worthington Biochemical Corporation, Freehold, NJ).

3. Devitellinization solution: 220 mM sodium aspartate, 10 mM KCl, 1 mM MgCl$_2$, 10 mM HEPES; adjust pH to 7.6 with NaOH.
4. Agarose, low melting point (Fisher, Pittsburgh, PA).
5. TEVC electrode filling solution: 1 or 2 M KCl and 10 mM HEPES, pH = 7.6.
6. Patch pipet filling solution (internal pipet solution [IPS]): 100 mM KCl, 1 mM CaCl$_2$, 1 mM MgCl$_2$, 5 mM EGTA, 5 mM HEPES. This is the normal solution we use for patching. For permeability measurements, this solution or another may be used, in which KCl is replaced with a different salt. For pH studies, a different buffer may be used, such as PIPES or MES, and the concentration may be increased.
7. *Xenopus laevis* frogs (Xenopus I, Ann Arbor, Michigan).
8. Qiagen PCR prep spin columns (Qiagen, Valencia, CA).
9. *Xenopus* Cx38 antisense oligonucleotide:5'-GCTTTAGTAATTCCCATCCTGC CATGTTTC-3', which is complementary to *Xen*Cx38 commencing at NT-5 with respect to the initiation codon. A 1 μM synthesis using phosphorothioate nucleotide analogs is resuspended at 8 pmol/μL concentration in RNase-free double-distilled H$_2$O (ddH$_2$O) and stored at −20°C. It is convenient to make higher concentration stocks of the antisense oligonucleotide and dilute them upon use.

2.2. Apparatus

1. Dissecting microscope.
2. 35- and 60-mm plastic Petri dishes (Falcon 3001 and 3002, Fisher).
3. Pasteur pipets for handling oocytes. During initial stages of oocyte isolation, the pipet is broken at the taper and used inverted, with the bulb attached at the break. For handling individual oocytes, the pipets are cut with a diamond knife 1 or 2 cm from the start of the taper (to make the tip wider) and firepolished to smooth the end.
4. Refrigerator set at 18°C.
5. Two-electrode voltage clamp (e.g., GeneClamp 500 from Axon Instruments, Foster City, CA).
6. Patch clamp (e.g., AxoPatch 200B from Axon Instruments).
7. Three-axis manipulators (e.g., MHW-3 from Narishige, Japan).
8. Pipet puller (e.g., P-97 from Sutter Instruments, Novato, CA or PP-83, Narashige).
9. Glass for rapid solution switching (thin-septum theta, cat. no. TGC150-10—tube glass from Warner Instrument, Hamden, CT).
10. Glass for patch pipets: 1.2 mm outer diameter (o.d.) (cat. no. 6020) or 1.5 mm o.d. (cat. no. 5968) filament glass from AM Systems, Everett, WA.
11. Piezoelectric actuator, amplifier/driver and mounting system (e.g., LSS-3000 from Burleigh, Fishers, NY).

3. Methods
3.1. Oocyte Isolation

Isolating oocytes from *Xenopus laevis* is a relatively straightforward process and has been reviewed elsewhere (*see* Skerrett et al., Chapter 13). We wish to stress some specifics relevant to hemichannels. There are a number of factors that can affect membrane properties and hemichannel expression, both of which will affect patch-recording success.

1. The use of collagenase with low ancillary protease activity was found to be best suited for studies of connexin hemichannels. Low protease activity appears to minimize the expression of endogenous *Xen*Cx38, consistently yields greater levels of exogenous hemichannel expression, and leaves the surface membrane amenable to obtaining gigaohm seals and excising patches in inside-out and outside out configurations. Otherwise, excised patches tend to form vesicles.

2. All collagenases have ancillary protease activity, the amount of which is determined by the purification procedure. Type I collagenase has the most ancillary protease activity, whereas type IV has the least. It is best to try a few different lots from different manufacturers before purchasing a collagenase. The trial lots are usually shipped with specification sheets that list the amounts of caseinase, clostripain, and tryptic activities. Usually, the lot with the lowest ancillary protease activities will give the best results for patching. Incubation time in collagenase as well as the collagenase concentration may be varied for each lot to optimize results. As a general rule, if the solution becomes too cloudy (as a result of cell rupture) during collagenase treatment or during the washing steps, try shortening the incubation time or lowering the collagenase concentration and washing more frequently.

3. The oocytes isolated using type IV collagenase from Worthington Biochemical do not adhere together very well, which is disadvantageous for promoting cell–cell coupling. Curiously, the use of collagenase with moderately higher ancillary protease activity yields better results in gap junctional coupling experiments. Very high protease activity tends to promote endogenous *Xen*38 coupling.

3.2. Preparation of cRNA

1. Expression of connexin proteins in *Xenopus* oocytes is usually achieved by injecting cRNA transcribed in vitro from cloned DNA. We will not discuss the details of making cRNA, as numerous protocols and commercial kits are available for this purpose (*see* Chapters 13 and 14). Usually, the cRNA is purified and concentrated by phenol-chloroform extraction and/or ethanol precipitation. However, resuspension of the cRNA pellet often is incomplete, and leads to small particles that can clog the injection pipets during loading. It is uncertain whether these particles are actually undissolved cRNA or salt, but our experience is that the level of expression is inversely proportional to the amount of uninjectable particulate matter in the cRNA suspension.

2. To achieve expression levels necessary for patching and avoid the extraction, precipitation, and resuspension steps, the completed cRNA transcription reaction can be purified on silica-gel spin columns, such as those supplied by Qiagen (Valencia, CA). The QIAquick polymerase chain reaction (PCR) purification kits are simple and provide excellent yield. Although sold as a means to purify DNA from the PCR reaction, the spin columns and reagents are ribonuclease free, and

the protocol is adapted for purification from an enzymatic reaction. The kits sold expressly for RNA isolation are designed for total RNA isolation from tissue, which is not ideal for our purposes. Connexin cRNA can be eluted from the columns with the Cx38 antisense oligonucleotide solution. The oligonucleotide is small enough not to bind to the silica gel, and passes through. Because the turnover rate of connexins is rapid, 2–3 h *(10)*, coinjection with Cx38 antisense almost eliminates Cx38 expression after a period of 1–2 d.

3.3. Patch Recording from Connexin Hemichannels

3.3.1. Expression Levels Required for Patching

1. To test for Cx46 expression levels suitable for patching, oocytes with intact vitelline membranes are voltage clamped to –60 mV and depolarized to +20 or +30 mV in XOS+Ca^{2+}. In our experience, oocytes that are considered "patchable" for Cx46 possess an outward current that exceeds 1 μA within 1–2 s of stepping the membrane voltage; these criteria will undoubtedly differ among connexins, depending on single-channel conductance and voltage activation properties. By "patchable" we mean that >80% of the patches will contain at least one functional hemichannel.

2. The oocytes meeting the expression criteria are devitellinized by placing them into a "discharged" 35-mm dish lid containing the hypertonic devitellinizing solution. Oocytes will adhere to a fresh plastic Petri dish, which becomes a serious problem when trying to remove them from the hypertonic media back to XOS+Ca^{2+} for storage or recording. To overcome this, the dishes can be "discharged" by rubbing distilled water around the surface with a finger. After ~10 min in the hypertonic solution, the cells shrink away from the vitelline membrane sufficiently and can be removed manually with forceps. Devitellinized cells are transferred to a 35-mm dish with the bottom coated with a 1% agarose/XOS+Ca^{2+} to keep the oocytes from adhering to the bottom.

3. Sometimes there is some loss of functional hemichannel expression after devitellinization. This became evident to us by an inability to see any hemichannel activity patch after patch from oocytes that had met or exceeded the criteria for expression. Subsequent testing of these devitellinized oocytes revealed significantly reduced currents compared to those just prior to devitellinization. The reason for this phenomenon is unclear, but appears to vary among batches of oocytes and collagenases. We find that if the oocytes are allowed to recover in XOS+Ca^{2+} for 2 h, expression levels often return. Overexpression can also a problem in that the cells become unhealthy and unresponsive to hypertonic shock. The best results for patching are obtained 24–60 h post-injection.

3.3.2. Recording Setup

3.3.2.1. RECORDING CHAMBER

1. A schematic view of a simple patch-recording chamber for oocytes is presented in **Fig. 1**. The chamber is constructed from Plexiglas and contains two compartments.

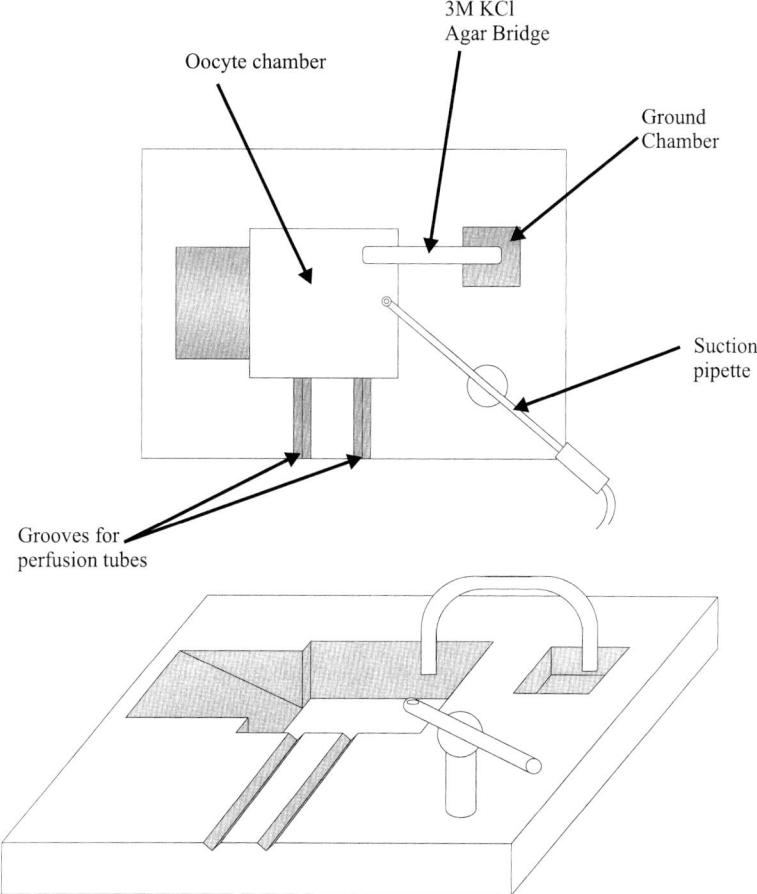

Fig. 1. A schematic diagram of a perfusion chamber suitable for studying connexin hemichannels in excised patches.

A glass slide affixed to the bottom with silicone adhesive forms the bottom of the larger compartment, which houses the oocyte. The angled entry (at the left of the larger compartment in the figure shown) allows access of a patch pipet to the oocyte at a shallow angle (beneficial when rapidly perfusing an excised patch with another electrode; *see* **Subheading 3.4.**). The smaller compartment serves as a ground reservoir. Two V-shaped grooves from the edge of the Plexiglas to the large compartment serve to restrict the movement of the bath perfusion pipets. These pipets are usually formed from patch pipet glass that has been bent to an S-shape, providing a wide stream parallel to the bottom of the chamber.

2. A suction pipet is formed from 1.2–1.5 mm inner diameter glass. One end is flame sealed, and then, upon reentering the flame, air is forced in to blow a hole

in the sealed end. At this stage, the thin glass cylinder that results from blowing the hole is broken off with forceps under a microscope. The jagged edges that remain should be fire polished to create a smooth hole. When used as a suction pipet, the hole will face up, and the solution level in the bath will be maintained at a level slightly below the hole, owing to capillary action. The suction pipet is in constant contact with the bath solution, eliminating the electrical noise generated from the breaking and reforming of contact with the bath solution that occurs with other types of suction pipets. A 3 *M* KCl–agarose bridge is constructed by bending a 3–4 cm piece of glass tubing into a U-shape. This is then filled with a 3 *M* KCl–1% agarose solution. The bridge can be reused multiple times if stored in 3 *M* KCl at 4°C.

3. Prior to recording, the bottom of the larger recording chamber is coated with 1% agarose–XOS. The ground electrode is placed in the smaller compartment, which is filled with the IPS. The 3 *M* KCl–agarose bridge connects the two compartments. The arrangement of a separate ground chamber with the IPS and a 3 *M* KCl bridge minimizes the differences in junction potentials between the perfused solutions and the bath.

3.3.2.2. ACQUISITION SOFTWARE AND HARDWARE

1. There are numerous software packages available for data acquisition. Most vendors offer a hardware–software combination that is sufficient for most electrophysiological experiments. The choice should depend on the type of computer system you are most familiar with (i.e., Apple or PC), the maximum number of data channels you will likely acquire, the maximum sample rate of the ADC board, and the number of waveform output channels, or DAC channels. Useful software features include automatic sampling of the gain and filter frequency settings of the amplifier and an easily configurable seal-test protocol. The software should have a variety of acquisition modes, including continuous, episodic, and event triggered. It should be able to sample multiple input channels simultaneously. In addition, it should have macro capabilities or the ability to alter the software through source code, as it is not likely that any single package will be flexible enough to perform every type of experiment. Some commercial software packages have limitations that are unsuitable for the special requirements of studies of gap junction channels (*see* **Notes 1** and **3**).

3.3.3. Analysis of Single Hemichannels

Cx46 hemichannels display complex voltage dependence that is best described by the action of two distinct voltage gates (*6*). Separation of the actions of voltage as resulting from two gates is suggested from the qualitative differences in channel behavior between positive and negative potentials. For negative potentials, the predominant gating behavior involves slow transitions between γ_{open} and γ_{closed} that are mediated by numerous short-lived substates. In contrast, at positive potentials the channel predominantly transits between

γ_{open} and γ_{sub}; these transitions are rapid and involve no intermediate substates. In fact, substates represent a significant component of connexin hemichannel gating, as well as cell–cell channel gating. Kinetic analyses of such complex gating behavior involving substates can be performed by utilizing a combination of existing methodologies that are particularly useful for channels with multiple substates, whether short-lived or long-lived.

3.3.3.1. Visualizing Short-Lived Substates

1. Although long-lived substates are easily discerned in a current time series and in all-points histograms, the substates characteristic of connexin hemichannel transitions between fully open and closed states are not generally visible in all-points amplitude histograms as they are too short-lived and only broaden the peaks of the open and closed states. Discerning these short-lived conductance levels from a view of the time series has also proved problematic. Thus, we employ mean-variance histograms (MVHs) of varying window widths as a means of visualizing the substates *(11)*.

2. A window consisting of a number of points (3–200) is moved along the sampled current from a single channel and the mean and variance within each window are calculated. A three-dimensional histogram is constructed by binning occurrences of mean-variance pairs. In the example of an MVH plot shown in **Fig. 2A**, the broad low-variance regions correspond to the open and closed current levels, at which the hemichannel spends most of the time, and the small low-variance regions correspond to the short-lived substates.

3. The substates that are visible in the MVH have lifetimes greater than or equal to the window width. An all-points histogram of the same current record is shown in **Fig. 2B** and is not informative in discerning substates.

3.3.3.2. Idealization with Substates

1. The first step in a kinetic analysis of single channel data is the construction of a noise-free idealized current time series. Although standard half-height threshold algorithms are useful for idealizing current records from channels that switch between zero and full conductance states, they deal poorly with transitions to multiple current sublevels. We employ a detection algorithm called the sublevel Hinkley detector (SHD) to idealize current records with multiple substates.

2. This algorithm examines all the possible current level jumps in parallel by calculating the velocity of increase to each level *(12)*. The velocity of increase has a maximum for a jump the correct level. For Cx46 hemichannel data, we identify current levels from the low-variance regions of the MVH. The SHD is then used to idealize the trace. The SHD, in addition to producing an idealized current trace, creates a transition matrix showing all possible transitions and their frequencies. An example of an idealization of a segment current record from a single Cx46 hemichannel is shown in **Fig. 2C**. When only two levels are selected for idealization, the algorithm behaves as a simple half-height crossing detector (*not shown*).

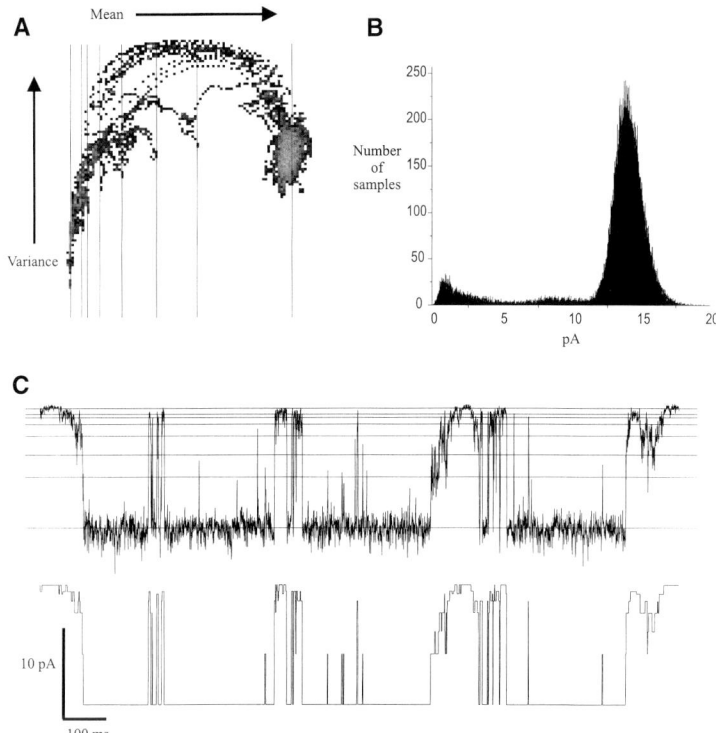

Fig. 2. A segment of a recording of a cell-attached patch containing a single active Cx46 hemichannel is analyzed by using mean-variance histograms (MVHs) and the sublevel Hinkley-detector (SHD). (**A**) A MVH of the patch current allows identification of the short-lived substates that constitute the transitions between the main open and closed states. (**B**) An all-points amplitude histogram is useful only for identifying the main open and closed states. (**C**) The SHD provides a noise-free idealization of the segment using the eight levels selected in **A**.

3.4. Rapid Perfusion of Excised Patches

The study of GJ channel sensitivity to chemical modulators in tissue or cell-pair preparations has been confounded by an inability to rapidly and uniformly change chemical concentrations, which is necessary to determine the kinetics of action and to avoid slower secondary effects. Single hemichannels in excised patches exposed to fast perfusion provide a means of examining the action of such modulators with millisecond time resolution.

In the late 1980s, a technology developed that enabled excised patches containing ion channels to be subjected to submillisecond solution switching. This was achieved by moving an excised patch through a "liquid filament," that is, an interface between laminarly flowing solutions *(13)*. The best way to achieve laminar flow between two solutions is to use theta tubing. Rapid switching can be achieved by a piezoelectric driven actuator.

3.4.1. Construction of Theta Tube Perfusion Pipets

1. The theta tube can be pulled on any standard microelectrode puller. To reduce oscillations caused by the elasticity of the theta pipet, it should be pulled with a short shank.
2. The tip diameter of the theta tube should be 100–200 μm. Because it is nearly impossible to consistently pull tips of this diameter, we pull the theta tubes to a smaller tip diameter (1–2 μm) and score the shank with a diamond knife. A micrometer mounted in the eyepiece of a dissecting scope is used to assess the diameter of the shank for scoring. Slight pressure with forceps to the scored shank produces a clean break leaving an interface perpendicular to the flow.
3. The pipet is mounted on a thin (1/8-in.) sheet of Plexiglas that is mounted on top of a larger sheet (same thickness). A groove milled in the upper Plexiglas sheet perpendicular to the direction of movement of the piezo-actuator helps maintain the same alignment from one experiment to another.
4. Once pulled to the desired tip, the theta tube is fastened to the Plexiglas with soft wax, placed under the microscope, and the septum is aligned so that it is straight up and down. Once adjusted, the perfusion pipet is permanently fixed with hot glue.
5. To supply solutions to each barrel, a biomedical needle is gently inserted into the back of each barrel until the tip of the needle lodges into the tapered portion of the shank. For the thin-walled theta glass we use, a 30-gauge stainless steel biomedical needle (Popper, New Hyde Park, NY) fits nicely into each barrel. The back ends of the barrels from which the needles emerge are sealed with sticky wax (Kerr, Emeryville, CA) to prevent leakage and the external portion of each needle is fastened to the bottom Plexiglas sheet.
6. Tubes connecting reservoirs with the desired solutions are then attached to the luer-end of the needles and the whole assembly is mounted onto the piezo-positioner.

3.4.2. Flow Adjustment

1. Flow rate through the theta tube barrels is easily controlled by gravity and can be adjusted by changing the heights of the reservoirs feeding each barrel.
2. Changes in flow rate change the shape of each solution stream and the width of mixing interface between the two streams.
3. Flow rates should be adjusted to achieve a thin interface between the two streams. Faster flow rates achieve thinner interfaces, but also tend to disrupt the patch seal. For a 200 μm diameter theta tube tip, we find an optimum flow rate of 0.2 mL/min with the patch pipet placed ~100 μm from the edge of the of flow tube.

3.4.3. Achieving a Rapid Transition Through the Interface

1. To achieve rapid-solution switching requires application of an appropriate voltage protocol to the amplifier/driver. The position of the piezo-actuator is nearly linearly related to the applied voltage. Programmable drivers are available that can be synchronized with electrophysiological data acquisition by means of a digital trigger input. If the driver is not programmable, an analog voltage waveform can be applied through an external input.

2. Software generation of a waveform through a DAC provides the most flexibility and ease of synchronization with electrophysiological data acquisition. Although, in principle, a voltage step would produce the most rapid movement of the piezo-actuator, it also introduces unwanted mechanical oscillations of the perfusion tip and may damage the piezo-device.

3. A voltage protocol consisting of three ramps, rather than a single step, provides the best results (*see* **Note 2** for discussion).

4. Pilot experiments using a high-resistance pipet and switching between two different salt solutions are useful in ascertaining the best flow rate, pipet placement, and piezo-voltage protocol. The speed of solution switching can tested by measuring the change in the holding current due to a change in tip potential of a high-resistance patch pipet when switching between two solutions differing in salt composition (solutions A and B in **Fig. 3A**). The 10–90% difference in tip potential (measured in voltage clamp as shift in holding current) was typically traversed in ~1 ms (**Fig. 3A**).

5. An example of the utilization of fast solution switching to examine the kinetics of pH action on a connexin hemichannel is shown in **Fig. 3B**. In this example, the pH at the intracellular face of an inside-out patch containing a single Cx46 hemichannel was rapidly switched from pH 7.5 to pH 6.0 by employing a piezo-actuator driven by the three-ramp protocol described previously. Three consecutive sweeps are shown in which the hemichannel was subjected to a 2-s application of pH 6.0. Switching was achieved in ~ 1 ms and very rapid compared to the time scale shown. The single hemichannel conductance can be seen to decrease almost immediately upon solution switching (middle trace), and after a brief latency the hemichannel closes and predominantly remains closed for the duration of the low-pH application *(14)*.

4. Notes

1. Special considerations are necessary for recording electrophysiological data from gap junction channels or hemichannels. For gap junction channels, continuous acquisition (no stimulus or gaps) requires at least two data channels to be acquired simultaneously. A commonly used software package, pClamp 6.0 (Axon Instruments, Foster City, CA), only allows acquisition of a single data channel when acquiring in continuous mode. A newer version, pClamp 7.0 (Axon Instruments), has remedied this problem, allowing up to 16 data channels to be sampled simultaneously. For hemichannel recordings, a single data channel may be all that is

A

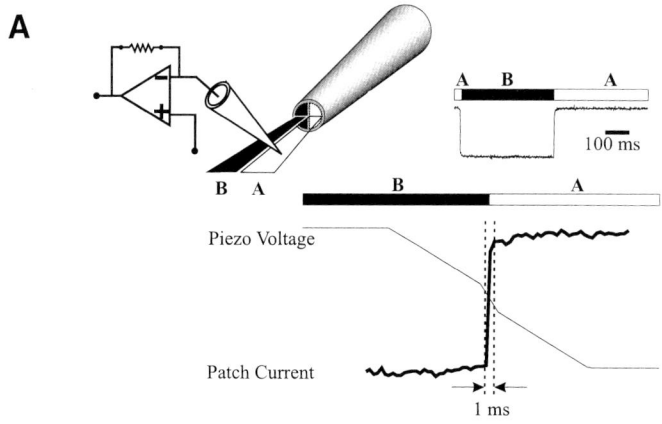

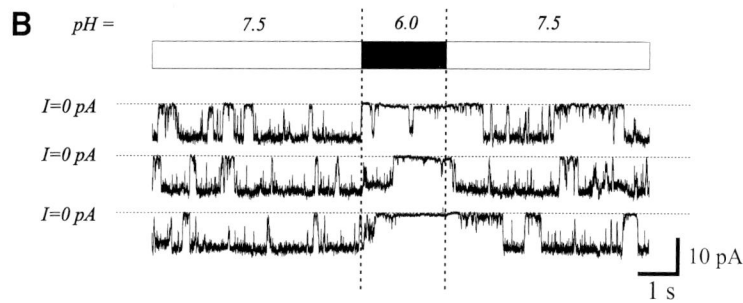

B

Fig. 3. Fast application of chemical modulators is possible with excised patches and a piezo-actuator. (**A**) An example of switching between two different salt solutions with a high-resistance pipet. The 10–90% difference in holding current is traversed in < 1 ms. (**B**) An example of multiple fast applications of pH 6.0 to the same inside-out patch using a piezo-actuator. *I* = 0 *pA* denotes the leak conductance of the patch.

necessary during continuous acquisition, unless the command voltage is also acquired. It is preferable to have two analog-out channels accessible from the software that will control the command voltage or a device such as a piezo actuator. The potential at each output should be modifiable during acquisition.

There are further considerations for pulsed data acquisition, data acquired during application of a stimulus waveform. Stimulus waveforms are divided into epochs that define the time at a given voltage, and whether that voltage should be reached by a single step or by ramping. pClamp 6 provides eight configurable epochs, whereas pClamp 7 provides 10, but both versions limit the number of samples per episode (or sweep) to 16,384 points. This means that when sampling at 5000 samples per second (5 kS/s), each waveform is ~3.3 s. If two channels are sampled at 5 kS/s, then the waveform time is halved. Both connexin cell–cell channels and hemichannels have mean open times on the order of hundreds of milliseconds to seconds. For Cx46 hemichannels, the opening and closing transitions are complex and full transitions require 10–100 ms for completion. Any acquisition program that limits the time per episode to just a few seconds with a sample rate such as 5 kS/s will be inadequate for studies of connexin channels.

2. To minimize the ringing of the theta tube assembly and piezo-actuator itself, a voltage protocol consisting of three ramps is ideal for a rapid transition across the interface between the two solutions flowing through the theta tube. The first ramp produces a slow approach (2 µm/ms) of the patch electrode from the center of one stream toward the interface. The second ramp rapidly (10 µm/ms) moves the patch pipet through the interface and the third ramp slowly (2 µm/ms) moves the patch pipet to the center of the second stream (*see* **Fig. 3A**). This waveform configuration minimizes ringing and resonance of the piezo-device and theta tube while achieving a rapid transition through the interface. The piezo-actuator we use has a maximum range of 70 µm; the corresponding input voltage range to achieve the full range of motion is 10 V. The voltage waveform we use consists of ramps from 0 to 4 V for 14 ms, 4–6 V for the next 1.4 ms, and 6–10 V for the last 14 ms. The distances moved during each of the three ramps are 28 µm, 14 µm, and 28 µm, respectively.

   Most commercially available stimulus/data acquisition software packages can output waveforms only when acquiring in episodic mode. As the "on" and "off" waveforms that position the actuator consist of three ramps, six epochs are necessary for piezo-movement. To record channel activity during the three periods before, during, and after exposure to the test solution, the software would have to provide at least nine configurable epochs (*see* **Note 1**).

   One workaround to allow software control of a piezo-actuator within software limitations is to modify the method by which the voltage is applied to the piezo-actuator. Most new commercial acquisition programs can acquire multiple data channels during a continuous acquisition. They also allow the voltage at either DAC channel to be modified without interrupting acquisition. The piezo-actuator can be repositioned by stepping the voltage between 0 and 10 V. To reduce the oscillations that an instantaneous voltage step will induce, filtering can be employed to slow the rise time. A simple RC circuit with a variable time constant of 10–100 ms will slow the rise time enough to prevent ringing without appreciably increasing the time spent crossing the interface between the two theta tube streams. Details of such conditioning circuits and the use of Clampex 7 to coordinate piezo movement and electrophysiological acquisition are provided by Kabakov et al. *(15)*. The exact time of interface crossing for each patch can be determined by first disrupting the patch and then flowing distilled water down the barrel that contained the test solution. The holding current of the patch will change as the piezo-actuator moves the interface across the patch pipet tip.

3. It may be desirable to write your own data acquisition software for the greatest flexibility in data acquisition. We have written our own software that controls

data acquisition boards from National Instruments (NI) (Austin, TX). NI hardware allows for a waveform output to start even if an acquisition is already running. On NI boards, separate timers drive acquisition and output. This means that a waveform output can operate at a different rate than the acquisition rate. With this flexibility, "on" and "off" waveforms can be applied at any time during a continuous acquisition of both patch current and piezo-voltage. Solution exchanges are thus synchronized with channel activity. Moreover, in episodic stimulation mode, waveforms can be of any length and any number of epochs, which is beneficial when compiling a large number of similar experiments for ensemble analysis.

References

1. Brink, P. R. and Fan, S. F. (1989) Patch clamp recordings from membranes which contain gap junction channels. *Biophys. J.* **56,** 579–593.
2. Phelan, P., Bacon, J. P., Davies, J. A., Stebbings, L. A., Todman, M. G., Avery, L., Baines, R. A., Barnes, T. M., Ford, C., Hekimi, S., Lee, R., Shaw, J. E., Starich, T. A., Curtin, K. D., Sun, Y. A., and Wyman, R. J. (1998) Innexins: a family of invertebrate gap-junction proteins [letter]. *Trends Genet.* **14,** 348–349.
3. Paul, D. L., Ebihara, L., Takemoto, L. J., Swenson, K. I., and Goodenough, D. A. (1991) Connexin46, a novel lens gap junction protein, induces voltage-gated currents in nonjunctional plasma membrane of *Xenopus* oocytes. *J. Cell Biol.* **115,** 1077–1089.
4. Ebihara, L. and Steiner, E. (1993) Properties of a nonjunctional current expressed from a rat connexin46 cDNA in *Xenopus* oocytes. *J. Gen. Physiol.* **102,** 59–74.
5. Ebihara, L., Berthoud, V. M., and Beyer, E. C. (1995) Distinct behavior of connexin56 and connexin46 gap junctional channels can be predicted from the behavior of their hemi-gap-junctional channels. *Biophys. J.* **68,** 1796–1803.
6. Trexler, E. B., Bennett, M. V. L., Bargiello, T. A., and Verselis, V. K. (1996) Voltage gating and permeation in a gap junction hemichannel. *Proc. Natl. Acad. Sci. USA* **93,** 5836–5841.
7. Li, H., Liu, T. F., Lazrak, A., Peracchia, C., Goldberg, G. S., Lampe, P. D., and Johnson, R. G. (1996) Properties and regulation of gap junctional hemichannels in the plasma membranes of cultured cells. *J. Cell. Biol.* **134,** 1019–1030.
8. DeVries, S. H. and Schwartz, E. A. (1992) Hemi-gap-junction channels in solitary horizontal cells of the catfish retina. *J. Physiol.* **445,** 201–230.
9. Pfahnl, A., Zhou, X. W., Werner, R., and Dahl, G. (1997) A chimeric connexin forming gap junction hemichannels. *Pflügers Arch.* **433,** 773–779.
10. Laird, D. W. (1996) The life cycle of a connexin: gap junction formation, removal, and degradation. *J. Bioenerg. Biomembr.* **28,** 311–318.
11. Patlak, J. B. (1993) Measuring kinetics of complex single ion channel data using mean-variance histograms. *Biophys. J.* **65,** 29–42.
12. Draber, S. and Schultze, R. (1994) Detection of jumps in single-channel data containing subconductance levels. *Biophys. J.* **67,** 1404–1413.
13. Franke, C., Hatt, H., and Dudel, J. (1987) Liquid filament switch for ultra-fast exchanges of solutions at excised patches of synaptic membrane of crayfish muscle. *Neurosci. Lett.* **77,** 199–204.

14. Trexler, E. B., Bukauskas, F. F., Bennett, M. V. L., Bargiello, T. A., and Verselis, V. K. (1999) Rapid and direct effects of pH on connexins revealed by the connexin46 hemichannel preparation. *J. Gen. Physiol.* **113,** 721–742.
15. Kabakov, A. Y. and Papke, R. L. (1998) Ultra fast solution applications for Prolonged gap-free recordings: controlling a Burleigh piezo-electric positioner with Clampex 7. *Axobits.* **22,** 6–8.

20

Exploring Hemichannel Permeability In Vitro

Andrew L. Harris and Carville G. Bevans

1. Introduction

The intercellular molecular signaling that occurs though gap junctions is determined by the particular molecules that can pass through the specific connexin channels involved. Differences in the biological roles played by different connexin isoforms are no doubt due, in part, to differences in the molecular selectivity of the channels they form. The molecular selectivity of connexins can be investigated in cells by monitoring intercellular movement of fluorescent or radiolabeled tracers. The tracers must be either injected into cells or loaded via acetoxymethyl (AM)-ester or similar chemistry. This approach has the advantage of working with native gap junction structures, but the disadvantage of being limited to the use as tracers of molecules that are tolerated well by cells and that do not affect connexin expression or function. In addition, enzymatic activities in cytoplasm can cause serious problems for interpretation of results, particularly for biologically active junction-permeant molecules such as second messengers.

The protocol described here was developed to assess in vitro the molecular selectivity of purified connexin channels (*see* **Note 1**). It has the advantages and disadvantages of a noncellular system in defining the molecular selectivity of junctional channels. In it, purified connexin is reconstituted into liposome membranes. The liposomes are then loaded with tracer molecules, and subjected to transport specific fractionation (TSF). In TSF the exchange of osmolytes (urea and sucrose) through a reconstituted channel increases the density of the liposome. The increase in density is detected as a change in position in a density gradient. A mixture of liposomes with and without functional channels forms two bands—liposomes *with* functional channels deep in the density gradient, and liposomes *without* functional channels nearer the top.

From: *Methods in Molecular Biology, vol. 154: Connexin Methods and Protocols*
Edited by: R. Bruzzone and C. Giaume © Humana Press Inc., Totowa, NJ

To assess molecular selectivity of the channels, the tracer(s) retained by the liposomes in the two populations are compared (*see* **Note 2**).

Channel *activity* is thus assessed by a change in density/position, while *permeability* is assessed by tracer loss. Liposomes that are permeable to the tracer rapidly lose it. The tracer per liposome retained in the upper band (composed of liposomes without functional channels) is an internal control for amount of tracer loaded into the liposomes, and for nonspecific leakage. Comparison of the tracer per liposome of the upper band with that of the lower band thus reveals whether the tracer is able to pass through the open channels. Owing to the small size of the liposomes, complete loss of tracer from a liposome occurs very rapidly, so each liposome either retains tracer or loses it completely. TSF has been used to identify connexin-specific differences in molecular selectivity for homomeric connexin32 (Cx32) channels and heteromeric Cx26/Cx32 channels *(1–4)*.

1.1. Stepwise Summary and Review of TSF

1. Liposomes are formed from connexin and lipid solubilized in detergent (including a trace amount of a rhodamine-labeled lipid).
2. Liposomes are formed in, and entrap, a urea-containing solution.
3. Liposomes are loaded with tracer.
4. Tracer is rapidly lost from liposomes that contain channels permeable to it.
5. Liposomes are centrifuged through an isoosmolar density gradient (TSF gradient) formed from the urea solution and a more dense solution in which the urea is osmotically replaced by sucrose (sucrose has greater density than urea).
6. Solutions and gradients are constructed so that the density of liposome membrane is that near the bottom of the gradient.
7. Liposomes that do not contain functioning connexin channels are not permeable to urea and sucrose. Their density remains low due to the entrapped urea solution, and they remain in the upper part of the gradient.
8. Liposomes that contain functioning connexin channels are permeable to urea and sucrose, and equilibrate these two osmolytes. These liposomes effectively have the (greater) density of the phospholipid membrane, and move to a lower position in the gradient.
9. Following centrifugation, the two populations of liposomes are recovered, and the ratio of retained tracer to lipid (assessed by rhodamine content) of each determined.

2. Materials

2.1. Reagents

1. Urea buffer (1XU): 10 mM KCl, 10 mM N-(2-hydroxyethyl) piperazine-N'-2-ethanesulfonic acid (HEPES), 0.1 mM EDTA, 0.1 mM EGTA, 3 mM sodium azide, 459 mM urea adjusted to pH 7.6 with KOH.
2. 10X urea buffer (10XU): Same as 1XU except concentration of urea is 10X (4.59 M).

3. Sucrose buffer (1XS): Identical to 1XU except that an osmotically equivalent concentration of sucrose, 400 mM, is substituted for the urea (*see* **Note 3**).
4. Detergent solutions: 80 mM octylglucoside (Calbiochem), 160 mM octylglucoside in 1XU.
5. Gel-filtration: Bio-Gel (A-0.5 m, 100–200 mesh, exclusion limit 500,000 daltons; Bio-Rad).
6. Lipid stock: 10 mg/mL of phosphatidylcholine (egg; Avanti) and phosphatidylserine (brain; Avanti) at a molar ratio of 2:1, also containing 1 mol% rhodamine-labeled phosphatidylethanolamine (Molecular Probes) (*see* **Note 4**)
7. Solvents (HPLC grade): Methanol, chloroform, DMSO
8. Miscellaneous: Compressed argon gas; 8-methoxypyrene-1,3,6-trisulfonic acid (MPTS, Molecular Probes); 3 M KCl.

2.2. Apparatus

2.2.1. Standard Items

1. Gel filtration (in cold room): glass columns (1.5×20 cm, 0.7×10 cm); peristaltic pump set at 9 mL/h; fraction collector.
2. Equilibration liposomes: capillary pipets (5–200 µL; Corning Pyrex); desiccator with vacuum pump, with high-adsorbency silica gel with indicator in the desiccator, always kept under vacuum; probe-tip sonicator, where probe can fit into 13×100 mm glass culture tubes and reach to within millimeters of the bottom (Branson); benchtop clinical centrifuge (in cold room); test tube vortex mixer (in the cold room); rotary shaker (in cold room); microcentrifuge.
3. TSF gradients: 37°C water bath; ring stand, magnetic stirplate; micro-stirbar (2×7 mm, Nalgene); 4–5-mL transparent ultracentrifuge tubes with greatest available height/diameter ratio (e.g., Beckman Ultra-Clear); gradient maker (Hoefer, SG Series); compressed air canister; vapor pressure osmometer (Wescor).
4. TSF centrifugation and analysis: ultracentrifuge at 37°C; swinging bucket rotor capable of developing 300,000g with 4–5-mL tubes (stored at 37°C); Wratten gelatin optical filters (75×75 mm; nos. 23A and 58; Kodak or Lee).

2.2.2. Homemade Items

1. For TSF gradients: Micrometer-Advance Gradient Pourer (*see* **Fig. 1**). Attached to the output fitting of the gradient maker is a set of short Tygon tubing segments that form an adaptor that accepts 2 mm outer diameter (o.d.) polyethylene (PE) tubing. The PE tubing is approx 25 cm long, and ends in a similar set of Tygon tubing segments attached to a plastic microliter pipet holder of the type usually included with boxes of microliter pipets. This set includes a 0.5 cm length of 1 cm o.d. tubing, which serves as a trap for small bubbles. The microliter pipet holder is affixed to a manual micrometer advance (e.g., Narishige) so that a 100-µL capillary pipet in the holder points along the axis of movement. The apparatus is oriented so that the axis of movement is vertical, with the pipet tip pointing downward. When forming gradients, the micropipet

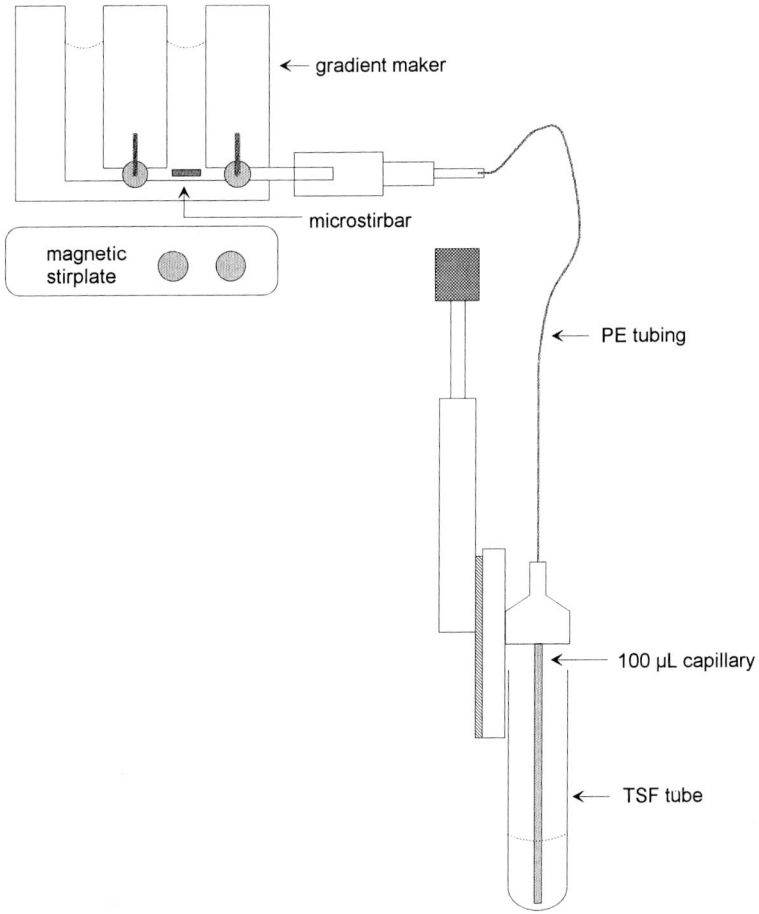

Fig. 1. Micrometer-Advance Gradient Pourer.

tip is at the bottom of the centrifuge tube. After the gradient is poured, the micrometer is used to smoothly remove the pipet from the centrifuge tube, upward
through the gradient.

2. For TSF gradients: Syringe–Tygon Tubing Device. This is a 5 or 10 mL plastic
   syringe with a short segment of large-bore (1 cm o.d., 0.6 cm inner diameter
   [i.d.]) Tygon tubing over the needle fitting. The Tygon tubing is selected so it can
   be snugly inserted into the top of the chambers of the Hoefer gradient maker, and
   also can accept Pasteur pipets. It is used in making gradients and in recovering
   the liposomes after TSF.
3. For TSF centrifugation and analysis: TSF viewing apparatus (*see* **Fig. 2**). This
   apparatus facilitates visualization and recovery of the bands of liposomes separated by TSF. A clamp on a ring stand holds a horizontal piece of 1/4-in. Lucite

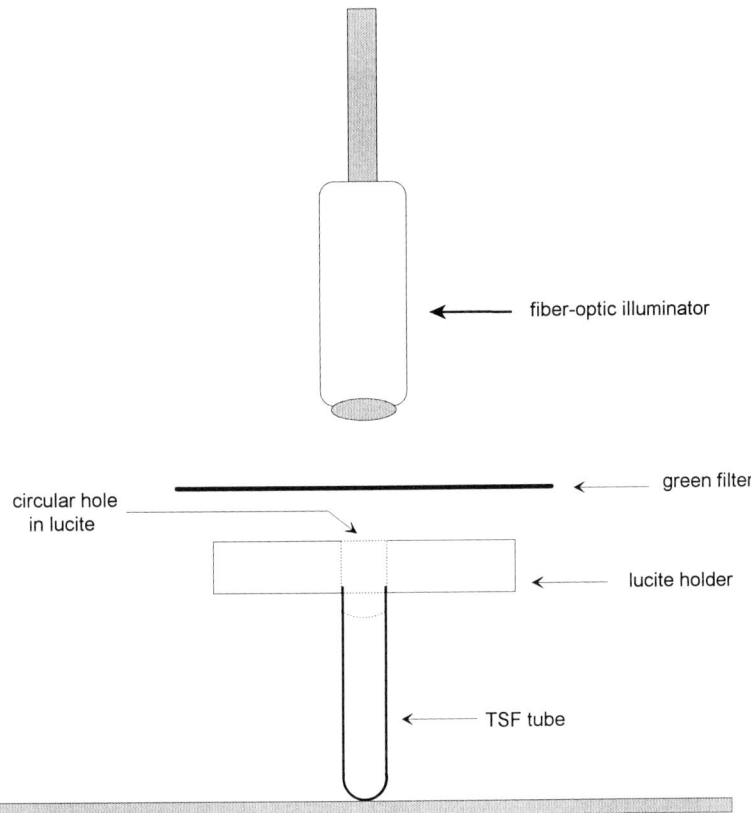

Fig. 2. TSF Viewing Apparatus.

into which has been drilled a smooth-walled hole the diameter of the TSF centrifuge tubes. The Lucite is positioned ~4.5 cm above the benchtop so that a TSF tube can be lowered through the hole from above and come to rest with only the top few millimeters of the tube within the Lucite, holding the tube vertical and visible almost its entire length. A fiber optic illuminator is then directed down onto the TSF tube from above. There is a black background behind the TSF tube. In this configuration, with the room completely dark, it is often possible to visualize the orange rhodamine-labeled band(s) of liposomes after TSF. Greater visual sensitivity can be achieved by positioning a no. 58 (green) Wratten gelatin filter between the illuminator and the TSF tube and viewing the TSF tube from the side through a no. 23A (red) Wratten gelatin filter. For optimal illumination, the fiber optic beam is focused to a wide angle and placed at some distance from the sample tube with a cardboard tube between the two to prevent stray light. The TSF tube remains stabilized in the Lucite holder while the liposome bands are recovered.

3. Methods

3.1. Purification

Connexin suitable for functional reconstitution can be obtained by any of several techniques. Although connexin extracted from purified preparations of junctional membranes has been used for this purpose by our group and others, it is not optimal as it tends to be insoluble and perhaps damaged. There are several other approaches (see also Chapters 3–6). Stauffer et al. used a combination of alkali conditions, detergent, and gel filtration to obtain hemichannels (5,6). Material purified in this way from transfected cells has been used in reconstitution studies (7). We routinely employ immunoaffinity purification from native tissues as a primary source of connexin (3). Connexin obtained in this way, whether from native tissues, primary cultures, or transfected cells, has the advantages of being solubilized in octylglucoside, and not exposed to harsh conditions. It is likely that in the future the use of epitope-tagged (cf. [8]) or polyhistidine-tagged (cf. [9]) constructs expressed in cultured cells will be the most efficient way to obtain purified connexins over which one has control of the amino acid sequence.

3.2. Reconstitution into TSF Liposomes

For optimal TSF, the connexin-containing liposomes should be unilamellar and of monodisperse size. Reconstitution of connexin into such liposomes may be achieved by any of several methods, including dialysis or gel filtration of a connexin–lipid–detergent mixture. Gel filtration is the more rapid and efficient method (1,3,10,11).

3.2.1. Gel-Filtration Column

In a coldroom prepare a glass 0.7×20 cm chromatography column with 17 to 19 cm of Bio-Gel A—0.5 m in 1XU according to the manufacturer's instructions. To avoid bubbles, the column should be formed using chilled 1XU. The gel must be degassed prior to use.

3.2.2. Pre-Equilibration of the Column

Before applying the connexin-containing liposome-forming solution to the column, nonspecific binding of lipid to the column matrix is minimized by pre-equilibrating the column with small liposomes that do not contain protein (equilibration liposomes [ELs]). These liposomes are prepared by sonication of a lipid–1XU mixture, usually 2 mL at 1 mg/mL of the same lipid stock used for the subsequent reconstitution. All work with lipid stocks in chloroform should be done in a fume hood. Care must be taken not to introduce surfactants or moisture into the lipid stocks or the liposome preparation.

1. Remove a tube of lipid stock (10 mg/mL) from –20°C freezer and allow to warm to room temperature for 30 min to avoid moisture condensation inside the tube (*see* **Note 5**).
2. Clean a 100×13 mm glass test tube by filling it with methanol, then with chloroform. Clean the containers used (Erlenmeyer flask, long Pasteur pipet) with each solvent prior to use (*see* **Note 6**).
3. Evaporate the residual chloroform in the tube under a stream of argon gas (*see* **Note 7**). **Caution:** Evaporation of the chloroform rapidly cools the glass and can cause moisture condensation inside the tube, which must be avoided. Continuously flowing argon keeps this to a minimum.
4. Open the lipid stock tube and immediately flow a small amount of argon gas into it to act as a barrier to moisture and oxygen while the tube is open.
5. Using a clean glass capillary pipet; remove 200 µL of the lipid stock and place it in the cleaned test tube.
6. Immediately regas the stock tube and seal. Wrap Parafilm around the lipid stock tube to form a moisture–condensation barrier when it is opened next, and return it to –20°C.
7. Gas the test tube with argon to evaporate the chloroform. While evaporating the chloroform, roll the tube at an angle between fingers to form an even coating of lipid on the sides of the tube up to about 2 cm from the bottom. Allow the gas to flow into the tube for about 2 min after the lipid is dry. Cap the tube.
8. Place the test tube with the dried lipid into a drying vacuum desiccator. Remove the tube cap and apply vacuum to the desiccator for 15 min (*see* **Note 8**).
9. After the vacuum pump has been turned off (remove the vacuum tube before turning off the vacuum pump to avoid having oil sucked back into the tube) attach the desiccator to the argon supply and flood the desiccator with argon to bring it back up to atmospheric pressure to open it.
10. Immediately recap the test tube and remove it to the coldroom.
11. Add 2 mL of 1×U buffer into the tube. Vortex-mix until the observable lipid is in solution. Place the capped tube in an ice bucket for 30 min.
12. Clean the sonicator probe tip by dripping ethanol down the tip and drying with a Kimwipe.
13. Using a clamp on a ring stand, position the test tube so that the probe tip is well into the liquid, but not touching the glass sides. Position a small flexible tube from the argon tank into the test tube about half way, and adjust the argon flow to about 2 psi. From below bring up on a magnetic stirplate a beaker containing an ice slurry and a magnetic stirbar so that the lipid tube is well cooled in the slowly stirring ice water.
14. Sonicate for 6 min, rest 2 min, and sonicate an additional 6 min. Replenish ice as required. Wear hearing protectors if you remain nearby. ELs have formed when the solution becomes clear—the solution must be completely clear before applying it to the gel column.
15. Recap the test tube and centrifuge it in a tabletop clinical centrifuge in a coldroom for 5–10 min. Let it coast to a full stop. Carefully remove and save as much of the

liposome solution as possible without disturbing the debris at the bottom of the tube (titanium from the sonicator probe).

16. Set the peristaltic pump attached to the Bio-Gel column to run at 9 mL/h.

17. Load the ELs onto the column after the meniscus just disappears into the top of the bead bed. Allow the ELs to flow into the bead bed until the meniscus just disappears again, then add 1XU to the top of the column to within about 1 cm of the top and replace the buffer reservoir cap and continue running. Be sure the column head is completely sealed (*see* **Note 9**).

18. Wait until the EL band is just completely off the bottom of the column before adding the liposome-forming solution. Keep the column running and proceed immediately to the reconstitution (*see* **Note 10**).

3.2.3. Reconstitution into Liposomes

The connexin to be reconstituted must be already solubilized in 80 mM octylglucoside. In addition, the final connexin-containing solution applied to the gel-filtration column must also contain 1 mg/mL of lipid and be composed of 1XU. For the protocol that follows, it is assumed that the stock of connexin is in 80 mM octylglucoside and 10 mM salt. Because different ways of obtaining pure connexin can yield it in different solutions (*see* **Subheading 3.1.**), some the specifics in the following protocol may need to be altered so that the final solution is as described above. For a typical reconstitution that yields 40–50% of the liposomes with functional channels, the protein/lipid ratio in the liposome-forming solution is approx 1:300 (w/w). This means that for the standard 1 mg/mL of lipid solution the connexin concentration should be 3–4 µg/mL (*see* **Note 11**). The liposomes formed by this method have a mean diameter of approx 900 Å *(12)*.

1. Clean and rinse a 13 × 100 mm test tube for lipid use as described in **Subheading 3.2.2.** (methanol followed by chloroform).

2. Warm the tube of stock lipid (10 mg/mL) to room temperature for 30 min before using.

3. Dry down 10 µL of stock lipid solution under argon as described in **Subheading 3.2.2.**, **steps 7–9** (*see* **Note 12**).

4. Place the tube with the lipid into a vacuum desiccator for about 15 min. After 15 min shut off the pump and bleed argon gas into the desiccator. Cap and remove the tube.

5. Add 90 µL of the solution of solubilized connexin to the lipid in the tube and replace the cap. Mix on a rotary shaker in a coldroom at 250 rpm until all lipid is dissolved (this is usually the time it takes to rinse the equilibration liposomes from the column, about 10 min). To avoid denaturing the protein, do not continue past the point when all the lipid is in solution.

6. Swirling the solution rapidly by hand, slowly drip in 10 µL of 10XU from a Pipetman P-20 (or similar) fitted with a long, thin, gel-loading tip into the dissolved lipid solution. The rapid mixing is to keep the urea concentration from becoming excessively high during pipetting (the 10XU contains 4.59 M urea).

7. Flush the tube with argon, cap with tight-fitting plug-tight caps (Elkay Plastics), and wrap with Parafilm. Optimally, let the solubilization mix sit in the dark in the cold room for 6–8 h, although shorter incubations can give acceptable results.

8. Remove the excess buffer from the top of the Bio-Gel column and allow the buffer meniscus to just run completely into the beads. With the column pump still on, carefully and slowly add the liposome-making solution using a long Pasteur pipet. Run the mixture entirely into the beads. As the meniscus just disappears, add (one drop at a time) 80 m*M* octylglucoside in 1XU until three drops have been added. Watch the column while adding the drops: After each drop, allow the meniscus to just go dry, then immediately add the next drop. After this the pink band of lipid will be observed to move down into the column bed. Slowly fill the column with regular 1XU. Replace the top to the buffer reservoir (*see* **Note 13**).

9. Make sure there is enough buffer in the reservoir, and turn on the fraction collector. Collect 1.45-min fractions (approx 0.5 mL).

10. The liposomes will elute in the void volume 40–60 min after addition (fraction nos. 25–35) at a pump rate of 9 mL/h.

11. When the liposomes have been collected, cap the collection tubes and label them.

12. Stop the column flow and turn off the stopcock at the bottom of the column. The column should be stripped the same day and washed overnight (*see* **Note 14**).

3.2.4. Stripping the Column

After each reconstitution, the column must be stripped of the residual lipid and protein remaining in the column matrix. If not removed, it builds up and oxidizes, degrading subsequent reconstitutions (*see* **Note 15**).

1. Prepare 0.5 mL of 160 m*M* octylglucoside in 1XU.

2. Turn on the pump of the gel-filtration column (9 mL/h). The buffer reservoir should contain at least 200 mL of 1XU (enough to run overnight).

3. Open the top of the column while it is running. Remove the excess buffer with a long Pasteur pipet, taking care not to disturb the beads. Allow the buffer to flow into the top of the beads until just before the meniscus disappears.

4. With a Pasteur pipet, load 0.5 mL of the 160 m*M* octylglucoside in 1XU onto the top of the column. The first few drops should be dropped lightly onto the middle of the bead bed. The rest of the solution can be flowed down the side of the glass wall. Do this slowly so the beads are not stirred up.

5. Allow the solution to completely flow into the beads and then with a new Pasteur pipet slowly begin adding fresh 1XU (first few drops onto the middle of the bead bed, then the rest down the glass wall). Continue adding one pipetful while washing down the walls of the column. After a few minutes to allow time for the concentrated detergent to diffuse away from the gel bed, remove the excess buffer and allow the rest to flow into the beads until the meniscus just disappears. Then use a new Pasteur pipet to slowly fill up the column with fresh 1XU. Stop when the fluid level is about 1.5 cm from the top. Replace the top of the column.

6. Rinse the column with at least ~40 mL of 1XU.

7. Proceed to equilibrating the column with ELs for the next reconstitution, if appropriate (*see* **Note 16**).

3.3. Loading the Liposomes with Tracers

The tracer molecules used to assess permeability are loaded into the formed liposomes (*see* **Note 17**). Tracers are loaded by incubation with the liposomes in the presence of 5% (v/v) DMSO for 30 min at room temperature. The DMSO and untrapped tracers are removed by gel filtration at 4°C (below the phase transition temperature of the lipids used). Radiolabeled, fluorescent, or spin-labeled tracers can be loaded by this method. This protocol should be practiced with liposomes that do not contain protein before applying it to the experimental samples.

3.3.1. Column to Remove DMSO and Untrapped Tracer(s)

Prepare a small (0.7 cm × 10 cm) desalting column with Bio-Gel A—0.5 m, 100–200 mesh, in 1XU.

3.3.2. Rinsing and Pre-Equilibration of Column (Before Each Use)

1. Degas the column.
2. Apply a 200-μL aliquot of 3 *M* NaCl in 1XU. Repeat (*see* **Note 18**).
3. Apply one pipetful of 1XU, let sit for a few minutes, then remove excess.
4. Wash with 25–50 mL of 1XU.
5. Apply 200 μL of ELs made as described in **Subheading 3.2.2.**
6. Let 25 drops pass to elute the ELs.

3.3.3. Calibration of the Column (Before Each Use)

It is essential to calibrate the column to determine the fraction in which the liposomes will elute, and to establish that unentrapped tracer does not contaminate the liposome fraction. This is done by a test trapping of a standard fluorescent tracer such as MPTS in gel-filtration-formed liposomes that do not contain connexin. These liposomes are not the ELs described (which are smaller and formed by sonication), but are formed on the gel-filtration column as described in **Subheading 3.2.3.**, except without the connexin (nonprotein liposomes [NPLs]).

1. At room temperature, add 10 μL of DMSO and 10 μL of 200 m*M* MPTS to 180 μL of NPL
2. Vortex-mix at high speed at room temperature for 30 min.
3. Cool the sample to 4°C for 15 min.
4. Apply the sample to the rinsed preequilibrated column. Let buffer run until the top of the bed is just dry, turn off column, and apply the 200-μL sample without disturbing beads. Turn column on and let flow until bed is just dry again. Turn column off, and apply a full head of buffer. Turn column on and begin counting drops from this point.

5. Collect each drop in a separate cuvet with enough phosphate-buffered saline (PBS) to bring the volume to 400 µL for fluorescence analysis.
6. From the rhodamine-PE fluorescence (560/590 nm) determine in which fractions the liposome elute (typically between drops 10 and 20). From the MPTS fluorescence (404/430) confirm that the conditions of the column completely separate the unentrapped MPTS from the liposomes (*see* **Note 19**). From the two measurements the efficiency of trapping can be calculated.

3.3.4. Tracer Loading

1. At room temperature, add tracer(s) in a 10-µL volume to 180 µL of protein-containing liposomes (formed as described in **Subheading 3.2.3.**), followed by 10 µL of DMSO.
2. Vortex-mix, cool, and apply the sample to the column as described in **Subheading 3.3.3., steps 2–4**. Processing of the sample must be done at room temperature—efficiency of trapping is much lower when done at 4°C.
3. Based on the calibrations, collect the drops that contain the peak of the eluted liposomes.
4. An aliquot of the combined peak fractions can be taken and the rhodamine and tracer fluorescence measured to confirm efficient trapping.

For determining the amount of tracer to use it is helpful to know how much will be loaded into the liposomes. For the 900 Å diameter liposomes produced by the previously described protocol, the aqueous volume of a liposome is 2.5×10^8 Å^3. If incubated with 1 mM of a tracer, approx 150 molecules of the tracer are entrapped in each liposome. In rough numbers, the intraliposomal volume in a peak fraction from the reconstitution column (lipid concentration of at least 1/3 mg/mL) is 10^{-5}–10^{-6} of the total volume (determined from MPTS trapping) (*see* **Note 20**).

When using radiolabeled tracers, the lowest detectable specific activity will depend on the counting conditions such as scintillation cocktail, scintillation counter used, etc. A rigorous analysis of this problem is presented in Rivkin and Seliger *(13)* (*see* **Subheading 3.5.3.1.**).

For fluorescent tracers, the amount to use depends on the overall quantum yield of the fluorescent compounds. A typical concentration for a bright compound such as MPTS is 10 mM.

3.4. Transport Specific Fractionation

3.4.1. Forming the TSF Gradients

3.4.1.1. THE NIGHT BEFORE

1. Place containers of urea (1XU) and sucrose (1XS) solutions into a 37°C water bath.
2. Set up a large water bath at 37°C. It will be used to keep the TSF tubes warm after pouring until they are centrifuged. Fill the water level of the water bath so that it comes to the level of the meniscus of filled TSF tubes.

3.4.1.2. SETTING UP THE GRADIENT MAKER

1. The gradient maker is clamped to a ringstand above a magnetic stirplate that has been raised to a suitable level on a jack platform.
2. The output of the gradient maker is connected to a length of 2 mm o.d. PE tubing. At the other end of the tubing is an adaptor for 100-µL micropipets. This adaptor is fixed to a vertically oriented micromanipulator (*see* **Subheading 2.2.2.2.** and **Fig. 1**). Be sure that the tubing between the gradient maker and the micropipet adaptor will not become stretched during gradient formation; test this by turning the micrometer to its fully lowered position.
3. Place 4.4-mL centrifuge tubes (Beckman Ultra-Clear) in a "fingered" test tube rack (Gilson 5 × 12 position rack for 13 × 100 mm culture tubes), or any rack that permits the tube to be visualized along its entire length. Wedge the tubes in place with a folded Kimwipe so they do not rattle in the rack and thus possibly mix the gradients after pouring.
4. Adjust the gradient maker to about 0.5 cm above the stirplate. The gradient maker should be level or very slightly sloped down to the right (output side). If the gradient maker is too close to the stirplate it will be difficult to turn the stopcocks.
5. Remove the micro-stirbar from storage in ethanol (using a stirbar retriever), rinse it carefully with deionized water, and dry it in a small Kimwipe. Place the stirbar in the right side chamber of the gradient maker.
6. Close the stopcocks of the gradient maker by turning them to the "up" position. Select a new 100-µL Corning Pyrex disposable micropipet. Check the top of the pipet for cracks or breaks that may break off or damage the micropipet holder.
7. Carefully place the pipet into the holder, pushing upwards and giving it a small twist once it is in. Be careful not to break off the pipet. Direct the pipet into a glass beaker to catch the rinsing solutions.

3.4.1.3. CLEANING/RINSING THE GRADIENT MAKER (BEFORE MAKING EACH TSF GRADIENT)

1. Place 4 mL of 1XS in the left chamber.
2. Open the left stopcock to allow the fluid to flow to the right chamber (*see* **Note 21**).
3. Turn on the stirplate and adjust the speed so that the stirbar turns freely as rapidly as possible without hitting the chamber sides.
4. Use a 5- or 10-mL plastic syringe a short segment of large-bore Tygon tubing over the end (*see* **Subheading 2.2.2.1.**) to force the fluid back and forth to clear air bubbles in the connecting tube between the chambers by placing it into the top of the left chamber and pushing the plunger up and down rapidly until all air bubbles have cleared the connection. Do not do this so forcibly that the fluid is pushed all the way through to the right side and even more air becomes trapped in the connection.
5. Open the right stopcock. If the solution does not begin to flow use the syringe–Tygon device as before, but place a finger tightly over the right chamber and at the same time push the fluid into and down the pouring tube until it flows out the end. Be careful not to introduce air bubbles into the connecting tube.

6. Close the right stopcock when the top of the meniscus in the left chamber reaches the top of the connecting tube between the chambers. Then close the left stopcock.
7. Carefully add 2 mL of 1XS into the left chamber.
8. Place 2 mL of 1XU in the right chamber.
9. Turn up the stirring speed, but be sure the stirbar is not bouncing off the sides. Adjust the position of the stirplate if necessary. Open the right stopcock.
10. When the solution nears the bottom of the right chamber (just before the meniscus travels through the channel), close the stopcock and fill it again with 2 mL of buffer. Do this three times. Be very careful that the level never goes low enough to let air into the output tubing. If, when pouring a gradient, even a single air bubble gets through the pouring apparatus and rises to the surface of the density gradient tube during pouring, the gradient is useless and must be repoured.
11. After adding the final 2 mL, turn the stirring speed down substantially and close the right stopcock as the meniscus top just falls to the upper level of the pouring tube exit. Be sure no air bubbles have entered into the pouring exit tube.
12. Fill the right chamber with 2 mL of 1XU. Turn the stirring speed back up to its previous level.
13. Slowly remove the pouring tube from the right side of the gradient maker. Avoid getting air bubbles trapped in the pouring tube connector.
14. Use a compressed air canister to blow through the tubing to force all the liquid out until no more drops fall from the glass capillary tube. Make sure water is not trapped at the tubing junctions.
15. Remove and discard the glass capillary and replace it with a new one.
16. Reconnect the pouring tube; the fluid should extrude a short distance into the pouring tubing with no trapped air bubbles.

3.4.1.4. PREPARING THE LIPOSOMES

1. Gently vortex-mix the tube containing the liposomes.
2. Remove 50–200 µL of liposomes and place into a microcentrifuge (e.g., Eppendorf) tube.
3. Place the tube into the 37°C water bath.

3.4.1.5. POURING THE GRADIENT

1. Remove the waste beaker and place a centrifuge tube in a rack into position. Advance the micrometer so the glass capillary just touches the middle of the bottom of the centrifuge tube. Adjust the micrometer so that the end of the capillary is just off the bottom of the tube.
2. Slowly open the left stopcock until it is horizontal to the left. Wait a few moments for the solutions to settle to equilibrium levels.
3. Now very slowly and evenly open the right stopcock. At some point fluid will begin to flow into the polyethylene pouring tube toward the centrifuge tube. Adjustment of the flow rate is critical, and difficult to control. The rate of flow will tend to be either too fast or too slow. For maximal control the best way to turn the stopcock is with an even pressure from your wrist. Do not try to turn

the stopcock by discrete amounts—this is virtually impossible. Turn as if the motion is going to be smooth and continuous; do not stop turning until the correct flow rate is achieved and, only then, let up on the turning pressure from your wrist (*see* **Note 22**).

4. As the meniscus flows down the polyethylene tube toward the centrifuge tube, the flow rate will increase. Be prepared to adjust the right stopcock to lessen the flow rate as this occurs. Use the gentle but firm pressure from the wrist as described previously. Make the last check on flow rate as the fluid falls through the glass capillary pipet. At this point any adjustment must be made quickly. Once the centrifuge tube has begun to fill it is not practical to adjust the flow rate without risking mixing and ruining the gradients (*see* **Note 23**).

5. With practice it should take about 45 min to pour six gradients. The rate can be monitored by checking for observable schlieren patterns in the right chamber of the gradient maker. If they are visible, the flow rate is too high (*see* **Note 24**).

6. The flow should be stopped when the fluid has flowed out of both chambers and is down to the level of the bubble trap. Stop the flow by turning off the right stopcock.

7. Using the micrometer, slowly raise the glass capillary out of the tube. Be careful not to cause gradient mixing by sudden movements. Particularly avoid vertical bumps.

8. When a gradient is completed, remove it from the test tube rack and place it in a 37°C water bath while the next gradient is poured. Rinse the gradient as described in **Subheading 3.4.1.3.** between each gradient.

9. After all the gradients are poured, adjust their fluid levels with a long Pasteur pipet. Use a clean pipet for each tube to avoid dropping heavier fluid into the gradient. Use the least-filled gradient as a guide and take the volumes of the others down to the same level. Be sure that the levels are exactly the same or the ultracentrifuge rotor will be unbalanced. The gradients are ready to have the liposomes applied to them and centrifuged.

3.4.2. Applying the Liposomes to the TSF Gradient

Remove 50–200 µL of liposomes from the tube in the 37°C water bath and gently layer onto the top of each gradient. All tubes must have exactly the same amount added so that the tubes balance.

3.4.3. TSF Centrifugation

Observing the usual precautions about balancing of tubes and good rotor practice, place the TSF tubes into the rotor buckets. Be particularly careful to avoid vertical shocks to the tubes (especially while putting the rotor on the ultracentrifuge spindle)—these are the most damaging to the gradients. Set the centrifuge controls for slow startup, and for minimal braking (braking should cut out below 800 and 1000 rpm).

Smooth transition of the buckets from vertical to horizontal and back should be confirmed with the machine in zonal mode with the door open for inspection of startup and stop (*see* **Note 25**). Centrifuge gradients at 300,000*g* (50K rpm in Sorvall TST 60.4 swinging bucket rotor). At 37°C the bands should be adequately separated in 2–3 h (*see* **Note 26**).

3.4.4. Cleaning the Gradient Maker (Typically Done After the TSF Tubes are Set Spinning)

1. Place a waste-water beaker underneath the pipet.
2. Pour distilled water into both compartments of the gradient maker. Fill them all the way up to the top and let water drain out through the pipet several times.
3. Disconnect the connecting tube and dry it using clean compressed air.
4. Remove the gradient maker from the ringstand and rinse it thoroughly in the sink using the distilled water.
5. Place the gradient maker back onto the ringstand and hang it upside down to allow water to drip out as it dries.

3.5. Analysis of TSF Gradients

3.5.1. Inspection of TSF Results

1. Remove the stopped rotor from the centrifuge, being careful not to disrupt the gradients.
2. Remove the TSF tubes and view them in the Viewing Apparatus described previously (*see* **Subheading 2.2.2.3.** and **Fig. 2**).
3. Put the TSF tube into place (by sliding it through the holder). Be careful not to let it bump on the table when lowering it into place.
4. Position the green (no. 58) Wratten filter above the tube.
5. Place the red (no. 23A) Wratten filter in front of the test tube holder.
6. Make the room completely dark and turn on the fiberoptic lamp.
7. Locate the bands, being careful to distinguish them from circular scratches on the outside of the centrifuge tube.
8. Typically, after a centrifugation of 2–3 h the band of liposomes without functional channels will be 3–6 mm into the gradient (*see* **Note 27**). This band will be somewhat diffuse (due to the distribution of sizes of the liposomes). If the liposomes have not entered the gradient, the TSF was not successful (*see* **Note 28**).
9. The band of liposomes with functional channels will be 8–10 mm below the band of liposomes without channels. This band should be much tighter than the one above it (*see* **Notes 29** and **30**).
10. Place the Wratten filters back into their envelopes (the UV light from the room lamps will photobleach them) (*see* **Note 31**).

3.5.2. Recovery of Bands

1. Place a long Pasteur pipet onto the end of the Syringe–Tygon Tubing device.
2. Carefully place the pipet into the tube, stopping just above the lower band. Carefully remove the entire band, and save it in an Eppendorf tube. The validity of

results depends on the efficiency and reproducibility of this procedure. Collection of the band should be a slow as possible and practical. The flow into the pipet will come from solution layers at and below the tip. Therefore to quantitatively recover the band in the smallest volume it is best to position the pipet tip just at or above the upper edge of the band. As the collection proceeds the band will thin until it is only visible when viewed exactly edge-on. At this point, while continuing to collect, move the pipet tip slightly up and down in the immediate vicinity of the last visible band to ensure complete collection. Before attempting this step on experimental tubes, the investigator must characterize his or her accuracy and reproducibility in collecting bands from multiple identically loaded tubes in the same TSF spin. Without such characterization, it is easy to unknowingly generate substantial and inconsistent errors.

3. Repeat for the upper band.

3.5.3. Measurement of Tracer Retention

For each recovered band, relative lipid amounts are determined by measurement of fluorescence (560 nm excitation, 590 nm emission) of the rhodamine-PE label in the bulk lipid phase and normalized to the volume in which the band was recovered. Retention of tracer can then be expressed on a per-liposome basis for each population of liposomes. The ratio of tracer to lipid in the upper band (composed of liposomes without functional channels) is an internal control for the amount of tracer loaded per liposome and for loss of tracer not due to channel activity. Comparison of this ratio with that of the lower band reveals whether the tracer is permeable through the functional channels.

3.5.3.1. SPECIAL CONSIDERATIONS FOR RADIOACTIVE TRACERS

Due to the low efficiency of liposome loading with tracer, TSF experiments will often yield low levels of radiolabel. Because of this, extra care must be taken to obtain statistically meaningful data, and the highest available specific nuclide activities should be used. In our experience the following protocol is optimal: The recovered bands are added to 5 mL of liquid scintillation cocktail (Ultima Gold, Packard), thoroughly mixed by vortex-mixing, and the activities measured in a liquid scintillation counter (counting window 0–18.6 keV). A blank control consisting of 0.75 mL of aqueous buffer and 5 mL of Ultima Gold is counted between each sample. Counting times are long enough to ensure that the Student's t-tests of net counting rates are significantly greater than background at the 95% confidence level. Counting rates are corrected for quench (external standards method) and background. The minimum detectable activity, the net signal that can be detected at the 95% confidence level, is approximately given by $\sqrt{8N/t}$ where N is the background noise (cpm) and t is the maximum counting time (min) *(13)*.

Quench correction is made by measuring the relative activities of a series of uniform, traceable standards quenched to various degrees with nitromethane. The calculated counting efficiencies (%) are plotted against the transformed spectral index of the external ^{137}Cs standard (tSIE) for each quenched sample in the series and the data are fit by a second-degree polynomial using a nonlinear least squares fit. Quenching correction of the data sets is made from the tSIE value reported for each sample. In addition, verification of direct measurements in dpm using a TriCarb 2250CA liquid scintillation counter (low background model, using preprogrammed Ultima Gold quench data from standards prepared by Packard) are made by comparison of data with the manual quench correction method outlined above.

4. Notes

1. The fundamentals of this technique were developed and first published by Goldin and Rhoden *(14)*.
2. The change in density of the liposomes and their separation on the basis of density occurs simultaneously in the TSF gradient. This occurs because the osmolyte exchange of the liposomal entrapped volume is very rapid compared with the movement of the liposome in the density gradient. TSF fractionates a set of liposomes into two populations—one that contains functional channels permeable to a given osmolyte and one that does not contain such channels. When the solute used is relatively small (sucrose, in the present protocol), TSF is also a way to assess the fraction of liposomes that contain channels that can open. Thus, in addition to its uses in permeability studies as outlined in this chapter, TSF can be used to explore the modulation of channel activity *(15–17)*.
3. The osmolality of the 1XU and 1XS buffers must be confirmed to be 500 ± 5 mOsm/kg. Since the pK_a of HEPES is temperature sensitive, the pH of the solutions must be adjusted at the temperature at which permeability is measured, typically 37°C.
4. There is nothing special about this particular lipid mixture. Essentially the same results are found with azolectin but reconstruction efficiency is poor in pure PC. However, use of defined lipids permits control over the lipid environment.
5. Lipid stocks are kept in glass tubes with Teflon-lined caps.
6. Tubes from some suppliers have a hydrophilic residue not removed by methanol. This is evident from inspection of the tube following the methanol rinse. These tubes must be rinsed with double-distilled water prior to use of the progressively less polar solvents.
7. Argon is used rather than nitrogen because it is more dense than air and so tends to remain in open tubes.
8. Use a small rotary oil-filled pump with an in-line vacuum filter. It should not be left running to the point that a limiting vacuum is reached because then oil vapor can diffuse back into the desiccator and contaminate the dried lipids.
9. If column runs dry, the gel must be removed to fresh 1XU buffer and completely degassed before repouring the column. The column will then need to be stripped with octylglucoside immediately to remove the lipid/protein.

10. If the column runs for an excessive period, the equilibration lipid will eventually wash off. Also, interrupting the flow by stopping the pump will cause the gel bed to expand as the pressure differential on the column dissipates. This can cause unreproducible reconstitution conditions and entrapment of air in the gel matrix.

11. Connexin concentration can be estimated from immunostained Western blots calibrated to a protein assay. However, it is difficult to calibrate a connexin sample because connexin is stained anomalously by most protein assays—calibration against known quantities of other proteins are usually artifactual. In addition, in our experience all dye-binding assays are inaccurate by up to an order of magnitude in the presence of octylglucoside, urea or sucrose, in spite of manufacturers' claims to the contrary.

To determine connexin concentration, connexin is quantitatively extracted from the sample, eliminating the lipid, detergent and aqueous solutes that interfere with protein determination. This is achieved using a methanol–chloroform extraction *(18)*. The protein sample can then be assayed by the *o*-phthalaldehyde (OPA; Pierce) fluorescence protein assay, in which the signal depends only on the type and amount of each amino acid present.

One approach is to hydrolyze the connexin into the constituent amino acids and apply the OPA assay to the product. A calibration curve is generated by applying the OPA assay to a stoichiometric mixture of individual amino acids corresponding to the composition of the connexin being assayed. To avoid doing a hydrolysis for each determination, a standard curve can be generated relating the OPA signals from the mixture of amino acids with those from intact polypeptides (the signal is lower from the intact peptides). Subsequently, the OPA assay can be performed directly on the intact connexin. An alternative to assembling a standard of known amino acids is to determine the connexin concentration by commercial quantitative amino acid analysis. The mixture of amino acids is useful as a permanent and invariant standard for calibration the OPA signal across subsequent purifications.

Once a standard connexin sample has been characterized, it can be Western blotted and immunostained. For connexin32, in our experience immunostains with monoclonal M12.13 primary antibody *(19)* and secondary alkaline phosphatase conjugate are reasonably linear between 0.5 and 10 ng per monomer band on PVDF membrane (the membrane saturates at higher levels). One can obtain a rough idea of connexin concentrations by inspection of immunostained blots.

12. The volumes given are for a final volume of 100 μL to be added to the 0.7 × 20 cm column. This can be scaled up to as much as 400 μL if a 1.5 × 20 cm column is used.

13. This entire step is the most crucial for successful and reproducible reconstitutions. It should be practiced several times with buffer alone to learn how to do it without running the top of the column dry. The pump should not be turned off at any time during this procedure because of the swelling of the bead bed that will result.

14. When the pump is not in use, the pressure heads on the silicone rubber pump tubing should be released to keep the tubing from being flattened and changing the nominal flow rate. Follow pump manufacturer's suggestions for replacing this tubing on a regular schedule or reconstitution conditions will become variable.

It is important to rinse the post-column tubing and fraction collectors with 95% ethanol weekly to remove buildup of lipid, protein, and detergent—especially at the fraction collector head outlet where buffer components tend to crystallize.

15. Excessively oxidized lipid is difficult to remove by this method and, in this case, the entire gel needs to be replaced.

16. For long-term storage the column should be infused with 1XU buffer (which contains 0.02% sodium azide). It is also important that the temperature be stable, and monitored with a recording thermometer. If the temperature varies by more than 1°C, the gel should be degassed.

17. An alternative is to load the liposomes with tracers during liposome formation. In this procedure, tracer is included in the liposome-forming solution and in the liposome-forming column. As the liposomes form the tracer is entrapped (cf. **ref. 3**). However, this method is wasteful of the tracer compounds, unless the unentrapped tracer can be recovered from the column eluent.

18. The high salt removes weakly bound residual tracer from the agarose.

19. With repeated tracer trappings these peaks tend to move together. If this occurs, the column must be stripped with 160 m*M* octylglucoside (*see* **Subheading 3.2.4.**) and then rinsed, pre-equilibrated, and calibrated as described in **Subheadings 3.3.2.** and **3.3.3.**).

20. Trapping is more efficient when the liposomes are concentrated. This is possible to achieve using 100-kDa cutoff UltraFree-MC cartridges (Millipore) at no greater than 3000–4000 rpm (lipid concentration up to about 3 mg/mL can be obtained).

21. Whenever turning the stopcocks, stabilize the gradient maker with the other hand.

22. Occasionally a gradient does not begin flowing. This may occur if there is an air bubble in the connecting tubes within the gradient maker. If this happens, close the left stopcock, carefully cover the right chamber with the fleshy part of an index finger and push up and down to begin the flow. As soon as flow begins turn off the right stopcock. Open the left stopcock and wait a few moments for the levels to equilibrate and the solutions to mix uniformly in the right chamber, then attempt pouring again by slowly opening the right stopcock, etc.

23. Occasionally the gradient stops flowing in mid-pour. If this happens, very carefully open the stopcock while looking at the centrifuge tube level—at first the rate will be too fast and the stopcock will have to be adjusted rapidly. This must be done without either filling too quickly or jockeying the level in the tube—if either happens one must toss out the tube and start over.

24. There can be no air bubbles in the gradients. Tiny air bubbles that do make it through the tubing are caught in the bubble trap which consists of a larger piece of Tygon tubing on the pourer head unit above the capillary.

25. This transition is critical. Sometimes bucket seatings can wear so that the reorientation is defective.

26. TSF centrifugations can be done at lower temperatures, but the time required for separation of the bands is longer. For example, at 4°C 10–12 h are required. Control studies have shown that the temperature does not affect the final result with regard to connexin channel permeability or activity.

27. During centrifugation the upper band continues to descend at a slow rate (due to the permeability of unmodified liposome membrane to urea and sucrose). After many hours (>24 h) it will eventually merge with the lower band.

28. This may occur for any of several reasons. Usually it indicates that one of the solutions was made incorrectly so that either the liposome-containing solution or liposomes are not dense enough to enter the gradient, or the gradient-forming solutions are too dense or viscous.

29. One may be tempted to estimate the fractional amount of the liposomes that are in the lower band, but this cannot be done by eye with any accuracy. The bands must be removed and the fluorescence measured.

30. If it is desired to take a photograph of the TSF bands, or to optimize visualization of them, make the following modifications to the Viewing Apparatus: To collimate the light illuminating the TSF tube, an opaque mask should be placed on the Lucite holder, with a hole permitting light to enter the Lucite only above the hole in which the TSF tube sits. The green filter can be placed on top of this mask. Collimating the beam minimizes the internal reflections from the TSF tube sides and boosts the signal/noise ratio. The distance between the fiberoptic lens and the TSF should be 20–30 cm. To keep stray light from hitting the TSF tube, an opaque tube (e.g., the cardboard core from a roll of paper towels) should enclose the light path from the fiberoptic lens down to the filter/mask/Lucite assembly.

31. If sucrose is spilled onto the filters, they can be cleaned with distilled water and a piece of lens cleaning paper. Do not use organic solvents such as ethanol.

References

1. Bevans, C. G., Kordel, M., Rhee, S. K., and Harris A. L. (1998) Isoform composition of connexin channels determines selectivity among second messengers and uncharged molecules. *J. Biol. Chem.* **273,** 2808–2816.

2. Harris, A. L. and Bevans, C. G. (1997) Molecular selectivity of homomeric and heteromeric connexin channels, in *Gap Junctions* (Werner, R., ed.), IOS Press, Amsterdam, pp. 60–64.

3. Rhee S. K., Bevans C. G., and Harris, A. L. (1996) Channel-forming activity of immunoaffinity-purified connexin32 in single phospholipid membranes. *Biochemistry* **35,** 9212–9223.

4. Sugawara, E. and Nikaido, H. (1994) OmpA protein of the *Escherichia coli* outer membrane occurs in open and closed channel forms. *J. Biol. Chem.* **269,** 17,981–17,987.

5. Stauffer, K. A., Kumar, N. M., Gilula, N. B., and Unwin, P. N. T. (1991) Isolation and purification of gap junction channels. *J. Cell Biol.* **115,** 141–150.

6. Stauffer, K. A. (1995) The gap junction proteins β_1-connexin (connexin-32) and β_2-connexin (connexin-26) can form heteromeric hemichannels. *J. Biol. Chem.* **270,** 6768–6772.

7. Buehler, L. K., Stauffer, K. A., Gilula, N. B., and Kumar, N. M. (1995) Single channel behavior of recombinant $\beta(2)$ gap junction connexons reconstituted into lipid bilayers. *Biophys. J.* **68,** 1767–1775.

8. Koster, J. C., Bentle, K. A., Nichols, C. G., and Ho, K. (1998) Assembly of ROMK1 (Kir 1.1a) inward rectifier K$^+$ channel subunits involves multiple interaction sites. *Biophys. J.* **74,** 1821–1829.

9. Clement, J. P., Kunjilwar, K., Gonzalez, G., Schwanstecher, M., Panten, U., Aguilar-Bryan, L., and Bryan, J. (1997) Association and stoichiometry of K$_{ATP}$ channel subunits. *Neuron* **18,** 827–838.

10. Mimms, L. T., Zampighi, G. A., Nozaki, Y., Tanford, C., and Reynolds, J. A. (1981) Phospholipid vesicle formation and transmembrane protein incorporation using octyl glucoside. *Biochemistry* **20,** 833–840.

11. Harris, A. L., Walter, A., and Zimmerberg, J. (1989) Transport-specific purification of large channels incorporated into lipid vesicles. *J. Membr. Biol.* **109,** 243–250.

12. Walter, A., Lesieur, S., Blumenthal, R. and Ollivon, M. (1993) Size characterization of liposomes by HPLC, in *Liposome Technology. Liposome Preparation and Related Techniques* (Gregoriadis, G., ed.), 2nd ed. CRC Press, Boca Raton, pp. 271–290.

13. Rivkin, R. B. and Seliger, H. H. (1981) Liquid scintillation counting for ^{14}C uptake of single algal cells isolated from natural samples. *Limnol. Oceanogr.* **26,** 780–785.

14. Goldin, S. M. and Rhoden, V. (1978) Reconstitution and "transport specificity fractionation" of the human erythrocyte glucose transport system. A new approach for identification and isolation of membrane transport proteins. *J. Biol. Chem.* **253,** 2575–2583.

15. Hong, E. J., Huh, K., and Rhee, S. K. (1996) Effect of ginseng saponin on gap junction channel reconstituted with connexin32. *Arch. Pharmacol. Res.* **19,** 264–268.

16. Bevans, C. G. and Harris, A. L. (1999) Direct high-affinity modulation of connexin channel activity by cyclic nucleotides. *J. Biol. Chem.* **274,** 3720–3725.

17. Bevans, C. G. and Harris, A. L. (1999) Regulation of connexin channels by pH: Direct action of the protonated form of taurine and other aminosulfonates. *J. Biol. Chem.* **274,** 3711–3719.

18. Wessel, D. and Flügge, U. I. (1984) A method for the quantitative recovery of protein in dilute solution in the presence of detergents and lipids. *Anal. Biochem.* **138,** 141–143.

19. Goodenough, D. A., Paul, D. L., and Jesaitis, L. (1988) Topological distribution of two connexin32 antigenic sites in intact and split rodent hepatocyte gap junctions. *J. Cell Biol.* **107,** 1817–1824.

21

Inducing *De Novo* Formation
of Gap Junction Channels

Feliksas F. Bukauskas

1. Introduction

Intercellular gap junction (GJ) channels arise from the association of two hemichannels, a hexameric assembly of connexin membrane proteins, between two cells (for review, *see* **refs. *1,2***). Connexins have the shortest half-lives of all known membrane channel proteins, approx 2–3 h, and consequently GJ channel formation and turnover is believed to be relatively fast (for review, *see* **ref. *3***).

Here we describe an induction method for the *de novo* formation of gap junction channels where coupling is observed by manipulating two individual cells into physical contact while recording the onset of electrical cell–cell coupling. This induction approach has been used to couple *Xenopus* blastulae *(4)*, *Xenopus* myoballs *(5,6)*, rat myocytes *(7)*, and insect cells *(8,9)*. We have refined and extended this technique to mammalian cells transfected with connexins *(10,11)*.

The formation of a functional GJ channel is a complex and multistage process. It is believed that the initial step in GJ formation is the establishment of a close juxtapositioning of the membranes in the junctional contact area, mediated by adhesion molecules *(12,13)*. Subsequent steps are likely to include recruitment of hemichannels to the contact area, alignment and docking of apposed hemichannels, the formation of a high-resistance seal isolating the channel pore from the extracellular space, and finally channel pore opening. The effective assembly of two hemichannels and insulation of the channel pore from the external medium is accomplished by electrostatic and hydrophobic noncovalent interactions between the E1 and E2 extracellular loops of the apposing hemichannels *(14)*.

From: *Methods in Molecular Biology, vol. 154: Connexin Methods and Protocols*
Edited by: R. Bruzzone and C. Giaume © Humana Press Inc., Totowa, NJ

Experiments with spontaneously preformed cell pairs has been the standard approach for examining cell–cell coupling for almost two decades, examining many different aspects of GJ channel regulation and gating at the macroscopic (many functional channels) and single channel levels. Usually, cell–cell coupling consists of hundreds of individual GJ channels arranged in junctional plaques *(15)*. In order to render cell pairs suitable for single channel analysis, the number of functional channels must be reduced. A widely used experimental approach consists of exposing the cells to submaximal doses of uncoupling agents, such as long-chain alkanols *(16)*, arachidonic acid *(17)*, etc. These treatments, however, may affect the parameters under investigation, for example, the single channel conductance and/or the open channel probability. These experimental limitations can be overcome by inducing two single cells into contact and analyzing initial events of *de novo* GJ channel formation. In most of our experiments, the period between the first and second channel formation lasts for minutes, which is more than adequate for single channel analysis.

The *de novo* GJ channel formation method, however, is more complex than the traditional technique, in that it involves additional steps and requires some experience in establishing contact between two individual cells. This method allows for junctional current analysis during the first GJ channel pore opening and for continuous single channel conductance measurements without the application of uncoupling factors and for the formation of heterotypic junctions without preloading cells with dye. We have used this technique on insect cell lines from *Aedes albobopictus* (cell line C6/36) *(18)* and *Spodoptera fruigiperda* (cell line Sf9) *(19)*; on mammalian cell lines such as N2A (mouse neuroblastoma cell line, ATCC No. CCL-131), HeLa (human cervix carcinoma cell line, ATCC No. CCL-2), and RIN (rat islet tumor cell line, ATCC No. CRL-2057) cells transfected with different connexins *(10,11)*; on a Novikoff hepatoma cell line (ATCC No. CRL-1604) endogenously expressing connexin43 (Cx43), on primary cultures of myocytes from neonatal rat heart *(20)* and Schwann cells from sciatic nerve of newborn rats *(21)*. We have observed that the lag time for the first GJ channel formation in mammalian cells is measured in tens of minutes *(11)*, and in few minutes for insect cells lines (C6/36 and Sf9) *(18)*. Lag times on the order of seconds were observed in primary cultures of insect hemocytes from *Periplaneta americana (22)* and epidermal cells from *Tenebrio molitor (8)*. Interestingly, in all cell types, after first channel opening coupling conductance (g_j) most often increases relatively rapidly and in a single sigmoidal fashion *(18)*. Presumably, *de novo* channel openings are proceeding the formation of junctional plaque, rather than docking of individual hemichannels. We attribute *de novo* channel opening to a special type of gating with slow channel opening kinetics (transition time takes up to tens of milliseconds) resembling a chemical- and V_m-sensitive gating *(18)*. The rapid single sigmoidal g_j rise suggests that this process occurs inside one plaque and, presumably, is cooperative.

2. Materials

2.1. Solutions

1. Ringer–Krebs solution for the perfusion of cells during the experimental procedure: 140 mM NaCl, 4 mM KCl, 2 mM CaCl$_2$, 1 mM MgCl$_2$, 5 mM glucose, 2 mM pyruvate and 5 mM *N*-(2-hydroxyethyl) piperazine-N'-2-ethanesulfonic acid (HEPES), pH 7.4.
2. Patch pipet solution. Pipets for whole cell voltage clamping are filled with different pipet solutions depending on the purpose of the experiment. The standard pipet solution consists of: 130 mM KCl, 10 mM NaCl, 1 mM MgCl$_2$, 3 mM MgATP, 1.4 mM CaCl$_2$, 5 mM EGTA, pCa ~7.5, and 5 mM HEPES, pH 7.2, filtered through 0.22-μm pores. For ionic permeability measurements, we use pipet solutions where the principal ions of the standard pipet solution, K$^+$ and Cl$^-$, are exchanged for cations and anions of differing size and molecular weight.
3. Solution for cell detachment from coverslips contained: 140 mM NaCl, 4 mM KCl, 1 mM MgCl$_2$, 5 mM EDTA, 5 mM glucose, 2 mM pyruvate, and 5 mM HEPES, pH 7.4.

2.2. Apparatus

2.2.1. Experimental Chamber

A schematic view of the experimental chamber is presented in **Fig. 1**. The experimental chamber is made of Perspex and contains three compartments. The coverslips are placed in the central compartment which has a thin glass bottom of about 100–200 μm thickness. Experiments for the induction of homotypic or heterotypic junctions require one and two coverslips, respectively. The central compartment has an inflow channel and is connected to the outflow reservoir. To prevent all possible mechanical microvibrations from the waves on the surface of the solution, created by the suction, the central chamber and outflow reservoir are connected through a relatively long and wide U-shaped channel. An outflow reservoir is connected through an agar bridge to the reservoir for the pipet solution. The former contains an Ag/AgCl electrode contact for the connection to the reference electrode of the clamp amplifier. The experimental chamber is mounted on the stage of an inverted microscope equipped with phase contrast optics (Olympus IX70).

2.2.2. Pipet Holders

For cell patching and double-voltage-clamp experiments, we use standard pipet holders from Axon Instruments (HL-1). For detachment of cells from the

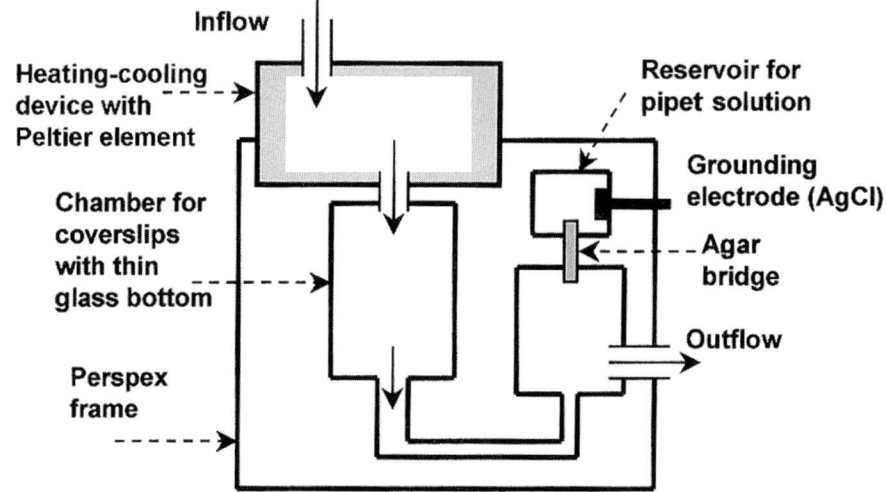

Fig. 1. Schematic view of an experimental chamber. *Continuous arrows* show perfusion solution flow.

coverslip, we use homemade pipet holders which allows us to create a local solution stream within the bulk perfusion of the whole chamber. This local solution stream is used to ensure gentle washout of individual cells from the coverslips and for their local perfusion with different solutions. A schematic view of this pipet holder is shown in **Fig. 2**. The pipet and thin inflow tube are inserted into the holder and sealed with resin rings. The solution in the pipet can be changed through the inflow tube which is connected to a reservoir of control or test solutions by hydrostatic pressure. By applying positive pressure through the connection on the right side of the pipet holder (*see* **Fig. 2**) we are able to create a local stream of the solution going out from the tip of the pipet. To prevent back flow of the solution through the microtube, we use an electronically regulated microvalve with remote control.

2.2.3. Patch Pipets

Patch pipets are pulled from glass capillary tubes with filaments (cat. no. 7052, A-M Systems) using a horizontal puller, Model P-97 (Sutter Instruments, USA). The pipets routinely have resistances of –5 MΩ (tip size ~2 µm). The pipets are filled with the appropriate pipet solution, immediately before use.

2.2.4. Pipets for Local Solution Streaming

These pipets have a tip diameter of about 5–10 µm and are made with the same horizontal puller and from the same glass as the patch pipets. The pipets

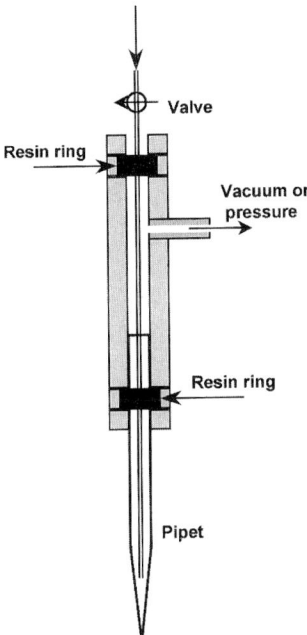

Valve

Resin ring

Vacuum or pressure

Resin ring

Pipet

Fig. 2. Schematic view of a pipet holder for creation of a local solution stream.

are filled with Ringer–Krebs solution. We use these pipets for detaching cells from coverslips and for local perfusion of cell pairs.

2.2.5. Heating-Cooling System

It is critical to allow cell–cell channel formation to occur at temperatures as close as possible to physiological conditions. We employ a custom made device, based on using a Peltier element (*see* Heating–cooling device in **Fig. 1**), that allows us to change the temperature of the perfusion solution in the range of 2–40°C. We constructed a sandwich-like system from the following three components: (1) a thermostat made from a thin rectangular metal box (–40 × 40 × 4 mm) which is perfused from inside by water thus keeping the temperature constant at one side of the Peltier element; (2) a Peltier element connected by two wires to the DC power supply (about 3 amp is required); and (3) a Perspex rectangular box (30 × 30 × 3 mm) containing a zigzag form channel, to maximize the surface area in contact with the Peltier element, and connected to the outside by two outputs; for the inflow and outflow of the perfusion solution. The temperature of the outflow depends on the amplitude and direction of the current through the Peltier element. At extreme currents, ice formation can stop the solution flow or damage the chamber when the temperature becomes too high. By using feedback control of the current through the Peltier element, the temperature in the experimental chamber can be automatically regulated.

2.2.6. Temperature Control

The temperature in the experimental chamber is not uniform and it is, therefore, important to measure the temperature as close as possible (0.5 mm or less) to the cells. The diameter of the temperature probe must be minimal to allow for such precise local measurement. In addition, the temperature probe can be influenced by the surrounding air temperature, through the wires which connect it to the measuring device, and their diameter should be minimal. We have tested different temperature probes and are currently using a thermocouple probe with a diameter of about 0.1 mm from Cole-Parmer.

We have empirically determined that physiological temperatures are important for gap junction formation; 37°C for mammalian cells and 27°C for insect cells (e.g., C6/36 and SL9 cell lines). Interestingly, it has been demonstrated that channel formation can occur even at 4°C in reaggregated Novikoff hepatoma cells *(15)*. Our data indicate that hypothermia causes an increase in the lag time of the first channel formation and slows down the subsequent rise in junctional conductance. In addition, temperature strongly affects single channel conductance with a Q_{10} of 1.25 *(11,23)*.

2.2.7. Setup for Electrical Recordings

1. Two amplifiers for whole cell recordings, Axopatch-1D (Axon Instruments, Foster City, USA).
2. Digital Data Recorder VR-100 (Instrutech, USA).
3. Low-Pass Bessel Filter 8 Pole (Warner Instrument, USA).
4. A/D converter with at least four input channels for signals and two output channels for stimulus (e.g., TL-1, Axon Instruments).
5. Data acquisition software and computer. Any version of pClamp (Axon Instruments) that runs on IBM compatible computers. It is advisable to consult a representative in the Technical Assistance Department at Axon Instruments to ensure compatibility of hardware and software.
6. Data analysis is performed using Pclamp software (Axon Instruments) and Sigma Plot (Jandel Corporation).

3. Method

The induction of *de novo* GJ channel formation involves aspects of cell biology, electrophysiology, and certain technically sensitive procedures. In brief, the method is based on the patching of two individual cells with two patch pipets and then bringing them into contact. This procedure involves plating cells on coverslips, their transfer to the experimental chamber, the identification of cells most suitable for GJ channel formation, cell detachment from the coverslip, whole cell voltage clamp patching of two individual cells at two different holding potentials, bringing these cells into contact and monitoring junctional current during *de novo* channel formation. Our empirical findings show that cells, that do not adhere to the glass coverslips or to multiwell plastic dishes do not exhibit GJ channel formation or at best do so inconsistently. Cells that exhibit adhesive properties usually show a GJ channel formation rate dependent on connexin type and expression level. Our preliminary data indicate that in HeLa cell transfectants, Cx43 GJ formation rate is more efficient than that of Cx46. We have determined that this discrepancy is not due to expression levels, as in spontaneously preformed HeLa cell pairs of both transfectants, grown under identical conditions, coupling conductance was similar. Interestingly, Cx43–Cx46 heterotypic junctions, show similar rates of *de novo* GJ channel formation as Cx43 homotypic junctions.

Different cell types possess different abilities to form GJ channels. For example, HeLa cell transfectants are more reliable than N2A cells for these gap junction channel formation experiments. There is an essential difference in the time required for GJ channel formation between invertebrate and mammalian cells. After manipulating two insect cells into contact, the lag time until the first GJ channel opening is on the order of tens of seconds (*8,22*) to a few minutes (*18*), while the time required for mammalian cells is tens of minutes (*11*).

Below, we will elaborate on these individual steps.

3.1. Cell Culture Conditions

Standard cell culture methods are used for all cell types. Usually, we passage the cells in multiwell culture dishes on square no. 1 coverslips (Clay Adams, USA) at a density of $1 - 2 \times 10^5$ cells/cm^2, which allows us to select individual cells for the coupling procedure. The optimal experimental time after cell passage depends on the cell type. For almost all cell lines we have tested, for example, HeLa, N2A, and RIN, transfected with different connexins, we found it to be optimal to perform the experiments on the second and third days after passage. The experiments on freshly isolated myocytes from neonatal rat heart, however, are conducted on the same day, 6–8 h after cell passage. After 2 d in culture, myocytes are so strongly adherent to the coverslips that it was difficult to separate them without damage. Other cells, such as freshly isolated fibroblast or Schwann cells from sciatic nerve of newborn rats, change they morphology with time, becoming very flat or projecting processes that makes their detachment from the glass coverslips nearly impossible. These cells should be used for experimentation on the first or second day after passage. We have found that insect cells from *Aedes albobopictus* (cell line C6/36) and *Spodoptera fruigiperda* (cell line Sf9) can be effectively detached 1–5 d after passage (*9,18,19*). If cells are too strongly attached to the coverslips they can be pretreated with a low Ca/Mg solution which weakens cell adhesion to the glass.

3.2. Induction of Cell–Cell Coupling

3.2.1. Mechanical Detachment of Cells from the Coverslips

The tissue culture coverslips are transferred to the central compartment of the experimental chamber (*see* **Fig. 1**) mounted on the stage of an inverted microscope that is equipped with phase-contrast optics. It is recommended that the transfer of coverslips be done as quickly as possible, while maintaining them in a horizontal position, since exposure of the cells to the air leads to cell lysis. Once individual cells are selected, the gentle flow of the local stream of Ringer–Krebs solution through the perfusion pipet is used to detach them from the coverslip. It should be emphasized that the flow intensity of the stream must be minimal to avoid mechanical cell damage. This cell separation procedure should be carried out slowly with maximum caution. Once the cell(s) is detached, we allow 1–2 min for cell recovery and visually evaluate the integrity of cell morphology. Damaged cells are usually more transparent than are viable cells when viewed by phase-contrast microscopy. In, addition, we have observed that damaged cells lose the ability to reattach to the coverslip, which is not true for viable cells. This is easy to assess by allowing the cells to recover for a few minutes and then reapplying the local stream.

3.2.2. Chemical Detachment of Cells from Coverslips

Cell lines differ in their ability to adhere to coverslips. For example, N2A and RIN cells have a rounder shape than HeLa cells when they are adherent to the coverslips and are relatively easier to separate by the gentle flow of the local stream. HeLa cells, however, are the cells of choice for GJ channel formation experiments, because they are easy to patch, have a relatively low level of nonjunctional current and at present they are one of the most reliable cell lines used for stable transfection with different connexins *(24)*. If the cells are too strongly attached to the coverslips, we pretreat them for 5 min with a low Ca/Mg Krebs–Ringer solution, which is an established method for cell dissociation *(15)*. After pretreatment with the low Ca/Mg Krebs–Ringer solution the cells start to retract taking on a rounder shape. The experimental chamber is then washed with control Krebs–Ringer solution and the local stream is applied to completely detach the cells from the coverslip.

3.2.3. Providing Cells into Contact

After recovery, the detached cell is patched with a patch-pipet, and after tight-seal formation whole cell clamp conditions are established. The same procedure is repeated with the second cell. The cells are mechanically manipulated into contact and are slightly pressed to each other to facilitate the formation of a cell–cell contact area. The stability of the micromanipulators at this stage of the experiment is critical to avoid any additional movement before the establishment of functional coupling. As described previously, the cells are believed to establish mechanical coupling through cell–cell adhesion. After the onset of electrical cell–cell coupling, slight movement of the cells does not have an effect on junctional conductance. Interestingly, we have observed in insect cell lines that after the establishment of functional coupling they can be moved away each from other up to distances of 10–30 µm with formation of thin processes and yet junctional conductance can still be measured. At greater distances the processes are disrupted but usually without changes in the holding current. In three experiments, we were successful in reestablishing electrical cell–cell coupling up to five times on the same two insect cells. We have not, however, been successful in reestablishing cell–cell coupling in mammalian cells after their separation.

3.3. Electrical Measurements

To study *de novo* channel formation and properties of single gap junction channels, a dual voltage-clamp method is used which has been previously described in detail *(25)*. After the establishment of whole-cell patch clamp conditions, membrane potentials of different amplitude (V_1, V_2) are clamped and nonjunctional membrane currents (I_{m1}, I_{m2}) are recorded individually for each cell. Subsequent to the opening of a newly formed GJ channel, the junctional current I_j will change I_{m1} and I_{m2} equally in opposite directions, and the total current in both pipets will be $I_{m1} + I_j$ and $I_{m2} - I_j$, respectively. The junctional conductance is determined as $g_j = I_j/(V_2 - V_1)$. Our data indicate that during the initial opening of *de novo* GJ channels there is no change in I_{m1} and I_{m2}. This indicates that a tight seal between two hemichannels must precede channel pore opening. After the first channel opening, spontaneous channel gating between the open and closed states can be observed. To evaluate the nonjunctional currents corresponding to the completely closed state ($I_j = 0$), we superimposed a voltage pulse, for example, of 50 mV amplitude and 100 ms duration on the basic holding potential of cell 1 or cell 2. The completely closed state was evidenced by the absence of an associated current of opposite polarity in the second cell.

Signals were recorded on videotape and for analysis were filtered at 2 kHz (8-pole Bessel; –3 dB) and digitized at 5 kHz with an A/D converter. Data acquisition and analysis were performed using Pclamp software and Sigma Plot.

4. Notes

1. The method of *de novo* GJ channel formation is invaluable for experiments where precise estimation of I_{m1} and I_{m2} at $I_j = 0$, that is, the closed state, are important. The short lag time just before the first channel opening provides an ideal window for such estimations. This has allowed us to show, for the first time, that the V_j-sensitive gating mechanism does not close the channel completely but to some

Table 1
Summary of Our Data on Homo- and Heterotypic Junction Formation
(+ or –; see Upper Left Corners) in HeLa Cells Transfected with Mouse (m) or Rat (r) Cxs

	Cx26m	Cx30.3m	Cx32r	Cx37m	Cx40m	Cx43r	Cx45m
Cx26m	+ (Xo+) 140[a]						
Cx30.3m		+ (XB+) 55					
Cx32r	+ (Xo+) 40–100[a]		+ (Xo+) 40[a]				
Cx37m	+ 85–35		+ (Xo–) 140–40	+ (Xo+) 380[a]			
Cx40m	– (Xo–)		– (Xo–)	+ (Xo+) 270–210	+ (Xo+) 165[a]		
Cx43r	– (Xo–)		+ (Xo+/–) 55–40	+ (Xo+) 45–100	– (Xo–)	+ (Xo+) 110[a]	
Cx45m	(Xo–)		(Xo–)	(Xo+)		+ (Xo+)	+ (Xo+) 40

	Cx26m	Cx30.3m	Cx32r	Cx37m	Cx40m	Cx43r	Cx45m	Cx46m	Cx50m
Cx46m	+ (Xo+) 100–135	+ 70–502	+ (Xo+) 100–30[b]		(Xo–)	+ (Xo+) 110–60[b]	+ (Xo+) 140[b]		
Cx50m	+ (Xo+)		(Xo+)		(Xo–)		+ (Xo+) 180–140	+ (Xo+) 200	
Cxs	Cx26m	Cx30.3m	Cx32r	Cx37m	Cx40m	Cx43r	Cx45m	Cx46m	Cx50m

Numerical entries represent single GJ channel conductance (g[open]) in pS. For heterotypic GJ channels, g(open) is given at $V_j = +100$ mV and $V_j = -100$ mV, respectively. We define V_j as positive when the cell expressing the connexin indicated in a column is more positive than the cell expressing connexin shown in a row. In our experiments, g(open) was measured at room temperature ($\sim$24°C) or at 36–37°C. All data were corrected for 24°C by using a temperature coefficient of $Q_{10} = 1.25$ (11). It is important to note that g(open) variations of $\pm 30\%$ as reported by different laboratories are likely due to using different pipet solutions (KCl, Kglutamate or Kaspartate) and in different concentrations, varying between 120 and 140 mM. (Xo+), (Xo–), or (Xo+/–) summarizes cell–cell coupling formation data in Xenopus oocytes as reported by different laboratories (28–33); (Xo+/–) indicates inconclusive data.

[a] Data obtained in collaboration with Dr. R. Weingart (*10,11,34*).

[b] Data obtained in collaboration with Dr. V. Verselis.

substate (residual state), which is different from the closed state *(26)*. Once two hemichannels dock and are tightly sealed to prevent current leakage between the channel pore and the intercellular space *(18)*, the junctional current rises which indicates the first channel pore opening. Usually this "formation" current is fragmented, indicating the existence of some fixed number of substates, and continues for tens of milliseconds *(11,18)*. The physiological meaning of these substates is unclear but it is postulated that they are directly related to the multimeric structure of the GJ channel (12 connexin molecules form the channel). We propose that the very first GJ channel opening is a special type of gating defined by a unique connexin conformation *(18)*. For a more precise analysis of substates it is very important to have low noise records. For this reason we often use different types of ion channel blockers, such as Cs^+, Ba^{2+}, TEA^+, Zn^{2+}, etc. In addition, it is recommended that the length of the pipet tip inserted into the bulk solution be minimized, by regulating solution outflow level, which substantially reduces the capacity of the input circuit and, consequently, high-frequency noise.

2. Our technique for *de novo* channel formation has been successfully applied to both homo- and heterotypic junction formation and their single channel analysis. Standard techniques described for heterotypic junction analysis involve loading one type of cell with dye and co-culturing unloaded and loaded cells for 1–2 d *(24)*. Thereafter, fluorescent microscopy is used to select mixed cell pairs for cell-to-cell dye diffusion or electrophysiological analysis. With our technique, two small coverslips (coverslips can be broken in small pieces) consisting of two different cell lines are positioned side by side in the experimental chamber. All free floating or weakly attached cells are washed out by strong bulk perfusion applied for the first few minutes of the experiment. Then, one cell from one of the coverslips is detached by using a gentle flow of solution as described previously, and patched to the pipet with the establishment of whole-cell recording. Then microscope stage with experimental chamber is moved to focus on other coverslip. The same procedure is repeated for the cell from the other coverslip and the two patched cells are manipulated into physical contact. Utilizing this method, we have observed formation of heterotypic junctions with two different types of insect cells, *Aedes albobopictus* (cell line C6/36) and *Spodoptera fruigiperda* (cell line Sf9) *(19)*. HeLa cells transfected with 9 different wild-type connexins have been observed to form 16 types of heterotypic junctions (e.g., Cx26/32, Cx43/46, etc.) of 21 tested. **Table 1** (*see previous page*) summarizes our data on homo- and heterotypic GJ channel formation in HeLa cell transfectants kindly provided to us by Dr. Willecke. In heterotypic junctions, for example, Cx26/40, Cx32/40, Cx40/43, etc., there was no GJ channel formation suggesting that these connexin pairs are not compatible. Most heterotypic junctions exhibit asymmetric $g_j - V_j$ dependence. For example, in Cx32/37 junctions this asymmetry is so high that at $V_j = 0$ most of the channels are closed. Therefore, g_j should be measured at different V_js or V_m because GJ channels can be functional but may be masked by voltage gating mechanisms. Our data on the formation of heterotypic

GJ channels in HeLa cells for most, but not all, types of heterotypic junctions are in agreement with the data obtained for *Xenopus* oocyte cell pairs expressing different connexins after injection of Cx mRNA (*see* Xo+ and Xo– in **Table 1**). We did however observed the formation of three heterotypic junctions (Cx32/Cx37 and Cx32/Cx43) which are not been reported to form in *Xenopus* oocytes. Compatibility of different connexins in HeLa transfectants as evaluated by using cell-cell dye diffusion methods have been extensively studied by Dr. Willecke and colleagues *(24,27)*

Acknowledgments

The technique of *de novo* formation of gap junction channel has developed over time in collaboration with Dr. R. Weingart, Department of Physiology, Bern University, Switzerland, Dr. C. Peracchia, Department of Physiology, Rochester University, USA and Dr. V. Verselis, Department of Neuroscience of Albert Einstein College of Medicine NY, USA. The author would like to thank Dr. K. Willecke, Institute of Genetics, Bonn University, Germany, for providing HeLa transfectants which were invaluable for our studies and Dr. Jak Kronengold for highly helpful comments on the manuscript.

References

1. Bennett, M. V. and Verselis, V. K. (1992) Biophysics of gap junctions. *Semin. Cell Biol.* **3,** 29–47.
2. Sosinsky, G. E. (1996) Molecular organization of gap junction membrane channels. *J. Bioenerg. Biomembr.* **28,** 297–309.
3. Laird, D. W. (1996) The life cycle of a connexin: gap junction formation, removal, and degradation. *J. Bioenerg. Biomembr.* **28,** 311–318.
4. Loewenstein, W. R., Kanno, Y., and Socolar, S. J. (1978) Quantum jumps of conductance during formation of membrane channels at cell-cell junction. *Nature* **274,** 133–136.
5. Chow, I. and Poo, M. M. (1984) Formation of electrical coupling between embryonic Xenopus muscle cells in culture. *J. Physiol.* **346,** 181–194.
6. Chow, I. and S. H. Young (1987) Opening of single gap junction channels during formation of electrical coupling between embryonic muscle cells. *Dev. Biol.* **122,** 332–337.
7. Rook, M. B., Jongsma, H. J., and van Ginneken, A. C. (1988) Properties of single gap junctional channels between isolated neonatal rat heart cells. *Am. J. Physiol.* **255,** H770–H782.
8. Churchill, D. and Caveney, S. (1993) Double whole-cell patch-clamp characterization of gap junctional channels in isolated insect epidermal cell pairs. *J. Membr. Biol.* **135,** 165–180.
9. Bukauskas, F. F. and Weingart, R. (1993) Multiple conductance states of newly formed single gap junction channels between insect cells. *Pflügers Arch.* **423,** 152–154.

10. Bukauskas, F. F., Elfgang, C., Willecke, K., and Weingart, R. (1995) Heterotypic gap junction channels (connexin26-connexin32) violate the paradigm of unitary conductance. *Pflügers Arch.* **429,** 870–872.

11. Bukauskas, F. F., Elfgang, C., Willecke, K., and Weingart, R. (1995) Biophysical properties of gap junction channels formed by mouse connexin40 in induced pairs of transfected human HeLa cells. *Biophys. J.* **68,** 2289–2298.

12. Keane, R. W., Mehta, P. P., Rose, B., Honig, L. S., Loewenstein, W. R., and Rutishauser, U. (1988) Neural differentiation, NCAM-mediated adhesion, and gap junctional communication in neuroectoderm. A study in vitro. *J. Cell Biol.* **106,** 1307–1319.

13. Meyer, R. A., Laird, D. W., Revel, J. P., and Johnson, R. G. (1992) Inhibition of gap junction and adherens junction assembly by connexin and A-CAM antibodies. *J. Cell Biol.* **119,** 179–189.

14. Ghoshroy, S., Goodenough, D. A., and Sosinsky, G. E. (1995) Preparation, characterization, and structure of half gap junctional layers split with urea and EGTA. *J. Membr. Biol.* **146,** 15–28.

15. Johnson, R., Meyer, R., and Lampe, P. (1989) Gap junction formation: A self-assembly model involving membrane domains of lipid and protein, in *Cell Interaction and Gap Junctions* (Sperelakis, N. and Cole, W. C., eds.), CRC Press, Boca Raton, FL, pp. 159–179.

16. Burt, J. M. and Spray, D. C. (1988) Inotropic agents modulate gap junctional conductance between cardiac myocytes. *Am. J. Physiol.* **254,** H1206–H1210.

17. Giaume, C., C. Randriamampita, and A. Trautmann (1989) Arachidonic acid closes gap junction channels in rat lacrimal glands. *Pflügers. Arch.* **413,** 273–279.

18. Bukauskas, F. F. and Weingart ,R. (1994) Voltage-dependent gating of single gap junction channels in an insect cell line. *Biophys. J.* **67,** 613–625.

19. Bukauskas, F. F., Vogel, R., and Weingart, R. (1997) Biophysical properties of heterotypic gap junctions newly formed between two types of insect cells. *J. Physiol. (London)* **499**(Pt 3), 701–713.

20. Valiunas, V., Bukauskas, F. F., and Weingart, R. (1997) Conductances and selective permeability of connexin43 gap junction channels examined in neonatal rat heart cells. *Circ. Res.* **80,** 708–719.

21. Bukauskas, F. F., Shrager, P., and Peracchia, C. (1998) Gating properties of gap junction channels in Schwann cells and fibroblasts isolated from the sciatic nerve of neonatal rats, in *Gap Junctions* (Werner, R., ed.), IOS Press, Amsterdam, pp. 25–29.

22. Churchill, D., Coodin, S., Shivers, R. R., and Caveney, S. (1993) Rapid de novo formation of gap junctions between insect hemocytes in vitro—A freeze–fracture, dye-transfer and patch-clamp study. *J. Cell Sci.* **104,** 763–772.

23. Bukauskas, F. F. and Weingart, R. (1993) Temperature dependence of gap junction properties in neonatal rat heart cells. *Pflügers Arch.* **423,** 133–139.

24. Elfgang, C., Eckert, R., Lichtenberg-Frate, H., Butterweck, A., Traub, O., Klein, R. A., Hulser, D. F., and Willecke, K. (1995) Specific permeability and selective formation of gap junction channels in connexin-transfected HeLa cells. *J. Cell. Biol.* **129,** 805–817.

25. Neyton, J. and Trautmann, A. (1985) Single-channel currents of an intercellular junction. *Nature* **317,** 331–335.
26. Weingart, R. and Bukauskas, F. F. (1993) Gap junction channels of insects exhibit a residual conductance. *Pflügers Arch.* **424,** 192–194.
27. Cao, F., Eckert, R., Elfgang, C., Nitsche, J. M., Snyder, S. A., Hulser, D. F., Willecke, K., and Nicholson, B. J. (1998) A quantitative analysis of connexin-specific permeability differences of gap junctions expressed in HeLa transfectants and Xenopus oocytes. *J. Cell Sci.* **111,** 31–43.
28. Barrio, L. C., Suchyna, T., Bargiello, T., Xu, L. X., Roginski, R. S., Bennett, M. V., and Nicholson, B. J. (1991) Gap junctions formed by connexins 26 and 32 alone and in combination are differently affected by applied voltage. *Proc. Natl. Acad. Sci. USA* **88,** 8410–8414.
29. Bruzzone, R., White, T. W., and Paul, D. L. (1994) Expression of chimeric connexins reveals new properties of the formation and gating behavior of gap junction channels. *J. Cell Sci.* **107,** 955–967.
30. Bruzzone, R., Haefliger, J. A., Gimlich, R. L., and Paul, D. L. (1993) Connexin40, a component of gap junctions in vascular endothelium, is restricted in its ability to interact with other connexins. *Mol. Biol. Cell* **4,** 7–20.
31. White, T. W., Bruzzone, R., Wolfram, S., Paul, D. L., and Goodenough, D. A. (1994) Selective interactions among the multiple connexin proteins expressed in the vertebrate lens: the second extracellular domain is a determinant of compatibility between connexins. *J. Cell Biol.* **125,** 879–892.
32. Werner, R., Levine, E., Rabadan Diehl, C., and Dahl, G. (1989) Formation of hybrid cell-cell channels. *Proc. Natl. Acad. Sci. USA* **86,** 5380–5384.
33. White, T. W., Paul, D. L., Goodenough, D. A., and Bruzzone, R. (1995) Functional analysis of selective interactions among rodent connexins. *Mol. Biol. Cell* **6,** 459–470.
34. Bukauskas, F. F. and Weingart, R. (1995) Gating properties of homo- and heterotypic gap junction channels formed by different mouse connexins, in Proceedings of the 1995 *Gap Junction Conference,* Ile des Embiez, France.

22

Recording and Analysis of Putative Direct Electrical Interactions in the Mammalian Brain

Taufik A. Valiante, Jose L. Perez Velazquez, and Peter L. Carlen

1. Introduction

The concept of electrical transmission in the mammalian central nervous system (CNS) and its functional implications for normal physiology and pathology has generated much controversy. While it was proposed as early as the mid-19th century that neurotransmission could be electrical and chemical (*1*), experimental evidence for chemical neurotransmission obtained by Loewi, Eccles, Katz, and colleagues (*2,3*), undermined the concept of electrical coupling. The first evidence for electrical transmission occurred in invertebrate preparations and was reported almost simultaneously by Watanabe in 1958 (*4*), and by Furshpan and Potter in 1959 (*5*), the former venturing that the electrical coupling promoted synchronous firing of neurons. Experimental evidence for electrical transmission in vertebrates was obtained by Bennett and colleagues in 1966–67 (*6,7*) while working on the supramedullary neurons of the pufferfish and the motoneurons of the toadfish, where high frequency synchronous contractions of the swim bladder muscle require electrically coupled neurons. Later in 1973, Henri Korn and colleagues found electrical interactions in mammals (*8*), and at the same time Baker and Llinas found evidence for electrical transmission in the rat mesencephalic nucleus (*9*).

Although it now appears that cellular communication in the mammalian CNS is mostly chemically mediated, the evidence gathered during the past two decades from many laboratories suggests that electrical interactions through gap junctions or field effects are relatively abundant. Thus the *functional role* of direct electrical interactions in normal mammalian CNS physiology and pathology (*10–13*) may be greater than first suspected (*14*).

From: *Methods in Molecular Biology, vol. 154: Connexin Methods and Protocols*
Edited by: R. Bruzzone and C. Giaume © Humana Press Inc., Totowa, NJ

Throughout this chapter we will introduce the reader to a few methods and tricks we have developed to study what we think represents electrical transmission, because, as it was half a century ago, the concepts and interpretation of observations are still very controversial. It has been our feeling that we have solved few and generated some new questions regarding the nature and implications of direct electrical interactions and their manifestations in the mammalian CNS. We stress, however, that different explanations to the experiments we describe here, in different circumstances, may be possible and certainly demand discrimination and an open mind as to their interpretations.

The best assay to determine electrical transmission involves simultaneous recording from two coupled neurons, which is not a trivial feat, at least in mammalian CNS preparations. Dual simultaneous recordings from clearly coupled mammalian neurons has been successfully achieved by MacVicar and Dudek *(15)*, in the rat hippocampus. We describe why this is difficult and present some alternatives that could help identify coupled neurons (*see* **Note 1**). A classical assay that does not involve dual intracellular recordings is the antidromic test *(16,17)* , but sometimes the results are very hard to interpret. In any case, these experiments gathered evidence for small amplitude events, termed fast prepotentials (FPPs) or spikelets, as manifestations of electrotonic interactions. A continuous debate as to the origin of these spikelets has developed. In the course of our investigations of gap junctional mechanisms in epilepsy, we observed spikelets in whole-cell recordings from pyramidal rat neurons in a variety of conditions *(11–13,18)*. In particular, when neurons are made hyperexcitable, the frequency of spikelet recordings increased to a reasonable value, 36% of the recordings, as compared with percentages using the antidromic test, 18% *(16)* or dual recordings *(15)*. As mentioned previously, the nature of these spikelets is controversial, but we have made some progress showing that they most likely represent direct electrical interactions and that their origin is external to the recorded neuron (*unpublished observations)*, and thus cannot be attributed to dendritic or calcium spikes *(12)*.

We describe the strategies that favor the appearance of spikelets and the methods we used to analyze and identify them. This is most important because some spikelets resemble excitatory postsynaptic potentials, but their sensitivity to membrane potential clearly separates them. The electrophysiological recordings involve the use of the patch-clamp technique, which provides relatively clean recordings making it possible to identify spikelets that have amplitudes sometimes on the order of a fraction of a millivolt.

2. Materials

The reagents and equipment mentioned below should be part of the standard repertoire of most electrophysiological laboratories.

2.1. Solutions

1. Artificial cerebrospinal fluid (ACSF), this can be any used in your laboratory. Ours is: 125 mM NaCl, 2.5–5 mM KCl, 1.25 mM NaH$_2$PO$_4$, 2 mM MgSO$_4$, 2 mM CaCl$_2$, 25 mM NaHCO$_3$, 10 mM glucose; the pH is 7.3–7.4 when the solution is aerated with 95% O$_2$, 5% CO$_2$. Osmolarity should be 300 ± 5 mOsm.
2. Internal solution for whole-cell patch clamp recordings includes: 150 mM potassium gluconate, 10 mM N-(2-hydroxyethyl) piperazine-N'-d-ethanesulfonic acid (HEPES), 1–2 mM Mg-ATP, 2–5 mM KCl, 0.1 mM EGTA, pH 7.2 adjusted with KOH, osmolarity 275 ± 5 mOsm.

2.2. Experimental Setup

1. For electrophysiological recordings, brain slices can be placed in any standard chamber; we have routinely used both interface and submerged type chambers. The submerged chamber is model PDMI-2, purchased from Medical System. The temperature should be set around 34–36°C, which is very important if, for example, epileptiform activity is required. A Faraday cage should surround the chamber to reduce electrical noise so that fast and small amplitude spikeletes can be reliably recorded.
2. The neuronal recordings are performed using the whole-cell configuration of the patch-clamp technique. Patch electrodes can be any standard borosilicate capillary tubing, such as TW150F-4 thin wall with filament sold by World Precision Instruments (WPI, New Haven, CT). The electrode tip resistance should range between 4 and 8 MΩ.
3. The amplifiers for the recordings can be any that allow for recordings in both bridge mode and voltage-clamp, such as the Axoclamp-2A from Axon Instruments (Foster City, CA).
4. Most of the spikelets recordings should be stored on videotape for later playback and analysis. Therefore a digital data recorder is needed, such as the VR-10 from Instrutech (New York).
5. For data acquisition software, we use any version of pClamp (Axon Instruments, Foster City, CA), which runs on IBM-compatible computers.
6. To reduce electrical noise, a Faraday cage will be necessary surrounding the vibration-free table on which the chamber and the microscope are placed. Our micromanipulators used for patching or extracellular recordings are water 3D manually driven manipulators (Model WR-60) purchased from Narishige Scientific (Japan).

3. Methods
3.1. Electrophysiological Recordings

1. The procedure obviously starts with the preparation of brain slices, 400 μm thick. We normally used young (17–30 d old) Wistar rats, which are anesthetized with Fluothane (Ayerst Laboratories, Montreal), and then decapitated. The brains are removed and transverse (or horizontal) brain slices are obtained using a vibratome

(Series 1000, Technical Products International). These procedures are carried in ice-cold ACSF. The slices are then kept at room temperature in a beaker for the duration of the experiments. In general, any other standard method to prepare brain slices could in principle be used. Most of our work took place in the hippocampus and cortex of rat brain, but any area of interest can be studied. A prerequisite is that neurons can be adequately patched such that recordings are of relatively low noise and of long duration.

2. We assume that spikelets represent electrically transmitted action potentials through gap junctions, or field coupling events. Thus, to reliably record spikelets, there is an absolute requirement of spontaneous or evoked activity. One method we used was to bathe hippocampal slices in calcium-free ACSF (with 1 mM EGTA added to eliminate residual calcium), which promotes spontaneous rhythmic activity recorded as field potential events synchronous with neuronal bursting *(11,12,18)*. During this epileptiform activity we could record spikelets in 35% of the patched pyramidal neurons.

3. To visualize spikelets during the recording, which normally are very fast events lasting 0.5–1 ms, and whose amplitudes are in the range 0.2–5 mV (*see* **Fig. 1**), all that is needed is to set the oscilloscope to relatively short sweep duration. We have routinely employed sweep durations of 0.5–2 ms with good sensitivity.

4. If calcium-free conditions are not desired, normal ACSF can be used, but increasing the potassium concentration (5–7 mM) increases the number of action potentials in response to extracellular stimulation. In **ref. *18*** and in **Fig. 1A** some recordings can be seen in normal ACSF. In these conditions an orthodromic or antidromic extracellular stimulus is required, and spikelets can be seen riding on the excitatory postsynaptic potential (EPSP) evoked in the patched cell. The efficiency of clear spikelet recordings under these conditions is lower than in calcium-free, approx 20%.

5. It is generally accepted that the amplitude of electrotonically transmitted spikes should be independent of the membrane potential (Vm) of the cell from which we record. Thus, to characterize the voltage dependency of spikelets the neuronal Vm has to be depolarized and hyperpolarized and the spikelet amplitudes measured. Then amplitude histograms can be constructed once spikelets have been identified for analysis (described below). We also recorded spikelets under voltage-clamp conditions *(12)*, but because they are rapid events the clamp will generally not be adequate.

Figure 1 shows the typical shapes of spikelets we recorded. While many of them will have the biphasic appearance shown in **Fig. 1B** (*upper* and *middle* traces), some will be long lasting, resembling EPSPs (*lower* trace). We think that the second type represents direct gap junctional communication, because of the low-pass filter characteristics (gap junctions act as low-pass filters, basically ohmic resistors connecting two cytoplasms), but the first type represents field coupling caused by a group of neuron possibly coupled through gap junc-

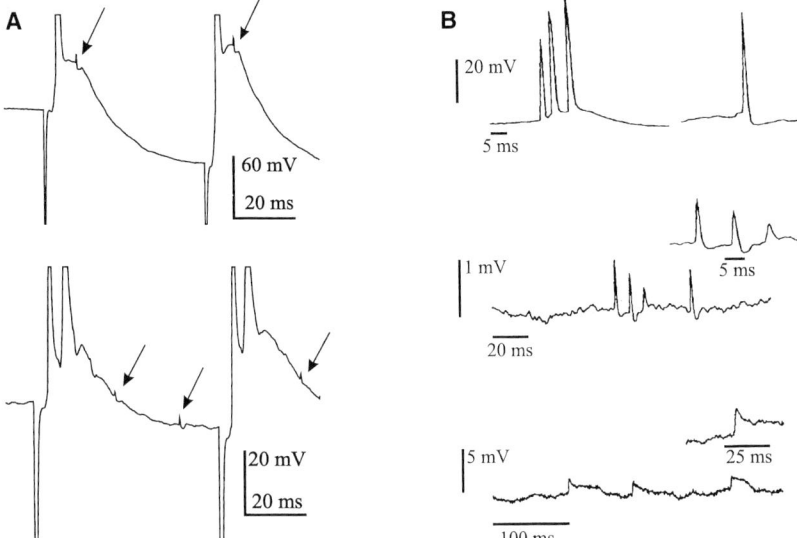

Fig. 1. Recording of spikelets in normal and calcium-free ACSF. **(A)** Extracellular stimulation of the Schaffer collaterals evokes an EPSP in the CA1 pyramidal neuron with superimposed action potentials and spikelets *(arrows)*. **(B)** Whole-cell recordings from two CA1 neuron in calcium-free ACSF reveal spikelets with different waveforms *(upper* and *middle* traces). All these spikelets disappear when gap junction blockers are bath applied *(lower* trace).

tions, firing synchronously in the neighborhood of the recording neuron *(13)*. The biphasic appearance of these events indicate differentiated spikes, characteristic of high-pass filters and capacitive coupling and therefore it is not suggestive of a direct low-pass gap junctional coupling. All these spikelets disappear (or their frequency decreases) when manipulations that block gap junctions are used *(11)*: acidification, octanol, or halothane *(see* **Note 2**). Conversely, spikelet frequency increases with manipulations that open gap junctions, such as alkalinization *(11)*.

3.2. Automated Identification and Analysis of Spikelets

Identification of spikelets posses several unique problems that arise from their small size and the high noise environment in which they exist. Spikelets are events that are on the order of millivolts and only several milliseconds long. Under the zero extracellular calcium conditions used to evoke synchronous neuronal discharge there are considerable membrane potential fluctuations most likely arising from intrinsic channel activity as well as electrical field effects.

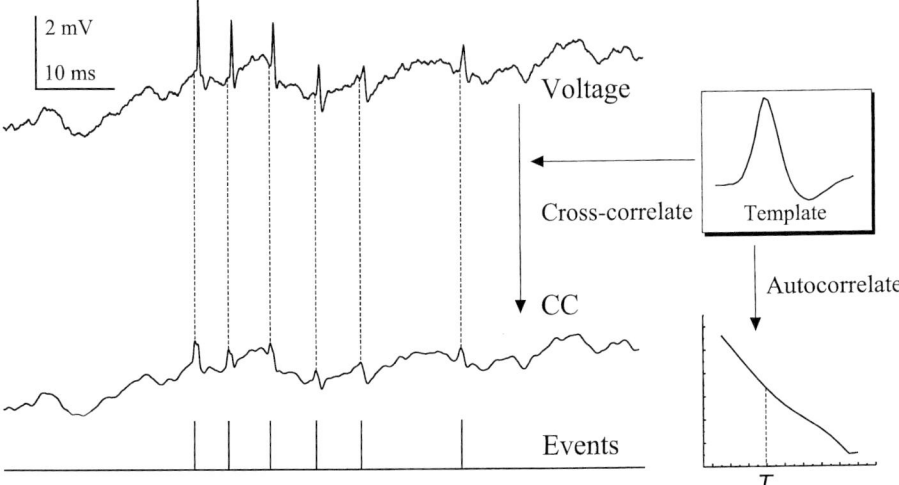

Fig. 2. Schematic of identification algorithm. Event detection begins with computation of cross- and autocorrelation functions. Representative data trace (voltage, 2048 points) along with a selected template (35 points) are shown. The cross-correlation is obtained (CC, bottom trace) as well as the autocorrelation of the template. Note how the high frequency noise is "filtered" out of the data through this mathematical manipulation. The *vertical dashed lines* are aligned to the peaks in the cross-correlation function considered to meet the criteria for an event. A user defined time interval (*T, see* also **Fig. 3**) is used to compute a slope from the autocorrelation function.

These fluctuations although morphologically dissimilar to spikelets may share a similar or larger amplitude than the waveform of interest. Under these conditions threshold detection algorithms fail, leaving the alternative, manual identification and analysis of spikelets as an arduous and daunting endeavour.

A unique detection algorithm first described for the analysis of synaptic currents *(19)* but more appropriately applied to low signal-to-noise ratio recordings was developed to identify spikelets. In essence the procedure involves searching for maxima within the correlation of a segment of data and a generic template (**Fig. 1**). Template here refers to a segment of data that typifies all or part of waveform for which one would like to search for within the data set (in this case a spikelet). Putative maxima are assessed according to four user defined criteria of which one arises from the autocorrelation function of the template (**Fig. 2**). Acceptable maxima are considered to demarcate the beginning of an event. A user-defined duration of data following the beginning of the event is taken to be a spikelet. The segment of data identified as a spikelet is saved to disc. Each spikelet can be reviewed and those that are spurious can be discarded by the user. The final set of spikelets with the times at which they occur in the data set are used for subsequent analysis.

3.2.1. Digitizing the Data

1. Long stretches of neuronal activity (20–120 min) were first recorded with a digital data recorder at a digitization rate of 88 kHz. To fully utilize the range of the analog-to-digital converter, the membrane potential was offset to zero. We then amplified this output from the Axoclamp 2A amplifier (Axon Instruments) by and additional 80 before recording to VCR tape.
2. During playback the amplifier gain was set to unity and the data digitized at 10 kHz using Fetchex (Axon Intruments) and saved to disk as a Fetchex formatted data file. This binary formatted data file was then read in as 2048-point data segments (**Fig. 2**, top trace), each being analyzed sequentially until the end of the file was reached.

3.2.2. Picking a Template

The entire data set was available for viewing and analysis in a graphical environment. The user was simply required to place two cursors anywhere within the entire data set to encompass a segment of data to be used as the template. This template was saved to disk and recalled when needed. The template could be any duration, and encompass any or all aspects of the spikelet to be detected.

3.2.3. Cross- and Auto-Correlations

Since the calculation of the auto- and cross-correlation functions requires arrays to be a power of two in length, two copies of the template were created each zero padded to the appropriate length. The first copy consisted of the template segment with trailing zeros padded to the next smallest power of two. The second copy was extended to the length of the data segment (albeit also a power of two). The cross-correlation function of the data and template (CC), and the autocorrelation function of the template were then obtained by standard numerical techniques *(20)*.

3.2.4. Criteria for Event Selection

From the autocorrelation function a single value was obtained. This value is the slope of the autocorrelation function over the interval from zero lag to a user specified time (T), which will be designated M_{auto} (*see* **Fig. 3A**). S_{min} and S_{max} are defined by the user and represent scale factors. We define M_{cross} as the slope over a specified interval within the cross-correlation function. A point in the cross-correlation function at time t was considered to be the start of an event or spikelet if it met all the following criteria:

A

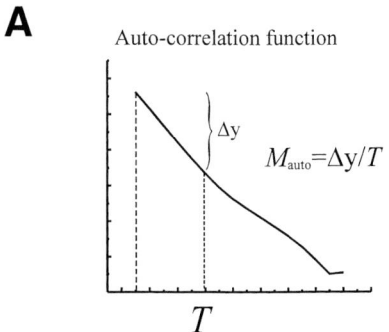

Auto-correlation function

$M_{auto}=\Delta y/T$

T

B

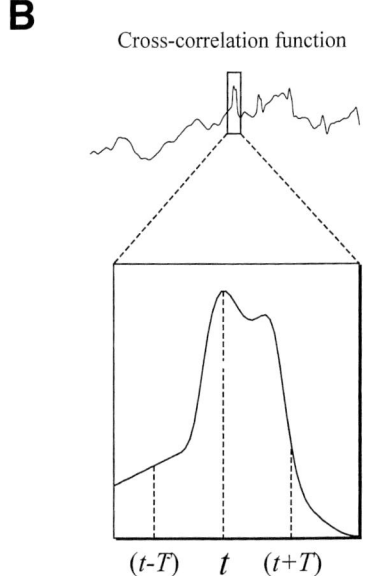

Cross-correlation function

$(t-T)$ t $(t+T)$

Fig. 3. Details of event selection. **(A)** Given the autocorrelation of the template, the slope over the user defined interval T is obtained (M_{auto}) **(B)** Segment of the cross-correlation function (same trace as in **Fig. 2**) and an expanded view of a short segment of this trace *(below)*. Two slopes are computed from the cross-correlation function, that over the interval $(t-T, t)$ and a second over the interval $(t, t+T)$. It is clear that the selection of T can significantly alter the fidelity of the search algorithm.

1. The absolute value of M_{cross} over the intervals $(t, t-T)$ and $(t, t+T)$ was greater than $S_{min}*M_{auto}$ (*see* **Fig. 3B**).
2. The absolute value of M_{cross} over the intervals $(t, t-T)$ and $(t, t+T)$ was less than $S_{max}*M_{auto}$.
3. Events could not be closer than a user specified time (usually 2 ms).

Once a putative event was identified in this fashion then a user specified amount of data prior to *t* time and following *t* was saved to disk as an "event." The time of occurrence of the event within the entire digitized segment was also saved to disk. This technique has several advantages and disadvantages which are described in **Notes 3** and **4**.

4. Notes

1. The main difficulties we found with dual recordings in the slice are the following: first, only a few of the neurons are coupled through gap junctions (15–35% in the CA1 area of the hippocampus in our conditions), therefore we should know to which neurons the one we patched is coupled before attempting a second patch. The visualization of the cells can be achieved using a 30× or 40× objective lens by infrared microscopy, a common method employed today. We have tried to identify coupled cells by including a fluorescent dye in the first patching electrode, hoping that the dye will spread to coupled neurons. We do not recommend this method due to the difficulty of visualizing clearly coupled cells under the patching conditions; for a typical dye coupling study, we must dehydrate and clear the slice with ethanol and methyl salycilate respectively *(11,12,18)* to clearly identify dye-coupled cells, which sometimes is unsuccessful with cleared tissue because the cells are to deep in the slice. Therefore, identifying unambiguously dye coupled cells under the patching setup is extremely difficult, and erroneous "coupling" can be seen frequently because of dye leakage from the patching electrode. In addition, fluorescent dyes are toxic for the cells, and more so when sustained UV irradiation is needed to perceive possible dye coupled neurons surrounding the patched cell. However, other dyes such as calcein may not be as toxic as Lucifer Yellow; keeping UV irradiation to a minimum helps in any case. For these reasons, we developed a trick that allows us to identify cells with direct cytoplasmic connections without using fluorescent markers. We employ an hyperosmolar internal solution, between 340 and 360 mOsm (the osmolarity raised with potassium gluconate or KCl), to patch the first cell. At 10–20 min after being patched the neuron is normally swollen, and one or two other cells surrounding the patched neuron can be seen to be clearly swollen, which indicates a direct cytoplasmic connection with the patched cell. This method avoids the leakage problem (extracellular hyperosmolar solutions cause the opposite effect, shrinkage), and neurons are relatively healthy for 15–20 min, which may allow for a second (swollen) cell to be patched, even though after this time cells become very leaky and may expire. As a note of caution, we have tried this method in immature rat neocortex, so far without success due to stability problems, for the tissue will move when the second patching electrode is positioned.

2. Several blockers of gap junction channels were successfully used in brain slices. Sodium propionate in the external ACSF, 20–25 m*M*, substituting the same amount of NaCl to maintain the osmolarity value, will acidify the cellular cytoplasm upon application in a few seconds, according to the experimental observations in muscle and nervous tissue. Another way to promote intracellular

acidification is to use normal ACSF but using 90% CO_2 to bubble the solution; also, washout of ammonium chloride (10 mM) will cause the same effect. Halothane was used at a concentration of 2% in the ACSF, and 0.2 mM octanol (directly added to the ACSF, do not make stock solution of octanol). Alkalinization is promoted by application of ammonium chloride, 10 mM (washout will cause acidification), trimethylamine (10 mM), or ACSF containing 60 mM bicarbonate. All these manipulations had an effect upon spikelet frequencies and spontaneous seizure-like events in 1–2 min upon application, were applied for not more than 7–10 min, and the effects were reversible in 2–10 min.

3. One of the major disadvantages of the recognition technique described above is the usage of computer memory. The original algorithm was written to run within the 640K memory limit of DOS-based personal computers. This does not pose a significant problem if the data is analyzed as 2048-point segments. However there is some data loss at the ends of each cross-correlation function. This, however, did not appear to pose a significant problem to analysis. The second major disadvantage is that the algorithm is rather slow as it involves three fast Fourier transforms and two inverse Fourier transforms. Nonetheless, a 66 MHz 80486 processor took only approx 5 minutes to analyze a 1 million point data set. This seems a small price to pay considering the manual alternative. Indeed some events were probably missed. However, by setting the four user specified variables defined above, the sensitivity can be increased at the expense of specificity. The putative events can then reviewed and those that are inappropriate (noise, action potentials, stimulus artifacts) discarded.

4. The algorithm presented above is a generic search algorithm. One needs only to selected a waveform of interest, and modify the search parameters. We have applied this algorithm to detection of spikelets *(12)* and miniature excitatory postsynaptic potentials. It has also been tested with field potentials and action potentials and has proven to be reliable. Under certain conditions this algorithm is clearly superior to threshold and slope detection algorithms. This is obvious by observing the data tracing in **Fig. 2**. A threshold detection algorithm would not be appropriate given the large and rapid fluctuations in baseline. We found this algorithm to be robust even in under poor recording conditions corrupted with high-frequency noise. This is most probably related to the fact that the cross-correlation filters out uncorrelated high-frequency noise (*see* cross-correlation trace in **Fig. 2**).

References

1. Brazier, M. A. B. (1959) The historical development of neurophysiology, in *Handbook of Physiology. Neurophysiology,* sect. 1, vol. I (J. Field, J., ed.), American Physiological Society, Washington, DC, pp. 1–58.
2. Eccles, J. C. (1964) *The Physiology of Synapses.* Springer-Verlag, New York, pp. 1–316.
3. Loewi, O. (1933) Problems connected with the principle of humoral transmission of nervous impulses. *Proc. R. Soc. B. Biol. Sci.* **118,** 299–316.

4. Watanabe, A. (1958) The interaction of electrical activity among neurones of lobster cardiac ganglion. *Jpn. J. Physiol.* **8,** 305–318.

5. Furshpan, E. J. and Potter, D. D. (1959) Transmission at the giant motor synapses of the crayfish. *J. Physiol.* **145,** 289–325.

6. Bennett, M. V. L., Nakajima, Y., and Pappas, G. D. (1967) Physiology and ultrastructure of electrotonic junctions. I. Supramedullary neurones. *J. Neurophysiol.* **30,** 161–179.

7. Pappas, G. D. and Bennett, M. V. L. (1966) Specialized junctions in electrical transmission between neurones. *Ann. NY Acad. Sci.* **37,** 495–508.

8. Korn, H., Sotelo, C., and Crepel, F. (1973) Electrotonic coupling between neurones in the rat lateral vestibular nucleus. *Exp. Brain Res.* **16,** 255–275.

9. Baker, R. and Llinas, R. (1971) Electrotonic coupling between neurones in the rat mesencephalic nucleus. *J. Physiol.* **212,** 45–63.

10. Draguhn, A., Traub, R. D., Schmitz, D., and Jefferys, J. G. R. (1998) Electrical coupling underlies high-frequency oscillations in the hippocampus. *Nature* **394,** 189–192.

11. Perez Velazquez, J. L., Valiante, T. A., and Carlen, P. L. (1994) Modulatuon of gap junctional mechanisms during calcium-free induced field burst activity: a possible role for electrotonic coupling in epileptogenesis. *J. Neurosci.* **14,** 4308–4317.

12. Valiante, T. A., Perez Velazquez, J. L., Jahromi, S. S., and Carlen, P. L. (1995) Coupling potentials in CA1 neurons during calcium-free-induced field burst activity. *J. Neurosci.* **15,** 6946–6956.

13. Vigmond E. J., Perez Velazquez, J. L., Valiante, T. A., Bardakjian, B. L., and Carlen, P. L. (1997) Mechanisms of electrical coupling between pyramidal cells. *J. Neurophysiol.* **78,** 3107–3116.

14. Traub, R. D. and Miles, R. (1991) *Neuronal Networks of the Hippocampus.* Cambridge University Press, Cambridge, UK.

15. MacVicar, B. A. and Dudek, F. E. (1981) Electrotonic coupling between pyramidal cells: a direct demonstration in rat hippocampal slices. *Science* **213,** 782–785.

16. MacVicar, B. A. and Dudek, F. E. (1982) Electrotonic coupling between granule cells of rat dentate gyrus: physiological and anatomical evidence. *J. Neurophysiol.* **47,** 579–592.

17. Taylor, C. P. and Dudek, F. E. (1982) A physiological test for electrotonic coupling between CA1 pyramidal cells in rat hippocampal slices. *Brain Res.* **235,** 351–357.

18. Perez Velazquez, J. L., Han, D., and Carlen, P. L. (1997) Neurotransmitter modulation of gap juctional communication in the rat hippocampus. *Eur. J. Neurosci.* **9,** 2522–2531.

19. Abdul-Ghani, M. A., Valiante, T. A., and Pennefather, P. S., (1996) Sr^{2+} and quantal events at glutamatergic synapses between mouse hippocampal neurones in culture. *J. Physiol.* **495,** 113–125.

20. Press, W. H., Flannery, B. P., Teukolsky, S. A., and Vetterling, W. T. (1988) *Numerical Recipes in C.* Cambridge University Press, New York.

23

Intercellular Calcium Signaling and Flash Photolysis of Caged Compounds

A Sensitive Method
to Evaluate Gap Junctional Coupling

Luc Leybaert and Michael J. Sanderson

1. Introduction

Communication is a fundamental paradigm of multicellular systems, and neighboring cells can exchange signals either by paracrine or juxtacrine communication *(1)*. In addition, cells that are coupled by gap junctions can communicate by the passage of electrical signals or by the diffusion of messenger molecules or ions through these junctions. Cardiac myocytes are extensively coupled by gap junctions to form a functional syncytium and are a good example of cells that communicate by electrical signals. Examples of nonexcitable, gap junctional coupled cells that communicate by the diffusion of intracellular messengers or ions are glial cells *(2,3)*, airway epithelial cells *(4,5)*, lens epithelial cells *(6,7)*, hepatocytes *(8,9)*, and endothelial cells *(10,11)*. In these cells, propagating increases in intracellular free Ca^{2+} concentration ($[Ca^{2+}]_i$) spread in all directions and over many rows of cells to form intercellular Ca^{2+} waves, and these are believed to form a major mechanism of cell communication *(12)*.

Intercellular Ca^{2+} waves were initially described in airway epithelial cells and astrocytes *(5,13)* but they have been observed in a wide variety of cells types and can be initiated either by mechanical *(5)*, chemical (focal application of agonists, e.g., glutamate, ATP, or ionomycin) *(14–17)* or electrical stimulation *(16,18,19)*. Evidence from a variety of cell types, including airway and lens epithelial cells and astrocytes, indicates that intercellular Ca^{2+} waves require functional gap junction channels but not extracellular Ca^{2+}, are trig-

From: *Methods in Molecular Biology, vol. 154: Connexin Methods and Protocols*
Edited by: R. Bruzzone and C. Giaume © Humana Press Inc., Totowa, NJ

gered by microinjection of inositol 1,4,5-trisphosphate (IP_3) or flash photoly-sis of caged IP_3, and are inhibited by the IP_3 receptor antagonist, heparin, and the phospholipase C inhibitor U-73122 *(5,7,16,17,20–26)*. As a result, inter-cellular Ca^{2+} waves are hypothesized to propagate between cells by the diffu-sion through gap junction channels of the intracellular Ca^{2+} mobilizing messenger IP_3. The molecular mass of IP_3 (437 daltons) is consistent with this hypothesis because it is below the permeation limit of gap junction channels (molecular mass 1000 daltons).

Although Ca^{2+} ions (molecular mass 40 daltons) can also permeate gap junc-tion channels, the contribution of the diffusion of Ca^{2+} ions to intercellular Ca^{2+} signaling appears to be limited because Ca^{2+} waves can be initiated with-out an increase in $[Ca^{2+}]_i$ in the stimulated cell *(12)* and can propagate in a manner that is independent of the magnitude of the $[Ca^{2+}]_i$ increase *(27)*. In addition, oscillations in $[Ca^{2+}]_i$ within individual cells do not generally initiate intercellular Ca^{2+} waves *(28–30)*. Similarly, increases in $[Ca^{2+}]_i$ induced by flash photolysis of caged Ca^{2+}, to concentrations equal or higher than those associated with Ca^{2+} waves, do not initiate intercellular Ca^{2+} changes *(26)*. Furthermore, the binding of Ca^{2+} to immobile buffers greatly reduces the effec-tive diffusion constant for Ca^{2+} in cytoplasm to about 20 times smaller than that for IP_3 *(31)*. As a result, IP_3 is more likely to reach and release Ca^{2+} from nearby IP_3 receptors before Ca^{2+}. However, under other conditions, Ca^{2+} may also act as an intercellular messenger. For example, Yule et al. *(32)* have dem-onstrated, in pancreatic acinar cells, that increasing $[Ca^{2+}]_i$ triggers intercellu-lar Ca^{2+} signaling when the cytosol has been transformed into an "excitable medium" by the priming of the IP_3 receptor to be more sensitive to Ca^{2+} by the binding of IP_3 *(33)*. Thus, in the presence of a uniform distribution of slightly elevated intracellular IP_3 concentration, Ca^{2+} may diffuse through gap junction channels to mediate Ca^{2+} waves. Another potential intercellular messenger is cyclic ADP-ribose (cADPr), an agonist that releases Ca^{2+} from internal ryanodine-sensitive Ca^{2+} stores *(34)*. cADPr has a molecular mass of 541 daltons which would be expected to permeate gap junction channels. When microinjected into lens epithelial cells, cADPr triggered the initiation of inter-cellular Ca^{2+} waves *(7)* and, as will be discussed, we have found that a Ca^{2+} wave can be initiated by the photolytic release of cADPr in glial cells. These results indicate that cADPr also acts as an intercellular messenger.

In addition to intracellular messengers, intercellular Ca^{2+} waves can also be propagated in a paracrine manner by the diffusion of an extracellular messen-ger. The propagation of Ca^{2+} waves in hepatocytes, pancreatic cells, mast cells, and glial cells can, under certain conditions, be biased in the direction of a superfusion of fluid and can cross a cell-free zone. In addition, in some cases, these Ca^{2+} waves can be attenuated by suramin, an ATP-receptor antagonist, or by apyrase, a hydrolytic enzyme for ATP. These results indicate that ATP may serve as an extracellular messenger *(9,19,32,35–39)*. The exact mechanism of cellular ATP release and the contribution to wave propagation of the regenerative release of ATP by the cells are not fully determined.

In summary, intercellular Ca^{2+} waves can propagate from cell to cell either by the diffusion of an intra- or extracellular messenger. However, the exact contribution of each mechanism is dependent both on the cell type and the kind of stimulation used to initiate the waves *(3,12,40)*.

1.1. Experimental Methods for Initiating Ca²⁺ Waves

The investigation of the mechanisms of intercellular Ca^{2+} waves requires a reliable method for their initiation from a single cell. Although both chemical and electrical stimulation trigger Ca^{2+} waves, these methods have the disadvantage that they can simultaneously stimulate adjacent cells. Microperfusion of an agonist is also a difficult technique to stimulate just a single cell because the applied messenger can diffuse to, and act on, neighboring cells. Electrical stimulation with an intra- or extracellular microelectrode has a similar disadvantage because current can flow to neighboring cells via gap junction channels or the extracellular fluid. By contrast, mechanical stimulation with a fine micropipet is an accurate protocol for single-cell stimulation. Obviously, the stimulus should not induce trauma but even gentle mechanical stimulation may induce microleaks in the plasmamembrane, and perhaps release small amounts of ATP. A disadvantage of mechanical stimulation is that is difficult to attain a reproducible stimulus strength.

A common disadvantage of all the stimulation protocols for Ca^{2+} wave initiation is that they act via a signaling cascade and consequently produce, in addition to an increase in IP_3, other second messengers, for example, diacylglycerol, that may influence cellular responses. Microinjection of the required intracellular messenger *(5,7)* is a good alternative to circumvent this nonspecificity, but this approach may be compromised by the need to impale the cell with a micropipet. A more sophisticated approach that avoids cell impalement is the liberation of a cellular messenger by flash photolysis of a "caged" compound.

1.2. Principles of Photolysis and Caged Compounds

Stimulation by flash photolysis involves the photocleavage by ultraviolet (UV) light of an inactive caged compound into an active messenger. The caged compound consists of a cellular messenger conjugated with a photolabile cage moiety, commonly a nitrobenzyl group, that renders the biologically relevant parts of the molecule inactive *(41)*. Because flash photolysis rapidly liberates active messengers, it is ideal for the study of kinetics of processes such as agonist–receptor interaction, excitation–contraction, and stimulus–secretion coupling.

We have used caged Ca^{2+}, IP_3, and cADPr to rapidly and locally initiate Ca^{2+} waves and investigate the mechanisms of intercellular wave propagation *(25,26)*. Caged Ca^{2+} consists of a Ca^{2+} binding molecule with a high affinity for Ca^{2+} that is transformed to a low-affinity state upon flash photolysis. Nitrophenyl-EGTA (NP-EGTA) is currently the most commonly used caged Ca^{2+} molecule because it has a high specificity for Ca^{2+} vs Mg^{2+}, it displays a large decrease (12,500-fold) in its Ca^{2+} affinity following photolysis *(42)*, and it is available as an acetoxymethyl (AM) ester, a form that permeates the cell membrane and allows easy loading of the cell by incubation in a solution of the ester molecule. Caged IP_3 is now available in two forms: the common form is IP_3 esterified with a nitrophenyl-ethyl (NPE) group at position P4 or P5 (mixture of the two isoforms). Another form has recently been synthesized and consists of a caged IP_3 derivative (methoxymethylene IP_3) rendered membrane permeant by substitution with propionyloxymethyl (PM) ester groups *(43)*. Caged cADPr consists of NPE esterified cADPr.

1.3. Photolysis and Caged Compounds to Study Intercellular Ca²⁺ Waves

To simultaneously perform flash photolysis and record changes in $[Ca^{2+}]_i$, an epifluorescence microscope system is required that is equipped with an extra light source to provide the UV photolysis light, and a camera sensitive to low light levels to capture the faint fluorescent Ca^{2+} reporter probe signals. An important principle to emphasize when combining photolysis with fluorescence imaging is that the fluorescent Ca^{2+} probe, in general, must have an excitation wavelength different from the UV excitation wavelength required for flash photolysis. If the excitation wavelengths overlap, photolysis can occur continuously during observation, and photobleaching of the Ca^{2+} probe may occur as a result of the high intensity of the photolysis flash. Ca^{2+} probes compatible with photolysis include fluo-3 (excitation 488 nm), rhod-2 (excitation 550 nm), or calcium green (excitation 488 nm). These are single excitation/emission wavelength probes, but $[Ca^{2+}]_i$ can be measured ratiometrically by using two single wavelength probes with different excitation or emission wavelengths, for example, fluo-3 (excitation 488 nm, emission 535 nm) and fura red (excitation 488 nm, emission 660 nm) *(44)*. Ratiometric Ca^{2+} indicators (e.g., fura-2 or indo-1) are not generally used although Kirby et al. *(45)* have described a method to use indo-1 in combination with UV photolysis.

Using flash photolysis of caged IP_3 and Ca^{2+}, we have demonstrated in astrocyte cultures and astrocyte-endothelial co-cultures that increasing IP_3, but not Ca^{2+}, in a single cell is sufficient to trigger intercellular Ca^{2+} waves (*see* **Fig. 1**).

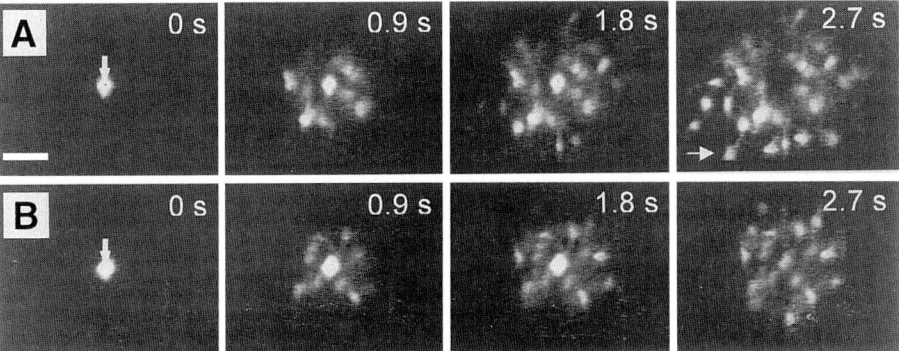

Fig. 1. Intercellular Ca^{2+} signaling in astrocytes induced by photolysis of microinjected caged IP_3. Changes in fluo-3 fluorescence are represented by a gray scale; an increase in gray scale intensity corresponds to an increase in $[Ca^{2+}]_i$. The time of each image, from the onset of the UV flash, is indicated in the *upper right*. (A) After the pressure injection of a single astrocyte *(vertical arrow)* with caged IP_3, the exposure of this cell to a UV flash for 2 s *(center white spot,* first three images) initiated a Ca^{2+} wave that propagated radially to surrounding astrocytes. Flash photolysis of microinjected or ester loaded caged Ca^{2+} (NP–EGTA) produced $[Ca^{2+}]_i$ changes only in the flashed cell (not shown). (B) After the Ca^{2+} wave shown in A had subsided, the culture was moved such that the astrocyte marked with the horizontal arrow *(right,* last image of A) was located in the center of the field of sequence B. Applying a UV flash to this noninjected astrocyte (located 100 μm away from the injected astrocyte) also initiated an extensive intercellular Ca^{2+} wave. Calibration bar = 40 μm. (Reprinted with permission from Leybaert et al., *Glia,* copyright ©1998 John Wiley & Sons) *(26)*.

These Ca^{2+} waves propagate to similar or dissimilar adjacent cells by the diffusion of IP_3 through gap junction channels *(see* **Figs. 2** and **3**). Similar results were obtained with caged cADPr in astrocytes (**Fig. 4**). Interestingly, Ca^{2+} waves propagated between cells that showed no evidence for dye coupling with Lucifer Yellow, 6-carboxyfluorescein, or biocytin *(46)*, but that displayed a certain degree of electrical cell coupling *(47)*. The importance of this result is that, although weakly coupled cells are difficult to detect by dye coupling techniques, these cells can communicate by the propagation of IP_3-mediated Ca^{2+} signals. These findings emphasize that flash photolysis of caged IP_3 and the observation of the resulting propagating Ca^{2+} changes is a very sensitive technique to demonstrate cell-to-cell coupling. The high sensitivity of the technique relies on the fact that small quantities of IP_3 diffusing between cells are greatly amplified by the cell into easily observable Ca^{2+} signals through Ca^{2+}-induced Ca^{2+} release at the IP_3 receptor *(48)*.

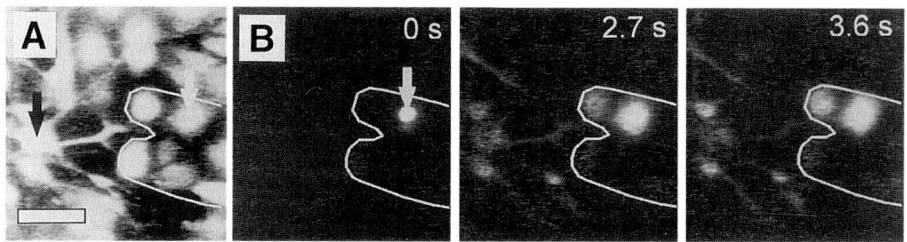

Fig. 2. Intercellular Ca^{2+} signals initiated by flash photolysis of caged IP_3 passing from endothelial cells to astrocytes. **(A)** Raw fluo-3 fluorescence image at the interface *(white line)* between endothelial cells and astrocytes in co-culture. An astrocyte *(black arrow)* located approx 45 µm left of the astrocyte–endothelial interface was injected with caged IP_3 and an endothelial cell *(white arrow)* was subsequently exposed to the UV flash. **(B)** Flashing the endothelial cell *(white arrow)* induced a Ca^{2+} increase that propagated to an adjacent endothelial cell and crossed the interface to continue to propagate through several astrocytes. Calibration bar = 40 µm. (Reprinted with permission from Leybaert et al., *Glia,* copyright © 1998 John Wiley & Sons) *(26).*

2. Materials
2.1. Cell Cultures

The experiments described in this chapter were performed on astrocyte cultures and on co-cultures of astrocytes and endothelial cells; all experiments were done on cultures grown to confluency. Astrocyte cultures were prepared from rat brain cortex based on the method described by McCarthy and de Vellis *(49)*. This method involves cell dissociation by trypsin incubation and mechanical trituration, filtration of the resulting cell suspension, growing the cells to confluency in culture flasks, and removal of overlaying nonastrocytic cells by a shaking procedure. For endothelial cells we used a spontaneously transformed endothelial cell line derived from human umbilical vein (ECV304; *see* **ref. 26**; obtained from European Collection of Animal Cell Cultures [ECACC], Salisbury, UK, http://fuseii.star.co.uk/camr/). Astrocyte–endothelial co-cultures were prepared by co-seeding both cells types onto glass-bottom microwell dishes (type P35G-0-14-C, MatTek, Ashwood, MA). Astrocyte identity was determined by immunostaining with Cy-3-conjugated antibodies to glial fibrillary acidic protein (GFAP) (Sigma, St. Louis, MO, http://www.sigma-aldrich.com); endothelial cell identification was based on their typical phase-contrast appearance and absence of GFAP immunostaining (*see* **Fig. 5**).

2.2. Solutions

1. General remarks. All experiments are done with the cells bathed in Hanks' balanced salt solution supplemented with 25 m*M N*-(2-hydroxyethyl) piperazine-*N'*-

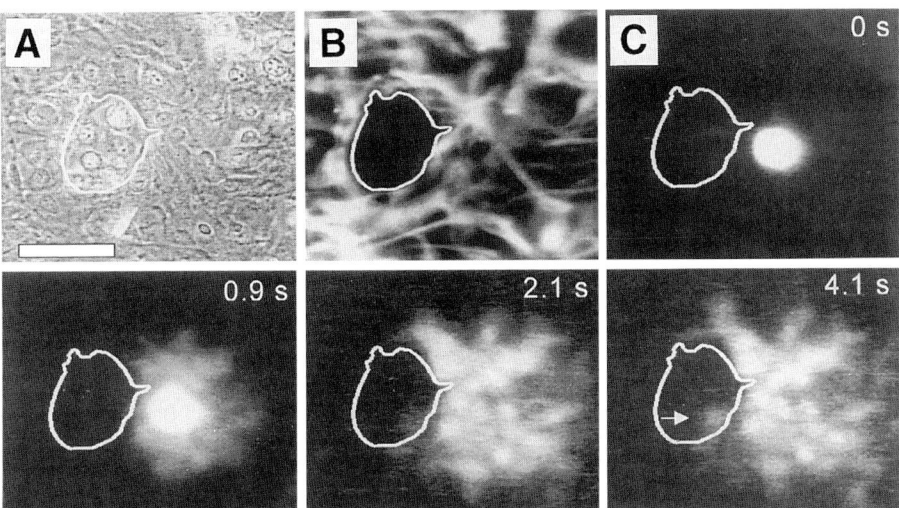

Fig. 3. Intercellular Ca^{2+} signals between astrocytes and endothelial cells initiated by flash photolysis of caged IP_3 loaded by electroporation. (**A**) Phase-contrast image of an astrocyte–endothelial co-culture with the endothelial cell region surrounded by the *white line*. (**B**) GFAP immunostaining showing GFAP-positive astrocytes surrounding GFAP-negative endothelial cells. (**C**) Flash photolysis of astrocytes (first two images) initiated an intercellular Ca^{2+} wave that propagated to a single endothelial cell *(white arrow)*. Calibration bar = 50 µm.

2-ethanesulfonic acid (HEPES) (HBSS–HEPES). It is important that solutions containing caged compounds are kept on ice during manipulation and that they are used in a darkened room with only low-power indirect illumination (no UV light, no day light).

2. Scrape loading buffer: 137 mM NaCl, 5.36 mM KCl, 0.81 mM $MgCl_2$, 5.55 mM D-glucose, and 25 mM HEPES, pH 7.4; store at 4°C.

3. Electroporation buffer: 300 mM sorbitol, 4.02 mM KH_2PO_4, 10.8 mM K_2HPO_4, 1 mM $MgCl_2$, and 2 mM HEPES, pH 7.20 (*20*); store at 4°C. For visualization of the electroporated zone, supplement with 2 µM Texas red dextran (molecular mass 3000 daltons, Molecular Probes Eugene, OR, http://www.probes.com) from stock of 75 µM Texas red dextran in electroporation buffer.

4. Caged Ca^{2+} solutions. Ester loading: stock solution of 1 mM o-nitrophenyl EGTA-AM (NP-EGTA-AM, Molecular Probes) in dimethyl sulfoxide (DMSO); store at –20°C. Microinjection: 50 mM K-HEPES, pH 7.2, with 10 mM o-nitrophenyl EGTA tetrapotassium salt (NP-EGTA, Molecular Probes) and 5 mM $CaCl_2$.

5. Caged IP_3 solutions. Microinjection: 50 mM K-HEPES, pH 7.2, with 1-2.5 mM D-*myo*-inositol 1,4,5-trisphosphate, P4(5)-(1-(2-nitrophenyl)ethyl)ester trisodium salt (NPE-caged IP_3, Molecular Probes or Calbiochem, La Jolla, CA, http://www.calbiochem.com), store at –20°C. Scrape loading: scrape loading buffer

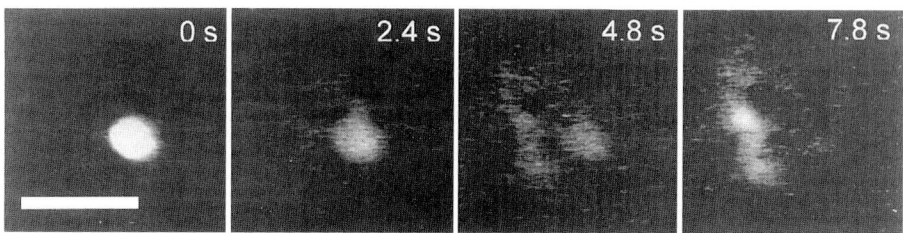

Fig. 4. Intercellular Ca^{2+} signals induced by flash photolysis in astrocytes loaded with caged cADPr by electroporation. Flash photolysis (*first image*) produced a Ca^{2+} increase in the flashed cell that slowly propagated to several neighboring cells. Calibration bar = 50 μm.

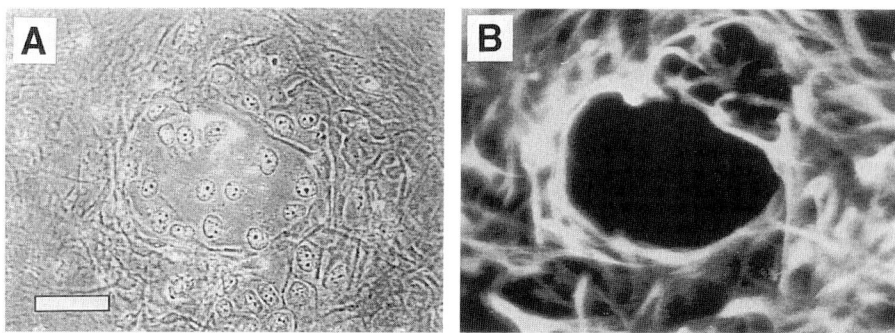

Fig. 5. Typical appearance of an astrocyte–endothelial co-culture. (**A**) Phase-contrast image illustrating a central region of endothelial cells surrounded by astrocytes. (**B**) GFAP immunostaining of the same culture region as shown in **A**, demonstrating GFAP-positive astrocytes surrounding GFAP-negative endothelial cells. Calibration bar = 50 μm.

with 200–400 μ*M* caged IP_3; store at –20°C. Electroporation loading: electroporation buffer with 200 μ*M* caged IP_3; store at –20°C.

6. Caged cADPr: Electroporation buffer with 200–400 μ*M* cyclic adenosine 5'-diphosphate ribose, 1-(1-(2-nitrophenyl)ethyl) ester (NPE-caged cADPr, Molecular Probes); store at –20°C.

7. Caged fluorescein-dextran: 1 m*M* 4,5-dimethoxy-2-nitrobenzyl (DMNB)-caged fluorescein, dextran, 3000 mol wt, anionic (Molecular Probes) in distilled water.

3. Methods

3.1. Ca^{2+} Imaging

Intercellular Ca^{2+} waves are monitored with the Ca^{2+}-sensitive fluorescent dye fluo-3 and the video imaging system illustrated in **Fig. 6**.

1. Cell cultures are loaded with fluo-3 by incubation for 1 h at 37°C in HBSS–HEPES containing 10 μM fluo-3-AM and 0.05% pluronic. One hour of de-esterification in HBSS–HEPES at room temperature is allowed before cells are used.

2. Cells are viewed with an inverted Nikon Diaphot 300 epifluorescence microscope using a ×40 phase-contrast oil immersion lens. Fluo-3 fluorescence images are obtained by excitation at 485 nm (e.g., band-pass filter type 485DF22 from Omega Optical, Brattleboro, VT, http://www.omegafilters.com) and observation at 535 nm (535DF35, Omega Optical). The excitation and emission wavelengths are separated by a dichroic mirror (DM-1) with a transmission cutoff at 505 nm (505DRLP02, Omega Optical).

3. Images are captured with a silicon intensified target (SIT) camera (Cohu, San Diego, CA, http://www.cohu.com) or an intensified CCD (ICCD) camera (Extended ISIS camera, Photonic Science, East Sussex, UK, http://www.photonic-science.ltd.uk) (the pros and contras of different camera types are discussed under **Note 1**) and are recorded on an optical memory disk recorder (OMDR; Panasonic TQ-2026F).

4. Recorded images are digitized with an image processor board (Data Translation, DT2861 or DT3155, Marlboro, MA; http://www.datx.com) within a PC, and are processed to display the change in fluorescence ($F_t – F_0$) by subtracting the image before UV exposure (F_0) from all subsequent images (F_t). Alternatively, images are processed to display the relative fluorescence change ($[F_t – F_0]/F_0$) using a pixel-by-pixel division performed offline (custom developed analysis routines written in C).

5. Intercellular Ca^{2+} waves are quantified by measuring the above threshold area of the fluorescence signal of the Ca^{2+} wave at its maximal expansion. Image processing packages such as NIH-image or Image Tool software are ideal for this purpose (NIH-image can be downloaded from: http://rsb.info.nih.gov/nih-image/ or http://www.scioncorp.com/frames/fr_scion_products.htm; a multiplatform Java version is available at http://rsb.info.nih.gov/ij/; Image Tool is available at http://ddsdx.uthscsa.edu/dig/itdesc.html). To reproducibly apply the same threshold level in all experiments, the conditions of dye loading and image recording must be kept as constant as possible.

6. The contribution of extracellular messengers to Ca^{2+} wave propagation can be explored by testing the effect of applying a fast superfusate flow over the cells on size and morphology of the Ca^{2+} wave. Superfusion of the cells is done by placing a source pipet and a suction pipet close to the borders of the imaged zone. Flow velocity can be determined by tracing the movement of suspended fluorescent beads (1 μm diameter Fluospheres, Molecular Probes; 1 μL of the 2% stock solution per 100 mL of HBSS-HEPES).

3.2. Flash Photolysis Setup

In our system, we use a continuously operating Hg arc lamp as a UV source rather than a flash lamp or UV laser (*see* **Note 2** for alternative UV light sources). Hg arc lamps have intense emission peaks in the UV region (at 334 and 365 nm) that can be selected for with optical bandpass filters.

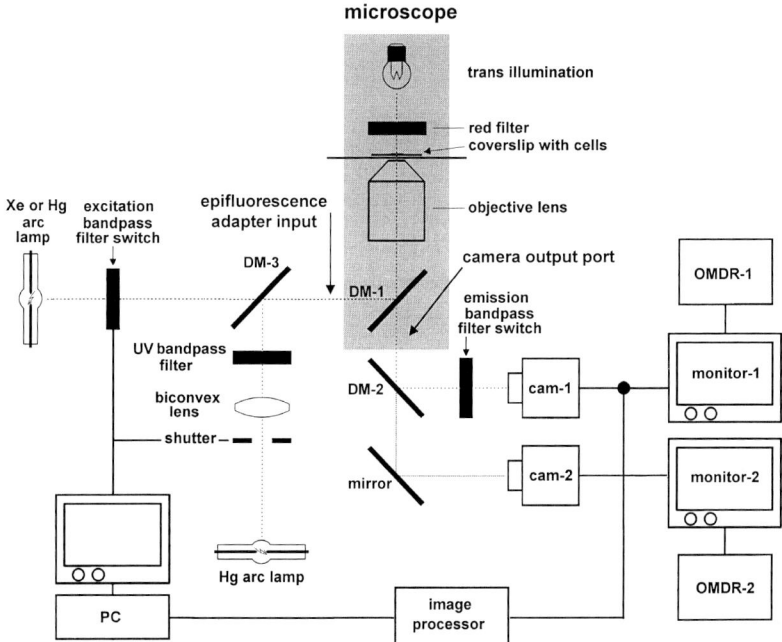

Fig. 6. The microscope system used for Ca²⁺ imaging and flash photolysis. The excitation light for imaging $[Ca^{2+}]_i$ is provided by either a Hg or Xe arc lamp (vertical bulb, *left*). The excitation light is band-pass filtered (485 nm for fluo-3), directed through a dichroic mirror (DM-3; transmission cutoff at 400 nm for fluo-3), and enters the microscope epifluorescence adapter to produce a uniform field of illumination. The UV excitation light for flash photolysis is provided by an Hg arc lamp (horizontal bulb) and is passed through a shutter, bandpass filtered (330 nm) and focused to a small spot at the epifluorescence field diaphragm with a biconvex lens. The UV light enters the epifluorescence adapter by reflection off DM-3. The shutter exposure time is controlled by computer software. The epifluorescence dichroic mirror, DM-1 (transmission cutoff at 505 nm for fluo-3) reflects the short-wavelength excitation light to the specimen. The long-wavelength fluorescence emitted by fluo-3 in the cells (wavelength ~535 nm) passes through DM-1, is reflected by DM-2 (transmission cutoff at 600 nm) to the camera port, and is imaged by a low-light sensitive camera (cam-1; ICCD or SIT camera). The video output of cam-1 is monitored, recorded on an OMDR, and connected to an image processor. A simultaneous visualization of the specimen is achieved by transillumination with red light (wavelength >600 nm) that passes through both DM-1 and -2 and is reflected by a mirror to cam-2 (CCD camera).

1. A small UV spot for flash photolysis is produced by imaging the arc of an Hg lamp (Osram type HBO 103W/2) with a biconvex fused silica lens (focal distance 250 mm, e.g., lens type F8068, Edmund Scientific, Barrington, NJ, http://

www.edmundscientific.com, or lens type SBX037, Newport Corporation, Irvine, CA, http://www.newport.com) onto the plane of the field diaphragm of the epifluorescence adapter of the microscope (**Fig. 6**).

2. The exposure time, ranging from 0.1 to 2 s, is controlled by a mechanical shutter (Uniblitz, Vincent Associates, Rochester, NY, http://www.uniblitz.com) that is activated by pulses coming from a PC. The UV spectrum of the Hg arc is selected with a bandpass filter (330 nm, e.g., 330WB80 from Omega Optical).

3. The UV beam is guided into the microscope optics by reflection off an external dichroic mirror (DM-3) with a transmission cutoff of 400 nm (400DCLP02, Omega Optical). The UV spot is focused and centered by adjusting the positions of the biconvex lens and Hg arc bulb.

4. The effective photolysis size of the UV spot is determined by quantifying the half-maximal energy diameter of the spot. This is achieved by exposing a thin layer of a fluorescent dye (e.g., fura-2 at 1 μM) on a microscope cover glass to the UV light and measuring the intensity profile from a recorded image of the fluorescence spot (using the image processing packages NIH-image or Image Tool referred to before). A thin fluorescent layer is made by pipetting 2.5–5 μL of fluorescent probe onto a coverslip and covering the droplet with a round 12-mm diameter coverslip. The intensity of the UV beam is attenuated with neutral density filters so that the brightest fluorescence at the center of the spot does not saturate the imaging system. The half-maximal energy diameter in our system is in the order of 7–10 μm.

5. The photolytic efficiency of uncaging, that is, the fraction of the caged probe released upon UV exposure, can be calculated by multiplying the measured photolytic efficiency of caged fluorescein–dextran (molecular mass 3000 daltons, Molecular Probes) by the ratio of the quantum efficiencies of the caged messenger (0.65 for caged IP_3) and caged fluorescein–dextran (~0.25). The photolytic efficiency of caged fluorescein-dextran can be determined by exposing a dried layer of 1 mM of the compound to UV flashes of increasing duration and measuring the intensity of the induced fluorescence. A plot of the fluorescence intensity vs UV exposure time is a saturating function from which the photolytic efficiency at various UV exposure times can be derived. The efficiency of a 2-s UV exposure is about 30% for caged fluorescein–dextran and about 80% for caged IP_3.

3.3. Loading Cells with Caged Messenger

Loading cells with caged messengers is performed after loading with the fluorescent Ca^{2+} indicator. The exact protocol depends on the caged messenger being used.

3.3.1. Loading Caged Ca²⁺

NP-EGTA is available as a membrane-permeant AM form and loading simply involves incubation of the cells in an NP–EGTA–AM-containing solution. It is, however, important that the intracellular concentration of caged Ca^{2+} does not rise too high as otherwise the buffering capacity of the cytosol will be greatly increased (*see* **Note 3**). While photolysis may produce a large increase in $[Ca^{2+}]_i$ in the flashed cell under these conditions, the caged Ca^{2+} will attenuate or even abolish Ca^{2+} responses in cells not exposed to UV light. We use very short incubation times to prevent overloading:

1. Dilute 10 µL of 1 mM NP-EGTA-AM to 5 µM just prior to loading with HBSS-HEPES.
2. Remove the HBSS–HEPES from the culture and add 2 mL of 5 µM NP–EGTA–AM per 35-mm culture dish.
3. Incubate for 10–15 min at 37°C.
4. Remove the loading solution and wash (*3*) with HBSS–HEPES.
5. Leave the cells for 30 min at room temperature to allow for deesterification before use.

By applying this protocol to confluent astrocyte cultures, we have demonstrated that flash photolysis of caged Ca^{2+} only produces an increase in $[Ca^{2+}]_i$ in the flashed cell without triggering any $[Ca^{2+}]_i$ changes in neighboring cells. The contribution of the caged Ca^{2+} to the cytosolic Ca^{2+} buffering capacity was negligible because mechanical stimulation of a single astrocyte still triggered propagating intercellular Ca^{2+} waves. Although fluo-3 fluorescence does not allow the quantification of $[Ca^{2+}]_i$ changes induced by flash photolysis, a comparison of the Ca^{2+} responses to a range of concentrations of the Ca^{2+} agonist ATP, measured with fluo-3 and the ratiometric dye fura-2, indicate that the peak values attained by photolysis are in the 500–750 nM range.

3.3.2. Loading Caged IP₃ or Caged cADPr

Caged IP₃ and cADPr are charged molecules that cannot freely diffuse through the plasma membrane. Li et al. (*43*) have described a membrane permeant ester form of caged IP₃ (caged methoxymethylene–IP₃–PM) but this molecule is not currently commercially available. Consequently, to load charged compounds into the cells, we use microinjection, scrape loading, or electroporation. Of these techniques, electroporation is particularly useful because it is relatively easy and has the advantage that it loads large fields of cells simultaneously.

3.3.2.1. MICROINJECTION

The technique consists of impaling a single cell with a glass micropipet and the subsequent pressure injection of a small amount of solution containing the caged messenger (*see* also Chapter 12).

1. Glass micropipets are pulled from filamented borosilicate capillaries (inner diameter 0.86 mm, outer diameter 1.5 mm; e.g., type 6030 from Clark Electro-

medical, Reading, UK) and have a resistance of 20–40 MΩ when filled with 1 M KCl. The tip size is < 1 μm and is hard to resolve with a 16× objective.

2. Pipets are back-filled with a small amount of caged IP$_3$ solution by placing their wide end in the filling solution and allowing capillary action to form a 2-mm column of solution at the pipet tip.

3. The pipets are mounted in a hydraulic micromanipulator and connected with polyethylene tubing to a Narishige pressure injector (IM-200).

4. The filled pipet is positioned above the cell to be injected and lowered to touch the plasma membrane and impale the cell. Several 100 ms pressure pulses of 50–100 kPa are then applied to inject a certain amount of pipet solution.

5. Injection is successful if the injected cell becomes obviously inflated. The injected volume can be estimated from the increase in cell shape. For astrocytes, this is normally in the order of a 25–50% increase in cell volume ($\Delta V/V = 2 \cdot \Delta r/r$ for a cylindrical cell with a volume V and radius r, or $= 3 \cdot \Delta r/r$ for a spherical cell with radius r).

6. Cell impalement and pressure injection triggers an intercellular Ca²⁺ wave. As a result, flash photolysis should not be initiated before the recovery of the Ca²⁺ signal, that is, approx 10 min following injection.

An advantage of microinjection of caged IP$_3$ and cADPr into gap junction-coupled cells is that large numbers of cells can be loaded by injection of a single cell, because these compounds have small molecular masses (635 and 690 daltons, respectively) enabling them to diffuse through gap junction channels to surrounding cells. However, a disadvantage of cell-to-cell diffusion is that it difficult to estimate the concentration of the compound in the injected cell. In astrocytes, which are extensively coupled by gap junctions, flash photolysis of noninjected cells, located up to 100 μm away from the cell injected with caged IP$_3$, still initiated large intercellular Ca²⁺ waves (**Fig. 1**). Flash photolysis applied 2 h after microinjection still triggered intercellular Ca²⁺ waves, indicating that the injected caged IP$_3$ does not leak from the cells and remains stable for a long period.

The injection of astrocyte–endothelial co-cultures with the gap junction permeable fluorescent dyes, Lucifer Yellow, 6-carboxyfluorescein, or biocytin did not result in any detectable dye spread between astrocytes and endothelial cells, suggesting that these cell types are weakly coupled (*46*). However, microinjection of an astrocyte with caged IP$_3$ and UV flash exposure of an endothelial cell did produce an endothelial Ca²⁺ increase that propagated back to the microinjected astrocyte (**Fig. 2**). This result demonstrates that caged IP$_3$ diffuses from astrocytes to endothelial cells, and that the IP$_3$ that is photolytically liberated in the endothelial diffuses back to the astrocytes.

3.3.2.2. SCRAPE LOADING

Scrape loading does not require sophisticated micromanipulation or instrumentation (*see* Chapter 12). This technique (adapted from **ref. 50**) loads cells with an exogenous compound by scratching a monolayer culture with a small needle *(51)* and has been extensively used to evaluate gap junction coupling of cells by measuring the intercellular spread of fluorescent dyes *(52–55)*.

1. For scrape loading, cells are placed in a nominally Ca^{2+}-free solution (scrape loading solution) containing the substance to be loaded. A nominally Ca^{2+}-free solution is used to avoid Ca^{2+}-dependent closure of gap junction channels in the injured cells. An EGTA-buffered Ca^{2+}-free solution is not used because this tends to detach cells after scratching. Remove the HBSS-HEPES from the cultures, wash (3×) with scrape loading solution, and add scrape loading solution containing 200–400 μ*M* caged IP_3. Leave for 10 min.
2. Apply a linear scratch across the cell culture with a thin syringe needle (27-gauge $^1/_2$-in) and leave for 10 min. After scratching, injured cells take up the compound in the loading solution, which diffuses via gap junction channels to neighboring cells to form a decreasing exponential concentration gradient of compound (*see* **Fig. 7**).
3. Remove the scrape loading solution and wash (3×) with HBSS-HEPES. Leave for 30 min before use to allow for recovery from trauma.

It is recommended to apply the photolysis light not too far from the scratch edge because the concentration of the scratch-loaded caged compound decreases exponentially away from the scratch. When astrocytes were scrape loaded with caged IP_3, flash photolysis successfully initiated Ca^{2+} waves when the UV light was directed up to 100 μm away from the scratch edge. Intercellular Ca^{2+} waves produced by photolysis and scrape loading were always smaller than Ca^{2+} waves generated with electroporation loading, indicating that the efficiency of scrape loading is less than that of electroporation loading.

3.3.2.3. ELECTROPORATION LOADING

Electroporation has been introduced to load cells with large molecular weight compounds such as DNA and antibodies *(56,57)* and consists of briefly exposing cells to a high strength electric field (~1000 V/cm) by applying a voltage pulse between two inert electrodes *(58)*. Electroporation transiently induces micropores in the plasma membrane and these allow substances, contained in the loading solution, to move into the cell *(59)*. To improve loading efficiency, the technique has been modified by applying an electric field that oscillates at a high frequency, rather than a single-voltage pulse, to generate numerous cycles of micropore formation *(60)*. Miniaturization of the electrodes has made it possible to electroporate a cell monolayer by positioning the electrodes close to the cell surface *(20)*.

1. Electroporation requires a specialized electrode composed of two parallel platinum/iridium wires (ϕ 25 μm, Johnson Matthey Catalog, Ward Hill, MA, type no. 10292), 1 mm or less apart, that are stretched and glued with cyanoacrylate across

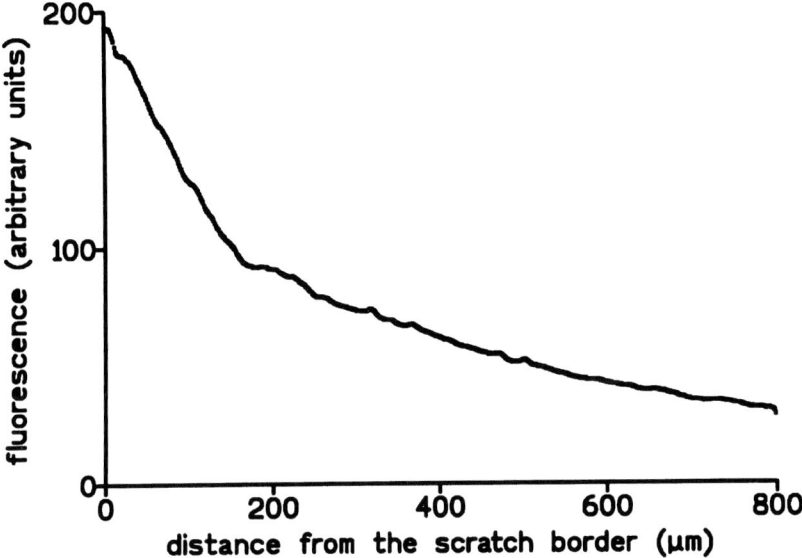

Fig. 7. Fluorescence diffusion profile following scrape loading of cells. Confluent astrocyte cultures were scrape loaded with 6-carboxyfluorescein by applying a linear scratch with a needle. An image was acquired 20 min after scraping, with the right border of the scratch positioned at the left border of the camera image. The graph shows the average line intensity profile that was calculated from the video image.

a hollow plastic cylinder (10 mm diameter, 10 mm high) (*see* **Fig. 8**). The cylinder is connected to a rod for attachment to a micromanipulator.

2. The train of electrical pulses applied to the electrode consists of four repeats of ten 1-ms bursts of positive going 50 kHz pulses of an amplitude of 30–50 V (modified from Chang [60] and Boitano et al. [20]). At an electrode wire separation of 1 mm, this corresponds to an electric field strength of about 300–500 V/cm.

3. Remove the HBSS–HEPES from the cultures and wash 3× with the electroporation buffer.

4. Remove the electroporation buffer as completely as possible (e.g., using a paper tissue) and add 25 μL of electroporation buffer containing 200 μM caged IP₃ (or caged cADPr), a quantity sufficient to cover the 1.4 cm² of cells on the glass bottom of the microwell dishes with a thin layer of buffer. If the electroporated cells need to be visualized, use the electroporation buffer supplemented with Texas red dextran.

5. Place the culture on the microscope and bring into focus with a ×10 lens. Move the lens 200 μm higher using the micrometer calibration scale on the microscope focusing dial.

6. Position the electroporation electrode with a micromanipulator so that the two parallel wires are parallel to the cell surface and just in focus (200 μm above the focal plane of the cells).

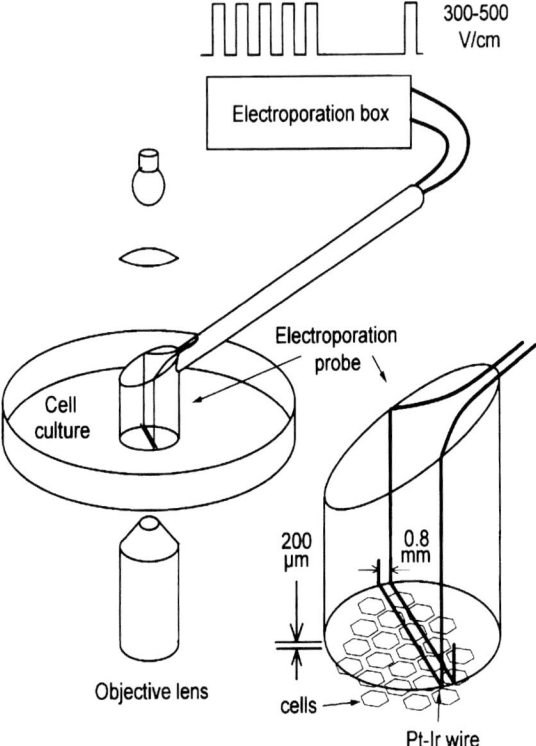

Fig. 8. A schematic of the electroporation setup. The electroporation electrode consists of two thin parallel Pt/Ir wires glued on a circular plastic probe that is positioned above the cells using a micromanipulator and a microscope. The electrode is connected to a high-frequency power supply that delivers a series of high-frequency oscillating voltage pulses.

7. Pipet another 25 μL of caged IP$_3$/electroporation buffer onto the electrode wires to fully immerse the electrodes in solution.
8. Apply the electroporation pulse train described under **step 2**.
9. Remove the electroporation solution and wash 3× with HBSS–HEPES.

This protocol uses a minimal quantity (50 μL) of caged IP$_3$-containing solution. Because the cell zone under the electrode often has an increased [Ca^{2+}]$_i$, after electroporation, a short recovery period is allowed. By using a low electric field strength (i.e., 300–500 V/cm), the electroporation-induced [Ca^{2+}]$_i$ elevation is minimized without affecting the loading efficiency (*see* **Note 4**) and cells can be used immediately after electroporation. In astrocyte cultures, the concentration of the loaded substance decreases exponentially away from the edge of the electroporation zone (the electroporation zone is the zone just

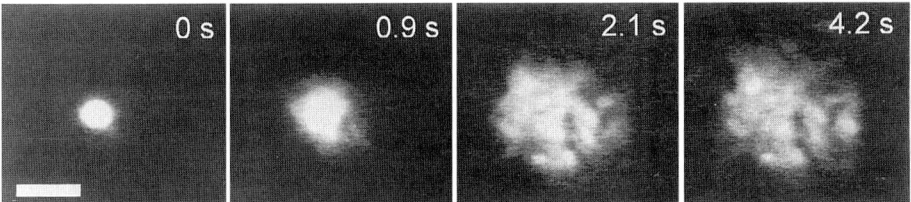

Fig. 9. An example of an intercellular Ca²⁺ wave initiated by flash photolysis in astrocytes loaded with caged IP₃ by electroporation. The UV flash is visible in the two first images. Calibration bar = 50 μm.

beneath the electrode wires–visualization of this zone is discussed in **Note 5**), just as observed with scrape loading, and the space constant of diffusional spread into the surrounding cells is not affected by electroporation. This indicates that gap junctional coupling in the zone just adjacent to the electroporation zone is preserved and not affected by electroporation. Flash photolysis experiments are best not done in the electroporation zone but are preferably performed in a region closeby (within the first 100 μm works well in astrocyte cultures).

An intercellular Ca²⁺ wave triggered in astrocytes by flash photolysis of caged IP₃ loaded into the cells by electroporation is shown in **Fig. 9**. A UV exposure of about 1 s was needed to produce intercellular Ca²⁺ waves of a substantial size and this exposure time was generally used in all our experiments. However, the area over which photolysis-induced Ca²⁺ waves spread is dependent on the duration of the UV exposure (*see* **Fig. 10**): this strongly indicates that the diffusion of IP₃ from a single cell can account for the generation of a Ca²⁺ wave and suggests that regeneration of IP₃ is not involved. Application of a fast superfusion of fluid (velocity about 400 μm/s) did not influence the shape of photolysis-induced intercellular Ca²⁺ waves, indicating that extracellular diffusing Ca²⁺ messengers are not contributing to the cell-to-cell propagation of a Ca²⁺ wave.

Electroporation was also used to load caged IP₃ into astrocyte–endothelial co-cultures. Flash photolysis of caged IP₃ in a single astrocyte triggered an intercellular Ca²⁺ wave that propagated through the astrocytes, crossed the astrocyte–endothelial interface and propagated into a single endothelial cell (**Fig. 3**). Wave propagation was also observed in the opposite direction, that is, from endothelial cells to astrocytes (not shown). The astrocytes did not appear to be dye coupled to endothelial cells, but did however show a low level of electrical coupling to endothelial cells, as previously mentioned in **Subheading 1.3**. We have also used electroporation to load astrocyte cultures with caged cADPr. Flash photolysis of caged cADPr in a single astrocyte produced Ca²⁺

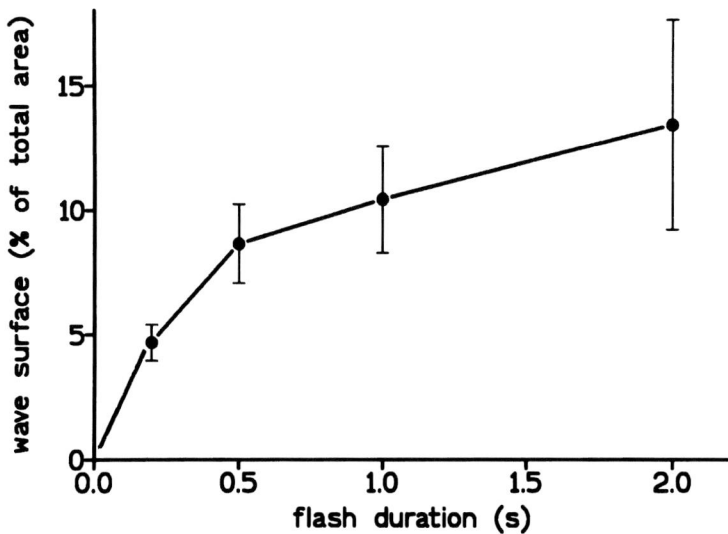

Fig. 10. Dose-response curve for the duration of the UV flash and the area of the induced Ca^{2+} wave. Astrocyte cultures were loaded with caged IP_3 by electroporation and UV flashes of different durations were applied at different locations within the culture. The area of the intercellular Ca^{2+} wave was calculated as the percentage of the total image area. The dose–response relation follows a saturating function.

changes in immediately adjacent cells (**Fig. 4**), indicating that cADPr is, as predicted by its molecular mass (541 daltons), able to diffuse through gap junction channels and trigger Ca^{2+} changes in adjacent cells. Thus, cADPr can, like IP_3, act as an intercellular Ca^{2+} messenger.

In conclusion, the use of flash photolysis to induce intercellular Ca^{2+} signals forms an alternative and highly sensitive method to demonstrate gap junctional coupling; the sensitivity of the method mainly derives from the amplification of the evoked Ca^{2+} responses by the process of Ca^{2+}-induced Ca^{2+} release via IP_3 or ryanodine receptors.

4. Notes

1. SIT cameras are intensified vacuum-tube cameras that are characterized by poor linearity, image nonuniformities, and a high time lag. Therefore, these cameras are currently not generally recommended for Ca^{2+} imaging. However, if appropriate corrections are made to eliminate nonlinearities and nonuniformities, these cameras are still useful to resolve the relatively slowly (10–20 µm/s) propagating intercellular Ca^{2+} waves. SIT cameras produce less noisy images as compared to noncooled ICCDs. ICCD cameras have superior linearity, image quality, response time and sensitivity, but the drawback of the high sensitivity is increased noise. ICCD cameras are indicated when speed is the ultimate goal and full video-rate

imaging speed is required, for example, as necessary to investigate the intracellular dynamics of intercellular Ca^{2+} wave propagation. Noise can be eliminated by averaging the recorded image sequence offline using image processing boards, but this obviously limits the rise time and time resolution. Alternatively, noise can be reduced at a more fundamental level by using a cooled ICCD, that is, ICCDs equipped with a liquid nitrogen or thermoelectrically cooled CCD chip. There are many good quality ICCD cameras on the market now; affordable noncooled ICCD cameras are available from Photon Technology International (PTI, Monmouth Junction, NJ, http://www.pti-nj.com). If time resolution is not the primary goal, slow-scan cooled CCD cameras (e.g., Photometrics, Tucson, AZ, http://www.photomet.com) offer another alternative to SIT or ICCD cameras. These cameras typically have thermoelectrically cooled CCD chips that result in reduced thermal noise levels. They are equiped with high-quality CCD chips that have a large fraction of the chip surface effectively exposed to light and a large pixel "depth" allowing the accumulation of many photons in the pixel "wells" on the chip before they are read out. These properties produce a camera with high dynamic range (12 or more bits pixel intensity digitization is possible, as compared to fewer than 8 bits per pixel for an ICCD) and allowing on-chip integration of the faint Ca^{2+} probe fluorescence signal. Owing to the integration, photon collection can occur over an extended period (e.g., 1 s) and this effectively omits the need for an image intensifier stage, which is the weakest part of an ICCD in terms of noise, linearity, and lifetime. Slow-scan cooled CCD cameras deliver superior image linearity and dynamic range but have the drawback of slow readout, that is, one or two images per second. Faster readout rates can often be obtained but at the expense of spatial resolution. If the fast intracellular Ca^{2+} dynamics of intercellular Ca^{2+} wave propagation are not the primary interest, slow-scan cooled CCD cameras deliver the best image quality; they are, however, more expensive than ICCDs.

2. In our setup, we use a Hg arc lamp in combination with a mechanical shutter to produce a small spot "flash" of UV light (*see* **Subheading 3.2.**). Owing to the limited energy, the exposure time with this kind of setup is quite long and in the range of 0.5–2 s. To make optimal use of the high-speed photolytical reactions and to fully resolve the dynamics of intercellular Ca^{2+} wave propagation, it can be advantageous to use UV sources of higher energy, that is, Xe flash lamps or pulsed lasers, so that the exposure time can be reduced to the millisecond or microsecond domain respectively. Xe flash lamps are specially designed Xe arc lamps (different from the classical Xe arc lamps used for epifluorecence excitation) that produce a short-lived ($\leq$1 ms) high-energy ($>$200 mJ between 300 and 400 nm) burst upon discharge of a capacitor over the bulb electrodes (*see* ref. *61*; flash lamp systems are available from Hi-Tech Scientific, Salisbury, UK, http://www.hi-techsci.co.uk). Pulsed lasers form another, more expensive, alternative as high-energy UV source. Compact, frequency tripled Nd:YAG lasers are now available that produce pulsed 355 nm light at 8 mJ per 5-ns pulse at a repetition frequency of up to 20 Hz (*see* ref. *62*; lasers available from, e.g., Continuum, Santa Clara, CA, http://www.continuumlasers.com).

3. Cells can be easily loaded with caged Ca^{2+} by using the membrane-permeable ester form (NP–EGTA–AM) but cells can be equally easily overloaded with the probe. Overloading the cells with NP–EGTA will buffer and attenuate $[Ca^{2+}]_i$ changes in all the cells not exposed to UV photolysis light as those cells contain the caged and high Ca^{2+} affinity form of the probe. It is imperative to perform control experiments to check the degree of cytoplasmic Ca^{2+} buffering introduced on NP–EGTA–AM loading, by carefully monitoring the $[Ca^{2+}]_i$ responses to known stimulation protocols or agonists.

4. The field strength of the electrical pulses used for electroporation loading that are given in the literature are often at the high end (order of 1000 V/cm). Our experience is that lower field strengths can be used with no obvious loss in loading efficiency and with significantly less cell damage (at least with the electrode configuration as described in **Subheading 3.3.2.3.**). With a field strength of 1000 V/cm, the cells right beneath the electroporation electrode often appear morphologically abnormal and often display poor or no $[Ca^{2+}]_i$ responses to known Ca^{2+} agonists or to flash photolysis of electroporation loaded caged IP_3. Field strengths of less than half this value result in comparable loading efficiencies, better morphological appearance, and preservation of agonist-induced $[Ca^{2+}]_i$ responses of the cells beneath the electroporation electrode.

5. The electroporation zone can be visualized by incorporating a high molecular weight (non-gap-junction-permeable) fluorescent dye in the loading solution with an excitation/emission wavelength that does not interfere with the fluo-3 Ca^{2+} measurements (e.g., Texas red dextran, excitation 596 nm, emission 620 nm). With the parallel wire electrodes described under **Subheading 3.3.2.3.**, the electroporation zone appears as a thin strip, 400–500 μm wide, across the monolayer. An additional advantage of introducing a fluorescent dye into the loading solution is that it provides some indication of the loading efficiency. By dividing the fluorescence intensity in the electroporated cells by the fluorescence intensity of a thin layer (ideally equal to the cell thickness) of the same probe, the loading efficiency was estimated to be about 30% for the described protocol. If the loading efficiency is known, an estimate of the cytoplasmic concentration of the caged messenger can be made. The concentration of messenger liberated by flash photolysis can be determined by multiplying the caged messenger concentration with the efficiency of the photolytic release process.

Acknowledgments

This work was supported by NIH Grant HL 49288 to Michael J. Sanderson and Belgian FWO grants 31508696, 3G002696, and 31552098 to Luc Laybaert.

References

1. Zimmerman, G. A., Lorant, D. E., McIntyre, T. M., and Prescott, S. M. (1993) Juxtacrine intercellular signaling: another way to do it. *Am. J. Respir. Cell Mol. Biol.* **9,** 573–577.
2. Giaume, C. and McCarthy, K. D. (1996) Control of gap-junctional communication in astrocytic networks. *Trends Neurosci.* **19,** 319–325.

3. Charles, A. C. (1998) Intercellular calcium waves in glia. *Glia* **24,** 39–49.

4. Sanderson, M. J., Chow, I., and Dirksen, E. R. (1988) Intercellular communication between ciliated cells in culture. *Am. J. Physiol.* **254,** C63–C74.

5. Sanderson, M. J., Charles, A. C., and Dirksen, E. R. (1990) Mechanical stimulation and intercellular communication increases intracellular Ca²⁺ in epithelial cells. *Cell Regul.* **1,** 585–596.

6. Goodenough, D. A. (1992) The crystalline lens. A system networked by gap junctional intercellular communication. *Semin. Cell Biol.* **3,** 49–58.

7. Churchill, G. and Louis, C. (1998) Roles of Ca²⁺, inositol trisphosphate and cyclic ADP-ribose in mediating intercellular Ca²⁺ signaling in sheep lens cells. *J. Cell Sci.* **111,** 1217–1225.

8. Spray, D. C., Bai, S., Burk, R. D., and Saez, J. C. (1994) Regulation and function of liver gap junctions and their genes. *Prog. Liver Dis.* **12,** 1–18.

9. Frame, M. K. and Defeijter, A. W. (1997) Propagation of mechanically induced intercellular calcium waves via gap junctions and ATP receptors in rat liver epithelial cells. *Exp. Cell Res.* **230,** 197–207.

10. Dejana, E., Corada, M., and Lampugnani, M. G. (1995) Endothelial cell-to-cell junctions. *FASEB J.* **9,** 910–918.

11. Demer, L. L., Wortham, C. M., Dirksen, E. R., and Sanderson, M. J. (1993) Mechanical stimulation induces intercellular calcium signaling in bovine aortic endothelial cells. *Am. J. Physiol.* **264,** H2O94–H2102.

12. Sanderson, M. J. (1996) Intercellular waves of communication. *News Physiol. Sci.* **11,** 262–269.

13. Cornell-Bell, A. H., Finkbeiner, S. M., Cooper, M. S., and Smith, S. J. (1990) Glutamate induces calcium waves in cultured astrocytes: long range glial signaling. *Science* **247,** 470–473.

14. Dani, J. W., Chernjavsky, A., and Smith, S. J. (1992) Neuronal activity triggers calcium waves in hippocampal astrocyte networks. *Neuron* **8,** 429–440.

15. D'Andrea, P. and Vittur, F. (1997) Propagation of intercellular Ca²⁺ waves in mechanically stimulated articular chondrocytes. *FEBS Lett.* **400,** 58–64.

16. Newman, E. A. and Zahs, K. R. (1997) Calcium waves in retinal glial cells. *Science* **275,** 844–847.

17. Venance, L., Stella, N., Glowinsky, J., and Giaume, C. (1997) Mechanisms involved in initiation and propagation of receptor-induced intercellular calcium signaling in cultured rat astrocytes. *J. Neurosci.* **17,** 1981–1992.

18. Nedergaard, M. (1994) Direct signaling from astrocytes to neurons in cultures of mammalian brain cells. *Science* **263,** 1768–1771.

19. Hassinger, T. D., Guthrie, T. D., Atkinson, P. B., Bennett, M. V. L., and Kater, S. B. (1996) An extracellular signaling component in propagation of astrocytic calcium waves. *Proc. Natl. Acad. Sci. USA* **93,** 13,268–13,273.

20. Boitano, S., Dirksen, E. R., and Sanderson, M. J. (1992) Intercellular propagation of calcium waves mediated by inositol trisphosphate. *Science* **258,** 292–295.

21. Hansen, M., Boitano, S., Dirksen, E. R., and Sanderson, M. J. (1993) Intercellular calcium signaling induced by extracellular ATP and mechanical stimulation in airway epithelial cells. *J. Cell Sci.* **106,** 995–1004.

22. Sanderson, M. J., Charles, A. C., Boitano, S., and Dirksen, E. R. (1994) Mechanisms and function of intercellular calcium signaling. *Mol. Cell. Endocrinol.* **98,** 173–187.

23. Sanderson, M. J. (1995) Intercellular calcium waves mediated by inositol trisphosphate. *Ciba Found. Symp.* **188,** 175–189.

24. Young, S. H., Ennes, H. S., and Mayer, E. A. (1996) Propagation of calcium waves between colonic smooth muscle cells in culture. *Cell Calcium* **20,** 257–271.

25. Sanderson, M. J., Paemeleire, K., Strahonja, A., and Leybaert, L. (1998) Intercellular calcium signaling between glial and endothelial cells, in *Gap Junctions* (Werner, R., ed.), IOS Press, Amsterdam, pp. 261–265.

26. Leybaert, L., Paemeleire, K., Strahonja, A., and Sanderson, M. J. (1998) Inositol trisphosphate dependent intercellular calcium signaling in and between astrocytes and endothelial cells. *Glia* **24,** 398–407.

27. Charles, A. C., Dirksen, E. R., Merril, J. E., and Sanderson, M. J. (1993) Mechanisms of intercellular calcium signaling in glial cells studied with dantrolene and thapsigargin. *Glia* **7,** 134–145.

28. Charles, A. C., Naus, C. C., Zhu, D., Kidder, G. M., Dirksen, E. R., and Sanderson, M. J. (1992) Intercellular calcium signaling via gap junctions in glioma cells. *J. Cell Biol.* **118,** 195–201.

29. Evans, J. and Sanderson, M. J. (1998) Intercellular calcium oscillations induced by ATP in airway epithelial cells. *Am J. Physiol.* **277,** L30–L41.

30. Stranhonja, A. and Sanderson, M. J. (1998) Intracellular Ca^{2+} oscillations induced in glia by intercellular Ca^{2+} waves. *Glia* **28,** 97–113.

31. Allbritton, N. L., Meyer, T., and Stryer, L. (1992) Range of messenger action of calcium ions and inositol **1,4,5**–trisphosphate. *Science* **258,** 1812–1815.

32. Yule, D. I., Stuenkel, E., and Williams, J. A. (1996) Intercellular calcium waves in rat pancreatic acini: mechanism of transmission. *Am. J. Physiol.* **271,** C1285–C1294.

33. Lechleiter, J. D. and Clapham, D. E. (1992) Molecular mechanisms of intracellular calcium excitability in X. laevis oocytes. *Cell* **69,** 283–294.

34. Lee, H. C. (1997) Mechanisms of calcium signaling by cyclic ADP-ribose and NAADP. *Physiol. Rev.* **77,** 1133–1164.

35. Osipchuk, Y. and Cahalan, M. (1992) Cell-to-cell spread of calcium signals mediated by ATP recptors in mast cells. *Nature* **359,** 241–244.

36. Cao, D., Lin, G., Westphale, E. M., Beyer, E. C., and Steinberg, T. H. (1997) Mechanisms for the coordination of intercellular calcium signaling in insulin-secreting cells. *J. Cell Sci.* **110,** 497–504.

37. Jorgensen, N. R., Geist, S. T., Civitelli, R., and Steinberg, T. H. (1997) ATP- and gap junction-dependent intercellular calcium signaling in osteoblastic cells. *J. Cell Biol.* **139,** 497–506.

38. Paemeleire, K., de Hemptinne, A., and Leybaert, L. (1998) Traumatic single cell injury to astrocytes causes ATP-dependent astrocyte-endothelial Ca^{2+} signalling. *Pflügers Arch.* **435,** R244.

39. Cotrina, M. L., Lin, J. H. C., Alves-Rodrigues, A., Liu, S., Li, J., Azmi-Ghadimi, H., Kang, J., Naus, C. C. G., and Nedergaard, M. (1998) Connexins regulate calcium signaling by controlling ATP release. *Proc. Natl. Acad. Sci. USA* **95,** 15,735–15,740.

40. Giaume, C. and Venance, L. (1998) Intercellular calcium signaling and gap junctional communication in astrocytes. *Glia* **24,** 50–64.

41. Kaplan, J. H. and Somlyo, A. P. (1989) Flash photolysis of caged compounds: new tools for cellular physiology. *Trends Neurosci.* **12,** 54–59.

42. Ellis-Davies, G. C. and Kaplan, J. H. (1994) Nitrophenyl-EGTA, a photolabile chelator that selectively binds Ca^{2+} with high affinity and releases it rapidly upon photolysis. *Proc. Natl. Acad. Sci. USA* **91,** 187–191.

43. Li, W., Llopis, J., Whitney, M., Zlokarnik, G., and Tsien, R. Y. (1998) Cell-permeant caged InsP3 ester shows that Ca^{2+} spike frequency can optimize gene expression. *Nature* **392,** 936–941.

44. Lipp, P., Luscher, C., and Niggli, E. (1996) Photolysis of caged compounds characterized by ratiometric confocal microscopy: a new approach to homogeneously control and measure the calcium concentration in cardiac myocytes. *Cell Calcium* **19,** 255–266.

45. Kirby, M. S., Hadley, R. W., and Lederer, W. J. (1994) Measurement of intracellular Ca^{2+} concentration using Indo-1 during simultaneous flash photolysis to release Ca^{2+} from DM-nitrophen. *Pflügers Arch.* **427,** 169–177.

46. Leybaert, L., Paemeleire, K., and Sanderson, M. J. (1998) Inositol trisphosphate dependent calcium signaling and gap junctional coupling in and between astrocytes and endothelial cells. *Soc. Neurosci. Abstr.* **24,** 266.

47. Leybaert, L., Paemeleire, K., D'Herde, K., and Sanderson, M. J. (1998) Inositol trisphosphate dependent calcium signaling and gap junctional coupling in and between astrocytes and endothelial cells in co-culture. *Pflügers Arch.* **436,** R29.

48. Berridge, M. J. (1993) Inositol trisphosphate and calcium signalling. *Nature* **361,** 315–325.

49. McCarthy, K. D. and de Vellis, J. (1980) Preparation of separate astroglial and oligodendroglial cell cultures from rat cerebral tissue. *J. Cell. Biol.* **85,** 890–902.

50. McNeil, P. L., Murphy, R. F., Lanni, F., and Taylor, D. L. (1984) A method for incorporating macromolecules into adherent cells. *J. Cell Biol.* **98,** 1556–1564.

51. El-Fouly, M. H., Trosko, J. E., and Chang, C. C. (1987) Scrape-loading and dye transfer. A rapid and simple technique to study gap junctional intercellular communication. *Exp. Cell Res.* **168,** 422–430.

52. Giaume, C., Marin, P., Cordier, J., Glowinski, J., and Prémont, J. (1991) Adrenergic regulation of intercellular communication between cultured astrocytes from the mouse. *Proc. Natl. Acad. Sci. USA* **88,** 5577–5581.

53. Vera, B., Sanchez-Abarca, L. I., Bolanos, J. P., and Medina, J. M. (1996) Inhibition of astrocyte gap junctional communication by ATP depletion is reversed by calcium sequestration. *FEBS Lett.* **392,** 225–229.

54. Bolanos, J. P. and Medina, J. M. (1996) Induction of nitric oxide synthase inhibits gap junction permeability in cultured rat astrocytes. *J. Neurochem.* **66,** 2091–2099.

55. Lavado, E., Sanchez-Abarca, L. I., Tabernero, A., Bolanos, J. P., and Medina, J. M. (1997) Oleic acid inhibits gap junction permeability and increases glucose uptake in cultured astrocytes. *J. Neurochem.* **69,** 721–728.

56. Neumann, E., Schaefer-Ridder, M., Wang, Y., and Hofschneider, P. H. (1982) Gene transfer into mouse lyoma cells by electroporation in high electric fields. *EMBO J.* **1,** 841–845.

57. Tsong, T. Y. (1991) Electroporation of cell membranes. *Biophys. J.* **60,** 297–306.

58. Weaver, J. C. (1993) Electroporation: a general phenomenon for manipulating cells and tissues. *J. Cell. Biochem.* **51,** 426–435.

59. Weaver, J. C. (1995) Electroporation theory. Concepts and mechanisms. *Methods Mol. Biol.* **55,** 3–28.

60. Chang, D. C. (1989) Cell poration and cell fusion using an oscillating electric field. *Biophys. J.* **56,** 641–652.

61. Rapp, G. (1998) Flash lamp-based irradiation of caged compounds. *Methods Enzymol.* **291,** 202–222.

62. Parker, I., Callamaras, N., and Wier, W. G. (1997) A high-resolution, confocal laser-scanning microscope and flash photolysis system for physiological studies. *Cell Calcium* **21,** 441–452.

24

Biochemical Analysis of Connexin Phosphorylation

Bonnie J. Warn-Cramer, Wendy E. Kurata, and Alan F. Lau

1. Introduction

The phosphorylation of connexins represents an important mechanism that regulates the biological activity of gap junctions. The methods described in this chapter to study connexin phosphorylation utilize [^{32}P]orthophosphate metabolic radiolabeling of intact cells which permits the subsequent direct identification of phosphorylated amino acids and phosphopeptides of connexin. These methods include: (1) the analysis of connexin phosphoisoforms by immunoprecipitation and sodium dodecyl sulfate-polyacrylamide gel electrophoresis (SDS-PAGE), (2) direct identification of the phosphorylated amino acid(s) by two-dimensional phosphoamino acid analysis, and (3) resolution of phosphorylated connexin peptides by two-dimensional phosphotryptic peptide analysis. Connexin43 (Cx43) is used as the primary experimental example in this chapter because of the authors' extensive experience with this connexin subtype.

2. Materials
2.1. Analysis of Phosphoisoforms of Connexins

1. Phosphate-deficient and methionine-free media (Gibco-BRL, Grand Island, NY).
2. Calf serum (Hyclone, Logan, UT).
3. Fetal bovine serum (Summit Biotechnology, Ft. Collins, CO).
4. ^{32}P$_i$ (NEN, Boston, MA), ^{35}S-Pro-Mix (Amersham, Arlington Heights, IL).
5. Cx43 antibody. Commercial sources of Cx43 antibodies include Zymed (South San Francisco, CA), Transduction Laboratories (Lexington, KY), Chemicon (Temecula, CA), etc.).
6. Phosphotyrosine antibodies (Transduction Labs, Lexington, KY).
7. Protein A–*S. aureus* (Sigma, St. Louis, MO) *(1)*.

From: *Methods in Molecular Biology, vol. 154: Connexin Methods and Protocols*
Edited by: R. Bruzzone and C. Giaume © Humana Press Inc., Totowa, NJ

8. Prestained molecular weight markers are from Sigma (St. Louis, MO). Unstained molecular weight markers are prepared in our laboratory from the following proteins obtained from Sigma: phosphorylase *b*—92.5 kDa (one part); bovine serum albumin—68 kDa (one part); catalase—60 kDa (two parts); ovalbumin—46 kDa (three parts); carbonic anhydrase—29 kDa (two parts); myoglobin—17 kDa (one part); cytochrome *c*—12.5 kDa (one part). All protein stock solutions are prepared at 2 mg/mL in 0.0625 *M* Tris-HCL, pH 6.8, and stored at –20°C. When needed, an aliquot of each stock solution is mixed with an equal volume of 2× SDS/PAGE (Laemmli) sample buffer (composition given in **step 14**), plus an equal volume of 1× gel sample buffer, and boiled for 3 min. The working unstained molecular weight marker mix is then prepared at the ratios indicated above in parentheses (to give approximately equal Coomassie blue staining intensities). The working mix is also stored at –20°C. Typically, we use 10 µL per well to give a good staining reaction in a 0.75-mm thick polyacrylamide gel.

9. Nonidet P-40 (NP-40) (Sigma, St. Louis, MO).

10. Ovalbumin (Sigma, St. Louis, MO).

11. RIPA buffer: 150 m*M* NaCl, 1% sodium deoxycholate, 1% Triton X-100, 0.1% SDS, 10 m*M* Tris-HCl (Fisher Scientific, Santa Clara, CA), pH 7.2. Prior to use, add phosphatase inhibitors to a final concentration of 160 µ*M* Na_3VO_4 (100 m*M* stock in deionized H_2O), 50 m*M* NaF (1 *M* stock in deionized H_2O), and 1 m*M* phenylmethylsulfonyl fluoride (PMSF), added as a dry powder. Sodium pervanadate and sodium fluoride stock solutions are stable and stored at 4°C. PMSF is not very soluble in H_2O and has a short half-life in aqueous solutions; however, it is still an effective protease inhibitor.

12. STE buffer: 150 m*M* NaCl, 20 m*M* Tris-HCl, pH 7.2, and 1 m*M* EDTA.

13. 2,5-Diphenyl-oxazole (PPO)–dimethyl sulfoxide (DMSO) solution for fluorography: PPO in DMSO, 22.2% w/v (*2*).

14. SDS–PAGE (Laemmli) sample buffer: 6.25 mL of stacking gel buffer (0.5 *M* Tris-HCl, 0.4% SDS, pH 6.8), 5 mL of glycerol, 0.75 mL of 0.1% bromphenol blue in H_2O, 7.5 mL of 20% SDS, and 25.5 mL of H_2O. Add 50 µL of β-mercaptoethanol to 950 µL of this buffer prior to use (5% β-mercaptoethanol).

15. All solutions are filtered through a 0.45-µm membrane and used at 4°C, unless indicated otherwise. Deionized H_2O at 10–18 mOhm is used to prepare all solutions.

2.2. Determination of Phosphoamino Acid Content

1. Phosphoamino acid standards (Sigma, St. Louis, MO).

2. Stoddard solvent (EM Science, Gibbstown, NJ).

3. Two-dimensional thin-layer cellulose sheets (MN Polygram CEL 300 thin-layer cellulose on plastic backing, 0.1-mm MN300 cellulose, Brinkmann Instruments, Postfach, Germany).

4. Immobilon-P (Millipore, Bedford, MA).

5. Glass microcapillary tubes (Fisher Scientific, Pittsburgh, PA).

6. Chromist laboratory sprayer (Gelman Sciences, Ann Arbor, MI) to spray pH 3.5 buffer.
7. Fluorescent glow paint (Polymark, Polymeric, Natick, MA).
8. Phosphoamino acid standards mix: Pser–Pthr–Ptyr at 1:1:1 in H_2O, each standard at 100 µg/mL in mixture. Store frozen at –20°C.
9. Phosphoamino acid analysis electrophoresis buffers: pH 1.9 buffer (2.5% formic acid v/v [25 mL of 88% stock/L, Fisher Scientific, Santa Clara, CA], 7.8% acetic acid) and pH 3.5 buffer (0.5% pyridine, 5% acetic acid).
10. CAPS transfer buffer: 10 mM CAPS (3-cyclohexylamino-1-propane sulfonic acid), pH 11, 10% methanol (v/v), 0.1% EDTA.

2.3. Phosphotryptic Peptide Mapping

1. Acetone (Fisher Scientific, Pittsburgh, PA).
2. Trichloroacetic acid (Fisher Scientific, Pittsburgh, PA).
3. NH_4HCO_3 (Sigma, St. Louis, MO).
4. TPCK-trypsin (Worthington Biochemicals, Lakewood, NJ): Trypsin treated with L-(tosylamido-2-phenyl) ethyl chloromethyl ketone to inhibit chymotrypsin.
5. Formic acid (99%) (Fisher Scientific, Pittsburgh, PA).
6. Hydrogen peroxide (30%) (Sigma).
7. Pancreatic RNase A (Sigma).
8. Thin-layer cellulose glass plates (20 × 20 cm, cat. no. 5716, EM Science, Gibbstown, NJ).
9. Trichloroacetic acid (TCA), 100%: Dissolve 500 g of TCA in 215 mL of H_2O. Store at 4°C in an amber bottle.
10. Pancreatic RNase A: 2 mg/mL in H_2O, boiled and stored in small aliquots at –20°C.
11. Performic acid: Prior to use mix nine parts of 99% formic acid with one part of 30% hydrogen peroxide and incubate for 60 min at room temperature. Transfer to ice.
12. 50 mM NH_4HCO_3, pH ~7.3–7.6: 80 mg of NH_4HCO_3 in 20 mL of H_2O.
13. Gel extraction buffer: 50 mM NH_4HCO_3 with 0.1% SDS and 5% β-mercaptoethanol.
14. Ascending chromatography buffer: 62.5% isobutyric acid, 4.8% pyridine, 2.9% acetic acid, and 1.9% *n*-butanol.
15. Tracking dye: 5 mg/mL of ε-dinitrophenyl lysine and 1 mg/mL of xylene cyanol FF in 50% of pH 4.72 buffer (5% *n*-butanol, 2.5% pyridine, and 2.5% acetic acid).
16. Most chemical reagents are from Sigma (St. Louis, MO), unless otherwise indicated.

2.4. Specialized Equipment

1. Hunter thin-layer electrophoresis system (HTLE-7000, CBS Scientific, Del Mar, CA).
2. Savant SpeedVac and Savant phosphoamino acid analysis tanks (Farmingdale, NY).
3. Beckman TL100 ultracentrifuge and Beckman LS5000TD scintillation spectrometer (Beckman Instruments, Palo Alto, CA).
4. Circulating cooling water bath (Model A81, Haake Instruments, Paramus, NJ).

3. Methods

3.1. Analysis of Phosphoisoforms of Connexins

This subsection describes the immunoprecipitation of radiolabeled Cx43 from cell lysates and the resolution of the common phosphoisoforms of connexins by SDS-PAGE (*see* Section *4.1.*, **Notes 1** and **2**).

1. On d 1, radiolabel freshly confluent cells plated on 100-mm cell culture dishes in phosphate-deficient culture medium, supplemented with 4% serum. Perform the $^{32}P_i$ radiolabeling using 0.5–3.0 mCi/mL in 3 mL of medium per dish for 2–4 h at 37°C in a humidified 5% CO_2 incubator. Radiolabeling with ^{35}S-Pro-Mix is performed with 50–100 μCi/mL in methionine-free medium, supplemented with 4% calf serum for 3–4 h at 37°C.

2. Harvest radiolabeled cell cultures by placing the dishes on ice and pipeting off the radioactive medium. Wash the dishes twice with ice-cold phosphate-buffered saline supplemented with phosphatase inhibitors (160 μM Na_3VO_4, 50 mM NaF, 1 mM PMSF). Freeze the cell cultures immediately without buffer at –70°C.

3. On d 2, thaw the frozen cells on ice and scrape the dishes using a rubber scraper in cold RIPA buffer plus phosphatase inhibitors (~2.5 mL of buffer/100-mm dish). Transfer the cell suspensions to 15-mL screw-capped plastic tubes and vortex-mix vigorously to aid lysis. Clarify the cell lysates by centrifugation at 541,000g (100,000 rpm) for 20 min at 4°C in a TLA 100.3 rotor using a TL100 table top ultracentrifuge.

4. Isolate Cx43 from the clarified lysates by adding immune serum that is capable of immunoprecipitating the protein. Generally, add 4–6 μL of Cx43 antiserum (*see* Section *4.1.*, **Note 3**) to a volume of lysate representing one-half to one-third of a 100-mm plate of confluent fibroblastic cells (e.g., vole cells or rat-1 cells, which express a fair amount of Cx43 protein). Mix the samples by vortex-mixing and incubate on ice for 1½–2 h. Add the same volume of nonimmune serum to an equivalent volume of cell lysate as a control for nonspecific immunoprecipitation.

5. During the incubation period prepare the Protein A-containing bacterial cells that will be used to collect the Cx43 immune complexes (*see* Section *4.1.*, **Note 4**). Pellet the Protein A-containing *S. aureus* at 3000g at 4°C for 15 min. Use a volume of *S. aureus* (at 10% suspension, w/v) that is 20× the total volume of antiserum used for the immunoprecipitations (i.e., 80 μL/tube for 4 μL of antiserum). Resuspend the bacteria to the original volume in STE–0.5% NP-40 buffer and incubate at room temperature for 15 min. Pellet the bacteria, wash in STE–0.05% NP-40 buffer, and resuspend in a final total volume of STE–0.05% NP-40 containing 1 mg/mL of ovalbumin to yield 0.1 mL of bacterial suspension per sample.

6. At the end of the antiserum-lysate incubation period, add the prepared Protein A-*S. aureus* to the samples, incubate on ice for 30 min, and pellet at 3000g at 4°C for 15 min. Wash the bacteria pellet containing the immune complexes 3× in 1.5 mL of RIPA buffer plus phosphatase inhibitors at 4°C and pellet as before. After the

third wash, transfer each sample in 1 mL of RIPA buffer plus phosphatase inhibitors to a 1.5-mL microcentrifuge tube and pellet the sample in a microcentrifuge at 4°C for ~20 s. Aspirate the supernatant off completely, add 30–40 µL of 1× SDS-PAGE sample buffer to each sample, resuspend the bacteria completely by vigorous vortex-mixing, and boil for 3–5 min. Pellet the bacteria for 2 min in a microcentrifuge. Collect the supernatant containing the released Cx43 using a micropipettor and store in a fresh microcentrifuge tube at –20°C overnight.

7. On d 3, resolve Cx43 from other phosphorylated proteins by electrophoresis on a 7.5–15% SDS-containing polyacrylamide gel poured during the previous day (*see* Section *4.1.*, **Note 5**). Directly compare the migration of Cx43 isoforms (phosphorylated and nonphosphorylated) by electrophoresing [^{32}P]Cx43 and [^{35}S]Cx43 immunoprecipitated from cells, respectively *(3)* (*see* Section *4.1.*, **Note 6**). The gel is fluorographed *(2)* and the radiolabeled proteins visualized by exposing the dried gel to Kodak X-Omat film with or without the aid of an intensifying screen at –70°C. Estimate the apparent molecular weights of the various ^{32}P-phosphoproteins by comparison to the migration of stained molecular weight standards. The phosphoisoforms of Cx43 are typically displayed as a ladder of ^{32}P-labeled bands (45K and 47K or P1 and P2, respectively) which migrate slower that the fastest migrating, nonphosphorylated Cx43 isoform (43K or NP) seen in the ^{35}S-labeled protein (*see* **Fig. 1**).

3.2. Determination of Phosphoamino Acid Content

This subsection describes the procedure for determining the identity of the amino acids in Cx43 that are phosphorylated in intact cells *(4)*. This procedure, which provides a direct and reliable determination of the content of the three major phosphoamino acids (serine, threonine, and tyrosine), requires the radiolabeling of intact cells with [^{32}P]orthophosphate.

1. On d 1 and 2 immunoprecipitate [^{32}P]-Cx43 as described in **Subheading 3.1.** and resolve it from other phosphoproteins by SDS-PAGE. Electrotransfer the proteins in the *unfixed* gel to an Immobilon-P membrane (*see* Section *4.2.*, **Note 1**) at 300 mAmp for 1 h at 4°C using CAPS transfer buffer. Expose the membrane to film overnight with or without the aid of an intensifying screen.

2. On d 3, localize the [^{32}P]-Cx43 bands by realigning the dried membrane with the developed X-ray film exposed with the aid of fluorescent glow paint spots (*see* Section *4.2.*, **Note 2**). Excise the [^{32}P]-Cx43 bands precisely from the Immobilon membrane, transfer each piece(s) into a 1.5-mL screw-capped microfuge tube, and determine the amount of ^{32}P-radioactivity by Cerenkov counting in a scintillation spectrometer. (Radioactive decay particles from β emitters, such as ^{32}P, generate light emissions that can be detected by spectrometry [Cerenkov radiation]. These energetic decay particles can be detected without the use of a scintillation cocktail to amplify the light signal, and thus the entire sample can be counted in an Eppendorf tube placed in a carrier vial and the sample recovered after determining the radioactivity). Wet the Immobilon membrane with methanol.

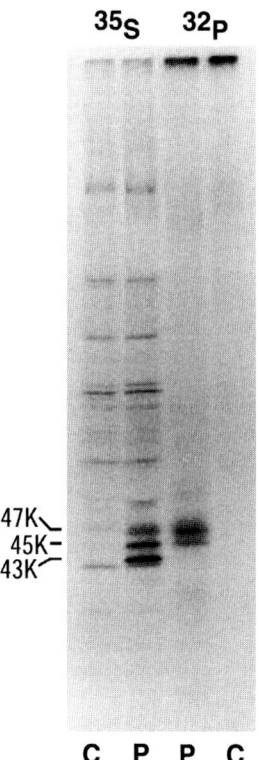

Fig. 1. Resolution of ³²P- and ³⁵S-labeled Cx43 immunoprecipitated from fibro-blast cells by SDS-PAGE. Separate cultures of vole fibroblast cells, an established rodent cell line *(3)* were labeled with either [³⁵S]methionine (50 μCi/mL for 4 h at 37°C) or ³²Pᵢ (0.5 mCi/mL for 4 h at 37°C). Cells were washed in phosphate-buffered saline, lysed in RIPA buffer, and clarified as described in **Subheading 3**. Clarified lysates were immunoprecipitated with either control nonimmune rabbit serum (*lanes C*) or Cx43 peptide antiserum (*lanes P*). Proteins were resolved by SDS-PAGE on a gradient gel and detected by autoradiography of the fluorographed gel. Positions of the nonphosphorylated 43K (NP) and the phosphorylated 45K (P1) and 47K (P2) Cx43 isoforms are indicated at the *left margin*. (Reprinted with permission from **ref.** *(3)*.

Wash the membrane twice with deionized H₂O and remove all H₂O after the last rinse. Acid hydrolyze the Cx43 protein from the membrane by adding 200 μL of 5.7 *N* HCl to each sample to cover all membrane pieces. Incubate samples in a preheated heating block at 110°C for 60 min (*see* Section *4.2.*, **Note 3**).

3. Chill the sample on ice for 5 min. Centrifuge it at 16,000*g* (14,000 rpm) in a microcentrifuge for 5 min, transfer the supernatant to a fresh microcentrifuge tube, add 300 of μL H₂O, and lyophilize it to dryness at the bottom of the tube

using a Savant SpeedVac (*see* Section *4.2.*, **Note 4**). Cerenkov count the sample in a scintillation spectrometer. Dissolve the sample completely in 5–10 µL of the phosphoamino acid standards mix.

4. Spot a sample volume that gives the appropriate number of cpm onto a 20×20 cm thin-layer cellulose (TLC) plastic-backed sheet using a 2–5-µL glass microcapillary pipet (*see* **Fig. 2** and Section *4.2.*, **Note 5**). Dry the sample spot between repeated applications with a gentle stream of cool filtered air. Typical total volumes spotted range from 4 to 6 µL. Up to four samples can be analyzed on a single 20×20 cm TLC sheet arranged as shown in **Fig. 2**.

5. Blot the spotted TLC sheet gently with a gauze pad wetted with pH 1.9 electrode buffer (*see* Section *4.2.*, **Note 6**). Avoid damaging the cellulose thin layer. Converge buffer around the sample origin in a concentric manner to concentrate each sample. Alternatively, the TLC sheet can be wetted with a blotter as described in **Subheading 3.3.** Mount the TLC sheet onto the Plexiglas holding rack and lower the rack slowly into the Savant electrophoresis tank through the Stoddard solvent layer and into the pH 1.9 electrode buffer. To prevent sample loss, avoid washing the electrode buffer over the sample origin. Electrophorese the sample toward the (+) electrode at 1000 V for 32 min at 4°C. Remove the TLC sheet, lay horizontally, and dry it thoroughly in a fume hood at room temperature.

6. Wet the TLC sheet minimally and evenly by spraying (Chromist laboratory sprayer) with pH 3.5 electrode buffer in a manner that avoids buffer pooling and runoff. Mount the TLC sheet in the rack rotated 90° from the first direction (*see* **Fig. 2**). Load the rack slowly into the pH 3.5 electrode buffer tank. Electrophorese the samples toward the (+) electrode at 1000 V for 21 min at 4°C. Lay the TLC sheet flat and dry it completely in a fume hood.

7. Spray the TLC sheet with 0.2% ninhydrin in ethanol. Avoid pooling of spray or buffer run off. Heat the sheet in an oven at 50°C for (3–5 min to develop the ninhydrin-stained spots. Outline the positions of the three phosphoamino acid standards lightly with a blunt no. 3 pencil. Visualize the radioactive spots by exposing the TLC sheet to Kodak X-Omat film with the aid of an intensifying screen at –70°C for 1–3 wk (*see* Section *4.2.*, **Note 7**).

8. The typical two-dimensional migration of the three major phosphoamino acids (pS, pT, and pY) yields an upside down, reversed L shape as shown in **Fig. 2B**. The appearance of a radioactive spot comigrating with one of the three phospho-amino acid standards indicates the phosphorylation of that particular phospho-amino acid in Cx43. Free inorganic phosphate appears as a radioactive spot migrating well ahead of the three phosphoamino acids in both dimensions. Any incompletely hydrolyzed phosphopeptide material will usually migrate as a smear just above and to the left of the origin.

3.3. Phosphotryptic Peptide Mapping

The procedures described in this subheading will produce a two-dimensional display of the phosphorylated peptides of Cx43 by the digestion of ^{32}P-labeled Cx43 with trypsin. This pattern gives an idea of the number of phosphorylated

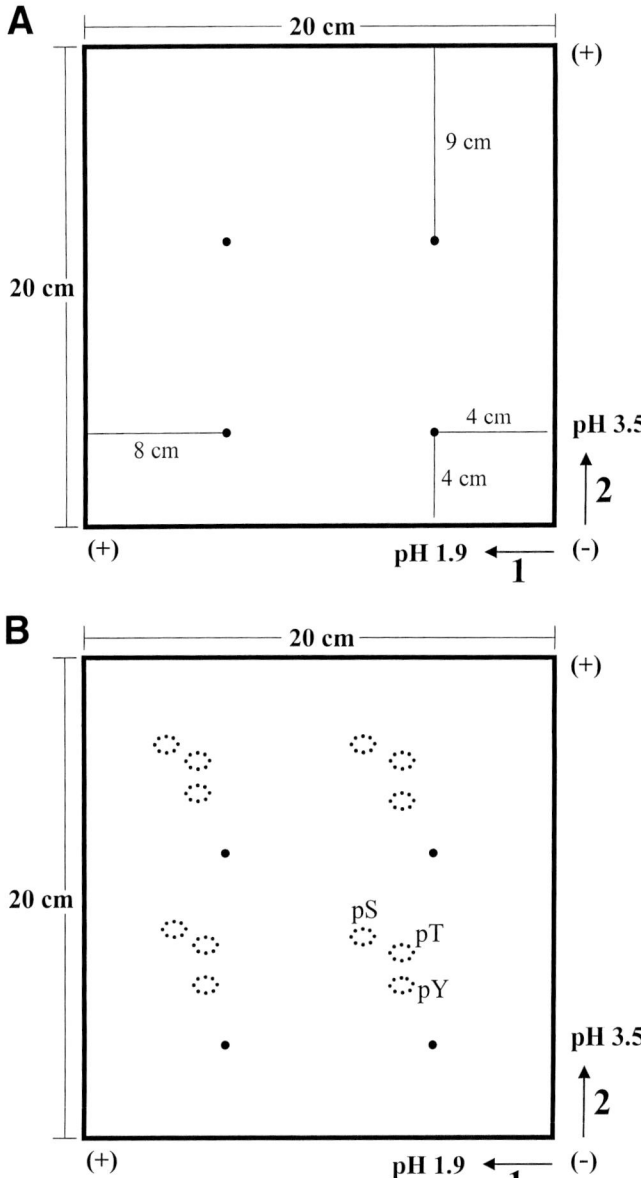

Fig. 2. Two dimensional resolution of phosphoamino acids from Cx43 on thin layer cellulose sheets. (**A**) Schematic representation of the 20 × 20 cm plastic-backed thin layer cellulose sheet. The origins for as many as four samples are shown by the closed circles (•). The directions of electrophoresis in the first dimension in pH 1.9 buffer and in the second dimension in pH 3.5 buffer are shown. Electrophoresis in both dimensions is toward the (+) electrode. (**B**) Schematic representation of a typical two-

sites in connexin under control conditions and an indication of changes in the phosphorylation status of Cx43 after an experimental treatment. The resulting phosphotryptic peptide pattern is reproducible and can serve as a "fingerprint" for phosphorylation under specific experimental conditions or by a particular protein kinase. This procedure is a key part of an approach to identify specific phosphorylation sites in connexin proteins. Because very small amounts of connexin are typically available for analysis, this procedure depends upon the high level ^{32}P-labeling of the connexin of interest. The procedure described here is patterned after Boyle et al. *(5)*.

1. On d 1–2, prepare [^{32}P]-Cx43 as described in **Subheading 3.1.** Because of the length of this procedure and the inherent losses of material, the radiolabeling of cells should be performed with 1–3 mCi/mL of [^{32}P]-orthophosphate using the necessary safety shielding and equipment (*see* Section *4.3.*, **Note 1**). We have also purified a larger amount of Cx43 starting material by radiolabeling more plates of cells and using a proportionally larger amount of Cx43 antibody.

2. On d 3, identify the Cx43 bands by autoradiography of the unfixed, hydrated, Saran-wrapped polyacrylamide gel. It is important to use Saran-Wrap brand of plastic-wrap. Precisely excise the appropriate band(s). Handle the gel pieces with gloves to avoid possible contamination of the sample. Cerenkov count the excised gel band. A minimum of 5000 cpm is generally required to complete these procedures successfully. This generally represents a very small amount of Cx43 protein. To extract the Cx43 protein, grind up the gel piece using a spatula in an Eppendorf tube or in a disposable tissue grinder tube (Kontes, Vineland, NJ). Transfer the gel pieces to a screw-capped microcentrifuge tube in 1 mL of freshly prepared extraction buffer (50 m*M* NH$_4$HCO$_3$ containing 0.1% SDS and 5% β-mercaptoethanol). The use of siliconized tubes for this and subsequent steps is recommended to prevent the loss of material caused by adsorption onto the sides of the tube. Boil the sample for 3–5 min and incubate for 2–3 h (or overnight) while shaking at room temperature.

3. On d 4, clarify the pooled supernatant by centrifuging in a microcentrifuge at 16,000*g* (14,000 rpm) for 10 min. Immediately remove the supernatant leaving behind any gel pieces. Extract gel pieces once more with 300 μL of extraction buffer (prepared as in **step 2**). Boil the sample again for 3–5 min and incubate for 2–3 h especially if significant cpm remain in the gel pellets (determine by Ceren-

---

(*Opposite page*) dimensional resolution of the three major phosphoamino acids from Cx43: phosphoserine (pS), phosphothreonine (pT), and phosphotyrosine (pY). The outlines correspond to the migration positions of the ninhydrin-stained phosphoamino acid standards. Autoradiography of the thin-layer sheet will reveal the identity of phosphorylated amino acid(s) present in the Cx43 protein isolated from the cells. Free ^{32}P$_i$ will migrate faster than the pS spot in both dimensions. Incompletely digested phosphopeptide material will migrate as a smear between the pY spot and the sample origin.

kov counting). Be very careful not to carry over any gel debris as this will lower the yield and produce a brown sample that electrophoreses poorly. Chill the clarified supernatant on ice. Add 20 µg of boiled carrier RNase A, followed by 250 µL of chilled 100% TCA. Mix completely and allow to precipitate on ice for 2–3 h. It is possible to precipitate the sample on ice overnight at 4°C. Pellet the precipitate in a microcentrifuge for 10 min at 16,000g (14,000 rpm). Wash the pellets once with 500 µL of cold acetone (–20°C) and air-dry the tubes. Cerenkov count the sample to determine the recovery, which should be approx 60% of the starting amount.

4. Dissolve the pellet in 50 µL of chilled performic acid and incubate on ice for exactly 1 h. Add 400 µL of chilled H_2O, freeze and lyphophilize to dryness in a SpeedVac (*see* Section *4.3.*, **Note 2**). During the performic oxidation step do not allow the sample to warm up. Dissolve the lyophilized sample in 50 µL of NH_4HCO_3, pH ~8.0–8.3. It is possible to use the NH_4HCO_3 solution prepared on d 3 because its pH will drift to ~8.0 overnight. Digest the sample with TPCK-trypsin (25 µg or 25 µL of 1 mg/mL solution) by incubating overnight at 37°C.

5. On d 5, remove tubes from water bath and centrifuge to collect volume. Continue the digestion by adding another aliquot of trypsin (10 µL). Vortex-mix the sample well and incubate for 3–5 h at 37°C. Add 400 µL of H_2O, freeze, and lyophilize in a SpeedVac to a small volume (~50 µL, ~2–3 h). Do not allow the sample to dry completely. Relyophilize twice with 300 µL of H_2O and then with 200 µL of pH 1.9 buffer. Add 100 µL of pH 1.9 buffer, vortex-mix well, and clarify the sample in a microcentrifuge at 16,000g (14,000 rpm) for 10 min at 4°C. Transfer the supernatant to a clean tube and lyophilize in the SpeedVac to a final volume of 5–10 µL. These steps, which remove NH_4HCO_3 and other debris, are essential to obtaining "nice", reproducible peptide maps (*see* Section *4.3.*, **Note 3**). Cerenkov count the sample.

6. On d 6, spot ~5 µL of sample at the origin on glass-backed TLC plates (EM Science) as described in **Fig. 3** (*see* Section *4.3.*, **Notes 4** and **5**). Keep the sample origin to a 1–2-mm diameter, drying with a stream of cool filtered air after each application. Apply tracking dye as indicated in **Fig. 3** (dye 1). The tracking dye is used to monitor the extent of electrophoresis. It is composed of two different dyes (ε-dinitrophenyl lysine and xylene cyanol FF) that migrate differently during electrophoresis, separating into a blue and a yellow spot. Prepare a blotter using a 20 × 20 cm double sheet of Whatman 3MM paper. Cut two ~1–2-cm diameter holes with a cork borer corresponding to the sample origin and dye spot. Wet the blotter by dipping it in a tray containing ~200 mL of pH 1.9 buffer and removing excess buffer by blotting it briefly with a second piece of Whatman 3MM paper. Wet the TLC plate by applying the damp blotter wetted with the pH 1.9 buffer. Try to wet and concentrate the sample first by pressing gently on the 3MM sheets surrounding the origin with a damp gauze to bring the buffer into the origin concentrically. Before loading the TLC plates, press the edges of the wicks in the Hunter apparatus to remove excess buffer. Load the TLC plate into the Hunter tryptic peptide apparatus with the left (+) and right (–) edges of the plate oriented

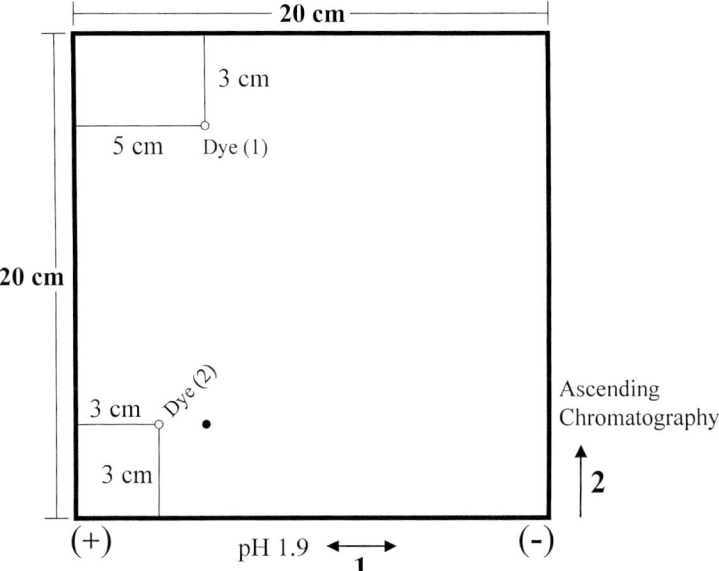

Fig. 3. Two-dimensional resolution of phosphotryptic peptides on thin layer cellulose. Schematic representation of the 20×20 cm thin-layer cellulose glass plate used for the separation of tryptic phosphopeptides of Cx43. The sample origin is indicated by the *closed circle* (•). The positions for the application of the tracking dye in the first (dye 1) and second (dye 2) dimensions are indicated by open circles (○). Electrophoresis in the first dimension in pH 1.9 buffer is toward the (+) and (–) poles. Resolution in the second dimension is by ascending chromatography in isobutyric acid buffer.

toward the correct electrode. Overlay ~1–1.5 cm of the edges of the plate with Whatman 3MM wicks leading from the buffer troughs (~400 mL of buffer in each trough). Place a sheet of Saran wrap over the TLC plate to protect the apparatus. Apply pressure and cool according to the manufacturer's instructions. Electrophorese in the pH 1.9 electrode buffer toward (+) and (–) electrodes for 65 min at 1000 V (*see* Section *4.3.*, **Note 6**). Peptides electrophorese toward the (–) electrode according to their charge to mass ratios *(5)*. Free phosphate migrates toward the (+) electrode. Remove the TLC plate and dry horizontally at room temperature.

7. Spot ~1 μL of tracking dye at the left edge of (+) of the TLC plate, at the same level as the origin (*see* **Fig. 3**, dye 2). Resolve phosphotryptic peptides in the second dimension (90° from first dimension and toward top of plate) (*see* **Fig. 3**) by ascending chromatography in a tank preequilibrated with isobutyric acid buffer at room temperature (*see* Section *4.3.*, **Note 7**). Chromatograph samples until the solvent front reaches the first dye spot near the top of the plate (approx 12–13 h).

8. On d 6, dry plates vertically in a fume hood. Expose the ^{32}P-labeled phosphotryptic peptides by autoradiography on Kodak X-Omat film at –70°C with the aid

of an intensifying screen for 1–3 wk depending upon the amount of sample radio-activity spotted (*see* Section *4.3.*, **Note 8**).

9. Peptides migrate in the electrophoretic direction according to their charge-to-mass ratios and in the chromatographic direction according to their hydrophobic-ity giving rise to a two-dimensional phosphopeptide map (*see* **Fig. 4** and **ref. 5**). Several features contribute to the complexity of peptide maps including: (1) incomplete cleavage by trypsin giving rise to multiple digestion products (i.e., lysine–aspartate/glutamate or arginine–aspartate/glutamate peptide bonds are not cleaved efficiently by trypsin; *see* **Fig. 4B** peptides d and e); (2) peptides that migrate differently due to different degrees of phosphorylation on the same peptide; and (3) altered migration in the chromatographic direction due to different phosphorylation sites on the same peptide (i.e., serine vs tyrosine sites) or different oxidation states of methionine or cysteine residues in a peptide. The addition of negatively charged phosphate groups to a peptide retards its migra-tion in both directions giving rise to a series of radioactive spots falling on a diagonal for a peptide with more than one phosphorylation state (*see* **Fig. 4B** peptides b and c). The peptide migrating furthest from the origin in both directions will then represent the lowest phosphorylation state of the peptide.

4. Notes

4.1. Analysis of Phosphoisoforms of Connexins

1. The relatively high level of $[^{32}P]$orthophosphate used in these methods requires careful sample handling, adequate Plexiglas shielding (≥ 0.75 in. thick), and a Geiger–Muller counter to monitor the radioactivity. If radiolabeled samples are not used immediately, they are best stored shielded at $-70°C$ as frozen cells on the dishes, rather than as scraped cell pellets or cell lysates.

2. The total profile of Cx43 can be displayed by immunoblotting cell lysates resolved on an SDS gel, if the cell line used contains sufficient amounts of Cx43. However, some phosphoisoforms of Cx43 may not be present at levels that can be readily detected on immunoblots and it may be necessary to immunoprecipi-tate radiolabeled Cx43 as described in **Subheading 3.1.** This will have to be determined empirically for the cell line and antibody preparation used. The phosphotyrosine content of Cx43 may be detectable by immunoblotting with an antiphosphotyrosine antibody. However, this approach is not amenable to the direct determination of the full spectrum of phosphoamino acid content of Cx43 and cannot be extended to tryptic peptide analysis.

3. We have found that approx 4–6 µL of our Cx43 peptide antiserum (raised against aa 368–382 of Cx43 (*3*) is adequate to immunoprecipitate the majority of Cx43 from one-half to one-third of a 100-mm dish of our rat-1 fibroblast cells. The amount of antibody used must be determined empirically for the particular anti-body and cell line used.

4. We have generally used Protein A-containing *S. aureus* instead of Protein A–Sepharose to collect the immune complexes because on a volume-to-volume basis

A   **B**

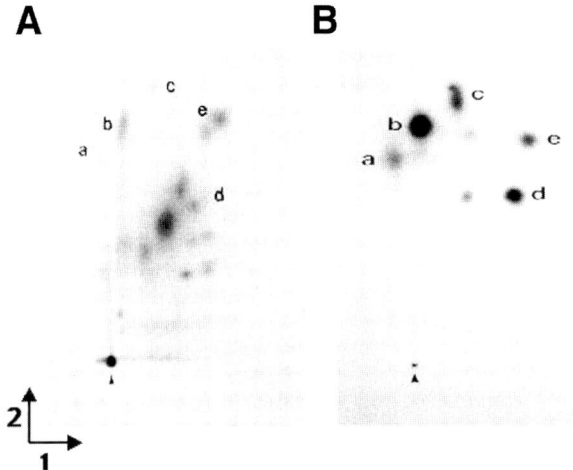

Fig. 4. Two-dimensional phosphotryptic peptide maps of ^{32}P-labeled Cx43. (**A**) Phosphotryptic peptides of Cx43 immunoprecipitated from ^{32}P-labeled, EGF-treated T51B rat liver epithelial cells. The sample origin is indicated by an *arrowhead*. Some phosphopeptides of interest are indicated by letters *a–e*. Electrophoresis in the first dimension and ascending chromatography in the second dimension are indicated at the *bottom left corner*. (**B**) Phosphotryptic peptides of the C-terminal tail of Cx43 phosphorylated in vitro by mitogen-activated protein (MAP) kinase. The carboxy-terminal half of Cx43, fused to glutathione-*S*-transferase (GST), was purified on glutathione–Sepharose beads, phosphorylated by purified MAP kinase, and purified by SDS-PAGE (*6*). The phosphotryptic peptides were prepared and resolved as described in **Methods**. The sample origin is indicated by an *arrowhead*. The letter designations correspond to peptides migrating similarly as those in **A**. Peptides d and e represent the same phosphorylation site with incomplete cleavage at a lysine–aspartate peptide bond giving rise to two digestion products. Peptides b and c represent different phosphorylation states of the same peptide and migrate on a diagonal in relation to each other. Peptide c represents the singly phosphorylated peptide and peptide b is doubly phosphorylated. Peptide a did not migrate in a diagonal relationship to peptides b and c on all chromatograms and has not been identified. (Reprinted with permission from **ref.** (*16*). *Chem.* **271,** 3779–3786, 1996.

the bacteria collect more immune complexes and they form smaller, firmer pellets which are easier to extract in small volumes.

5. Cx43 is resolved well on a SDS-containing 12% polyacrylamide gel or on a 7.5–15% polyacrylamide gradient gel (Rainin Minipuls 2 gradient mixer, Rainin/Gilson, Woburn, MA). The gel dimension is 0.5 cm thick and 12–13 cm long. We routinely pour the separating gel the day before use and polymerize the gel overnight at room temperature.

6. For the side-by-side comparison of [³⁵S]- vs [³²P]-Cx43, use approx 3× more ³⁵S cpm than ³²P cpm to give equivalent exposures. This gel should be fluorographed with PPO–DMSO to maximize detection of the ³⁵S radioactivity *(2)*.

4.2. Determination of Phosphoamino Acid Content

1. The recovery of acid-hydrolyzed Cx43 sample is most quickly and efficiently accomplished from Immobilon-P rather than the extraction of a gel piece followed by the subsequent TCA precipitation and acid hydrolysis of the sample. In addition, the use of Immobilon in the acid hydrolysis step avoids possible contamination of the samples by hydrolyzed acrylamide, which may occur when a gel piece is ground up and extracted. Immobilon membrane is highly acid- and base-resistant and does not bind free phosphoamino acids after acid hydrolysis, which makes it an ideal support for acid extraction.
2. The exposed X-ray film is accurately realigned with the Immobilon membrane by the use of strategically placed, small fluorescent dots on the membrane, which make corresponding exposed dots on the film.
3. The use of screw-capped microcentrifuge tubes in the acid hydrolysis step at 110°C is essential to prevent loss of sample. Excessive pressure buildup in the tube may cause the inadvertent popping of the caps and loss of sample volume.
4. A Savant SpeedVac is desirable because it lyophilizes samples to the very bottom of the microcentrifuge tube, which permits the subsequent resuspension of the sample in the small volume (4–6 µL) required for spotting on the TLC sheet.
5. We have obtained readable phosphoamino acid maps from the spotting of samples containing as few as 50–100 cpm of ³²P Cerenkov radiation. However, to detect a less prevalent phosphoamino acid or to rule out the presence of a particular phosphoamino acid, it may be necessary to spot more cpm.
6. Care must be taken not to gouge the TLC sheet during the application of the sample and the pH 1.9 electrode buffer. The buffers must be applied minimally to produce TLC sheets that are dull in appearance, not shiny, indicating an overapplication of buffer. Correct the latter event by wicking away the excess buffer carefully with a Kimwipe. Pooling of buffer or buffer runoff can ruin a chromatogram. A circulating water bath is used to cool the electrophoresis tanks.
7. It is advisable to run the phosphoamino acid standards first in a test run to determine that all of the standards are visible with the ninhydrin stain and that they are resolved well from each other.

4.3. Phosphotryptic Peptide Mapping

1. The time frame given for these procedures is approximate and it is possible to do several steps in one day and complete the entire procedure in < 6 d.
2. An aliquot of the performic acid treated sample can be removed at this step, lyophilized, and acid hydrolyzed for phosphoamino acid analysis if sufficient sample radioactivity is available.
3. The key to the production of reproducible, clear peptide maps is the removal of all NH_4HCO_3 and debris material from the sample after the trypsin digestion

through the described rounds of lyophilization and centrifugation and transfer of the sample to a fresh tube *(5)*. Also, it is important not to let the sample go dry during the final washes and lyophilization. This may result in the irreversible loss of hydrophobic peptides on the walls of the tube. Carryover of any gel material from the extraction step will produce a brownish insoluble pellet that may contain the bulk of the phosphopeptides (and therefore radioactivity). After application, the sample should wet evenly and quickly when the electrode buffer migrates into the sample origin. If this does not happen, the sample may be contaminated with foreign material and it is not likely to run well.

4. Each TLC sheet must be examined prior to use to determine the direction of cellulose flow, which should be chosen as the direction for electrophoresis. Cellulose plates are poured in one direction and the cellulose allowed to flow to the other end of the plate. By holding the plates up to the light it may be possible to see spaces on the edges or grains within the cellulose that indicate the direction of flow. Gloves should be used when handling the plates.

5. Loading an excessive amount of protein on the plates may cause the phosphopeptides to smear or streak.

6. The pH 1.9 buffer used in the first dimension resolves Cx43 peptides well *(6)*. It is a frequently used starting buffer because most peptides are soluble in it and it provides good resolution. A circulating water bath is used to cool the apparatus during electrophoresis. We have found that the isobutyric acid buffer gives good resolution of Cx43 peptides in the second ascending chromatography dimension. Resolution of other connexins may require different buffer types, which must be determined empirically *(5)*.

7. We perform the ascending chromatography (second dimension) in a large Pyrex glass tank ($30 \times 30 \times 60$ cm) that has been equilibrated with the chromatography buffer overnight. The tanks should be opened only for the brief time that it takes to insert the TLC plates. Smaller glass "brick" type tanks ($10 \times 27 \times 30$ cm) can also be used, but they must be lined with Whatman 3MM paper on four sides to help maintain the buffer-equilibrated gas phase in the tanks. The glass lid should be tightly sealed on the tank top with silicone grease. The chromatography tank should be used in a quiet corner of the laboratory to minimize vibrations that disturb chromatography. Different batches of plates and new batches of chromatography buffer may affect the time required for the ascending chromatography run. Do not allow the solvent front to run to the top of the plate.

8. Use of a phosphoimaging device may increase the sensitivity of detecting lower amounts of radiolabeled phosphopeptides.

References

1. Kessler, S. W. (1975) Rapid isolation of antigens from cells with a staphylococcal Protein A-antibody adsorbent: parameters of the interaction of antibody-antigen complexes with protein A. *J. Immunol.* **115,** 1617–1624.
2. Bonner, W. M. and Laskey, R. A. (1974) A film detection method for tritium-labelled proteins and nucleic acids in polyacrylamide gels. *Eur. J. Biochem.* **46,** 83–88.

3. Crow, D. S., Beyer, E. C., Paul, D. L., Kobe, S. S., and Lau, A. F. (1990) Phosphorylation of connexin43 gap junction protein in uninfected and Rous sarcoma virus-transformed mammalian fibroblasts. *Mol. Cell. Biol.* **10,** 1754–1763.
4. Kamps, M. P. and Sefton, B. M. (1989) Acid and base hydrolysis of phosphoproteins bound to Immobilon facilitates analysis of phosphoamino acids in gel-fractionated proteins. *Anal. Biochem.* **176,** 22–27.
5. Boyle, W. J., Van Der Geer, P., and Hunter, T. (1991) Phosphopeptide mapping and phosphoamino acid analysis by two-dimensional separation on thin-layer cellulose plates. *Methods Enzymol.* **201,** 110–149.
6. Warn-Cramer, B. J., Lampe, P. D., Kurata, W. E., Kanemitsu, M. Y., Loo, L. W. M., Eckhart, W., and Lau, A. F. (1996) Characterization of the mitogen-activated protein kinase phosphorylation sites on the connexin43 gap junction protein. *J. Biol. Chem.* **271,** 3779–3786.

25

How to Close a Gap Junction Channel

Efficacies and Potencies of Uncoupling Agents

Renato Rozental, Miduturu Srinivas, and David C. Spray

Alladin found a magical lamp and rubbed it. The genie appeared and told him that he had one wish. Alladin did not think twice. He said: "I need a potent and selective uncoupling agent". The genie responded "I said a wish, not a miracle".

1. Introduction

There are several reasons that one might want to selectively do away with gap junction channels. Of critical importance to electrophysiologists, coupling interferes with isopotentiality, a requirement for voltage-clamping cells. Second, by measuring the function of cell groups or tissues in which gap junctions have been eliminated, it may be possible to infer the roles that gap junctions normally play ("negative" physiology: **refs. *1–5***). Finally, there are pathological conditions in which gap junction overexpression might be an underlying cause (or problematic consequence) of the pathology, and therefore gap junction blockers might be therapeutically useful. Despite the desirability of finding agents that would specifically block intercellular communication, however, there is as yet no "silver bullet" that will close gap junction channels without side effects and, as considered later, some of the most commonly used gap junction blockers do not totally close the channels, but only partially impair their conductance.

For other types of ion channels, pharmacological experiments have provided seminal evidence defining their properties and physiological roles. For example, the current understanding of interactions of neurotransmitters with their receptors and the behavior of voltage-gated ionic channels on excitable membranes,

From: *Methods in Molecular Biology, vol. 154: Connexin Methods and Protocols*
Edited by: R. Bruzzone and C. Giaume © Humana Press Inc., Totowa, NJ

leading to channel activation, inactivation, and desensitization, owes much to the development of pharmacological agents and the discovery of highly specific toxins. Examples include tetrodotoxin (TTX) *(6,7)* and saxitoxin for Na^+ channels *(8,9)*, D-tubocurarine *(10–13)* and α-bungarotoxin *(14–16)* for the nicotinic acetylcholine receptor ion channel complex (nAChR), tetraethylammonium (TEA) for K^+ channels *(17)*, and ω-conotoxins and ω-agatoxins *(18–20)* for voltage-activated Ca^{2+} channels all of which have helped to define these channels, as discrete entities and to clarify their roles in function of specific tissues.

For gap junction channels, no such high-affinity and highly selective inhibitor has yet been identified. The lack of naturally occurring gap junction toxins might be due in part to the inaccessibility of gap junction proteins to extracellular space. Whereas other ion channels have large exposed domains, and such exposure is even crucial for the operation of ligand-gated channels, the lumen of gap junction channels is continuous between connexons contributed by each cell and is thus inaccessible to extracellular agents. The lack of discovery and development of synthetic uncoupling agents is likely ascribable to insufficient motivation, as well as the lack of simple assays by which large number of potentially active compounds could be screened. Because there are currently only a few correlations of gap junction channel dysfunctions or gene mutations with diseases, the pharmaceutical industry has not actively supported the development of uncoupling agents as a strategy to uncover potential therapeutic targets. By and large, the uncoupling agents now in general use were discovered as side effects of drugs that had been shown to affect other ion channels. Little attention has been given to optimization of specificity through pharmacokinetic manipulation, and it is not surprising that these tools are lacking. An exception is the recent use of antibodies directed toward the relatively conserved extracellular loop domains required for the process of formation of gap junction channels by docking of two hemichannels, as well as the use of connexin-mimetic peptides. These strategies offer the possibility of blocking junctional channels under long-term conditions *in situ,* perhaps ultimately combining high specificity, potency, and efficacy. The applicability of these approaches for both short- and long-term studies is highlighted in this chapter.

Although a moderately large number of agents has been shown to reduce intercellular coupling via gap junctions (**Table 1**), only a very few are in general use. The modes and possible sites of actions of the most useful of these are reviewed in historical order of appearance in Results and Discussion. Among these, we discuss alterations in pH_i *(21,22)*, long-chain alcohols (i.e., heptanol and octanol) *(23,24)*, halothane and ethrane *(25,26)*, glycyrrhetinic acid derivatives *(27,28)*, and antibodies against external loop domains of Cx43 (aEL1–42, Dr. Nedergaard, unpublished results; aEL2–186, ref. *29*) and connexin mimetic peptides corresponding to these regions *(30–34)*.

Table 1
Agents that Reduce Coupling

Lipophiles	Antibodies, peptides, antisense RNA
Arachidonic acid	Extracellular loop antibodies
Doxyl stearic acids	Connexin mimetic peptides
Heptanol, octanol	Connexin antisense RNA
Halothane	Other molecules
Oleic Acid	Glycyrrhetinic acids
2,3 butanedione monoxime	Anandamide
Oleamide and derivatives	Diamide
Acidifiers	Tricaine (MS222)
Weak acids (CO_2, lactate, acetate, etc)	Insulin
Nitrobenzyl esters	Gossypol
Nigericin	Testosterone
Dinitrophenol	Tamoxifen
Tumor promoters	
Dieldrin	
Phorbol esters	
DDT	

2. Material and Methods

2.1. Intracellular Acidification (See Note 1)

Junctional conductance (g_j) can be rapidly and reversibly reduced by acidification of the cytoplasm *(21)*; extracellular acidification does not affect g_j. Intracellular acidification is most easily achieved by bubbling the extracellular solution with CO_2 (to produce H_2CO_3, carbonic acid) or by adding weak acids such as acetate, proprionate, or acetate as Na^+ salts. Because the nondissociated form of weak acids are membrane permeant, the efficacy of weak acids as uncoupling agents is governed by both the extracellular concentration of the weak acid and by the extracellular pH, as described by the Henderson–Hasselbalch equation:

$$pH = pK_a + \log [BA]/[HA]$$

where [HA] is the concentration of a weak acid, [BA] the concentration of its membrane-permeant salts, and K_a is the apparent dissociation constant of the acid, i.e., the tendency to give rise to free hydrogen ions.

2.2. Halothane (Fluothane) (See Note 2)

Halothane (Halocarbon Lab, North Augusta, SC), 2-bromo-2-chloro-1,1,1-trifluoroethane ($C_2HBrClF_3$; mol wt 197.39) *(see* **Fig. 1**), is a nonflammable

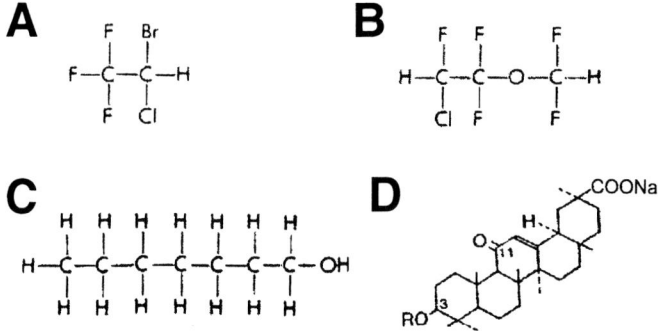

Fig. 1. Structure of some of the most commonly used gap junction inhibitors. (**A**) Halothane (fluothane); (**B**) Enflurane (ethrane); (**C**) heptanol; (**D**) glycyrrhetinic acid derivatives: 18α-glycyrrhetinic acid and 18β-glycyrrhetinic acid (R=H); carbenoxolone (R=NaOOC(CH$_2$)$_2$CO–).

highly volatile liquid routinely used as a general anesthetic; halothane is the standard volatile anesthetic to which others are compared. This compound is sensitive to light and may be stabilized with 0.01% thymol; solubility in water is 0.35% (17 m*M*). Halothane is a potent and readily reversible uncoupling agent when applied at concentrations ranging from 2–3.5 m*M*; however, its action is not at all specific for gap junction channels. A halothane concentration of 43 μ*M* (approx 1 MAC [*minimum alveolar concentration*]) has been shown to potently suppress acetylcholine-induced currents in the α$_3$β$_2$ subunit composition of the neuronal nicotinic acetylcholine receptor (*35*). Recent studies have demonstrated that this volatile agent may bind to discrete sites on a diverse number of protein targets, altering both local protein dynamics and global protein stability. In general, halothane tends to depress neuronal firing and excitatory synaptic transmission and to potentiate synaptic inhibition. These actions are mediated by inactivation of a variety of both voltage-dependent and agonist-triggered currents; in addition to action on gap junctions, halothane has been shown to:

(1) Alter inhibitory postsynaptic currents (IPSCs) in hippocampal neurons through a direct postsynaptic action (*36,37*).
(2) Suppress Ca^{2+} influx into presynaptic terminals, thereby accounting for the depression of excitatory synaptic transmission (*38*).
(3) Block both TTX-resistant and TTX-sensitive Na$^+$ channels (*39,40*).
(4) Inhibit the function of certain glutamate receptor subtypes (e.g., GluR3), but enhance kainate (GluR6) receptor function (*41,42*) (results with chimeric receptors implicate that Gly-819 in the transmembrane region [TM4] of GluR6 is critical in the action of halothane).

(5) Antagonize glutamatergic neurotransmission at the *N*-methyl-D-aspartate (NMDA) receptor *(43)*.
(6) Inhibit the function of muscarinic receptors (M1) through activation of protein kinase C *(44,45)*.
(7) Activate K^+ channels *(46,47)*.
(8) Inhibit thromboxane A_2 (TXA_2) signaling at the membrane receptor *(48)*.
(9) Modulate γ-aminobutyric acid (A) (GABA[A]) receptor function *(36–38)*.

2.3. Heptanol and Octanol (See Note 3)

Higher alcohols such as heptanol (mol wt 116.2) and octanol (mol wt 130.2) block junctional conductance in a wide range of cell types and tissues. These aliphatic alcohols are miscible with ether, benzene, chloroform, and alcohols; solubility in water is 0.096 mL/100 mL (~ 1 mM). The potencies of these compounds to block intercellular communication are inversely related to the length of the aliphatic chain, the dose required for blockage increasing as the aliphatic chain is shortened (i.e., 20 mM ethanol *[49]*, 2–3 mM heptanol, and 1 mM octanol *[50–52]*). Because of their low solubility in aqueous solutions, alchohols longer than eight carbons are rarely used. As with ethanol, these aliphatic alcohols (Sigma Chemical, St. Louis, MO) unspecifically perturb the function of ion channels and other proteins.

2.4. 18-Glycyrrhetinic Acid and Carbenoxolone (See Note 4)

The α- and β-stereoisomers of 18-glycyrrhetinic acid (3-hydroxy-11-oxo-18,20-olean-12-en-29-oic acid; FW 470.7) and carbenoxolone (3-hydroxy-11-oxoolean-12-en-30-oicacid 3-hemisuccinate; disodium salt; FW 614.7) are available from Sigma Chemical. The structure–activity relationships for glycyrrhetinic acid derivatives are illustrated in **Fig. 1**. In addition to their actions on gap junctions, licorice and its analogs have been shown to interact with mineralocorticoid receptors and to inhibit the enzyme 11-dehydrogenase *(53–55)*, resembling symptoms caused by excess mineralocorticoid secretion (i.e., pseudo-aldosteronism). These side effects have led to use of carbenoxolone in trials for the treatment of gastric ulcer *(56)*. Carbenoxolone has also been shown to shift the reversal potential of the opioid current in locus coeruleus neurons to the K^+ equilibrium potential.

2.5. Arachidonic Acid and Oleamide (See Note 5)

Arachidonic acid (5,8,11,14-eicosatetraenoic acid, FW 304.5) is a precursor of prostaglandins and leukotrienes and has been shown to activate protein kinase C and to interfere with the V_m-sensitive gating mechanism of gap junction channels *(50,57)*. Oleamide (cis-9-octadecenamide; oleylamide; FW 281.5) is an endogenous fatty acid primary amide that possesses sleep-inducing properties in animals and has been shown to potentiate the function of serotoninergic (5-HT$_{1A}$ and 5-HT$_{2A}$) receptors and benzodiazepine-sensitive GABA systems *(58–62)*. Both arachidonic acid and oleamide have been shown to block gap junctional communication at 10 μM and 100 μM, respectively. Whereas arachidonic acid uncouples cells of all types, effects of oleamide have been reported to be cell type specific, working more effectively on astrocytes than on cardiac myocytes *(62,63)*.

2.6. Closure of Gap Junctional Channels by Transjunctional Voltage (See Note 6)

The dependence of junctional conductance on transjunctional voltage was first demonstrated in amphibian embryonic cell pairs using the dual voltage clamp technique *(64)*. In these and all mammalian gap junctions thus far studied, application of large V_j steps of either polarity caused a strong decline of the junctional current to a nonzero steady-state level, whereas small V_j steps produced no appreciable change in junctional conductance.

2.7. Targeted Gap Junction Disruptors

Recently, several laboratories have contributed to set new uncoupling strategies by developing antibodies against external loop domains of connexins and mimetic peptides corresponding to these regions.

1. Rabbit polyclonal antibodies and Fab fragments were prepared by Maiken Nedergaard against peptides corresponding to the amino acid residues 46–76 of Cx43 gap junctional protein Before use, the prepared Fab fragments were dialyzed against serum-free phosphate-buffered saline (PBS), then concentrated by centrifugation through Centricon-30 filters. According to Dr. Nedergaard (*personal communication*), EL1–46 antibodies should be used at a concentration of 60 mg/mL and cells should be incubated for at least 3 h before testing (3–18 h) (*see* **Fig. 2**). *Efficacy:* Coupling among astrocytes can be reduced up to 70% by 60 mg/mL of EL1-46 antibodies.

2. An affinity-purified polyclonal antibody (EL2-186) which immunoreacts with positions 186–206 of Cx43 was produced in rabbits by Hofer and Dermietzel *(29)* and shown to yield a robust signal on Western blots. The EL2-186 antibody has the ability to recognize external loop domains of hemichannels and recognizes additional bands at positions corresponding to 30 kDa and 32 kDa in brain homogenates (presumably Cx32 and perhaps another connexin species). This cross-reactivity is expected, as sequence homology of the peptide for immunization is >70% with the corresponding Cx32 sequence. According to the authors, EL2-186 antibodies should be used at a concentration of 30 mg/mL and the cells should be incubated for at least 12 h before testing (*see* **Fig. 3**; for more details, *see* ref. *29*). *Efficacy:* Coupling among astrocytes can be reduced up to 50% by 30 mg/mL of EL2-186 antibodies.

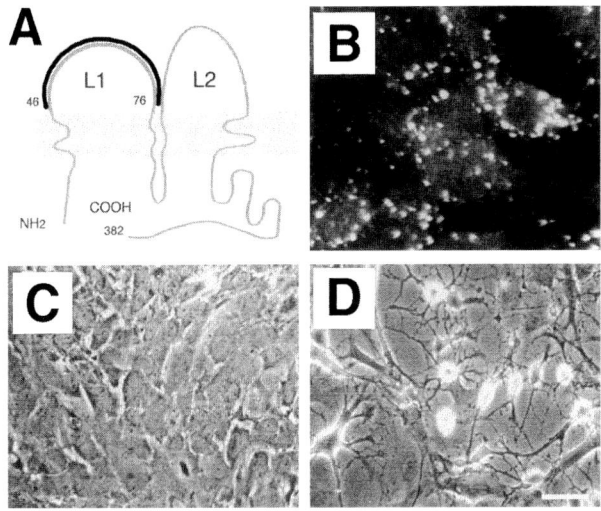

Fig. 2. EL1-46 (extracellular loop) Antibodies directed against Cx43 reduces astrocytic cell-to-cell contact. (**A**) Predicted topology of Cx43 with four transmembrane segments, two external loops (EL1 and EL2), and the N- and C-termini located at the cytoplasmic side of the membrane. To analyze the function of Cx43 in astrocytes, a polyclonal antibody was raised against the sequence that covered position 46–76 at EL1 (indicated by outer black line). (**B**) Immunolabeling of confluent culture of cortical astrocytes with the antibody directed against EL1. Cx43 immunoreactive plaques were located in areas of cell-to-cell contacts. Eighteen-hour incubation of a confluent astrocytic culture in the antibody Fab fragments (60 mg/mL) led to loss of their native flat epithelioid phenotype (**C**) and transformation into loose spindle shaped cells with few cellular contacts as previously noted (**D**). Incubation in antibodies to 8D9 or L1 or vehicle (PBS) had no such effect. Immunocytochemistry and functional coupling assay: Cells were plated on 12-mm uncoated coverslips (0.5–1 × 10^5 cells/mL) and fixed 1–2 d later with 4% paraformaldehyde. Cultures were permeabilized with 0.1% Triton X-100 and blocked with 10% normal goat serum. The dye transfer technique was adapted from Goldberg et al. *(128)*. Cells were loaded with CDCF diacetate for 5 min, washed, and trypsinized. After resuspension, cells were labeled in suspension with 10 *M* DiIC$_{18}$ *(5)* (Molecular Probes) for 10 min and mixed with unlabeled cells at a 1:250 ratio. One hour after plating on polylysine-coated dishes, dye transfer from the CDCF/DiIC$_{18}$ labeled (donor) cells to unlabeled (recipient) cells was evaluated using confocal scanning microscopy. These results suggest that gap junction formation is a regulator of astrocytic phenotype. (Figure provided by Dr. Maiken Nedergaard, New York Medical College, Valhalla, NY, USA.)

3. Connexin-mimetic polypeptides possessing sequence homology to selected amino acid sequences of rat connexins were synthesized. Among them, Gap 27 peptide (amino acid sequence: SRPTEKTIFII; possesses conserved sequence

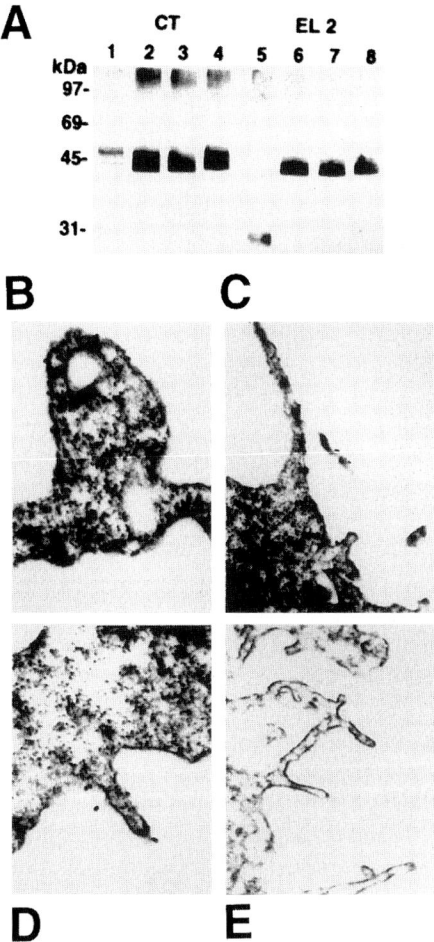

Fig. 3. EL2-186 (extracellular loop) antibodies inhibit LY transfer among primary astrocytes. **(A)** Western blots of homogenates from different brain regions and heart tissue using the antibodies CT-360 and EL2-186 (*lanes 1* and *5,* heart controls; *lanes 2* and *6,* cerebral cortex; *lanes 3* and *7,* hippocampus; *lanes 4* and *8,* cerebellum) (for more details, *see* **ref.** *29*). **(B–E)** Immunogold labeling of astrocytes with EL2-186. Gold label is concentrated at cytoplasmic processes **(B)** or filopidia **(C,D)**. **(E)** Control with preabsorbed antibody. Magnification: ×85,000 **(B,D,E)**, ×53,000 **(C)**. (Figure provided by Dr. Rolf Dermietzel, Ruhr University Bochum, Bochum, Germany.)

homology to a portion of the second extracellular loop leading to the fourth transmembrane connexin segment) and Gap 20 peptide (EIKKFKYGC; possesses sequence homology with the intracellular loop of Cx43) has been mostly used in functional assays; the latter is used as a biologically inactive control when applied extracellularly. At present, the mechanism of action(s) of the connexin-mimetic peptides is unknown. They: (1) have been shown to delay the assembly of connexons into functional gap junctions *(32)*; and (2) have been suggested to act directly by destabilizing the interactions between the extracellular loops in pre-formed gap junction channels *(33)*. Typically, the Gap 27 peptide is used within the range of concentrations of 300 μM up to as high as 10 m*M*. *Efficacy:* Coupling among cells incubated overnight can be reduced up to 80% by 10 m*M* Gap27 peptide; the IC_{50} is 300 μM. Irrespective of the mechanism of action, synthetic peptides have been reported to preferentially inhibit one type of gap junction in cells expressing multiple connexin types *(34)*.

4. Polyclonal antibodies directed to synthetic oligopeptides corresponding to the amino acid sequences corresponding to residues 363–382 (the carboxy[C]-terminus) of rat Cx43 were prepared by immunizing rabbits with the peptide YPSSRASSRASSRPRPDDLEI, synthesized according to Merrifield *(66)*. *Efficacy*: Diffusional communication in microinjected cultured neonatal rat heart cells is promptly reduced in about 50% by 0.8 mg/mL a363–382; the failure to observe an effect in a higher proportion of assays has been correlated with the difficulty of controlling the volume of microinjected fluid; the presence of junctional channels made up of other cardiac connexins, such as Cx40 and Cx45, could provide another explanation. It is suggested that the blocking of junctional communication by these antibodies results from interference with a regulatory domain of Cx43 *(67)*.

3. Results and Discussion
3.1. The Many Definitions of "Uncoupling"

Strength of functional coupling can be measured in several ways, and each type of measurement has a different threshold for detection and may depend on different sets of variables and have different limitations. The most rigorous type of measurement is that of junctional conductance, which is most simply achieved on mammalian cell pairs using the dual whole cell voltage clamp technique. Although this method is exquisitely sensitive at low coupling strengths (being able to resolve junctional currents flowing through single channels), modifications are required to measure junctional conductances above about 10 nS, owing to series resistance errors. What is measured here is the total current carried by small, mobile, intracellular ions. The simplest method for determining coupling strength is to measure diffusion of a tracer molecule such as Lucifer Yellow or neurobiotin after their intracellular injection, or loading from extracellular space either by scrape-loading or loading one population of cells with dye esters and then mixing these with unloaded cells (*see* Chapter 12). These methods are most sensitive to large changes in well coupled systems, and we generally see little or no dye coupling in Cx43-expressing cells when junctional conductance is 2 nS or less. What is measured using this technique is permeability to molecules of specific charge and of large size, although the coculture method is also dependent on rate of gap junction formation.

Decreases in functional coupling can be due to altered effective size or charge selectivity of the channel, decreased percentage of time the channel is open, or decreased number of channels functionally present. For few uncoupling agents have detailed enough studies been done thus far to conclude just how coupling strength is reduced, although as considered below, it is clear that certain effectors (such as transjunctional voltage and intracellular acidification or halothane treatment) act through largely independent mechanisms.

Quantitative analysis of connexin-specific permeabilities to fluorescent dyes and to ions of different sizes have supported the concept that connexins show differential permeabilities that cannot be predicted only on size considerations, but strongly depend on the charge of the probe, cell volume, cell input resistance, junctional capacity and binding of the dye to axoplasmic elements (Lucifer Yellow [LY] binds irreversibly to organelles). In addition, since pore diameters vary significantly among homotypic channels and pore diameter can directly affect ionic permeability depending on the relative strength of the electrostatic surface charge of the pore and the distance to the permeant ion or molecule, precise determinations should be performed for each cell type in each specific condition. Several issues that remain to be rigorously determined include: (1) whether subconductance states differ in selective permeability when compared to the main state; and (2) what number of channels represents the threshold for dye permeation? Without these answers, experimental data regarding the lack of detection of dye transfer and uncoupling effects may be misinterpreted.

If a few channels effectively modulate a specific effect, partial blockage of the total number of channels would prevent dye transfer among neighboring cells and decrease the magnitude but not prevent the expression of the effect. For example, we have recently showed that bystander killing of adjacent cells by 20 μ*M* ganciclovir (GCV) is dependent on cell communication via gap junctions; HSV-*tk* transfected cells are killed in vitro within 5 d of treatment with GCV and transmit this toxicity to adjacent cells lacking HSV-*tk*. In short, the extent of cell death and sensitivity to GCV depend on the degree of connexin expression in transfectants and also depend on which connexin is expressed in the cocultured cells. Our results support the notion of the bystander effect is dependent on the strength of cell communication via gap junctions; even when the transfer of the dye LY could not be detected among weakly coupled cells, cell death was significantly assessed by the fluorescence-activated cell sorting (FACS) technique *(68–70)*. Thus, the lack of detection of dye transfer could have been misinterpreted as lack of involvement of gap junctions in the bystander effect. Similarly, the results obtained by treating cells with partial uncouplers would be misinterpreted if evaluated only by dye transfer assays.

The use of the terms *efficacy* and *potency* has been confused in our field. Efficacy (or intrinsic activity) is related to the magnitude of the effect (i.e., blockade). Potency represents the dependency of the uncoupling effect on its concentration, the concentration of the drug at which it is half-maximally effective. Thus, the efficacy of a specific uncoupler is not necessarily related to its potency as an antagonist . For example, 2–3 mM halothane and 1 mM octanol inhibit g_j by 100%, in contrast to 50 µM oleic acid that inhibits g_j by 80%; the efficacies and potencies in regard to their uncoupling effects are octanol ~ halothane >>>> oleic acid (*efficacy*) and oleic acid >>> octanol > halothane (*potency*).

As summarized in **Table 1**, agents that have been found to reduce junctional conductance may to some extent be grouped according to hypothesized mode of action. For example, intracellular acidification sufficient to close gap junction channels may be achieved by membrane permeant weak acids, by application of nigericin in the presence of high K$^+$ and low extracellular pH, by metabolic inhibition with dinitrophenol and by sufficiently high concentrations of nitrobenzyl esters. Likewise, several lipophilic compounds uncouple cells, including doxylstearic acids, heptanol and octanol, halothane, and oleic acid. Arachidonic acid may exert is uncoupling effect in certain types through direct action or through cyclooxygenase or lipoxygenase metabolites. For other agents, such as dieldrin, DDT and phorbol esters (TPA), there is a phenomenological correlation with their action as tumor promoters, but whether TPA (and the endogenous protein kinase C activator diacylglycerol) exert their action through an accompanying phosphorylation of the connexin proteins remains controversial. Finally, many of the agents used to close gap junction channels have no known mechanism of action, and it remains to be determined whether effects are directly on the channel protein or through other effects on thc cell. In the sections that follow, we have briefly summarized some of the information available regarding modes of action of uncoupling agents.

3.2. Modes of Action of Uncoupling Agents

3.2.1. pH

Gap junction channels composed of different connexins seem to have different pH sensitivities. For example, Cx45 channels expressed in SKHep1 cells are mostly closed at an intracellular pH value of 6.7 (*71*), implying a pK value near pH 7 and a high Hill coefficient, whereas the apparent pK measured for Cx43 in a number of cell types is about 6.5, with a lower Hill coefficient (*51,71,72*). In early quantitative studies of gap junction sensitivity to pH, it was hypothesized that the Hill coefficient might reflect the cooperative titration of sites on the connexin molecules, within the connexon (*73*). Based on the nearly neutral pK of the pH sensitivity, histidine residues initially seemed likely candidates for the pH sensor, and it was proposed that perhaps those residues located within the cytoplasmic loop of the connexin molecules might participate *(74)*. Mutagenesis and expression of altered Cx43 and Cx32 sequences in *Xenopus* oocytes have revealed that this region may participate in some connexins but not in others. For Cx43, the most amino (N)-terminal histidine residue within the cytoplasmic loop seems to participate in acidification-induced channel closure, acting as a binding site for a region within the C-terminus *(see* ref. *75)*.

A different mechanism has been proposed for Cx32, where positively charged residues are hypothesized to be the pH responsive portions of the protein *(76)*. Coexpression of low pH sensitivity Cx32 mutant proteins along with wild-type Cx32 has been reported to rescue wild-type pH sensitivity, implying that cooperativity between the titration of charges among the connexin subunits is not required for channel closure *(77)*.

3.2.2. Voltage

The sensitivity of junctional conductance to transjunctional voltages was found to be well described by the Boltzmann equation:

$$g_j = (g_{max} - g_{min})/\{1 + \exp[A(V - V_0)]\} + g_{min}$$

where g_j is steady-state junctional conductance, g_{max} and g_{min} are the maximal and minimal conductances obtained at lowest and highest V_j, V_0 is the voltage at which the voltage-sensitive component of g_j $(g_{max} - g_{min})$ is reduced by 50%, and A is a slope factor from which the equivalent number of gating changes, n, can be calculated *(22)*. Channels of each connexin type display distinct Boltzmann parameters but for most, steady-state conductance is highest at 0 mV and decreases with voltages of each polarity.

Single-channel studies reveal that voltage sensitivity involves interconverting transitions between the fully open state (O) and the voltage-insensitive or residual conductance substate (O_s) *(78,79)*. The ratio of the unitary conductances of the main open state and the subconductance state has in all cases been found to be similar to the g_{min}/g_{max} ratio, thus indicating that the residual conductance g_{min} seen at high V_j arises from channel transitions occurring from the main state to the residual subconductance state *(79,80)*. Single-channel open probability measurements further indicate that the voltage sensitivity of the macroscopic conductance is due to the ensemble activity of identical and independent channels *(78,80)*.

Molecular mechanisms of voltage dependence gating remain to be fully determined. Connexins do not contain a highly charged helical motif upon which the voltage gradient is likely to act, and voltage sensitivity seems to depend on several regions of the protein, including a proline residue in the second transmembrane domain *(81)* and charged residues in the amino terminus and at the M1–E1 margin *(82)*. Nevertheless, conceptual understanding of just which residues are the voltage sensors and how sensing of the voltage field is transduced into conformational change resulting in channel closure remain to be refined.

The mechanism of pH-induced gap junction channel closure differs from that caused by transjunctional voltage in that there is no detectable residual conductance or substate associated with acidification-induced gating. Single-channel studies further suggest that acidification produces slow transitions between open and closed states, in contrast to the rapid transitions between O and O_s caused by V_j *(83)*.

3.2.3. Ca²⁺ as an Uncoupling Agent

More than a century ago, Engelmann *(84)* discovered that when a small region of myocardium was cut in normal physiological salt solution, adjacent tissue initially rendered quiescent eventually recovered its responsiveness and contractility. Subsequent studies found that recovery occurred less rapidly when Ca^{2+} was removed from the medium *(85,86)*, although it was accelerated in low Ca^{2+} solution if pH was lowered *(87)*. These findings have been interpreted as indicating that both Ca^{2+} and H^+ ions can close gap junction channels, thereby allowing injured cells to repolarize when uncoupled from the damaged cells.

Cells do uncouple when Ca^{2+} levels are sufficiently elevated, but it has remained somewhat controversial whether concentrations of Ca^{2+} required for closure of gap junction channels are obtained under physiological or pathological situations and whether this ion acts independently or by changing pH or by binding to an accessory molecule. Several studies have examined responses of junctional conductance to either changing pH or changing internal Ca^{2+} while concentration of the other ion was held constant. Such studies on internally perfused fish embryonic cells indicated that pH sensitivity was high (apparent pK about 7.3, Hill coefficient about 4), whereas Ca^{2+} sensitivity was low (apparent pK about 3.5, Hill coefficient about 1: refs. *73,88*). Other workers, using different preparations and different methods, have obtained results that are at odds with each other and with the studies cited above. In a early study using cardiac myocyte pairs, in which one myocyte was broken open to allow direct manipulation of intracellular pH and Ca^{2+}, Noma and Tsuboi *(89)* measured an apparent pK for uncoupling by H^+ (6.1) that was independent of Ca^{2+}, whereas the pK for uncoupling by Ca^{2+} was pH-sensitive (6.6 to 5.6 at pH values of 7.4 and 6.5). Firek and Weingart *(90)* directly examined pH and Ca^{2+} sensitivity in neonatal cardiac myocytes in which ionic concentrations of each were well buffered. Their values for the apparent pK of Ca^{2+}-induced channel closure were about 3.5 and for H^+ ions were near 6; when concentrations of both Ca^{2+} and H^+ ions were increased, junctional conductance was reduced further, supporting an additive interaction of these cations in inhibition of junctional communication. White et al. *(91)*, by contrast, reported the lack of appreciably uncoupling of cardiac myocytes with either low pH or high Ca^{2+} when the cytoplasmic concentration of the other ion was maintained at low levels.

If there is synergism between H^+ and Ca^{2+} ions in the closure of gap junction channels, this might arise from competition for the same sites on the connexin molecules, with different affinities for the two ions. Alternatively, an intermediary molecule with binding sites for both ions might be involved, such as calmodulin *(92,93)*.

3.2.4. Sensitivity to Glycyrrhetinic Acid Derivatives

Glycyrrhetinic acid derivatives have been shown to reduce intercellular coupling *via* gap junctions *(27,28,94–102)*. The effects of glycyrrhetinic acid and carbenoxolone on junctional conductance are illustrated in **Figs. 4** and **5**. Basically, 18α-glycyrrhetinic acid and 18β-glycyrrhetinic acid and carbenoxolone mediate concentration-dependent inhibition of junctional conductance by 60% and 80%, respectively; complete blockade of gap junction channels is not observed even with concentrations up to 100 μM. These results that 18α-glycyrrhetinic acid and related compounds do not totally uncouple even though dye transfer may be undetectable are consistent with reports from others *(102)*. At concentrations above 75 μM, cytotoxicity and irreversibility become an issue with these agents.

3.2.5. Modes of Action of Lipophiles: Lipid Chaos or Protein Binding Sites?

Alcohols, specifically heptanol and octanol, were discovered by Fidel Ramon and his colleagues to uncouple crayfish axons *(24)*. These compounds

---

Fig. 4. (*Opposite page*) Effects of glycyrrhetinic acid derivatives on intercellular junctional communication of cultured mouse hippocampal neural progenitor cells *(129)* **(A–J)** and mouse cardiac myocytes **(K,L)** as shown with the scrape-loading Lucifer Yellow transfer and dual whole-cell voltage-clamp techniques. The methods for the scrape loading technique were adapted from El-Fouly et al. *(130)*. **A–J.** Cells were incubated for 30 min with 18α-glycyrrhetinic acid (75 μ) **(C,D)**, carbenoxolone (75 μM) **(E,F)**, heptanol (2 mM; positive control) **(G,H)** and glycyrrhizic acid (75 μM; negative control) **(I,J)** prior the experiments were performed for each condition. Control **(A,B)**. **(K,L)** Junctional conductance measurements revealed that coupling

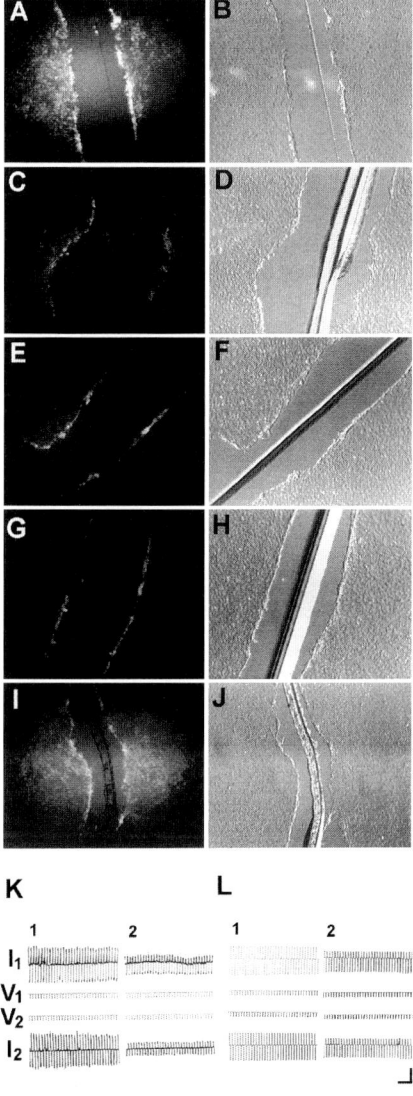

strength between myocytes is reduced by 50–60% by glycyrrhetinic acid derivatives. Cells were treated for 30 min with 75 μ*M* 18α-glycyrrhetinic acid (**K2**). (**L2**) Cells treated for 30 min with 75 μ*M* 18β-glycyrrhetinic acid. Command voltage pulses were applied alternately to cells 1 and 2. Currents in the same cell in which the voltage step is applied represent the sum of conductances of junctional and nonjunctional membranes; junctional currents are recorded in the other cell. Calibration bars: horizontal, 15 s vertical, 30 mV (V_1, V_2), 250 pA (I_1, I_2). Note that adjacent cell remained dye and electrically coupled in the presence of the above glycyrrhetinic acid derivatives. Similar results have been described by Martin et al. (*102*).

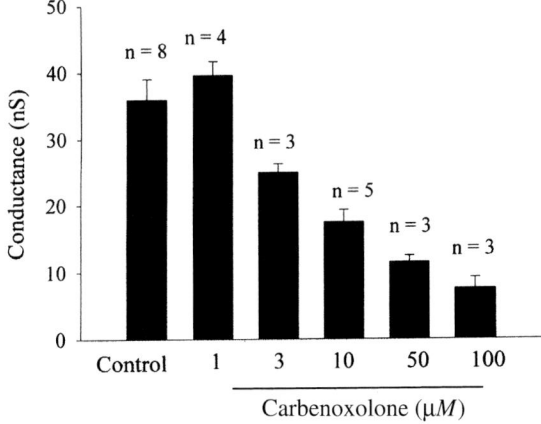

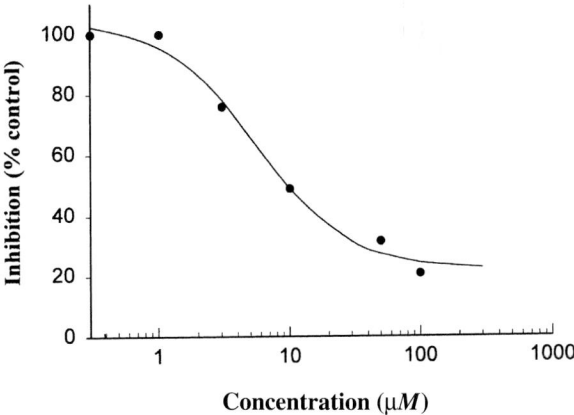

Fig. 5. Carbenoxolone reduces the strength of coupling in cardiac myocytes in a concentration dependent-manner. However, as described for 18α-glycyrrhetinic acid and 18β-glycyrrhetinic acid, complete blockade was never observed. A 100 μM carbenoxolone concentration caused a inhibition of gap junction channels by 80% and no further inhibition was observed even when the carbenoxolone concentration was increased to 300 μM.

have subsequently been shown to be effective in every mammalian tissue examined, and the reduction in junctional conductance is generally rapid, complete, and reversible at concentrations in the range 0.1–3 mM (e.g., **103**). Halothane, a volatile anesthetic with arrhythmogenic properties, also totally and reversibly uncouples all mammalian cell types that have been examined. This action can occur at concentrations lower than those affecting excitability (**Fig. 6G**), implying relative specificity of their action on junctional channels *(26,104)*. It has been proposed that these agents may act through membrane fluidity changes, either due to increased bulk fluidity *(105)* or decreased fluidity of cholesterol-rich domains *(106)*, although an alternative mode of action involving domains at specific depths within the lipid bilayer was suggested by experiments using doxyl-stearic acid probes *(107)* and might be inferred from studies on astrocytes using a wide variety of oleamide derivatives *(63)*. Like acidification-induced uncoupling, lipophile uncoupling does not induce substates in gap junction channels, but rather completely closes the channels to zero conductance.

A major question regarding the mechanism of action(s) of pharmacological uncoupling agents is whether they act through binding to specific sites in the connexins or exert an action through changing properties of the lipids in which connexons are imbedded. For anesthesia in general this issue is unresolved; the widely varying structures of the lipid soluble substances to which gap junctions are sensitive or resistant (*see* **Fig. 1**) suggest that the uncoupling agents interact directly with specific binding sites within or nearby the junctional channel although this is by no means proven.

The state of general anesthesia is a drug-induced absence of perception of all sensations *(38,108)*. A large number of structurally unrelated agents are capable of producing anesthesia, including inert gases, simple inorganic and organic compounds (i.e., nitrous oxide), and more complex organic molecules (i.e., halogenated alkanes and ethers); although they all induce anesthesia, no specific chemical group is required for this activity *(109)*. However, their wide structural differences confer limitations and disadvantages of their utilization in each specific situation. Because of their nonspecific effects on several different ionic channels, it is not surprising that a subgroup of the general anesthetics also uncouple cells. Particularly, oleamide has been suggested to be a possible endogenous analogue of general anesthesia *(109)*. This concept is worth considering because it would represent a new form of modulation of anesthesia.

Evidence that general anesthetics bind to specific sites on proteins has come from studies showing that the activity of the firefly luciferase, a soluble lipid-free enzyme, can be modulated by a diverse range of anesthetic agents at IC_{50} concentrations similar to their EC_{50} values for producing anesthesia. In addition, the strongest indication of the existence of specific binding sites for general anesthetics comes from studies based on physicochemical characteristics and stereoselectivity, potency and efficacy. For example, although isoflurane and enflurane are isomers (**Fig. 1**), they differ in their pharmacological properties. More strikingly, stereoisomers differ in their anesthetic potency two-fold (e.g., the optical isomers of thiopental and pentobarbital; **Table 1**); several

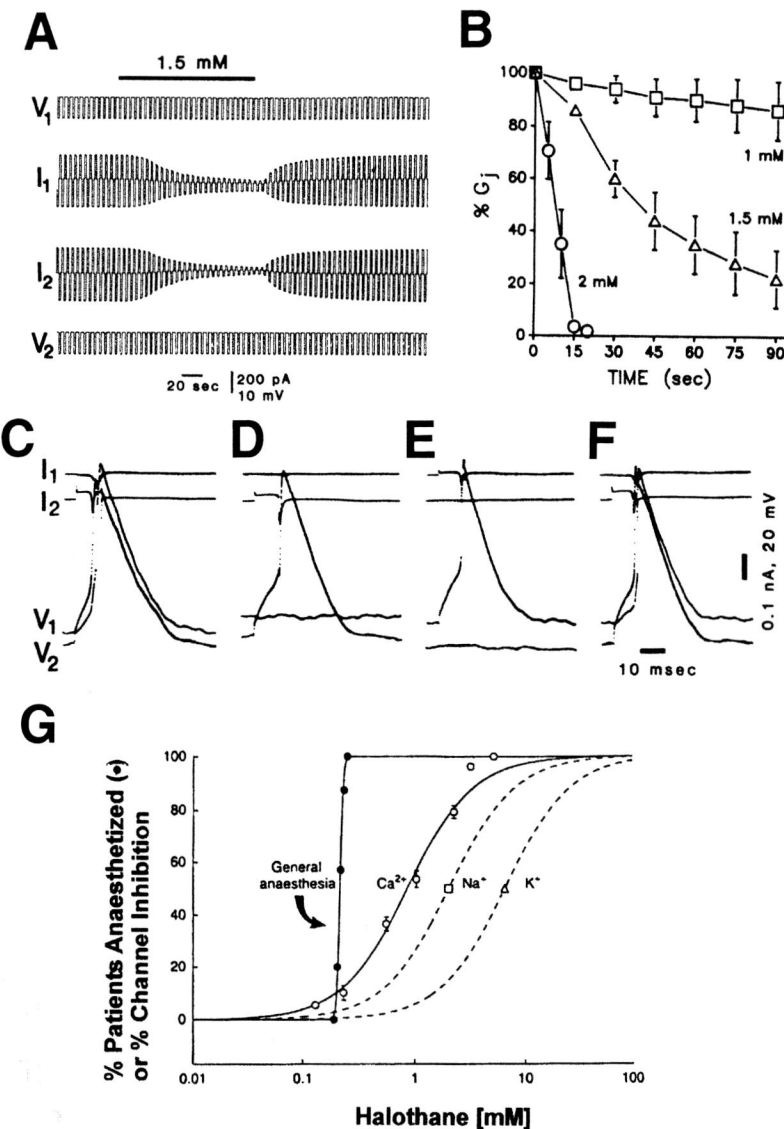

Fig. 6. Halothane blocks different types of ionic channels in a concentration-dependent manner. (**A**) Uncoupling of cardiac myocytes by 1.5 m*M* halothane under dual whole-cell voltage-clamp conditions. Note that both uncoupling and its reversal are rapid processes. Cells were clamped at 0 mV and command pulses of –10 mV were delivered to one cell and currents were recorded in both cells. Junctional currents are upward transitons above baseline in I_1 and I_2. (**B**) Time course for uncoupling by halothane (1, 1.5, and 2 m*M*). (**C–F**) Influence of halothane on action potential configuration and propagation under current clamp configuration. Cell 2 is stimulated

studies show that at least for the barbiturates, the *S* isomer is more potent than the *R* isomer *(110,111)*.

Although halothane is not very potent, about 2–3 m*M* being required to inhibit junctional currents completely, this agent currently may be one of the least toxic uncoupling agents available. One potentially useful direction for the field would be to determine whether available or novel derivatives of halothane would exert more potent and selective actions on gap junction channels than the parent compound. Furthermore studies of potency and efficacy of stereoisomers should be revealing as to mode of action.

3.2.6. Uncoupling by Design: Engineered Uncoupling Agents

A novel approach recently pursued by several laboratories has been to apply connexin antibodies *(112,113)* or polypeptides corresponding to extracellular domains (e.g., ref. *29* and Dr. Nedergaard, *unpublished results*), in the hope of occupying binding sites on the extracellular aspect of connexons, thereby inhibiting channel formation. The block of functional coupling achieved by these agents has been remarkable, as has been the use of antisense oligonucleotides corresponding to Cx43 and Cx40 sequences *(114–117)*. The use of dominant-negative strategies, in which a connexin construct is expressed that inhibits functional channel formation by other connexins, has also been successful in studies of *Xenopus* and mouse development *(118–120)*, and improved potency and efficacy are current goals of several laboratories that should prove useful in future studies of transgenic animals and of cultured cells.

Astrocytes are connected by gap junctions *in situ* and in culture. The identity and properties of astrocytic gap junction channels have been characterized by immunocytochemical, molecular biological, and electrophysiological techniques *(121–126)*. These studies have demonstrated that although Cx30, Cx40, Cx43, Cx45 and Cx46 are expressed in rodent cortical astrocytes, these glial cells are ~ 90–95% connected by Cx43; the residual 5–10% coupling contributed by the other Cx types were estimated from dual whole-cell recordings between astrocytes from Cx43 knockout mice *(127)*. Thus, one would expect

---

(*Continued from opposite page*) in **C** (before halothane), **D** (after uncoupling), and **F** (following recovery). In **E**, cell 1 is stimulated while cells are uncoupled. **(G)** Voltage-gated ion channels are also sensitive to halothane. Note that general anaesthesia in humans (•), as measured by lack of response to a surgical incision, occurs at concentrations of halothane 4–30 times lower than the EC_{50} concentrations needed to half-inhibit peak currents through L-type Ca^{2+} channels from the pituitary cells, Na^+ channels from the squid giant axon, delayed rectifier K^+ channels from the squid giant axon, and gap junction channels from rat cardiac myocytes (Figures **A–F** adapted from **ref. *26***; **G** from **ref. *108***).

coupling efficiency by dye spread to be virtually absent among primary astrocytes treated with specific antibodies against Cx43.

Polyclonal antibodies recognizing the extracellular loops I and II of Cx43 have been suggested to inhibit junctional coupling completely with EC_{70} and EC_{50} values of 60 mg/mL and 30 mg/mL, respectively (*see* **Figs. 2** and **3**). Both authors suggest that the cells should be incubated with their antibodies for a 12-h period prior to the performance of the functional assays. The time for the duration of their treatment was derived from densitometric measurements of Cx43 over a chase period of 8 h; accordingly, Dermietzel has shown that Cx43 half-time is between 2 and 3 h *(29)* and exceeding 4× the half-time should fully inhibit gap junction mediated intercellular communication. In addition, both authors state that their antibodies also recognize and block other Cx types. Nevertheless, the drawback regarding the use of the above EL1 and EL2 antibodies is the large residual dye coupling (30% and 50%, respectively) in treated astrocytes and the additional blockage of other Cx types. Thus, although these preliminary results are encouraging, the efficacy, toxicity and other side effects mediated by the EL1 and EL2 antibodies have yet to be compared with those of the "conventional" uncoupling agents discussed in this chapter.

Short synthetic peptides possessing the amino acid motifs QGP and SHVR of extracellular loop 1 and the SRPTEK motif of extracellular loop 2 have been shown to inhibit intercellular communication via gap junctions composed of Cx43. A short peptide termed Gap 27 has been shown to be a potent and reversible inhibitor of the gap junction channels formed of Cx43 and has been reported as a specific uncoupling agent of both hetero- and homocellular gap junctional communication. However, like most of the "uncoupling" agents available, this peptide does not totally block intercellular communication. For example, although the vascular attenuation of the response induced by cholinergic agents by Gap 27 peptide is concentration dependent, ranging from 50% (300 µM) to 80% (10 mM), it does not reach 100% blockade *(33)*. Thus, although Gap 27 appears to be a potent partial antagonist of junctional communication, with potency higher than for the external loop domain antibodies, it does not exceed the efficacies described for the antibodies or for uncoupling agents such as heptanol, halothane, or glycerrhetinic acid derivatives.

3.3. Conclusion

In general terms, there are three ways of depressing cell-to-cell signaling: (1) binding to the channel and affecting g_j; (2) alterations of the lipid environment around the channel, and (3) indirect effect(s) on the channel through a change in pH_i or $[Ca^{2+}]_i$. Decrease of P_o of the junctional channel by compounds with no apparent structural relationship is postulated to occur by particular combinations of the aforementioned events. However, given the lack of specific antagonists, the functional distinction of specific binding site within the channel is difficult at the present time. To date, the pharmacological agents used to block gap junction channels have not been very discriminatory. Halothane, heptanol, octanol, 18α- and 18β-glycyrrhetinic acid, and carbenoxolone rapidly inhibit coupling among cells but their mechanism of action is not known. In contrast to halothane, heptanol, and octanol, the blockade mediated by 18α- or 18β-glycyrrhetinic acid and carbenoxolone is not complete or fully reversible. In contrast, the potency of these compounds in inhibiting gap junction permeability is much higher than that shown by the volatile agents: carbenoxolone >18α-glycyrrhetinic acid and 18β-glycyrrhetinic acid >>> octanol > heptanol > halothane. Neither 18α-glycyrrhetinic acid nor 18β-glycyrrhetinic acid or carbenoxolone completely inhibits intercellular communication via gap junctions up to concentrations of 100 µM; increasing concentrations are not indicated because of the cytotoxicity mediated by these agents. Affinity-purified antibodies (EL1-46 and EL2-186) prepared against extracellular loops of Cx43 and peptides corresponding to these regions presumably provide a more selective blockade of junctional channels than the drugs described in this review. However, we suggest caution in using these agents because of their limited efficacy. Future experiments are clearly necessary to explore the mechanisms by which junctional conductance can be reduced and to design better tools with which to block these channels.

4. Notes

1. Although all gap junction channels appear to be pH sensitive, some are more sensitive than others; however, all gap junction channels seem to be closed when pH_i is lowered to values near 6. Such low values can be achieved by substituting 20 mM of the Na^+ salts of lactate, propionate, or acetate for NaCl in the extracellular solution and buffering the solution to pH 6.5 with the low-pK buffer 4-morpholinepropanesulfonic acid (MOPS). Alternatively, cells can be superfused with external solutions and equilibrated with air/CO_2 mixtures (100% to 50% CO_2). To ensure that sufficient intracellular acidification is achieved, pH should ideally be monitored using ratiometric fluorescent pH indicators or intracellular pH-sensitive microelectrodes.

2. Halothane is supplied in amber bottles with 0.01% (w/w) thymol. The recommended storage conditions are to keep the bottle securely closed and to avoid excessive heat. Stock halothane solutions are prepared in physiological solutions (e.g., Tyroide's or Ringer's) (1:1 v/v) and kept at room temperature (21–22°C); the concentration of the anesthetic in the saturated stock solution was determined by gas chromatography of extracted samples as 17 ± 1.5 mM *(26)*. Serial dilutions are made immediately prior to each experiment, and optimal concentrations of halothane to uncouple cell pairs range from 2 to 3 mM; this procedure provides a highly reproducible uncoupling effect.

3. Heptanol and octanol are freshly diluted from 200 mM stock solution in 95% ethanol. Addition of 3–5 mM ethanol alone has no effect on junctional conductance, but should be applied as a control for unspecific effects mediated by alcohol. Hexanol, which is also inactive at concentrations below 5 mM, provides an additional control.

4. 18α- and 18β-glycyrrhetinic acid (αGA, βGA) are insoluble in water; they are miscible with dimethyl sulfoxide (DMSO), chloroform, and ethanol. Stock solutions of up to 100 mM can be prepared in DMSO or a mixture of DMSO and ethanol (2:3 v/v). Final solvent concentration should be 0.5% or less and appropriate solvent controls should be included in each experiment. Glycyrrhizic acid (Sigma Chemical, St. Louis, MO), a glycyrrhetinic acid analog with no uncoupling effects, should be used as a *control* for the nonspecific effects of vehicle and acid. Carbenoxolone is water soluble.

5. Arachidonic acid is available from Sigma Chemical (St. Louis, MO) both as a sodium salt and as free acid. Stock solutions should be prepared in ethanol (10 mg/mL) or in DMSO (10 mg/mL, but needs sonication); be aware that its initial decomposition at 37°C may result in a purity loss of 10% over a 3-d period. Oleamide is available from Sigma Chemical (St. Louis, MO). Special considerations should be given when mixing these agents with albumin/globulin containing media, which have high affinity for these compounds, thereby changing effective concentration.

6. Both transient and sustained uncoupling by voltage require large V_j gradients that are difficult to generate except under conditions where space clamp is achievable (as in voltage-clamped cell pairs). Moreover, the completeness of channel closure is limited at high voltages by the residual conductance of the substate. Nevertheless, V_j gradients have been maintained adequately long to allow measurement of junctional permeability *(65)*.

Acknowledgments

The authors thank Profs. Maiken Nedergaard and Rolf Dermietzel for generously contributing **Fig. 2** (unpublished data) and **Fig. 3** (Hofer and Dermietzel, 1998), respectively. In addition, we thank Ms. Fran Andrade for secretarial assistance. This work was supported by N.I.H. grants to Renato Rozental and David C. Spray.

References

1. Nicholson, S. M. and Bruzzone, R. (1997) Gap junctions: getting the message through. *Curr. Biol.* **7,** R340–344.
2. Scherer, S. S., Xu, Y. T., Nelles, E., Fischbeck, K., Willecke, K., and Bone, L. J. (1998) Connexin32-null mice develop demyelinating peripheral neuropathy. *Glia* **24,** 8–20.
3. White, T. W. and Paul, D. L. (1999) Genetic diseases and gene knockouts reveal diverse connexin functions. *Annu. Rev. Physiol.* **61,** 283–310.

4. Lo, C. W. (1999) Genes, gene knockouts, and mutations in the analysis of gap junctions. *Dev. Genet.* **24,** 1–4.

5. Spray, D. C., Kojima, T., Scemes, E., Suadicani, S. O., and Gao, Y. (1999) Negative physiology: what connexin-deficient mice reveal about the functional roles of individual gap junction proteins, in *Biophysics of Gap Junction Channels* (Peracchia, C. ed.), Academic Press, Orlando.

6. Tahara, Y. (1910) Uber das tetrodongift. *Biochem. Z.* **10,** 255–275.

7. Narahashi, T., Deguchi, T., Urakawa, N., and Okhubo, Y. (1960) Stabilization and rectification of muscle fiber membrane by tetrodotoxin. *Am. J. Physiol.* **198,** 934–938.

8. Nakamura, Y., Nakajima, S., and Grundfest, H. (1965) Analysis of spike electrogenesis and depolarizing K inactivation in the electroplaques of Electrophorus electricus. *J. Gen. Physiol.* **49,** 321–349.

9. Narahashi, T., Haas, H. G., and Therrien, E. F. (1967) Saxitoxin and tetrodotoxin: Comparison of nerve blocking mechanism. *Science* **157,** 1441–1442.

10. Bernard, M. C. (1857) *Leçons sur les Effets des Substances Toxiques et Médicamenteuses.* Ballière et Fils, Paris, pp. 238–306.

11. Manalis, R. S. (1977) Voltage-dependent effect of curare at the frog neuromuscular junction. *Nature* **267,** 366–368.

12. Katz, B. and Miledi, R. (1978) A re-examination of curare action at the motor endoplate. *Proc. Natl. Soc. Lond. (Biol.)* **203,** 119–133.

13. Colquhoun, D., Dreyer, F., and Sheridan, R. E. (1979) The actions of tubocurarine at the frog neuromuscular junction. *J. Physiol.* **293,** 247–284.

14. Lee, C. Y. (1972) Chemistry and pharmacology of polypeptide toxins in snake venoms. *Annu. Rev. Pharmacol.* **12,** 265–286.

15. Mathews-Bellinger, J. and Salpeter, M. M. (1978) Distribution of acetylcholine receptors at frog neuromuscular junctions with a discussion of some physiological implications. *J. Physiol.* **279,** 197–213.

16. Albuquerque, E. X., Barnard, E. A., Porter, C. W., and Warnick, J. E. (1974) The density of acetylcholine receptors and their sensitivity in the postsynaptic membrane of muscle endplates. *Proc. Natl. Acad. Sci. USA* **71,** 2818–2822.

17. Hagiwara, S. and Saito, N. (1959) Voltage-current relations in nerve cell membrane of *Onchidium verruculatum. J. Physiol.* **148,** 161–179.

18. McCleskey, E. W., Fox, A. P., Feldman, D. H., Cruz, L. J., Olivera, M., Tsien, R., and Yoshikami, D. (1987) ω-conotoxin: direct and persistent blockade of specific types of calcium channels in neurons but not muscle. *Proc. Natl. Acad. Sci. USA* **84,** 4328–4331.

19. Aosaki, T. and Kasai, H. (1989) Characterization of two kinds of high-voltage-activated Ca-channel currents in chick sensory neurons, differential sensitivity to dihydropyridines and ω-conotoxin GVIA. *Pflügers Arch.* **414,** 150–156.

20. Plummer, M. R.D., Logothetis, E., and Hess, P. (1989) Elementary properties and pharmacological sensitivities of calcium channels in mammalian peripheral neurons. *Neuron* **2,** 1453–1463.

21. Turin, L. and Warner, A. E. (1980) Intracellular pH in early *Xenopus* embryos: its effects on current flow between glastomeres. *J. Physiol.* **300,** 489–504.

22. Spray, D. C., Harris, A. L., and Bennett, M. V. L. (1981) Gap junctional conductance is a simple and sensitive function of intracellular pH. *Science* **211,** 712–715.
23. Délèze, J. and Hervé, J. C. (1983) Effect of several uncouplers to cell-to-cell communication on gap junction morphology in mammalian heart. *J. Membr. Biol.* **74,** 203–215.
24. Johnston, M. F., Simon, S. A., and Ramon, F. (1980) Interactions of anesthetics with electrical synapses. *Nature* **286,** 498–500.
25. Terrar, D. A. and Victory, J. G. G. (1988) Influence of halothane on electrical coupling in cell pairs isolated from guinea-pig ventricle. *Br. J. Pharmacol.* **94,** 509–514.
26. Burt, J. M. and Spray, D. C. (1989) Volatile anesthetics block intercellular communication between neonatal rat myocardial cells. *Circ. Res.* **65,** 829–837.
27. Davidson, J. S., Baumgarten, I. M., and Harley, E. H. (1986) Reversible inhibition of intercellular junctional communication by glycyrrhetinic acid. *Biochem. Biophys. Res. Comm.* **134,** 29–36.
28. Davidson, J. S. and Baumgarten, I. M. (1988) Glycyrrhetinic acid derivatives: a novel class of inhibitors of gap-junctional intercellular communication. Structure–activity relationships. *J. Pharmacol. Exp. Ther.* **246,** 1104–1107.
29. Hofer, A. and Dermietzel, R. (1998) Visualization and functional blocking of gap junction hemichannels (connexons) with antibodies against external loop domains in astrocytes. *Glia* **24,** 141–154.
30. Dahl, G., Werner, R., Levine, E., and Rabadan-Diehl, G. (1992) Mutational analysis of gap junction formation. *Biophys. J.* **62,** 172–182.
31. Dahl, G., Nonner, W., and Werner, R. (1994) Attempts to define functional domains of gap junction proteins with synthetic peptides. *Biophys. J.* **67,** 1816–1822.
32. Warner, A., Clements, D. K., Parikh, S., Evans, W. H., and DeHaan, R. L. (1995) Specific motifs in the external loops of connexin proteins can determine gap junction formation between chick heart myocytes. *J. Physiol.* **488,** 721–728.
33. Chaytor, A. T., Evans, W. H., and Griffith, T. M. (1998) Central role of heterocellular gap junctional communication in endothelium-dependent relaxations of rabbit arteries. *J. Physiol.* **508,** 561–573.
34. Kwak, B. R. and Jongsma, H. J. (1999) Selective inhibition of gap junction channel activity by synthetic peptides. *J. Physiol.* **516,** 679–685.
35. Narahashi, T., Aistrup, G. L., Lindstrom, J. M., Marszalee, W., Nagata, K., Wang, F., and Yeh, J. Z. (1998) Ion channel modulation as the basis for general anesthesia. *Toxicol. Lett.* **100–101,** 185–191.
36. Banks, M. I. and Pearce, R. A. (1999) Dual actions of volatile anesthetics on GABA(A) IPSCs: dissociation of blocking and prolonged effects. *Anesthesiology* **90,** 12–134.
37. Jones, M. V. and Harrison, N. L. (1993) Effects of volatile anesthetics on the kinetics of inhibitory postsynaptic currents in cultured rat hippocampus neurons. *J. Neurophysiol.* **70,** 1339–1349.
38. Krnjevic, K. (1992) Cellular and synaptic actions of general anaesthetics. *Gen. Pharmacol.* **23,** 965–975.
39. Scholz, A., Appel, N., and Vogel, W. (1998) Two types of TTX-resistant and one TTX-sensitive Na$^+$ channel in rat dorsal root ganglion neurons and their blockade by halothane. *Eur. J. Neurosci.* **10,** 2547–2556.

40. Sirois, J. E., Pancrazio, J. J., III, C. L., and Bayliss, D. A. (1998) Multiple ionic mechanisms mediate inhibition of rat motoneurons by inhalation anaesthetics. *J. Physiol.* **512,** 851–862.

41. Dildy-Mayfield, J. E., Eger, E. I., 2nd, and Harris, R. A. (1996) Anesthetics produce subunit-selective actions on glutamate receptors. *J. Pharmacol. Exp. Ther.* **276,** 1058–1065.

42. Minami, K., Wick, M. J., Stern-Bach, Y., Dildy-Mayfield, J. E., Brozowski, S. J., Gonzales, E. L., Trudell, J. R., and Harris, R. A. (1998) Sites of volatile anesthetic action on kainate (glutamate receptor 6) receptors. *J. Biol. Chem.* **273,** 8248–8255.

43. Beirne, J. P., Pearlstein, R. D., Massey, G. W., and Warner, D. S. (1998) Effect of halothane in cortical cell cultures exposed to *N*-methyl-D-aspartate. *Neurochem. Res.* **23,** 17–23.

44. Nietgen, G. W., Honemann, C. W., Chan, C. K., Kamatchi, G. L., and Durieux, M. E. (1998) Volatile anesthetics have differential effects on recombinant m1 and m3 muscarinic acetylcholine receptor function. *Br. J. Anaesth.* **81,** 569–577.

45. Minami, K., Vanderah, T. W., Minami, M., and Harris, R. A. (1997) Inhibitory effects of anesthetics and ethanol on muscarinic receptors expressed in *Xenopus* oocytes. *Eur. J. Pharmacol.* **339,** 237–244.

46. Patel, A. J., Honore, E., Lesage, F., Fink, M., Romey, G., and Lazdunski, M. (1999) Inhalational anesthetics activate two-pore-domain background K$^+$ channels. *Nat. Neurosci.* **2,** 422–426.

47. Lopes, C. M., Franks, N. P., and Lieb, W. R. (1998) Actions of general anaesthetics and arachidonic pathway inhibitors on K$^+$ currents activated by volatile anaesthetics an FMRFamide in molluscan neurones. *Br. J. Pharmacol.* **125,** 309–318.

48. Honemann, C. W., Nietgen, G. W., Podranski, T., Chan, C. K., and Durieux, M. E. (1998) Influence of volatile anesthetics on thromboxane A2 signaling. *Anesthesiology* **88,** 440–451.

49. Abou Hashieh, I., Mathieu, S., Besson, F., and Gerolami, A. (1996) Inhibition of gap junction intercellular communications of cultured rat hepatocytes by ethanol: role of ethanol metabolism. *J. Hepatol.* **24,** 360–367.

50. Weingart, R. and Bukauskas, F. F. (1998) Long-chain *n*-alkanols and arachidonic acid interfere with the V_m-sensitivie gating mechanism of gap junction channels. *Pflügers Arch.* **435,** 310–319.

51. Spray, D. C. and Bennett, M. V. L. (1985) Physiology and pharmacology of gap junctions. *Annu. Rev. Physiol.* **47,** 281–303.

52. Burt, J. M. (1990) Modulation of cardiac gap junction channel activity by the membrane lipid environment. Biophysics of Gap Junction Channels, C. Peracchia (ed)., CRC Press, pp. 75–93.

53. Walker, B. R. and Edwards, C. R. (1994) Licorice-induced hypertension and syndromes of apparent mineralocorticoid excess. *Endocrinol. Metab. Clin. North Am.* **23,** 359–377.

54. Sewell, K. J., Shirley, D. G., Michael, A. E., Thompson, A., Norgate, D. P., and Unwin, R. J. (1998) Inhibition of renal 11–beta-hydroxysteroid dehydrogenase

in vivo by carbenoxolone in the rat and its relationship to sodium excretion. *Clin. Sci.* **95,** 435–443.

55. Monder, C., Stewart, P. M., Lakshmi, V., Valentino, R., Burt, D., and Edwards, C. R. (1989) Licorice inhibits corticosteroid 11-beta-dehydrogenase of rat kidney and liver: in vivo and in vitro studies. *Endocrinology* **125,** 1046–1053.

56. Fraser, P. M., Doll, R., Langman, M. J., Misiewicz, J. J., and Shawdon, H. H. (1972) Clinical trial of a new carbenoxolone analogue (BX24), zinc sulphate, and vitamin A in the treatment of gastric ulcer. *Gut* **13,** 459–463.

57. Schmilinksy-Fluri, G., Valiunas, V., Willi, M., and Weingart, R. (1997) Modulation of cardiac gap junctions: the mode of action of arachidonic acid. *J. Mol. Cell. Cardiol.* **29,** 1703–1713.

58. Boger, D. L., Patterson, J. E., and Jin, Q. (1998) Structural requirements for 5-HT2A and 5-HT1A serotonin potentiation by the biologically active lipid oleamide. *Proc. Natl. Acad. Sci. USA* **95,** 4102–4107.

59. Yost, C. S., Hampson, A. J., Leonoudakis, D., Koblin, D. D., Bornheim, L. M., and Gray, A. T. (1998) Oleamide potentiates benzodiazepine-sensitive gamma-aminobutyric acid receptor activity but does not alter minimum alveolar anesthetic concentration. *Anesth. Analg.* **86,** 1294–1300.

60. Lees, G., Edwards, M. D., Hassoni, A. A., Ganellin, C. R., and Galanakis, D. (1998) Modulation of GABA(A) receptors and inhibitory synaptic currents by the endogenous CNS sleep regulator *cis*-9,10-octadecenoamide (cOA). *Br. J. Pharmacol.* **124,** 873–882.

61. Thomas, E. A., Carson, M. J., and Sutcliffe, J. G. (1998) Oleamide-induced modulation of 5-hydroxytryptamine receptor-mediated signaling. *Ann. NY Acad. Sci.* **861,** 183–189.

62. Boger, D. L., Patterson, J. E., Guan, X., Cravatt, B. F., Lerner, R. A., and Gilula, N. B. (1998) Chemical requirements for inhibition of gap junction communication by the biologically active lipid oleamide. *Proc. Natl. Acad. Sci. USA* **95,** 4810–4815.

63. Guan, X., Cravatt, B. F., Ehring, G. R., Hall, J. E., Boger, D. L., Lerner, R. A., and Gilula, N. B. (1997) The sleep-inducing lipid oleamide deconvolutes gap junction communication and calcium wave transmission in glial cells. *J. Cell. Biol.* **139,** 1785–1792.

64. Spray, D. C., Harris, A. L., and Bennett, M. V. L. (1979) Voltage dependence of junctional conductance in early amphibian embryos. *Science* **204,** 432–434.

65. Verselis, V., White, R. L., Spray, D. C., and Bennett, M. V. (1986) Gap junctional conductance and permeability are linearly related. Science **234,** 461–464.

66. Merrifield, R. (1963) Solid-phase peptide synthesis. 1: The synthesis of a tetrapeptide. *J. Am. Chem. Soc.* **85,** 2149–2154.

67. Bastide, B., Jarry-Guichard, T., Briand, J. P., Deleze, J., and Gros, D. (1996) Effect of antipeptide antibodies directed against three domains of connexin43 on the gap junction permeability of cultured heart cells. *J. Membr. Biol.* **150,** 243–253.

68. Vrionis, F. D., Wu, J. K., Qi, P., Waltzman, M., Cherington, V. S., and Spray, D. C. (1997) The bystander effect exerted by tumor cells expressing the herpes sim-

plex virus thymidine kinase (HSVtk) gene is dependent on connexin expression and cell communication via gap junctions. *Gene Ther.* **4**, 577–585.

69. Elshami, A. A., Saavedra, A., Zhang, H., Kucharczuk, J. C., Spray, D. C., Fishman, G. I., Amin, K. M., Kaiser, L. R., and Albelda, S. M. (1996) Gap junctions play a role in the "bystander effect" of the herpes simplex virus thymidine kinase/ganciclovir system in vitro. *Gene Ther.* **3**, 85–92.

70. Andrade-Rozental, A. F., Rozental, R., Hopperstad, M. G., Wu, J. K., Vrionis, F., and Spray, D. C. (2000) Gap junctions: the "kiss of death" and the "kiss of life". *Brain Res. Rev.* **32**, 308–315.

71. Hermans, M. M., Kortekaas, P., Jongsma, H. J. and Rook, M. B. (1995) pH sensitivity of the cardiac gap junction proteins, connexin 34 and 43. *Pflügers Arch.* **431**, 138–140.

72. Morley, G. E., Taffet, S. M., and Delmar, M. (1996) Intramolecular interactions mediate pH regulation of connexin43 channels. *Biophys. J.* **70**, 1294–1302.

73. Spray, D. C., Harris, A. L., and Bennett, M. V. (1981) Equilibrium properties of a voltage-dependent junctional conductance. *J. Gen. Physiol.* **77**, 77–93.

74. Spray, D. C. and Burt, J. M. (1990) Structure–activity relations of the cardiac gap junction channel. *Am. J. Physiol.* **258**, LC195–C205.

75. Ek-Vitorin, J. F., Calero, G., Morley, G. E., Coombs, W., Taffet, S. M., and Delmar, M. (1996) pH regulation of connexin43: molecular analysis of the gating particle. *Biophys. J.* **7**, 1273–1284.

76. Wang, X. G. and Peracchia, C. (1997) Positive charges of the initial C-terminus domain of Cx32 inhibit gap junction gating sensitivity to CO_2. *Biophys. J.* **73**, 798–806.

77. Wang, X. G. and Peracchia, C. (1998) Chemical gating of heteromeric and heterotypic gap junction channels. *J. Membr. Biol.* **162**, 169–176.

78. Bukauskas, F. F., Elfgang, C., Willecke, K., and Weingart, R. (1995) Biophysical properties of gap junction channels formed by mouse connexin40 in induced pairs of transfected human HeLa cells. *Biophys. J.* **68**, 2289–2298.

79. Moreno, A. P., Rook, M. B., Fishman, G. I., and Spray, D. C. (1994) Gap junction channels: distinct voltage-sensitive and insensitive conductance states. *Biophys. J.* **67**, 113–119.

80. Srinivas, M., Costa, M., Gao, Y., Fort, A., Fishman, G. I., and Spray, D. C. (1999) Voltage dependence of macroscopic and unitary currents of gap junction channels formed by mouse connexin50 expressed in rat neuroblastoma cells. *J. Physiol.* **517**, 673–689.

81. Suchyna, T. M., Xu, L. X., Gao, F., Fourtner, C. R., and Nicholson, B. J. (1993) Identification of a proline residue as a transduction element involved in voltage gating of gap junctions. *Nature* **365**, 847–849.

82. Verselis, V. K., Ginter, C. S. and Bargiello, T. A. (1994) Opposite voltage gating polarities of two closely related connexins. *Nature* **368**, 348–351.

83. Bukauskas, F. F. and Peracchia, C. (1997) Two distinct gating mechanisms in gap junction channels: CO_2-sensitive and voltage-sensitive. *Biophys. J.* **72**, 2137–2142.

84. Engelmann, T. W. (1877) Ueber die Leitung der Erregung im Herzmuskel. *Pflügers Arch.* **11**, 465–480.
85. De Mello, W. C., Motta, G. E., and Chapeau, M. (1969) A study on the healing-over of myocardial cells of toads. *Circ Res.* **24**, 475–487.
86. Deleze, J. (1970) The recovery of resting potential and input resistance in sheep heart injured by knife or laser. *J. Physiol.* **208**, 548–562.
87. De Mello, W. C. (1983) The influence of pH onthe healing-over of mammalian cardiac muscle. *J. Physiol.* **339**, 299–307.
88. Spray, D. C., Stern, J. H., Harris, A. L., and Bennett, M. V.L. (1982) Gap junctional conductance: comparison of sensitivities to H and Ca ions. *Proc. Natl. Acad. Sci. USA* **79**, 441–445.
89. Noma, A. and Tsuboi, N. (1987) Dependence of junctional conductance on proton, calcium and magnesium ions in cardiac paired cells of guinea pig. *J. Physiol.* **382**, 193–211.
90. Firek, L. and Weingart, R. (1995) Modification of gap junction conductance by divalent cations and protons in neonatal rat heart cells. *J. Mol. Cell. Cardiol.* **27**, 1633–1643.
91. White, R. L., Doeller, J. E., Verselis, V. K., and Wittenberg, B. A. (1990) Gap junctional conductance between pairs of ventricular myocytes is modulated synergistically by H$^+$ and Ca^{++}. *J. Gen. Physiol.* **95**, 1061–1075.
92. Peracchia, C. and Bernardini, G. (1984) Gap junction structure and cell-to-cell coupling regulation: is there a calmodulin involvement? *Fed. Proc.* **43**, 2681–2691.
93. Peracchia, C. and Wang, X. C. (1997) Connexin domains relevant to the chemical gating of gap junction channels. *Braz. J. Med. Biol. Res.* **30**, 577–590.
94. Yamamoto, Y., Fukuta, H., Nakahira, Y., and Suzuki, H. (1998) Blockade by 18-β-glycyrrhetinic acid of intercellular electrical coupling in guinea pig arterioles. *J. Physiol.* **511.2**, 501–508.
95. Eugenin, E. A., Gonzalez, H., Saez, C. G., and Saez, J. C. (1998) Gap junctional communication coordinates vasopressin-induced glycogenolysis in rat hepatocytes. *Am. J. Physiol.* **274**, G1109–G1116.
96. Goldberg, G. S., Moreno, A. P., Bechberger, J. F., Hearn, S. S., Shivers, R. R., MacPhee, D. J., Zhang, Y.-C., and Naus, C. G. (1996) Evidence that disruption of connexon partical arrangements in gap junction plaques is associated with inhibition of gap junctional communication by a glycyrrhetinic acid derivative. *Exp. Cell Res.* **222**, 48–53.
97. Seseke, F. G., Gardemann, A., and Jungermann, K. (1992) Signal propagation via gap junctions, a key step in the regulation of liver metabolism by the sympathetic hepatic nerves. *FEBS Lett.* **301**, 265–270.
98. Guan, X., Wilson, S., Keith, K., Schlender, K., and Ruch, R. J. (1996) Gap-junction disassembly and connexin43 dephosphorylation induced by 18-β-glycyrrhetinic acid. *Mol. Carcinog.* **16**, 157–164.
99. Munari-Silem, Y., Lebrethon, M. C., Morand, I., Rousset, B., and Saez, J. M. (1995) Gap junction-mediated cell-to-cell communication in bovine and human

adrenal cells. A process whereby cells increase their responsiveness to physiological corticotropin concentrations. *J. Clin. Invest.* **95,** 1429–1439.

100. D'Andrea, P., and Vittur, F. (1996) Gap junctions mediate intercellular calcium signalling in cultured articular chondrocytes. *Cell Calcium* **20,** 389–397.

101. Frame, M. K. and DeFeijter, W. (1997) Propagation of mechanically induced intercellular calcium waves via gap junctions and ATP receptors in rat liver epithelial cells. *Exp. Cell. Res.* **230,** 197–207.

102. Martin, W., Zempel, G., Hulser, D., and Willecke, K. (1991) Growth inhibition of oncogene-transformed rat fibroblasts by cocultured normal cells: relevance of metabolic cooperation mediated by gap junctions. *Cancer Res.* **51,** 5348–5354.

103. Burt, J. M. and Spray, D. C. (1988) Single channel events and gating behavior of the cardiac gap junction channel. *Proc. Natl. Acad. Sci. USA* **85,** 3431–3434.

104. Niggli, E., Rudisuli, A., Maurer, P., and Weingart, R. (1989) Effects of general anesthetics on current flow across membranes in guinea pig myocytes. *Am. J. Physiol.* **256,** C273–281.

105. Takens-Kwak, B. R., Jongsma, H. J., Rook, M. B., and Van Ginneken, A. C. (1992) Mechanism of heptanol-induced uncoupling of cardiac gap junctions: a perforated patch-clamp study. *Am. J. Physiol.* **262,** C1531–C1538.

106. Bastiaanse, E. M., Jongsma, H. J., van der Laarse, A., and Takens-Kwak, B. R. (1993) Heptanol-induced decrease in cardiac gap junctional conductance is mediated by a decrease in the fluidity of membranous cholesterol-rich domains. *J. Membr. Biol.* **136,** 135–145.

107. Burt, J. M. (1989) Uncoupling of cardiac cells by doxl stearic acids: specificity and mechanism of action. *Am. J. Physiol.* **256,** C913–924.

108. Franks, N. P. and Lieb, W. R. (1994) Molecular and cellular mechanisms of general anaesthesia. *Nature* **367,** 607–614.

109. Lerner, R. A. (1997) A hypothesis about the endogenous analogue of general anesthesia. *Proc. Natl. Acad. Sci. USA* **94,** 13,375–13,377.

110. Richter, J. A. and Holtman, J. R., Jr. (1982) Barbiturates: their in vivo effects and potential biochemical mechanisms. *Prog. Neurobiol.* **18,** 275–319.

111. Andrews, P. R. and Mark, L. C. (1982) Structural specificity of barbiturates and related drugs. *Anesthesiology* **57,** 314–320.

112. Chaytor, R. T., Evans, W. H., and Griffith, T. M. (1997) Peptides homologous to extracellular loop motifs of connexin 43 reversibly abolish rhythmic contractile activity in rabbit arteries. *J. Physiol.* **503,** 99–110.

113. Boitano, S., Dirksen, E. R., and Evans, W. H. (1998) Sequence-specific antibodies to connexins block intercellular calcium signaling through gap junctions. *Cell Calcium* **23,** 1–9.

114. Moore, L. K. and Burt, J. M. (1994) Selective block of gap junction channel expression with connexin-specific antisense oligodeoxynucleotides. *Am. J. Physiol.* **267,** C1371–1380.

115. Makarenkova H. and Patel, K. (1999) Gap junction signalling through connexin-43 is required for chick limb development. *Dev. Biol.* **207,** 380–392.

116. Becker, D. L., McGonnel, I., Makarenkova, H. P., Patel, K., Tickle, C., Lorimer, J., and Green, C. R. (1999) Roles for alpha 1 connexin in morphogenesis of chick embryos revealed using a novel antisense approach. *Dev. Genet.* **24,** 33–42.

117. Minkoff, R., Bales, E. S., Kerr, C. A., and Struss, W. E. (1999) Antisense oligonecleotide blockade of connexin expression during embryonic bone formation: evidence of functional compensation within a multigene family. *Dev. Genet.* **24,** 43–56.

118. Goliger, J. A., Bruzzone, R., White, T. W. and Paul, D. L. (1996) Dominant inhibition of intercellular communication by two chimeric connexins. *Clin. Exp. Pharmacol. Physiol.* **23,** 1062–1067.

119. Paul, D. L., Yu, K., Bruzzone, R., Gimlich, R. L., and Goodenough, D. A. (1995) Expression of a dominant negative inhibitor of intercellular communication in the early *Xenopus* embryo causes delamination and extrusion of cells. *Development* **121,** 371–381.

120. Sullivan, R. and Lo, C. W. (1995) Expression of a connexin 43/beta-galactosidase fusion protein inhibits gap junctional communication in NIH3T3 cells. *J. Cell Biol.* **130,** 419–429.

121. Dermietzel, R. and Spray, D. C. (1993) Gap junctions in the brain: where, what type, how many and why? *Trends Neurosci.* **16,** 186–192.

122. Giaume, C., and McCarthy, K. D. (1996) Control of gap-junctional communication in astrocytic networks. *Trends Neurosci.* **19,** 319–325.

123. Wolff, J. R., Stuke, K., Missler, M., Tytko, H., Schwarz, P., Rohlmann, A., and Chao, T. I. (1998) Autocellular coupling by gap junctions in cultured astrocytes: a new view on cellular autoregulation during process formation. *Glia* **24,** 121–140.

124. Nagy, J. I., Patel, D., Ochalski, P. A., and Stelmack, G. L. (1999) Connexin30 in rodent, cat and human brain: selective expression in gray matter astrocytes, co-localization with connexin43 at gap junctions and late developmental appearance. *Neuroscience* **88,** 447–468.

125. Dermietzel, R., Hertzberg, E. L., Kessler, J. A., and Spray, D. C. (1991) Gap junctions between cultured astrocytes: immunocytochemical, molecular, and electrophysiological analysis. *J. Neurosci.* **11,** 1421–1432.

126. Kunzelmann, P., Schroder, W., Traub, O., Steinhauser, C., Dermietzel, R., and Willecke, K. (1999) Late onset and increasing expression of the gap junctions protein connexin30 in adult murine brain and long-term cultured astrocytes. *Glia* **25,** 111–119.

127. Scemes, E., Dermietzel, R., and Spray, D. C. (1998) Calcium waves between astrocytes from Cx43 knockout mice. *Glia* **24,** 65–73.

128. Goldberg, G. S., Bechberger, J. F., and Naus, C. G. (1995) A pre-loading method of evaluating gap junctional communication by fluorescent dye transfer. *BioTechniques* **18,** 490–497.

129. Rozental, R., Morales, M., Mehler, M. F., Urban, M., Kremer, M., Dermietzel, R., Kessler, J. A., and Spray, D. C. (1998) Changes in the properties of gap junctions during neuronal differentiation of hippocampal progenitor cells. *J. Neurosci.* **18,** 1753–1762.

130. El-Fouly, M. H., Trosko, J. E., and Chang, C. C. (1987) Scrape-loading and dye transfer. A rapid and simple technique to study gap junctional intercellular communication. *Exp. Cell Res.* **168,** 422–430.

Index

A

Amphotericin B, dual perforated
  patch technique, 285, 286, 289
Antisense oligodeoxynucleotide,
  connexin knockdown,
  brain slices and cultured neurons,
    183, 184
  chicken embryo studies, 180,
    181, 183, 184
  controls, 179, 180
  gene knockout comparison,
    advantages, 176
    disadvantages, 177
  materials, 176–178
  oligodeoxynucleotide design and
    synthesis, 178, 179, 183
  Pluronic gel delivery system,
    175, 176, 179
  principle, 175, 183
  rat brain lesion studies, 181, 182
Arachidonic acid, gap junction
  uncoupling, 451, 452, 468
Assembly assays, gap junctions,
  cell-free translation/integration
    system assays,
    coimmunoprecipation assay,
      109, 112, 113
    sucrose density gradient
      centrifugation, 107, 108
  cell–cell coupling, 380
  chloroform/methanol
    precipitation, 129, 133
  crosslinking, 120, 125–127, 131

de novo formation,
  apparatus,
    electrical recordings, 384
    experimental chamber, 381
    heating-cooling system, 383,
      384
    patch pipets, 382
    pipet holders, 381, 382
    solution streaming pipets,
      382, 383
    temperature control, 384
  cell culture, 385
  cell lines, 380, 385
  coupling of cells, 386, 387
  detachment of cells from
    coverslip,
    chemical, 386
    mechanical, 386
  electrical measurements, 387,
    390
  homotypic versus heterotypic
    junction formation, 390, 391
  lag time and cooperativity,
    380, 381
  overview, 384, 385
  solutions, 381
immunoprecipitation, 120, 121,
  128, 129, 132, 133
purification for studies,
  cell lines, 118, 119
  labeling medium, 119
  materials, 119, 120

Triton X-100 solubilization,
cell cultures, 123, 130, 131
overview, 122, 123
rat liver tissue, 124
rat myometrial tissue, 124, 125
steps in formation, 379
sucrose density gradient
centrifugation, 120, 127,
128, 131, 132
Western blot analysis, 121, 129, 133

B

Baby hamster kidney (BHK) cell,
Cx43 truncated gap junction
expression,
detergent extraction, 60, 67–69,
74
DotMetric protein assay, 79,
85, 86
induction of expression, 83
materials, 78–80, 85
membrane isolation, 83, 84
methotrexate selection, 81
scaleup, 81, 82, 86
storage of stocks, 82, 83
subculture, 81
two-dimensional crystallization,
see Two-dimensional
crystallization
thawing of stocks, 80, 81
tissue culture, 80
BCECF, gap junction permeability,
204
BHK cell, *see* Baby hamster kidney
cell
Brain slice electrophysiology,
antisense oligodeoxynucleotide
connexin knockdown, 183, 184
dual recording from coupled
neurons, 396, 403

historical perspective of electrical
transmission studies, 395
reverse transcriptase polymerase
chain reaction combination
studies, 146, 147, 154, 155
spikelet recording,
automated identification and
analysis,
cross- and auto-correlations,
402
digitization of data, 401
event selection criteria,
401–404
overview, 399–401
template selection, 401
blockers of gap junctions, 399,
403, 404
characteristics, 396
data acquisition, 397, 398
materials, 396, 397
shapes, 398, 399

C

Caged compounds, *see* Calcium wave
Calcein, gap junction permeability, 204
Calcium phosphate, transfection of
HeLa cells, 190, 191, 195, 196
Calcium wave,
cell type distribution, 407
experimental initiation,
caged compound loading,
calcium NP-EGTA-AM, 418
electroporation, 420–424,
426, 427
microinjection, 419, 420
scrape loading, 420
flash photolysis of caged
compounds,
caged compounds, 410
cell coupling, 411

cell culture, 412
flash photolysis setup, 417, 426
principle, 409, 410
solutions, 413, 414
mechanical stimulation limitations, 409
imaging,
data analysis, 415
dye loading, 415
fluorescent calcium probes, 410
instrumentation, 415, 425
inducers,
extracellular messengers, 408, 409
intracellular messengers, 407, 408
intracellular calcium role, 408
propagation, 410, 411
Carbenoxolone, gap junction uncoupling, 441, 468
6-Carboxyfluorescein, gap junction permeability, 203
5(6)-Carboxyfluorescein diacetate, fluorescence recovery after photobleaching,
concentration determination, 318
dye loading, 315, 318, 319
reloading of cells, 319
stock solution, 316
washing off excess dye, 319
gap junction permeability, 204
Cell-free translation/integration, connexins,
advantages, 91, 92
combined transcription/ translation systems, 102
connexon assembly in microsomes, assembly assays,
coimmunoprecipitation assay, 109, 112, 113

sucrose density gradient centrifugation, 107, 108
membrane integration, 107
overview, 106
translation, 106, 107, 112
equipment, 93, 94
full-length, membrane-integrated connexin synthesis, 100–102
materials, 92–94
membrane integration, 92, 93, 96–99, 110, 111
membrane integration/trans-membrane orientation assays,
alkali extraction, 103, 104, 111
N-glycosylation tagging assay, 105, 106
overview, 102, 103
protease protection assays, 104, 105, 111, 112
optimization of translation, 99
polyacrylamide gel electrophoresis, 109, 110
principle, 91, 94
transcription, 92, 94–96, 110
translation, 92, 93, 96
Connexon, *see* Gap junction
Crosslinking,connexons, 120, 125–127, 131
Cysteine scanning mutagenesis, *see* Mutagenesis, structure–function analysis of gap junctions

D

Dextrans, gap junction permeability, 203
2',7'-Dichlorofluorescein, gap junction permeability, 203
Digoxygenin-labeled probes, *see In situ* hybridization, messenger RNA

E

Electron microscopy, *see also*
    Freeze-fracture, gap junctions;
    Two-dimensional
    crystallization,
    gap junction purity assessment,
        60, 71
    tracer microinjection analysis,
        212
Electrophysiology, *see* Brain slice
    electrophysiology; Patch-clamp
Electroporation, caged compounds,
    420–424, 426, 427

F

Flash photolysis, *see* Calcium wave
Fluorescence-activated cell sorting,
    transjunctional metabolite
    capture, 336, 339
Fluorescence recovery after
    photobleaching (FRAP),
    advantages in diffusion studies,
        313
    apparatus,
        controls for light source, 317,
            318
        filters, 317
        fluorescence microscope, 317
        image acquisition, 318
        light source, 317
        technical requirements, 316,
            317
    5(6)-carboxyfluorescein
        diacetate,
        concentration determination,
            318
        dye loading, 315, 318, 319
        reloading of cells, 319
        stock solution, 316

        washing off excess dye, 319
    cell injury, 313, 325
    cell-to-cell transfer of cytosolic
        molecules, measurement, 314
    comparison with other methods,
        dye injection, 323, 324
        electrophysiology, 324
    controls,
        cell damage, 322, 323
        correction for spontaneous loss
            of fluorescence, 322
    limitations, 324, 325
    principle, 314
    rate constant of fluorescence
        recovery,
        determination, 319, 325
        normalization for number of
            corrected cells, 319, 320,
            322
    solution preparation, 316
    transfer constant determination
        for solute across gap
        junctions, 315
FRAP, *see* Fluorescence recovery
    after photobleaching
Freeze-fracture, gap junctions,
    fracture plane, 34, 42, 44
    principle, 33, 34, 41, 42, 44, 45
    sodium dodecyl sulfate-digested
        freeze fracture replica
        labeling,
        antibodies,
            description, 37
            monoclonal antibody
                preparation, 41
            polyclonal antibody
                preparation, 40
        apparatus, 37
        detergent treatment, 37, 39

dissection, 38
freeze-fracture and replication, 38, 39, 49, 50
immunolabeling, 37, 39, 40, 50, 51
materials, 34, 36, 37
mounting, 37, 41
overview, 34, 45, 47
precautions with propane, 47, 49
quick freezing, 34, 36, 47, 49
thawing, 39

G

Gap junction,
assembly stages, 117
calcium signaling, *see* Calcium wave
cell-free assembly of connexons in microsomes,
assembly assays,
coimmunoprecipitation assay, 109, 112, 113
sucrose density gradient centrifugation, 107, 108
membrane integration, 107
overview, 106
translation, 106, 107, 112
connexons, 33
de novo formation, *see* Assembly assays, gap junctions
definition, 33, 57
electrophysiology, *see* Brain slice electrophysiology; Patch-clamp
expression systems, *see* Xenopus oocyte, gap junction expression
freeze-fracture, *see* Freeze-fracture, gap junctions
homotypic versus heterotypic, 187

mutagenesis, *see* Mutagenesis, structure–function analysis of gap junctions
permeability assays, *see* Permeability, gap junctions
purification, *see also* Assembly assays, gap junctions,
baculovirus-insect cell expressed gap junctions containing Cx32 and Cx26, alkali extraction, 60, 69, 74
connexin distribution in tisses, 117, 118
Cx43 truncated gap junctions from transfected BHK cells, detergent extraction, 60, 67–69, 74
electron microscopy assessment of purity, 60, 71
heart gap junction purification, detergent extraction, 59, 60, 65–67, 73
liver gap junction purification, alkali extraction, 59, 64, 65, 73
detergent extraction, 59, 62–64, 71–73
split gap junction preparation, 60, 69–71, 74
materials, 59–62, 71, 72
overview, 58
structure, 44, 45
transjunctional metabolite capture, *see* Transjunctional metabolite capture, gap junctions
turnover, 135, 379
uncoupling, *see* Uncoupling, gap junctions
X-ray crystallography, 57, 58

N-Glycosylation tagging,
connexin assay for membrane
integration, 105, 106
sugar scanning for structure–
function analysis of gap
junctions, 255
18-Glycyrrhetinic acid, gap junction
uncoupling, 451, 460, 468
Green fluorescent protein-connexin
chimeras,
applications, 135–137
construction,
engineering, 138, 140
expression, 138, 140
materials, 137–140
overview, 135, 136
transfection, 138, 140
imaging in live cells, 139–142

H

Halothane, gap junction uncoupling,
miscellaneous actions, 450, 451
mode of action, 462, 463, 465
properties, 449, 450, 467
HeLa cell, connexin expression,
calcium phosphate transfection,
190, 191, 195, 196
cell culture, 190
comparison with *Xenopus* system,
187, 188
lipofection, 190–192, 196
materials, 188–190
overview, 188
screening of cell clones,
immunofluorescence, 194, 195
Northern blot, 193, 194, 197
selection of stably transfected
cells, 190, 192, 193, 196, 197
vector construction, 190, 195

Heptanol, gap junction uncoupling,
451, 460, 462, 468
High-performance liquid
chromatography (HPLC),
transjunctional metabolite
capture, 336, 337, 339
HPLC, *see* High-performance liquid
chromatography

I

Immunofluorescence microscopy,
connexins in retrovirus
expression systems, 170,
171, 173
HeLa cell, connexin expression
screening, 194, 195
Immunolabeling,
In situ hybridization, *see* In situ
hybridization, messenger
RNA
sodium dodecyl sulfate-digested
freeze fracture replica
labeling, *see* Freeze-fracture,
gap junctions
Immunoprecipitation,
assembly of gap junctions, assay,
coimmunoprecipitation, 109,
112, 113
immunoprecipitation, 120, 121,
128, 129, 132, 133
phosphorylated connexins, 434,
442, 443
Xenopus oocyte gap junctions,
230–233, 244, 245, 248
In situ crystallization, *see* Two-
dimensional crystallization
In situ hybridization, messenger
RNA,
apparatus, 8–10

historical perspective for connexins, 2, 3
microscopy and analysis of sections, 23, 28, 29
quantitative analysis, 23, 29, 30
radioactive detection on sections, detection,
    counterstaining, 22, 28
    development of signal, 22, 28
    photographic emulsion preparation, 21, 27, 28
    slide preparation, 22
    hybridization,
        hybridization conditions, 21, 27
        mix preparation, 20, 27,
        post-hybridization washes, 21
        solutions, 7, 8
    materials, 6–8, 24
    nonradioactive detection comparison, 1, 2
    pretreatment of paraffin sections, 7, 20, 27
    principle, 3
    riboprobe radiolabeling,
        hydrolysis, 19, 26
        materials, 6, 7, 24
        purification, 26, 27
        sulfur-35 calibration slide preparation, 18
        transcription, 18, 19, 26
    tissue processing, 6, 16–18, 26
specificity, 23, 28, 29
whole-mount,
    anti-digoxygenin antibody incubation, 13, 14, 25
    dissection, 12
    embedding and sectioning, 5, 6, 15, 16, 24
    fixation of embryos, 4, 12, 13, 24, 25
    hybridization, 4, 5, 13, 25
    immunodetection of riboprobe, 5, 14, 15, 26
    materials, 3–6, 23, 24
    prehybridization, 4, 13, 25
    principle, 3
    requirements, 10
    riboprobe preparation,
        digoxygenin incorporation, 11, 12
        plasmid DNA linearization and purification, 11, 25
        synthesis, 4, 11, 12, 23–25
    sectioning comparison, 2

L

Lipofection, HeLa cells, 190–192, 196
Lucifer Yellow CH, gap junction permeability, 202

M

Membrane integration, *see* Cell-free translation/integration, connexins
Messenger RNA (mRNA),
    antisense inhibition, *see* Antisense oligodeoxynucleotide, connexin knockdown
    expression analysis, *see* reverse transcriptase polymerase chain reaction, connexin expression in single cells
    localization, *see In situ* hybridization, messenger RNA
Metabolic cooperation, gap junction permeability assay, 217, 218, 221

Microinjection,
  caged compounds, 419, 420
  membrane-impermeant tracers for
      gap junction permeability
      studies,
    advantages and limitations,
        208
    apparatus, 204, 205, 218–220
    cell buffer, 206, 220
    cell preparation, 209, 210, 220,
        221
    electrode preparation and
        mounting, 210, 211
    electron microscopy analysis,
        212
    fixation solution, 207
    injection, 211, 212
    injection solution, 207
    light microscopy analysis, 212,
        221
    penetration, 211
    recording of injection, 212
    tracer solution, 206
  retrovirus in chick embryos, 168,
      169
  RNA in *Xenopus* oocytes,
    materials, 228, 229, 245, 246
    oocyte preparation, 237, 238
    setup, 237
mRNA, *see* Messenger RNA
Mutagenesis, structure–function
      analysis of gap junctions,
  applications, 251, 252
  strategies,
    accessibility test for localizing
        channel gates, 255, 256
    cysteine scanning mutagenesis,
        254
    deletion mutagenesis, 253, 254

domain swapping, 256
evolutionary guidance, 252
site-directed mutagenesis, 252,
    253
sugar scanning, 255
substituted cysteine accessibility
    method,
  cassette mutagenesis, 258
  materials, 256–258
  overview, 254
  plasmid preparation, 261
  polymerase chain reaction,
    cassette generation, 258,
        259, 264
    fusion of fragments, 259, 260
    screening for mutant clones,
        260, 264, 265
  transcription, 261, 262
  Xenopus oocyte
      electrophysiology,
    oocyte preparation, 262
    patch-clamp, 262–266
    solutions, 257
    two-electrode voltage clamp,
        262, 265

N

Neurobiotin, gap junction
    permeability, 203
Northern blot, HeLa cell screening
    for transfected connexins, 193,
    194, 197
Nystatin, dual perforated patch
    technique, 285, 289

O

Octanol, gap junction uncoupling,
    451, 460, 462, 468
Oleamide, gap junction uncoupling,
    451, 452, 468

P

Patch-clamp,
  direct patching onto earthworm
    junctional membrane, 341
  dual patch-clamp,
    cell preparation, 271
    corrections for nonjunctional
      resistances,
      cytoplasmic access
        resistance, 280
      membrane resistance, 280–
        282, 285
      series resistance, 277, 278,
        280, 282, 285
      time-sharing technique, 288,
        289
    ionic permeability
      measurements,
      calculation of relative ionic
        permeability ratios,
        304–306
      equipment, 295, 296, 304, 307
      external solutions, 296
      internal pipet solutions,
        297–299, 307–309
      ionic reversal potential
        measurements from
        homotypic gap
        junctions, 299, 300,
        302, 303, 309, 310
      materials, 294–299
      overview, 293, 294
    macroscopic gap junctional
      conductance
      determination, 272, 274
    macroscopic voltage sensitivity
      determination, 274, 275
    materials, 269–271, 286, 287
    overview, 269, 272, 287

perforated patch technique,
  advantages and limitations,
    286
  amphotericin B-perforated
    patch, 285, 286, 289
  nystatin-perforated patch,
    285, 289
  pipet solutions, 270, 271
  setup, 269, 270, 286, 287
  single-channel conductance
    determination, 275, 276,
    287, 288
  single-channel kinetics, 276, 288
reverse transcriptase polymerase
  chain reaction combination
  studies,
  apparatus, 145, 146
  brain slice recordings, 146,
    147, 154, 155
  expelling patch pipet contents
    into polymerase chain
    reaction tubes, 147, 148
  harvesting of cell content, 147
  overview, 143
  solutions, 145
single connexin hemichannel
  recording in *Xenopus*
  oocytes,
  acquisition software and
    hardware, 347, 351–354
  apparatus, 343, 345–347
  Cx46 hemichannels, 347, 348
  idealization with substrates, 348
  overview, 341, 342
  rapid perfusion of excised
    patches,
    achieving rapid transition
      through interface, 351,
      353

complementary RNA
   preparation, 344, 345
expression levels required
   for patching, 345
flow rate adjustment, 350
oocyte isolation, 343, 344
overview, 349, 350
recording chamber, 345–347
theta tube perfusion pipet
   construction, 350
reagents, 342, 343
sublevel Hinckley detector, 348
visualizing short-lived
   substrates, 348
Xenopus oocyte, substituted
   cysteine accessibility
   method for gap junction
   analysis, 262–266
PCR, *see* Polymerase chain reaction
Peptide mapping, *see*
   Phosphorylation, connexins
Permeability, gap junctions,
   endogenous molecule
      permeability assays,
      coupling assessment, 217
      loading cells with radiolabeled
         precursos, 217
      metabolic cooperation, 217,
         218, 221
      overview, 216, 217
   fluorescence recovery after
      photobleaching, *see*
Fluorescence recovery after
   photobleaching
ionic permeability measurements
   with dual whole cell patch-
   clamp,
   calculation of relative ionic
      permeability ratio, 304–306

equipment, 295, 296, 304, 307
external solutions, 296
internal pipet solutions, 297–
   299, 307–309
ionic reversal potential
   measurements from
   homotypic gap junctions,
   299, 300, 302, 303, 309,
   310
materials, 294–299
overview, 293, 294
membrane-permeant fluorogenic
   ester incorporation,
   analysis, 215, 216
   apparatus, 205
   dye loading, 215
   overview, 215
   solutions, 208
microinjection of membrane-
   impermeant tracers,
   advantages and limitations, 208
   apparatus, 204, 205, 218–220
   cell buffer, 206, 220
   cell preparation, 209, 210, 220,
      221
   electrode preparation and
      mounting, 210, 211
   electron microscopy analysis, 212
   fixation solution, 207
   injection, 211, 212
   injection solution, 207
   light microscopy analysis, 212,
      221
   penetration, 211
   recording of injection, 212
   tracer solution, 206
overview, 201, 202
scrape-loading experiments,
   analysis, 214

apparatus, 205, 220
culture preparation, 213
overview, 213
recording, 214
scrape-loading, 213, 214
solutions, 207, 220
tracers,
apparatus, 359–361
connexin purification, 362
features, 202
loading liposomes with tracers,
366, 367, 375
membrane-impermeant tracers,
202–204
membrane-permeant
fluorogenic esters, 204,
357
overview, 357, 358
radioactive tracers, 202, 372, 373
reagents, 358, 359, 373
reconstitution with gel
filtration,
column preparation, 362
pre-equilibration of column,
362–364, 373, 374
reconstitution, 364, 365,
374, 375
stripping of column, 365,
366, 375
tracer retention measurement,
372, 373
transport specific fractionation,
band recovery, 371, 372
centrifugation, 370, 371,
375
formation of gradients, 367–
370, 375
gradient maker, 368, 369, 371
inspection of results, 371, 376

liposome application to
gradient, 370
transport specific fractionation,
reconstituted connexin assay
in liposomes,
Phosphorylation, connexins,
phosphoamino acid content
analysis,
blotting of gels, 435, 444
materials, 432, 433
sample preparation, 435–437,
444
thin-layer chromatography,
437, 445
phosphoisoform analysis,
gel electrophoresis, 435, 443,
444
immunoprecipitation, 434, 442,
443
materials, 431, 432
radiolabeling, 434, 442
phosphotryptic peptide mapping,
band extraction from gel, 439,
440
data analysis, 442
desalting, 440, 444, 445
materials, 433
overview, 437–439
performic acid oxidation, 440,
444
thin-layer chromatography,
440–442, 448
regulation, 431
Polymerase chain reaction (PCR),
see also Reverse transcriptase
polymerase chain reaction,
connexin DNA fragments in
retrovirus construct
preparation, 162–164, 172

substituted cysteine accessibility
method,
cassette generation, 258, 259, 264
fusion of fragments, 259, 260
screening for mutant clones,
260, 264, 265
Potassium channel, crystal structure
verification of models, 251
Protease protection assay, connexins in
membranes, 104, 105, 111, 112

R

Retrovirus expression systems,
connexins,
applications in gap junction
studies, 159, 160
construct preparation,
adapter plasmid preparation, 164
cloning into RCAS(A), 164, 165
overview, 162
polymerase chain reaction of
connexin DNA
fragments, 162–164, 172
detection of expression,
immunofluorescence staining,
170, 171, 173
immunostaining of tissue
sections, 171, 172
Western blot of crude
membranes, 171, 173
expression in chick embryo,
microinjection, 168, 169
overview, 167, 168
materials, 160–162
overview of vector, 159
sectioning of chick embryo, 169,
170, 173
viral stock preparation,
chick embryo fibroblast packaging cells, 165, 166, 172

concentrating, 166, 172, 173
titering, 166, 167, 173
Reverse transcriptase polymerase
chain reaction (RT-PCR),
connexin expression in single
cells,
amplification reactions, 144, 148,
149, 154, 155
applications, 143
complementary DNA synthesis,
144, 148
connexin identification,
confirmation,
multiplex polymerase chain
reaction, 154, 155
sequencing, 154
Southern blot analysis, 145,
154
specific primers for
polymerase chain
reaction, 153, 154
degenerate primers, 149, 151,
152, 155
plaque identification, 153
product subcloning into M13,
144, 152
sequencing, 153
materials, 144–146
oligonucleotide radiolabeling, 144
patch-clamp combination studies,
apparatus, 145, 146
brain slice recordings, 146,
147, 154, 155
expelling patch pipet contents
into polymerase chain
reaction tubes, 147, 148
harvesting of cell content, 147
overview, 143
solutions, 145

precautions, 146
RT-PCR, *see* Reverse transcriptase polymerase chain reaction

S

Scrape loading,
  calcium wave initiation, 420
  gap junction permeability assays,
    analysis, 214
    apparatus, 205, 220
    culture preparation, 213
    overview, 213
    recording, 214
    scrape-loading, 213, 214
    solutions, 207, 220
Sodium dodecyl sulfate-digested freeze fracture replica labeling, *see* Freeze-fracture, gap junctions
Southern blot, connexin identity confirmation, 145, 154
Spikelet, *see* Brain slice electrophysiology
Substituted cysteine accessibility method, *see* Mutagenesis, structure–function analysis of gap junctions
Sucrose density gradient centrifugation,
  connexon assembly assays, 107, 108, 120, 127, 128, 131, 132

T

Thin-layer chromatography (TLC),
  transjunctional metabolite capture, 337, 338
  phosphoamino acid content analysis, 437, 445
  phosphotryptic peptide mapping, 440–442, 448

TLC, *see* Thin-layer chromatography
Transjunctional metabolite capture, gap junctions,
  co-culture conditions, 334, 335, 339
  donor cell labeling for fluorescence-activated cell sorting,
    fluorescent labeling of donors, 334
    overview, 333
    radioactive labeling of donors, 333, 334, 339
  filtration, 336, 339
  fluorescence-activated cell sorting, 336, 339
  high-performance liquid chromatography, 336, 337, 339
  labeling time and metabolic turnover, 329, 330
  materials, 330, 332, 338
  preloading support protocol, 332, 333
  principle, 329, 330
  releasing cells from culture dish, 335, 336
  thin-layer chromatography, 337, 338
Transjunctional voltage, gap junction uncoupling, 452, 458, 459, 468
Transport specific fractionation, *see* Permeability, gap junctions
Triton X-100, connexon solubilization,
  cell cultures, 123, 130, 131
  overview, 122, 123
  rat liver tissue, 124

rat myometrial tissue, 124, 125
Two-dimensional crystallization,
  in situ crystallization of Cx43
      truncated gap junctions
      expressed in BHK cells,
    crystallization conditions, 85
    enrichment, 84, 85
    factors in resolution
        improvement, 86, 87
    materials, 78–80
  in vitro versus *in situ*, 77, 78

U

Uncoupling, gap junctions,
  acidification agents,
    intracellular acidification, 449
    mode of action, 457, 458
    pH sensitivity of channels,
        457, 458, 467
  antibody targeting, 452, 455, 465,
      466
  applications, 447
  arachidonic acid, 451, 452, 468
  bystander effects, 456
  calcium uncoupling mechanism,
      459, 460
  carbenoxolone, 441, 468
  comparison with other channels,
      448
  connexin-mimetic peptides, 453–
      455, 465, 466
  efficacy of agents, 457, 467
  engineering of agents, 465, 466
  18-glycyrrhetinic acid, 451, 460,
      468
  halothane,
    miscellaneous actions, 450, 451
    mode of action, 462, 463, 465
    properties, 449, 450, 467
  heptanol, 451, 460, 462, 468

measurement,
    quantitative analysis, 456
    tracer diffusion, 455, 456
  octanol, 451, 460, 462, 468
  oleamide, 451, 452, 468
  overview of agents, 448, 449,
      457
  potency of agents, 457, 467
  transjunctional voltage and
      closure, 452, 458, 459, 468

W

Western blot,
  connexon assembly assay, 121,
      129, 133
  retrovirus-expressed connexins,
      171, 173

X

X-ray crystallography,
  gap junctions, 57, 58
  potassium channel, crystal
      structure verification of
      models, 251
Xenopus oocyte, gap junction
      expression,
  advantages, 225, 226
  applications, 225, 226
  comparison with HeLa cell
      system, 187, 188
  cytoplasmic perfusion of paired
      oocytes,
    electrical recording, 243, 244
    materials, 229, 232, 246
    oocyte preparation, 243
    perfusion chamber preparation,
        241–243
  endogenous Cx38 expression, 226
  immunoprecipitation, 230–233,
      244, 245, 248

junctional conductance
measurement after pairing,
229, 238–241, 243, 244,
247, 248
limitations, 226, 227
materials, 227–233, 245, 246
oocyte,
extraction and preparation,
227, 233, 234, 246
handling buffer, 231
pairing, 238
RNA,
DNA template preparation,
227, 228, 234, 235, 246
microinjection,
materials, 228, 229, 245, 246
oocyte preparation, 237, 238
setup, 237
recovery and quantification,
227, 228, 235, 236, 246
transcription, in vitro, 227,
228, 235
single connexin hemichannel
patch-clamp recording,
acquisition software and
hardware, 347, 351–354
apparatus, 343, 345–347
Cx46 hemichannels, 347, 348
idealization with substrates,
348

overview, 341, 342
rapid perfusion of excised
patches,
achieving rapid transition
through interface, 351,
353
complementary RNA
preparation, 344, 345
expression levels required
for patching, 345
flow rate adjustment, 350
oocyte isolation, 343, 344
overview, 349, 350
recording chamber, 345–347
theta tube perfusion pipet
construction, 350
reagents, 342, 343
sublevel Hinckley detector, 348
visualizing short-lived
substrates, 348
substituted cysteine accessibility
method, electrophysiology
of gap junctions,
oocyte preparation, 262
patch-clamp, 262–266
solutions, 257
two-electrode voltage clamp,
262, 265
translation, in vitro, 228, 236,
237